Foundations of Mathematics

A Quantitative Reasoning Approach

Barbara L. Miller · L. Robin Hendrix

Foundations of Mathematics

A Quantitative Reasoning Approach

Executive Project Manager
Kimberly Cumbie

Vice President, Research and Development
Marcel Prevuznak

Editorial Assistants
Kimberly Cumbie, Susan Fuller, Nina Waldron

Review Coordinator
Lisa Young

Senior Graphic Designer
Jennifer Moran

Art and Cover Design
Jennifer Moran

A division of Quant Systems, Inc.

546 Long Point Road, Mount Pleasant, SC 29464

Printed in the United States of America

ISBN:

Student Workbook: 978–1–938891–56–4

Student Workbook and Software Bundle: 978–1–938891–57–1

Acknowledgements

First and foremost, we would like to thank our former students for inspiring us to write this workbook. Their victories and their struggles have helped us to understand what students need to succeed in mathematics. We would also like to thank everyone at Hawkes Learning, especially the book content and software content teams, for their support and faith in us during the creation of this workbook.

We would like to thank our reviewers for giving us their time and feedback throughout the editorial process.

Alan Hayashi, Oxnard College
Amy Young, Navarro College
Beverly Meyers, Jefferson College
Bonny Rainforth, Central Community College
Brian Leonard, Southwestern Michigan College
Donald Ransford, Edison State College
Emily Hantsch, Mohawk Valley Community College
Jack Rotman, Lansing Community College
Jordyn Nail, East Georgia College
Kristen Campbell, Elgin Community College
Michael Kirby, Tidewater Community College
Mike McComas, Mountwest Community College
Patricia Rhodes, Treasure Valley Community College
Shakir Manshad, New Mexico State University

We would also like to thank our friends who helped us develop the Math@Work sections.

Rachel Sulyok
Karen Hunter
J Darby Smith
Jonathan H. Hafner

In addition, we would like to thank our families and friends for their support and understanding throughout this project.

Letter to the Student

Foundations of Mathematics takes a different approach than the mathematics textbooks you may have used in the past. This workbook focuses more on the thorough understanding of concepts and the use of reasoning skills to solve problems. By showing you where the mathematics you are learning can be applied in real-life contexts and in a variety of career paths, we have addressed the question that many of our students have asked us, "When will I ever use this?"

Accompanying Software Usage

The software that accompanies the *Foundations of Mathematics* workbook has three modes: Learn, Practice, and Certify. For each section of the workbook, you should first go to the software and complete the Learn mode, which provides the basics of the concepts covered in the corresponding section of the workbook. After reading through Learn, you should complete the Understand Concepts section of the workbook. Then, proceed to the Practice mode of the Hawkes Learning software to sharpen your math skills before attempting the Skill Check and Apply Skills sections of the workbook. If your instructor requires you to complete the Certify mode to demonstrate mastery of the material presented in the section, this should be done after finishing the entire workbook section. Keep in mind that there are no directions in the workbook concerning Certify mode since an instructor may decide not to use this mode of the software.

How to Use This Book

Before starting the first chapter, be sure to read through the First Day of Class Resources located at the beginning of the workbook. These have been especially prepared to help you succeed in this math course. Many of the resources and skills presented in this section can be used in your other classes as well. After the first day of class, be sure to fill out the Important Information for Success page and keep this as a handy reference throughout the semester. Once you are ready to start the first chapter, here is how you should proceed through the workbook.

1. Look over the Table of Contents for the chapter to get an idea of the topics presented in the section. Read through the Math@Work Introduction to get a quick overview of a career that uses the mathematics from the chapter.
2. Read through the Study Skill at the beginning of the chapter and refer back to this as needed as you work through the workbook.
3. Read through the Objectives listed at the start of the section and the Success Strategy that is designed to help you be successful with the material in the section.
4. Skim through the section to get an idea of what you will be learning, paying attention to titles, headings, definitions, and diagrams.
5. Read through the Learn mode for the corresponding lesson in the software.
6. Complete the Understand Concepts part of the workbook for the section.
7. Use the Practice mode of the software to reinforce the math skills and concepts learned.
8. Complete the Skill Check and Apply Skills parts of the workbook for the section.
9. Complete the Certify mode, if assigned by your instructor.
10. At the end of a Chapter, be sure to read through the Math@Work to see how the mathematics learned in the chapter would actually be used in a particular career.
11. Complete one or more of the Chapter Projects, as assigned by your instructor.
12. Complete the Foundations Skill Check before proceeding to the next chapter to ensure that you are prepared. If necessary, review the skills from the current and previous chapters, as noted by the section headings above each set of exercises.

Key Features of the Workbook

We have designed the features of this workbook to help you be more successful in learning the concepts and skills presented in the course and enhance your ability to apply these concepts and skills to real-world problems. Here is an overview of some of the key features of the workbook.

Math@Work

At the beginning of each chapter, the Math@Work Introduction presents a brief description of a career or field of study that uses the mathematics contained in the current and previous chapters. This career or field of study is explained in greater depth at the end of the chapter along with an "on the job" situation involving mathematics. This feature helps answer the question often asked by students about where they will ever use the mathematics they are learning and gives you exposure to a variety of careers you may not be aware of.

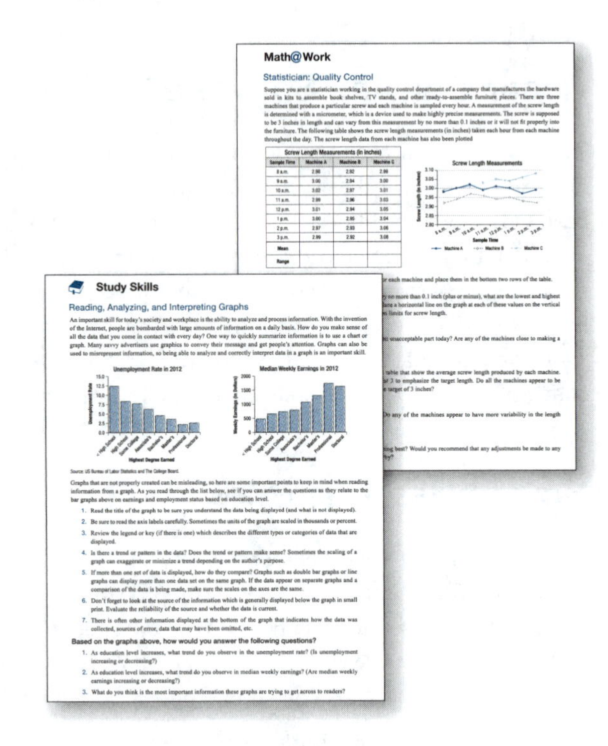
Math@Work

Statistician: Quality Control

Study Skills

Reading, Analyzing, and Interpreting Graphs

Study Skills

At the beginning of the workbook, we have included a set of useful information and study skills to get you off to a good start. New study skills are also included at the beginning of each chapter so you will continue to be successful throughout the course. Many of these study skills can be applied to your other college courses as well, not just math courses.

Objectives & Success Strategy

Objectives

The objectives provide you with a clear and concise list of the main skills taught in each section, which will enable you to focus your time and effort on the most important topics. You should review these objectives after finishing the section to make sure you understand the skills listed and can successfully use them.

Success Strategy

At the beginning of each section of the workbook, we have included a success strategy that is specific to the topics and skills taught in the section. This strategy is designed to help you be more successful in learning, understanding, or applying a particular skill or concept.

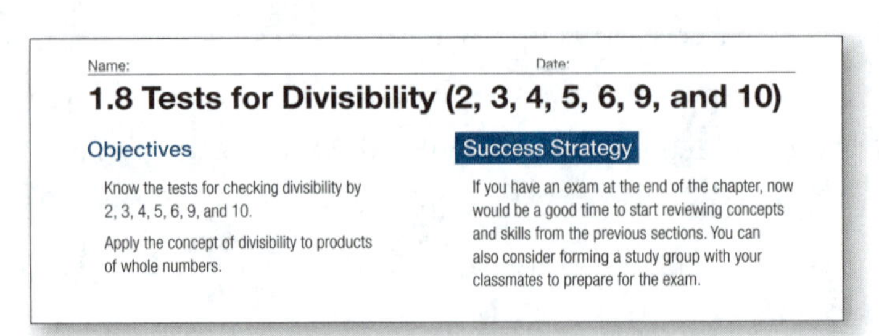
Name: Date:

1.8 Tests for Divisibility (2, 3, 4, 5, 6, 9, and 10)

Objectives

Know the tests for checking divisibility by 2, 3, 4, 5, 6, 9, and 10.

Apply the concept of divisibility to products of whole numbers.

Success Strategy

If you have an exam at the end of the chapter, now would be a good time to start reviewing concepts and skills from the previous sections. You can also consider forming a study group with your classmates to prepare for the exam.

Understand Concepts

This portion of each section focuses on building an understanding of new concepts and skills. This feature includes definitions, formulas, procedures, and other important information that will enhance your understanding of the material presented in the Learn mode of the software. There are also problems that guide you through the steps involved in learning the new concepts and skills presented in the section.

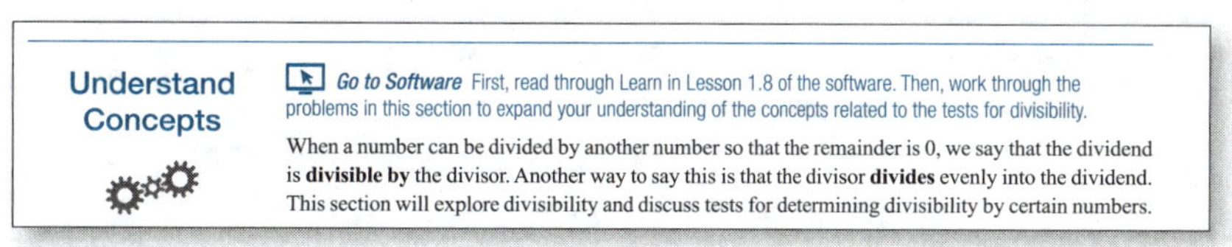
Understand Concepts

Go to Software First, read through Learn in Lesson 1.8 of the software. Then, work through the problems in this section to expand your understanding of the concepts related to the tests for divisibility.

When a number can be divided by another number so that the remainder is 0, we say that the dividend is **divisible by** the divisor. Another way to say this is that the divisor **divides** evenly into the dividend. This section will explore divisibility and discuss tests for determining divisibility by certain numbers.

Skill Check

This portion of each section checks your ability to perform the skills presented in the Understand Concepts part of the workbook or in the Learn mode of the software. Only a few skill problems are presented in each section. The Practice mode of the Hawkes Learning software should be used for more practice.

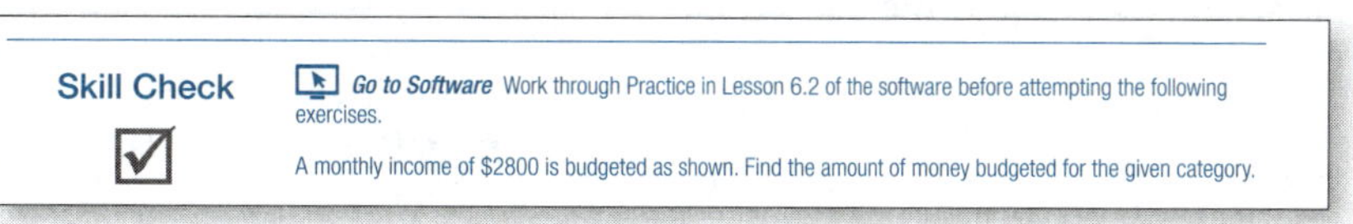
Skill Check

Go to Software Work through Practice in Lesson 6.2 of the software before attempting the following exercises.

A monthly income of $2800 is budgeted as shown. Find the amount of money budgeted for the given category.

Apply Skills

This portion of each section uses the concepts and skills presented earlier in the section and applies them to real-world contexts. The focus of this feature is to build problem-solving skills, which are essential in higher-level math courses and in real life.

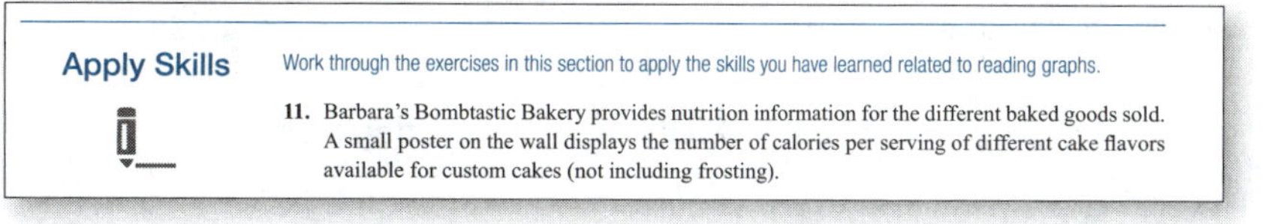
Apply Skills

Work through the exercises in this section to apply the skills you have learned related to reading graphs.

11. Barbara's Bombtastic Bakery provides nutrition information for the different baked goods sold. A small poster on the wall displays the number of calories per serving of different cake flavors available for custom cakes (not including frosting).

Quick Tip & Lesson Link

Quick Tip

These tips appear in the margin throughout each section to provide you with just-in-time information, such as definitions, formulas, or other resources that should help you better understand the material being presented in the section. Quick Tips are also used to clarify terms or concepts, provide help with answer entry or simplification, and provide additional guidance.

Lesson Link

This feature is used to help you recall previously learned skills that are relevant to the current topic or to point out when the current topic will be reintroduced in a later section. Topics that are noted multiple times in the workbook should be considered very important and you should spend extra time studying and understanding them.

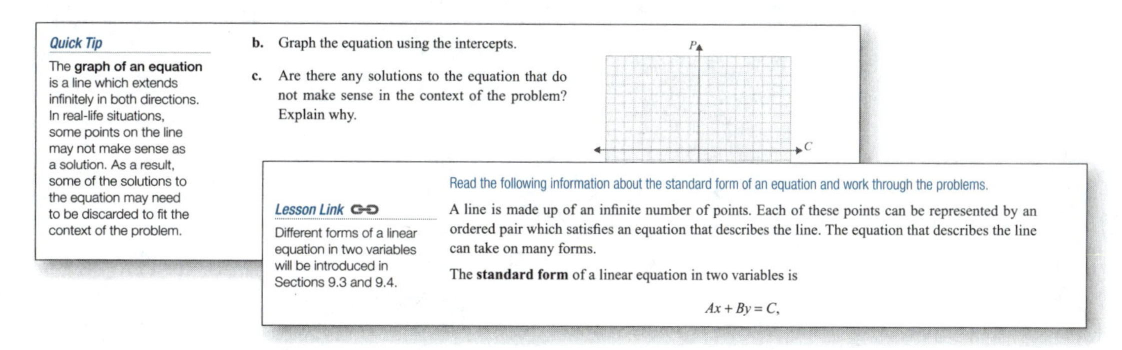
Quick Tip

The **graph of an equation** is a line which extends infinitely in both directions. In real-life situations, some points on the line may not make sense as a solution. As a result, some of the solutions to the equation may need to be discarded to fit the context of the problem.

b. Graph the equation using the intercepts.

c. Are there any solutions to the equation that do not make sense in the context of the problem? Explain why.

Lesson Link

Different forms of a linear equation in two variables will be introduced in Sections 9.3 and 9.4.

Read the following information about the standard form of an equation and work through the problems.

A line is made up of an infinite number of points. Each of these points can be represented by an ordered pair which satisfies an equation that describes the line. The equation that describes the line can take on many forms.

The **standard form** of a linear equation in two variables is

$$Ax + By = C,$$

Projects

There are two projects at the end of each chapter, including the appendix, and two final projects at the end of the workbook. Instructors may use these as individual projects or as group projects, depending on how they have structured the course. The chapter projects incorporate the ideas learned in the current and previous chapters. The final projects incorporate a subset of the important skills learned throughout the course.

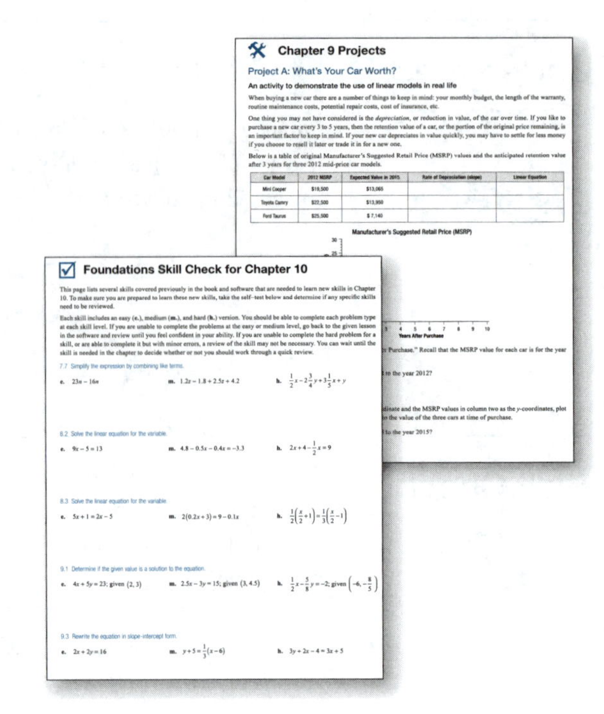

Chapter 9 Projects

Project A: What's Your Car Worth?

An activity to demonstrate the use of linear models in real life

When buying a new car there are a number of things to keep in mind: your monthly budget, the length of the warranty, routine maintenance costs, potential repair costs, cost of insurance, etc.

One thing you may not have considered is the *depreciation*, or reduction in value, of the car over time. If you like to purchase a new car every 3 to 5 years, then the retention value of a car, or the portion of the original price remaining, is an important factor to keep in mind. If your new car depreciates in value quickly, you may have to settle for less money if you choose to resell it later or trade it in for a new one.

Below is a table of original Manufacturer's Suggested Retail Price (MSRP) values and the anticipated retention value after 3 years for three 2012 mid-price car models.

Car Model	2012 MSRP	Expected Value in 2015	Rate of Depreciation (slope)	Linear Equation
Mini Cooper	$19,500	$13,065		
Toyota Camry	$22,500	$13,950		
Ford Taurus	$25,500	$7,140		

Foundations Skill Check for Chapter 10

This page lists several skills covered previously in the book and software that are needed to learn new skills in Chapter 10. To make sure you are prepared to learn these new skills, take the self–test below and determine if any specific skills need to be reviewed.

Each skill includes an easy (**e.**), medium (**m.**), and hard (**h.**) version. You should be able to complete each problem type at each skill level. If you are unable to complete the problems at the easy or medium level, go back to the given lesson in the software and review until you feel confident in your ability. If you are unable to complete the hard problem for a skill, or are able to complete it but with minor errors, a review of the skill may not be necessary. You can wait until the skill is needed in the chapter to decide whether or not you should work through a quick review.

7.7 Simplify the expression by combining like terms.

e. $23n - 16n$ **m.** $1.2z - 1.8 + 2.5z + 4.2$ **h.** $\frac{1}{2}x - 2\frac{3}{4}y + 3\frac{1}{5}x + y$

8.2 Solve the linear equation for the variable.

e. $9x - 5 = 13$ **m.** $4.8 - 0.5x - 0.4x = -3.3$ **h.** $2x + 4 - \frac{1}{2}x = 9$

8.3 Solve the linear equation for the variable.

e. $5x + 1 = 2x - 5$ **m.** $2(0.2x + 3) = 9 - 0.1x$ **h.** $\frac{1}{2}\left(\frac{x}{2} + 1\right) = \frac{1}{3}\left(\frac{x}{2} - 1\right)$

9.1 Determine if the given value is a solution to the equation.

e. $4x + 5y = 23$; given $(2, 3)$ **m.** $2.5x - 3y = 15$; given $(3, 4.5)$ **h.** $\frac{1}{2}x - \frac{5}{8}y = -2$; given $\left(-6, -\frac{8}{5}\right)$

9.3 Rewrite the equation in slope-intercept form.

e. $2x + 2y = 16$ **m.** $y + 5 = \frac{1}{3}(x - 6)$ **h.** $3y + 2x - 4 = 3x + 5$

Foundations Skill Check

This feature is a self-test for you to check your understanding of previously learned skills that will be needed in the next chapter. An easy, medium, and hard version of each skill is included to more accurately test your knowledge. If you have difficulty with any particular skill in the foundations skill check, be sure to review the section associated with that skill.

Icons

This icon indicates that access to the Internet is required to complete the problem or project.

This icon indicates that you should work through the specified mode of the software before completing the section of the workbook.

This icon indicates that the Quick Tip is about using technology, such as a calculator or software.

Our vision while creating this workbook was to more efficiently and effectively prepare students like you for the mathematics courses you will take in college, and to hopefully change your way of thinking about mathematics as a whole. We hope that after completing this course, you will no longer think of mathematics as something you must endure in order to reach your educational goals, but instead as a subject you need to master in order to make your life and career more productive and rewarding.

Wishing you much success,

Barbara L. Miller

L. Robin Hendrix

Foundations of Mathematics
A Quantitative Reasoning Approach

Table of Contents

Letter to the Student iii
First Day of Class Resources xii
First Day of Class Projects xix

Chapter 1: Whole Numbers 1
Study Skills 2
1.1 Reading and Writing Whole Numbers 3
1.2 Addition and Subtraction with Whole Numbers 7
1.3 Multiplication with Whole Numbers 11
1.4 Division with Whole Numbers 17
1.5 Rounding and Estimating with Whole Numbers 21
1.6 Exponents and Order of Operations 27
1.7 Problem Solving with Whole Numbers 33
1.8 Tests for Divisibility (2, 3, 4, 5, 6, 9, and 10) 39
1.9 Prime Numbers and Prime Factorizations 43
Chapter 1 Projects 47
Math@Work 51
Foundations Skill Check for Chapter 2 52

Chapter 2: Fractions and Mixed Numbers 53
Study Skills 54
2.1 Introduction to Fractions and Mixed Numbers 55
2.2 Multiplication and Division with Fractions and Mixed Numbers 61
2.3 Least Common Multiple (LCM) 67
2.4 Addition and Subtraction with Fractions 73
2.5 Addition and Subtraction with Mixed Numbers 79
2.6 Order of Operations with Fractions and Mixed Numbers 83
Chapter 2 Projects 89
Math@Work 93
Foundations Skill Check for Chapter 3 94

Chapter 3: Decimal Numbers 95
Study Skills 96
3.1 Introduction to Decimal Numbers 97
3.2 Addition and Subtraction with Decimal Numbers 103
3.3 Multiplication with Decimal Numbers 107
3.4 Division with Decimal Numbers 111
3.5 Decimal Numbers and Fractions 115
Chapter 3 Projects 121
Math@Work 125
Foundations Skill Check for Chapter 4 126

Chapter 4: Ratios and Proportions, Percent, and Applications 127

Study Skills 128
4.1 Ratios and Proportions 129
4.2 Solving Proportions 133
4.3 Decimal Numbers and Percents 139
4.4 Fractions and Percents 143
4.5 Solving Percent Problems Using the Proportion $P/100 = A/B$ 147
4.6 Solving Percent Problems Using the Equation $R \cdot B = A$ 151
4.7 Applications: Discount, Sales Tax, Commission, and Percent Increase/Decrease 155
4.8 Applications: Profit, Simple Interest, and Compound Interest 161
Chapter 4 Projects 167
Math@Work 171
Foundations Skill Check for Chapter 5 172

Chapter 5: Geometry 173

Study Skills 174
5.1 Angles 175
5.2 Perimeter 181
5.3 Area 185
5.4 Circles 189
5.5 Volume and Surface Area 193
5.6 Triangles 199
5.7 Square Roots and the Pythagorean Theorem 205
Chapter 5 Projects 209
Math@Work 213
Foundations Skill Check for Chapter 6 214

Chapter 6: Statistics, Graphs, and Probability 215

Study Skills 216
6.1 Statistics: Mean, Median, Mode, and Range 217
6.2 Reading Graphs 223
6.3 Constructing Graphs from Data Sets 231
6.4 Probability 237
Chapter 6 Projects 241
Math@Work 245
Foundations Skill Check for Chapter 7 246

Chapter 7: Introduction to Algebra 247

Study Skills ... 248
7.1 The Real Number Line and Absolute Value ... 249
7.2 Addition with Real Numbers ... 255
7.3 Subtraction with Real Numbers ... 259
7.4 Multiplication and Division with Real Numbers ... 263
7.5 Order of Operations with Real Numbers ... 269
7.6 Properties of Real Numbers ... 273
7.7 Simplifying and Evaluating Algebraic Expressions ... 277
7.8 Translating English Phrases and Algebraic Expressions ... 283
Chapter 7 Projects ... 287
Math@Work ... 291
Foundations Skill Check for Chapter 8 ... 292

Chapter 8: Solving Linear Equations and Inequalities 293

Study Skills ... 294
8.1 Solving Linear Equations: $x + b = c$ and $ax = c$ ... 295
8.2 Solving Linear Equations: $ax + b = c$ ... 301
8.3 Solving Linear Equations: $ax + b = cx + d$ ... 305
8.4 Applications: Number Problems and Consecutive Integers ... 309
8.5 Working with Formulas ... 313
8.6 Applications: Distance-Rate-Time, Interest, Average ... 317
8.7 Linear Inequalities ... 323
Chapter 8 Projects ... 329
Math@Work ... 333
Foundations Skill Check for Chapter 9 ... 334

Chapter 9: Linear Equations and Inequalities in Two Variables 335

Study Skills ... 336
9.1 The Cartesian Coordinate System ... 337
9.2 Graphing Linear Equations in Two Variables: $Ax + By = C$ ... 343
9.3 The Slope-Intercept Form: $y = mx + b$ ... 349
9.4 The Point-Slope Form: $y - y_1 = m(x - x_1)$ ... 355
9.5 Introduction to Functions and Function Notation ... 361
9.6 Graphing Linear Inequalities in Two Variables ... 367
Chapter 9 Projects ... 373
Math@Work ... 379
Foundations Skill Check for Chapter 10 ... 380

Chapter 10: Systems of Linear Equations 381

Study Skills 382
10.1 Systems of Linear Equations: Solutions by Graphing 383
10.2 Systems of Linear Equations: Solutions by Substitution 389
10.3 Systems of Linear Equations: Solutions by Addition 395
10.4 Applications: Distance-Rate-Time, Number Problems, Amounts, and Costs 399
10.5 Applications: Interest and Mixture 403
Chapter 10 Projects 409
Math@Work 413
Foundations Skill Check for Chapter 11 414

Chapter 11: Exponents and Polynomials 415

Study Skills 416
11.1 Exponents 417
11.2 Exponents and Scientific Notation 423
11.3 Introduction to Polynomials 429
11.4 Addition and Subtraction with Polynomials 433
11.5 Multiplication with Polynomials 437
11.6 Special Products of Binomials 441
11.7 Division with Polynomials 447
Chapter 11 Projects 451
Math@Work 455
Foundations Skill Check for Chapter 12 456

Chapter 12: Factoring Polynomials and Solving Quadratic Equations 457

Study Skills 458
12.1 Greatest Common Factor and Factoring by Grouping 459
12.2 Factoring Trinomials with Leading Coefficient 1 465
12.3 Factoring Trinomials with Leading Coefficient Not 1 469
12.4 Special Factoring Techniques 473
12.5 Additional Factoring Practice 477
12.6 Solving Quadratic Equations by Factoring 481
12.7 Applications of Quadratic Equations 485
Chapter 12 Projects 489
Math@Work 493
Foundations Skill Check for Chapter 13 494

Chapter 13: Rational Expressions 495

Study Skills 496
13.1 Multiplication and Division with Rational Expressions 497
13.2 Addition and Subtraction with Rational Expressions 503
13.3 Complex Fractions 509
13.4 Solving Equations with Rational Expressions 513
13.5 Applications of Rational Expressions 517
13.6 Variation 521
Chapter 13 Projects 525
Math@Work 529
Foundations Skill Check for Chapter 14 530

Chapter 14: Radicals 531

Study Skills 532
14.1 Roots and Radicals 533
14.2 Simplifying Radicals 537
14.3 Addition, Subtraction, and Multiplication with Radicals 543
14.4 Rationalizing Denominators 549
14.5 Equations with Radicals 555
14.6 Rational Exponents 559
14.7 Functions with Radicals 563
Chapter 14 Projects 567
Math@Work 571
Foundations Skill Check for Chapter 15 572

Chapter 15: Quadratic Equations 573

Study Skills 574
15.1 Quadratic Equations: The Square Root Method 575
15.2 Quadratic Equations: Completing the Square 579
15.3 Quadratic Equations: The Quadratic Formula 583
15.4 Applications 587
15.5 Quadratic Functions 591
Chapter 15 Projects 597
Math@Work 601

Appendix 603

Study Skills 604
A.1 US Measurements 607
A.2 The Metric System 611
A.3 US to Metric Conversions 615
A.4 Absolute Value Equations and Inequalities 621
A.5 Synthetic Division and the Remainder Theorem 627
A.6 Graphing Systems of Linear Inequalities 631
A.7 Systems of Linear Equations in Three Variables 635
A.8 Introduction to Complex Numbers 641
A.9 Multiplication and Division with Complex Numbers 645
A.10 Standard Deviation and z-Scores 649
A.11 Mathematical Modeling 655
Appendix Projects 669

Final Projects 673

Index 679

First Day of Class Resources

Contents

- Important Information for Success
- Where Do I Go For Help?
- Dispelling Myths About Math
- How You Learn Best: Left-Brained vs. Right-Brained
- Organization and Time Management
- The Basic Steps of Problem Solving
- First Day of Class Projects

Important Information for Success

You should read the Study Skill *Where Do I Go for Help?* to understand why the information below is important for your success in this course. Return here after your first day of class to fill in the information.

Instructor's Name: ______________________ Instructor's Office Location: ______________________

Instructor's Office Hours: ______________________ Instructor's E-mail Address: ______________________

Class days and times: ______________________

Hours of operation of the academic success center: ______________________ Hours of operation of the library: ______________________

Contacts from my math class: ______________________

____

____

Web links for my math e-book, courseware, or tech support: ______________________

____

____

Internet resources I have found: ______________________

____

____

Where Do I Go for Help?

All students need academic help at some point in their college career, so it is important to be aware of the resources that are available to you. You should take advantage of these resources and locate them in advance, so that when you do need help, you know where to go immediately.

1. Your Instructor At most colleges, instructors are required to hold a minimum number of office hours. The days and times of the instructor's office hours can typically be found on their syllabus. This is the time the instructor has set aside to be available to students who may need additional help. If you have questions that you did not get answered in class or concerns about an upcoming test, this is a great time to get one-on-one help. If the instructor's office hours do not fit with your schedule, be sure to talk to the instructor after class and see if he or she is available to meet with you at a time that is more convenient for you. Even if you don't have any immediate questions for the instructor, it is a good idea to find out where their office is and just drop in to say hello and get to know them a little better.

2. Tutoring Most colleges have a Learning Center or Student Success Center where they offer tutoring in most subject areas, often free of charge. The center may offer small group tutoring, one-on-one tutoring, and online tutoring. You may be able to just walk in and get help, but often you will need to set up an appointment. You should find out the hours of operation of the tutoring center and note them in the front of this workbook. If online tutoring is available, find out the hours of operation, as it may differ from that of the tutoring center on campus. **For more information on tutoring, see the Study Skill for Chapter 10.**

3. Establish Contacts in Class Going to the tutoring center is not a substitute for attending class. If you miss a class, you are responsible for learning the information presented in class and making up any missed assignments. It is a good idea to exchange contact information with one or two classmates in each of your classes. Then, if you are unable to attend class and want to know what material was covered or if any work was assigned during your absence, you have someone you can contact besides the instructor.

4. Study Groups If the professor allows or encourages group work in or out of class, you might want to consider forming a study group with some of your classmates. The hardest part in establishing a study group is finding a time that is convenient for everyone. Typically, three to six people are ideal for a study group. If you include too many people in the group, it will be hard to find a time that works best for everyone to meet and it may be hard to get any work done. **For information on how to work effectively in a group, see the Study Skill for Chapter 12.**

5. Library Most college campuses have a library that is open with extended hours for students to access books and reference materials. Libraries typically have study areas available for students to study in between classes and often provides access to computers for typing up papers or working online. You may find other versions of textbooks by different authors or study aids that may explain the material you are learning in class differently or in a way that you understand better. Most college libraries also have a website where you can access thousands of books, reference materials, and journals through their online databases.

6. Publisher Supplemental Materials Many textbooks, especially math and science textbooks, have CDs or a website address included with the textbook that contain additional practice problems, chapter projects or activities, practice tests, and review materials. They may also have a companion website where you can find videos, solution manuals to purchase, and other resources. The publisher may also provide courseware to accompany the textbook where you can complete homework assignments, create your own practice tests, or use interactive activities to explore concepts.

7. Internet Resources There are numerous videos on YouTube that explain how to do common types of math problems. There are other reputable websites such as Purplemath, Khan Academy, Math is Fun, and others that publish free math resources for anyone to use. Your college may also have free math resources posted on its website or may have references to reputable sources to obtain help. **For more information about online resources, see the Study Skill for Chapter 14.**

Be sure to go back and fill out the information on the previous page for future reference!

Dispelling Myths about Math

Have you ever said to someone, "I'm just not good at math" or "My brain just isn't wired to do math"? Many of us have heard about the left-brain vs. right-brain dominance theory. According to this theory, people who think with the left part of their brain are considered to be more logical and analytical, and therefore better at math. People who think with the right part of their brain are considered to be more creative and intuitive, and therefore better at subjects like English, art, and music. There is actually some truth to this theory but don't fear, there's more to the story.

Knowing which side of your brain is dominant is helpful in that it will help you understand how you learn best. Are you a visual learner (right-brain) or an auditory learner (left-brain) or somewhere in between? Do you have to perform tasks in a sequential manner (left-brain) or are you able to multitask and do many things at once (right-brain)? Knowing how you learn best will help you determine how to approach learning in the classroom and help you develop good study habits that coincide with your learning style. You can find an array of psychological inventories on the Internet to help you determine your brain dominance.

So, if it turns out that you are right-brained, does this mean you are doomed in math? No, not at all! Did you know that you can actually increase your brainpower and become smarter? Many people think you are either born with certain abilities or you're not. In other words, some people believe that you are either good at math or you aren't, and that can't be changed. However, research has shown that to be incorrect. According to Stanford University psychology professor Carol Dweck, the human brain is like a muscle. The more you exercise it, the stronger it becomes. In fact, the more you struggle with a concept or activity, the more likely you are to develop more brainpower!

Within the last 10 years, scientists have learned that the brain can actually generate new neurons throughout one's entire life through a process called "neurogenesis." Neurons are special cells that are responsible for sending, receiving, and interpreting information from all parts of the body. Scientists now believe that learning itself actually stimulates neurogenesis. So, this means that your intelligence level is not a static trait. Likewise, your mathematical abilities are not static and they can be improved.

You should look at learning as a challenge, especially learning math. Understand that when you struggle with a tough subject, such as math, you are providing your brain with the opportunity to expand and grow intellectually. Don't give up easily. Perseverance is the key to math success!

Source: Carol Dweck, Mindset (New York: Ballantine Books, 2006), 221.

How You Learn Best: Left-Brained vs. Right-Brained

The brain is one of the most complex and remarkable organs of the body. It is divided into two hemispheres with the left hemisphere controlling the right part of the body and the right hemisphere controlling the left part of the body. You can tell a lot about a person by knowing which hemisphere is dominant. People who are right-brain dominant tend to be more intuitive and emotional. People who are left-brain dominant tend to be more logical and analytical. Also, people who are right-brain dominant tend to excel in sports, art and music, whereas, people who are left-brain dominant tend to excel in areas like math, law, and accounting.

You can find an array of psychological inventories or questionnaires on the Internet that will help you determine your brain dominance. Do some research on the Internet using the keywords "brain dominance" or "left-brain vs. right-brain," and find one or more inventories on brain dominance. Fill out a couple of the inventories to find out which brain dominance you exhibit the most.

If you are right-brained, don't think that you can't be good at math. Remember the information on *Dispelling Myths about Math?* You can actually improve, or "grow," your brain, therefore improving your math skills. In this resource, we are going to give some tips on improving your study skills based on your brain dominance.

Right-brained people are very visual and have the ability to focus on patterns. Much of math is about patterns. Instead of trying to memorize rules or formulas, look for patterns in the mathematics. Also, being a visual learner, pay attention to figures and diagrams in the textbook. Although it may be difficult at first, read through each step of the textbook examples thoroughly and see if you can mimic the steps and explain each of them. When having difficulty completing a homework problem, find an example problem similar to it in the textbook. Use the steps in the example as a pattern or guide to work through the problem you are having difficulty with. Look for video clips on the Internet that may help you in solving particular types of problems.

Left-brained people tend to be more verbal, focusing on words and symbols. As a result, they tend to be auditory learners, which means they learn by listening. Auditory learners typically do not like to take notes, which may hinder them when trying to recall what an instructor said in class a few weeks earlier. If an instructor allows it, recording the instructor's lectures on tape would help with recalling the information later on. Video clips on the Internet can also be helpful in recalling concepts you may have forgotten, since most contain detailed step-by-step explanations of the math work presented in the video.

People with right-brain dominance have a tendency to be disorganized. If this applies to you, then this is something you definitely need to work towards improving. Organizational skills are necessary when taking any college course, not just math. It is important that you keep your notes, homework assignments, and any graded work that has been returned to you in the same binder or folder for easy reference. Mathematics courses typically build upon concepts learned at the beginning and expand on these concepts as you progress through the course. As a result, final exams are often cumulative, which means you will be tested on all the major concepts learned throughout the course. Don't throw away any notes or assignments, as you may need them later on.

People who are left-brain dominant tend to be highly organized and, in some cases, too organized. They may spend so much time organizing their work that they have less time to study and prepare for exams or presentations. Try to find a balance and make sure that you have ample time to prepare for major assignments.

People who are right-brain dominant often have difficulty setting priorities, which means they often procrastinate and fail to make deadlines. This is an important skill for anyone attending college as you often have multiple assignments due in different courses at the same time. Keep a calendar, either on paper or electronically, on which you place all assignments for all courses as well as any personal commitments. Print it out on a weekly basis and carry it with you to avoid having to hastily throw an assignment together at the last minute. Last-minute assignments typically don't earn the highest grade!

People with left-brain dominance tend to be very logical and detail-oriented, but sometimes can miss the "big picture" by focusing on lesser important parts of an assignment or project. Sometimes it is necessary to step back from the details and make sure the overall goal or key point has been adequately addressed. Also, keeping a calendar with too much detail can be stressful and time-consuming, taking valuable time away from more important assignments.

Attending class on a regular basis is a must for both left and right-brained learners. As mentioned earlier, math concepts tend to build upon each other. Missing class may result in you missing an important concept or procedure upon which the next class or section of material is based.

Keep in mind that brain-dominance is only one factor that may affect how you learn, and that many people exhibit behaviors or traits of both left- and right-brained thinkers. Understanding how you learn best and realizing your shortcomings and how to overcome them is the primary goal of this study skill.

Summary of Traits	
Left-Brain Dominant	**Right-Brain Dominant**
Logical	Intuitive
Organized	Disorganized
Detailed	Sees "big picture"
Analytical	Creative
More detached	Emotional
Visual	Auditory

Sources:

http://www.personalityquiz.net/profiles/leftrightbrain.htm

http://www.scholastic.com/teachers/article/left-brainright-brain

Organization and Time Management

In college, it is very important to be well-organized and keep a schedule or calendar of all assignment due dates. You should write down all class times on your calendar, as well as any personal or family events so that you have everything in one place. If you have a job, you should also put your work hours on the schedule. Be sure to also mark time on your calendar for routine studying and homework. Keep in mind that you may need extra study time right before quizzes and tests. In college, a good rule of thumb is to spend 2 to 3 hours studying outside of class for every hour in class. For a full-time schedule of 12 hours, this would result in 24 to 36 hours a week spent on studying.

Keeping a calendar with all your time constraints on it will help to reduce stress since you won't have any surprise assignments to "blindside" you at the last minute. Also, don't take on more personal commitments than you can handle while in college. Learn to say "no" to people and remember what your main focus and goals are. Remember to also give yourself time to eat well and get plenty of rest.

To keep work organized, some students like to keep a separate binder for each class, while other students like to keep one large binder with everything in it. Use what works for you, as long as you are able to keep up with your notes and assignments. If you use one large binder and it starts getting too big, then you may want to remove some of the older notes and put them in a separate place. Do not throw away any work or notes until the class is over. Instructors sometimes make errors in their gradebook and online gradebooks may get corrupted or lost, so keep all work until final grades are determined. You should also consider keeping your notes as reference for a more advanced course that you need to take in the future.

Don't spend so much time organizing your notes and assignments that it consumes all your study time. You need to make sure you have plenty of time to study for quizzes and tests, and are not rushing through your study time because you spent too much time organizing and highlighting your notes. It is also important to organize your study area so you have quick access to the materials you need and don't have to spend a lot of time searching for materials to do your work. Be sure to set up your study area to minimize distractions, so don't create a study area in front of the TV. Turn off your phone and your radio when you study so that you can concentrate.

During the week, establish an evening routine to allow time for family, meals, exercise, studying, and plenty of rest. You need to also establish a morning routine that gives you time for a good breakfast and possibly some time to do some last-minute studying before class. Make sure you have a dependable alarm clock and that you give yourself plenty of time to get to class and not be late. Be sure to check the expected weather conditions the night before, so you can plan to get up a little earlier if necessary. Always plan for the unexpected by getting to campus ahead of time, so you won't feel rushed and panicky, especially on days that you have a test or major assignment due. Some instructors have very strict late policies, so be sure to be on time if this is the case. Also, some instructors may require an entry ticket or assignment that must be shown to gain access to the classroom. Make sure you have this assignment ready the night before it is due and have it safely put away in your notebook so it isn't forgotten if you have to leave in a hurry the next morning.

The Basic Steps of Problem Solving

Throughout this workbook, you will be asked to solve problems involving real-world applications. It won't be an easy process to start with, but it will get easier over time. This page provides you with a tool called *Pólya's four-step problem-solving approach* to help you get started on the right path to becoming an expert problem solver. Later on, in Chapter 12, we will introduce you to some other problem-solving strategies that you might like better and that may be easier to remember.

According to a Hungarian mathematician named George Pólya, the reasoning for solving math problems is the same as solving everyday problems that come up in real life. As a result, you shouldn't need a different method of problem solving just for math.

There are four steps to this problem-solving technique.

1. **Understand the problem.**
    - **a.** Read the problem.
    - **b.** Understand all the words.
    - **c.** If it helps, restate the problem in your own words.
    - **d.** Be sure that there is enough information.
2. **Devise a plan.**
    - **a.** Guess, estimate, or make a list of possibilities.
    - **b.** Draw a picture or diagram.
    - **c.** Formulate a mathematical expression or equation to answer the problem.
3. **Carry out the plan.**
    - **a.** Try all the possibilities you have listed.
    - **b.** Study your picture or diagram for insight into the solution.
    - **c.** Evaluate your expression or solve any equation that you may have set up.
4. **Look back over the results.**
    - **a.** Can you see an easier way to solve the problem?
    - **b.** Does your solution actually work? Does it make sense in terms of the original wording of the problem? Is it reasonable?
    - **c.** If there is an equation, check your answer in the equation.

Go to http://en.wikipedia.org/wiki/How_to_Solve_It for more information on Pólya's problem-solving approach and examples of questions to ask when solving problems.

Name: Date:

First Day of Class Projects

"What we hope to do with ease, we must learn first to do with diligence." -Samuel Johnson

Perseverance Project

1. Write down some of your thoughts about the subject of math.

2. Now, read the information on *Dispelling Myths about Math.*

3. Did you learn anything from reading this article that changed any of your thoughts from Problem 1?

4. When working on a tough problem (math or otherwise), how would you rate your perseverance in finding the solution to the problem on a scale from 1 to 10 (with 1 being "I give up easily" and 10 being "I don't stop till I get the right answer")?

5. Write a goal statement for yourself with regard to your level of perseverance for this math course.

6. If you start having difficulty with the material presented in the course, list at least three resources that you can use for help.

7. What do you think the quote at the top of this page by Samuel Johnson means and how does it relate to perseverance?

Left-Brained vs. Right-Brained Project

Do an Internet search to find one or more inventories on brain dominance. Take at least two of the inventories you found and compare the results.

Here are some resources you can use:

http://www.personalityquiz.net/profiles/leftrightbrain.htm

http://www.scholastic.com/teachers/article/left-brainright-brain#quiz

1. Based on the inventories do you exhibit more left-brained or more right-brained characteristics?

2. Do you agree with this assessment? Explain why or why not?

3. List at least three traits or behaviors associated with your brain-dominance, as determined from the inventories, that you believe apply to you the most.

4. If you haven't already read the Study Skill *How You Learn Best: Left-Brained vs. Right-Brained* in this section, read it now. List at least three things that you learned from your reading that you can to do to increase your success in this math course.

Chapter 1: Whole Numbers

Study Skills

1.1 Reading and Writing Whole Numbers

1.2 Addition and Subtraction with Whole Numbers

1.3 Multiplication with Whole Numbers

1.4 Division with Whole Numbers

1.5 Rounding and Estimating with Whole Numbers

1.6 Exponents and Order of Operations

1.7 Problem Solving with Whole Numbers

1.8 Tests for Divisibility (2, 3, 4, 5, 6, 9, and 10)

1.9 Prime Numbers and Prime Factorizations

Chapter 1 Projects

Math@Work

Foundations Skill Check for Chapter 2

Math@Work

Introduction

If you plan to go into business management, you can choose to run an already established business or start your own company. Either way, you will be responsible for the day to day operations of the business, from hiring employees to determining which new products to produce and sell. You will need to constantly analyze various aspects of the company, from sales records to employee time sheets, and determine if the company is running as smoothly as possible. This analysis requires strong math skills and the ability to communicate your findings and decisions to the owner of the company and the employees.

As the manager of a business, one part of your job may be to manage the inventory of the company you work for, which is the amount of each product the company has in stock and ready to sell. Some companies design and sell a product but have it manufactured by another company. When managing the inventory of a company, there are several questions that you will need to know the answers to. How do you determine how much of which products to order? How much will it cost to reorder the needed products and ship them to the warehouse? Finding the answers to these questions (and many more) requires several of the skills covered in this chapter. At the end of the chapter, we'll come back to this problem and explore how the inventory for a business is managed.

Study Skills

Doing Research on the Web

Throughout this workbook, you may be asked to do some research or look up information on the Internet. The Internet is a useful tool, but you have to be careful. Not everything you read on the Internet is factual and some websites are more reliable than others. Here are some tips on using the Internet to make sure the information you are finding is quality information and to help you more efficiently navigate the Internet to save you time.

First, there is some basic terminology that you need to understand:

Browser: The software program used to access the Internet. Typical browsers include Internet Explorer, MSN Explorer, AOL, Netscape, Firefox, Google Chrome, Opera, and Safari.

Search engine: This is the Internet tool that allows you to search for a particular topic or website using a word or phrase typed into the browser's search bar. The two most popular search engines are Google and Bing.

URL: This acronym stands for Universal Resource Locator and it is the address of a website. For example, http://www.wikipedia.org is the URL for a free online encyclopedia website.

Tips on Using the Internet

1. The three letters at the end of a URL give you some indication as to the reliability of the information you might obtain from the website. The sites near the beginning of the list tend to be more reliable than those at the end, although there can be exceptions.

   a. .edu: colleges and universities (K–12 public schools use xx.us where xx is the state abbreviation.)
   b. .gov: US Federal Government sites
   c. .mil: US military sites
   d. .org: nonprofit organizations
   e. .com: commercial
   f. .net: Internet companies
   g. .biz: businesses

2. Always look up your information on at least two different websites and compare it. If the two agree, then the information has some factual basis. If they do not, then look for a third and possibly fourth website to confirm the information.

3. Determine who owns or supports the website to make sure it is not biased in its opinions or favors a particular group's beliefs or views.

4. For larger research projects and papers, try to find peer-reviewed articles, which are articles reviewed by an impartial panel of the author's peers who are experts in the field. These are often available in online databases through your college library or a nearby local library. These articles usually contain citations and references to other sources that may also be helpful in your research.

5. Always evaluate the author's credentials of any article, blog, or research study that you find on the Internet. What degrees or titles do they hold? Where was the research conducted? Did the results agree with other similar studies? If this level of detail is not provided, it is probably not a reliable source.

6. Online encyclopedias such as Wikipedia can be used as a starting point to obtain general information about a topic. They will often provide you with links to more reliable information that is cite-worthy. Wikipedia provides a list of citations at the bottom of the entry and will post a warning if something is not very well cited.

7. Stay away from websites that allow anyone to publish or edit information on the site. You want to use a website whose content is written by experts in the field and can be documented as to the source of the information.

8. Check the date on the information to be sure that you have the most recent facts on a topic. Older articles may provide some background for your research, but always look for the most current information available.

Name: Date:

1.1 Reading and Writing Whole Numbers

Objectives

Understand the place value system.

Write a whole number in standard and expanded notation.

Write a whole number as its English word equivalent.

Success Strategy

As you work through the Hawkes Learning software, write down the key terms and their definitions in a notebook or in the margins of this workbook.

Understand Concepts

Go to Software First, read through Learn in Lesson 1.1 of the software. Then, work through the problems in this section to expand your understanding of the concepts related to reading and writing whole numbers.

Whole Numbers and Natural Numbers

The **whole numbers** are the **natural numbers** (also known as the **counting numbers**) and the number 0.

Natural numbers: {1, 2, 3, 4, 5, 6, ...}

Whole numbers: {0, 1, 2, 3, 4, 5, 6, ...}

1. In previous math courses, you've worked with several different types of numbers besides the natural numbers and whole numbers. Write down the names of these other types of numbers that you remember and describe the key features of each type. We'll check back on this later to see how your current knowledge matches the information given in this course.

Quick Tip

A **digit** is a single piece of a number like a letter is a single piece of a word. Numbers are made from digits in a specific order like words are made from letters in a specific order.

2. The numerical system we use depends on a place value system to determine the value of a digit in a number. Each place in the place value system represents a multiple of 10. In the table, the top row represents the period name.

a. Fill in the second row with the place value name for each position.

b. Fill in the third row with the multiple of ten that each place value represents.

Millions period			Thousands period			Ones period			Decimal
								Ones (or units)	
								1	.

3. Why do you think it is important to learn how to properly read and write all types of numbers in both their numerical form and their English word equivalent?

Place Value System

The **decimal system** (or **base ten system**) is a place value system that depends on three things:

1. the **ten digits**: 0, 1, 2, 3, 4, 5, 6, 7, 8, 9
2. the **placement** of each digit; and
3. the **value** of each place.

While searching for the answers to the following questions, refer to the Doing Research on the Web information at the beginning of the chapter.

Quick Tip

While searching on the Internet, refer to the *Doing Research on the Web* information at the beginning of the chapter.

4. The number system we use is called a "base ten" system because it uses ten digits (0, 1, 2, 3, 4, 5, 6, 7, 8, 9) and its place value system depends on the number 10. Perform an Internet search using the key words "base number systems" to find answers for the following questions.

   a. Do other base number systems exist? If so, how do they compare to the base ten system?

   b. Do any groups of people or professions use different base number systems? If so, what base number system do they use and how?

Lesson Link

In Section 3.1 we will discuss the proper use of the word "and" in mathematics.

5. The use of the word "and" when writing whole numbers in their English word equivalent is considered to be improper or incorrect. However, this usage occurs a lot in conversation and media. For example, the title of the story *One Hundred and One Dalmatians* by Dodie Smith uses the "and" improperly according to mathematical rules, but we understand which number is intended by the context.

   a. Think of two other improper uses of the word "and" where the meaning is clear by context.

   b. Why do you think the meaning of each phrase is clear?

Name: ______________________ Date: ______________

Skill Check

Go to Software Work through Practice in Lesson 1.1 of the software before attempting the following exercises.

Write the following whole numbers in expanded notation.

6. 634

7. 7512

Quick Tip

Hyphens are used when writing the English word equivalent of two-digit words larger than 20. An example of this is "twenty-five."

Write the following whole numbers in their English word equivalent.

8. 414,646

9. 94,840

Write the given number in standard notation.

10. one hundred five thousand, six hundred seventeen

11. fifty-eight thousand, two hundred fifty-nine

Apply Skills

Work through the problems in this section to apply the skills you have learned related to reading and writing whole numbers.

12. During a talk show, a guest says that his website has four hundred eighty thousand one hundred five views per day. The transcriptionist writes this number as "408,105". Is this the correct standard notation? If not, what should it be?

13. A jewelry maker made one thousand five hundred dollars during one week. She keeps track of her income and writes "$1500" in a spreadsheet. Is this the correct standard notation? If not, what should it be?

14. Kady, a pharmacist, calls a pharmaceutical company to order more medicine of a certain type. While ordering she says she needs seventeen thousand eight hundred fifty tablets. The sales representative tells Kady that the order will cost one thousand two hundred forty-nine dollars. How would Kady and the sales representative write these numbers on a sales invoice (in standard form)?

Quick Tip

Problem 15 highlights some of the "improper" ways that people say numbers that don't match their standard English word equivalents. Through experience, the meanings of the phrases are understood.

15. Nikola is an intern at a warehouse where his job is to listen to voice mails and fill out order forms. He listens to the following voice mail.

"Hello. This is Ethan Taylor calling from store number one four nine five. I'm calling in my weekly order to be delivered on June fifteenth. We need fifty-nine of item one thousand one. We need one hundred forty-five of item fifteen oh six. We also need seven hundred seventy-five of item nine fifty-two. That's all. Thank you."

a. How would Nikola fill out the order form?

Name:	
Delivery Date:	
Store Number:	
Item Number	**Quantity**

b. Which numbers were said in a non-standard English word equivalent? How should these be written in their English word equivalents?

Name: Date:

1.2 Addition and Subtraction with Whole Numbers

Objectives

Add whole numbers.

Subtract whole numbers.

Recognize and use the properties of addition.

Calculate the perimeter of geometric figures.

Success Strategy

Drawing a diagram or picture may be helpful when solving word problems. Remember, "a picture is worth a thousand words."

Understand Concepts

Go to Software First, read through Learn in Lesson 1.2 of the software. Then, work through the problems in this section to expand your understanding of the concepts related to adding and subtracting whole numbers.

1. Label the parts of an addition and a subtraction problem.

$$\begin{array}{r} 123 \\ +\,278 \\ \hline 401 \end{array}$$ ________ ________ ________

$$\begin{array}{r} 500 \\ -\,209 \\ \hline 291 \end{array}$$ ________ ________ ________

Read the following paragraph about a different way to think of subtracting numbers with borrowing and work through the problems.

When simplifying subtraction problems such as 1000 – 276, the minuend and the subtrahend can be decreased by 1 to make calculations easier and remove the need for borrowing. In this case, the resulting subtraction problem is 999 – 275. In both situations, the difference will be 724.

2. Calculate the following and determine whether or not the alternate method would be useful for the given calculation to eliminate the need for borrowing.

a. 1200 – 624 **b.** 206 – 190

Properties of Addition		
Property	**Examples**	
Commutative Property	**Algebraic**	**Numerical**
The order of the numbers in the addition statement can be reversed.	$a+b=b+a$	$14+6=6+14$
Associative Property	**Algebraic**	**Numerical**
The grouping of numbers in the addition statement can be changed.	$(a+b)+c=a+(b+c)$	$(1+2)+3=1+(2+3)$
Identity Property	**Algebraic**	**Numerical**
Addition by 0 leaves the original number unchanged.	$a+0=a$	$5+0=5$

Perimeter

The **perimeter** of a geometric figure is the distance around the figure.

3. You may recall from previous math courses that you need a formula to find area. However, to determine the perimeter of a figure such as a rectangle you really don't need to use a formula. Why is this? (**Hint:** Think about how you would determine the perimeter of different figures such as a triangle or trapezoid.)

True or False: Determine whether each statement is true or false. Rewrite any false statement so that it is true. (There may be more than one correct new statement.)

Quick Tip

Try to set up a few addition or subtraction problems that match the situations described in the True or False problem statements.

4. When you borrow a 1 from the hundreds place, you can think of it as borrowing 10 tens.

5. When you carry a 1 from the tens place and add it to the hundreds place, you are really adding 10 to the entire sum.

6. When you borrow a 1 from the millions place, you can think of it as borrowing 100 hundred thousands.

7. When finding the sum of $128 + 153$, you need to carry a 1. This 1 is added to the ones place.

Skill Check

Go to Software Work through Practice in Lesson 1.2 of the software before attempting the following exercises.

Add.

8. $\begin{array}{r} 7527 \\ +\,2050 \\ \hline \end{array}$

9. $\begin{array}{r} 4826 \\ +\,9375 \\ \hline \end{array}$

Name: Date:

Find the perimeter of each figure.

10.

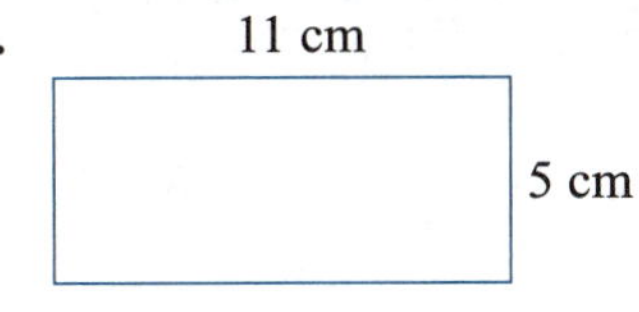

11.

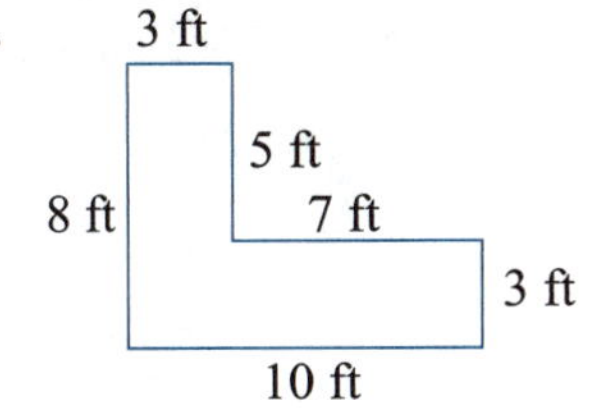

Subtract.

12. $\begin{array}{r} 294 \\ -181 \\ \hline \end{array}$

13. $\begin{array}{r} 4500 \\ -3476 \\ \hline \end{array}$

Apply Skills

Quick Tip

Key words for addition include *add*, *plus*, *increase by*, *sum*, and *total*.

Key words for subtraction include *subtract from*, *minus*, *decreased by*, *less*, and *difference*.

Work through the problems in this section to apply the skills you have learned related to adding and subtracting whole numbers. Circle any key words or phrases that suggest that either addition or subtraction should be performed.

14. Theo is keeping track of how many visitors his website receives each day. The website receives 858 visitors on the first day, 723 visitors on the second day, and 1055 visitors on the third day. How many visitors did Theo's website receive during those three days?

15. The Smith family is creating a simple monthly budget to keep track of their spending habits. Each month they pay $1085 for their mortgage. Their average monthly utility bills are $104 for electric, $49 for water & sewer, and $109 for a cable TV, Internet, and mobile phone bundle. They budget $550 per month for food. They also have a car payment of $295 per month, insurance payment of $65 per month, and they estimate $225 for fuel.

a. Fill in the table with the missing information.

Category	Expenses
Mortgage	$1085
Utilities	
Food	$550
Car Costs	

b. How much money does the Smith family need to budget for these expenses each month?

c. If their combined family income is $3500 per month (after taxes), how much money do they have for other bills and expenses each month?

16. Three friends rent a local party room for a holiday party. Due to the local fire code, the maximum occupancy of the party room is 125 people. Manuel invites 43 people, George invites 52 people, and Ali invites 27 people to the party.

a. If everyone accepts the invitation and attends the party, how many people will be at the party?

b. If everyone accepts the invitation and attends the party, will they be within the maximum occupancy?

Quick Tip

When solving problems in real life, the original solution may need to be adjusted based on new information. Problem 17 describes such a situation.

17. Daria has a small garden in her backyard. Lately she has had trouble with rabbits and deer eating her vegetables, so she wants to put a fence around the garden. The garden is in the shape of a rectangle with a length of 12 feet and a width of 8 feet.

a. How many feet of fencing will Daria need to buy?

b. After determining how much fencing she needs to buy, Daria realizes that she doesn't need to put any fencing on one side of the garden because it is along the side of her garage. The side next to the garage has a length of 12 feet. With this new information, how many feet of fencing will Daria need to buy?

c. While at the home improvement store, a salesperson suggests that Daria install a gate in the fence so she doesn't have to climb over the fence to work in her garden. The gate that she picks out is 4 feet wide and will go in place of one section of fencing. Not including the length of the gate, how many feet of fencing will Daria need to buy?

Name: Date:

1.3 Multiplication with Whole Numbers

Objectives

Identify the parts of a multiplication problem.

Multiply whole numbers.

Recognize and use the properties of multiplication.

Recognize and use the distributive property.

Calculate the area of a rectangle.

Success Strategy

Practice mental math to increase your speed at solving problems and decrease your dependency on calculators.

Understand Concepts

Quick Tip

Multiplication can be time consuming so learning the multiplication tables for whole numbers up to 12 can save a lot of time. To practice multiplication of numbers 1 through 12, visit: http://gadgets.hawkeslearning.com/flashcards

Go to Software First, read through Learn in Lesson 1.3 of the software. Then, work through the problems in this section to expand your understanding of the concepts related to multiplying whole numbers.

1. Label the parts of a multiplication problem.

45 ________

$\times$ 5 ________

225 ________

2. The following table lists the properties that are common for both addition and multiplication. The property and its statement are given. Fill in the columns for addition and multiplication with an example of each property.

Properties of Addition and Multiplication		
Property	**Example**	
Commutative Property	**Addition**	**Multiplication**
The order of the numbers in the operation can be reversed.		
Associative Property	**Addition**	**Multiplication**
The grouping of numbers can be changed.		
Identity Property	**Addition**	**Multiplication**
The operation by this number leaves the original number unchanged.		
Zero-Factor Law	**Addition**	**Multiplication**
The product of 0 and a number is always equal to 0.	Not applicable	

The Distributive Property

For any whole numbers a, b, and c, $a(b+c)=a\cdot b+a\cdot c$.

3. In math, there are often different ways to do things. This problem demonstrates two different methods of multiplying. The method you use will depend on which you like better and which is easiest for you.

a. Compare the results of multiplying $42 \cdot 15$ and finding the sum of $42 \cdot 10$ and $42 \cdot 5$.

b. How would you rewrite this relationship using the distributive property?

c. Show how you can extend this idea to multiplying three digit numbers such as $256 \cdot 425$.

4. While no formulas are given yet in this course, you may remember some formulas for area from previous math classes. If so, write them down and what they represent. We'll check back on this later to see how your current knowledge matches the information given in this course.

Read the following paragraph about the word "identity" and work through the problems.

Addition and multiplication both have something called the "identity". For a particular operation the identity is the number that always leaves the value it is operating on unchanged. The value of the identity depends on the operation being performed: addition or multiplication.

5. First, let's look at the additive identity.

a. If you are using the operation of addition, what value would you need to add to the number 5 to leave it unchanged?

In other words, put a number in the following blank that makes this equation true: $5 + \_\_\_ = 5$.

Quick Tip

A letter in an equation or expression is called a **variable**. A variable is a symbol which is used to represent an unknown number or any one of several numbers.

b. Does the value you determined in part **a.** work for all whole numbers a in this equation: $a + \_\_\_ = a$?

c. Based on your results from parts **a.** and **b.**, the additive identity is _____.

Name: ______________________ Date: ______________________

6. Next, let's look at the multiplicative identity.

a. If you are using the operation of multiplication, what value would you need to multiply the number 5 by to leave it unchanged?

In other words, put a number in the following blank that makes this equation true: $5 \cdot \_\_\_ = 5$.

b. Does the value you determined in part **a.** work for all whole numbers a in this equation: $a \cdot \_\_\_ = a$?

c. Based on your results from parts **a.** and **b.**, the multiplicative identity is _____.

7. In math, the word *unique* means that only one exists.

a. Is the additive identity unique? That is, is your answer from Problem 4 part **c.** the only number that we can use to get the same result as Problem 4 parts **a.** and **b.**? If it is not unique, what other number(s) can be used as the additive identity?

b. Is the multiplicative identity unique? If not, what other number(s) can be used as the multiplicative identity?

Read the following information about square units and work through the problems.

In math we often use words that have different meanings in different situations. The word *unit* is one of these words. It has two different meanings which are important in relation to area. One definition of unit is "the first and least counting number" or "a single number which represents a whole." Another important definition of unit is "a defined quantity (of length, time, etc.) adopted as a standard of measurement."

8. What number does the first definition refer to?

9. What are some different units of length?

In area, a *square unit* is something which combines both definitions of unit. You may remember that a square unit is what we use to cover up a shape to determine its area. A square unit has a side length of 1 and the length is measured in a specific unit of measurement. The unit of measurement used will vary depending on the situation. In general, it looks like this:

1 unit

1 unit

(This figure of a square unit is general and the unit can be changed to inches, feet, miles, etc., depending on the area you are measuring.)

Quick Tip

The **area of a rectangle** is equal to its length multiplied by its width.

A square is a rectangle, so the area of this square unit is equal to its length times its width. The area of a square inch is calculated as follows:

$$\text{Area} = 1 \text{ inch} \cdot 1 \text{ inch} = (1 \cdot 1) \cdot (\text{inch} \cdot \text{inch}) = 1 \text{ inch}^2$$

10. What properties allow us to change the order and the grouping of numbers and units when multiplying to find area?

11. The figure shown is divided into square inches.

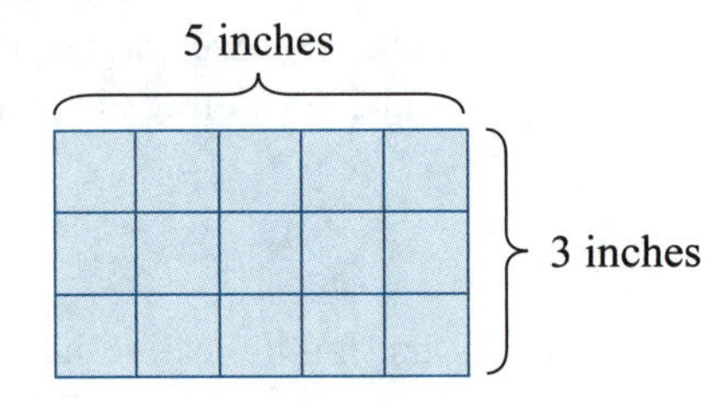

Quick Tip

Using units in mathematical calculations is important because it keeps track of measurements. This becomes especially important in Section A.3 when **dimensional analysis** is used. (Dimensional analysis is the process of finding an equivalent value of a measurement expressed in a different unit of measurement.)

a. Find the area of the figure by counting the square units.

b. Find the area by multiplying the side lengths together (be sure to include the units).

c. Do the areas from parts **a.** and **b.** match? Explain why or why not.

Skill Check

Go to Software Work through Practice in Lesson 1.3 of the software before attempting the following exercises.

Find each product.

12. $\begin{array}{r} 437 \\ \times\ 7 \\ \hline \end{array}$

13. $\begin{array}{r} 420 \\ \times 104 \\ \hline \end{array}$

Name: ______________________ Date: ______________________

Evaluate the expression.

14. $4(5+9)$

15. $9(8+3)$

Find the area of the following figures. Be sure to label your answers with the correct units.

16.

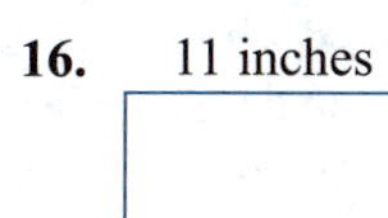

17.

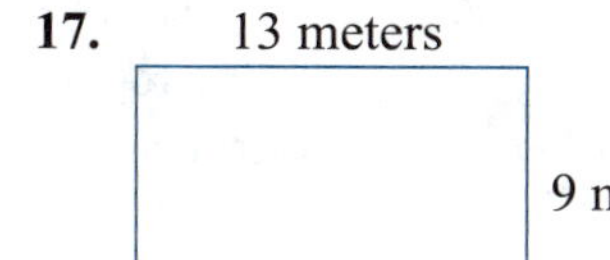

Apply Skills

Work through the problems in this section to apply the skills you have learned related to multiplying whole numbers.

Since these problems are in a section covering only multiplication, it may be obvious to you which operation to use to solve the problem. However, if these problems were not given in this section, it may require more thought to figure out which operation to use. Think about how you would know which operation to use for each problem. Are there keywords for multiplication? Does the context of the problem indicate multiplication? After solving each problem, give a reason for using multiplication. Circle any key words or phrases that suggest multiplication.

Quick Tip

Key words that indicate multiplication are *multiply*, *times*, *each*, *per*, *of*, *double*, *triple*, *twice*, *factors*, and *product*.

18. The United States Postal Service sells sheet of stamps which have 4 rows of 5 stamps each. How many stamps are sold in a sheet?

19. The International Space Station (ISS) is a large spacecraft designed for humans to live and do research in. It is currently in orbit around the Earth and manned by six astronauts from different countries. According to NASA, the ISS travels at an average speed of 18,000 miles per hour. How far does the ISS travel in one day?

20. While looking for a part-time summer job, you get two job offers. With the first job you would make \$9 an hour and work 32 hours per week. With the second job you would make \$11 per hour and work 24 hours per week.

a. How much would you make per week at each job (before taxes)?

b. Which is the better job offer? Give a reason to support your answer.

21. A pharmacy typically dispenses 520 tablets of a certain medicine every day, seven days per week. To make sure there is enough of this medicine in stock, the head pharmacists orders more every two weeks. How many tablets of the medicine does the head pharmacist need to order every two weeks?

22. The maintenance crew at an office building is tasked with laying new tile in the reception room. The room is a 12 foot by 8 foot rectangle. The tile to be laid is in the shape of a 1 foot by 2 foot rectangle. The maintenance crew buys 3 boxes of tile which contain 18 tiles each.

a. What is the area of the reception room?

b. What is the area of each tile?

c. What is the total area that the purchased tiles can cover?

d. Did the maintenance crew buy enough tiles for the reception room?

Name: Date:

1.4 Division with Whole Numbers

Objectives

Divide whole numbers.

Know how division is related to multiplication.

Know how to divide with 0.

Learn the long division process.

Success Strategy

Checking your work whenever possible is a good habit to develop. Try to always check your work on in-class and homework assignments. Also, if you finish a test or exam early, be sure to use the extra time to check over your test.

Understand Concepts

Go to Software First, read through Learn in Lesson 1.4 of the software. Then, work through the problems in this section to expand your understanding of the concepts related to dividing whole numbers.

1. Label the parts of a division problem.

$$\underline{\qquad\qquad} \quad 4\overline{)20}^{\;5} \quad \begin{matrix}\underline{\qquad\qquad} \\ \underline{\qquad\qquad}\end{matrix}$$

Read the following paragraph about the term *undefined* and work through the problems.

In math, the term **undefined** is very important since everything we learn is based on solid and clear definitions. In a mathematical context, *undefined* means that something does not have meaning and no interpretation has been assigned to it. All of the operations in math work because they are clearly defined, which means that the result is easily determined and there is only one result. For instance, no one who understands division will argue against you when you say $10 \div 2 = 5$. This result can be verified by the equation $10 = 2 \cdot 5$. Using this idea of checking division with multiplication, let's see what makes division by zero *undefined*.

2. We will start by claiming that 7 divided by 0 is equal to some number n. Writing this as an equation gives us $7 \div 0 = n$.

a. Pick any value for n and substitute it into the equation.

b. We are able to check division by using multiplication. To check the division equation from part **a.**, we use the expression

$7 = \_\_\_\_ \cdot \_\_\_\_$

c. What do we know about multiplication by zero (the zero-factor law)?

Lesson Link

The zero-factor law was covered in Section 1.3.

d. Is the multiplication statement from part **b.** true?

3. Think about what would happen if we pick a different number for n in the equation $7 \div 0 = n$.

a. Can you find any number which would make the multiplication check $7 = 0 \cdot n$ a true statement?

b. What does this suggest about the simplification of $7 \div 0$?

4. Will the outcome of the simplification change if we use a different number other than 7 to divide by zero? Provide an example to support your answer.

Quick Tip

While searching on the Internet, refer to the *Doing Research on the Web* information at the beginning of the chapter.

5. The number zero is important in mathematics and has a very interesting history. In fact, it wasn't part of the original counting systems developed thousands of years ago. Perform an Internet search using the keywords "history of zero" to research the history of the number zero and answer the following questions.

a. What are the names of two civilizations that developed the concept of zero independently?

b. Did these civilizations have a symbol for zero? If so, what did they use for zero?

True or False: Determine whether each statement is true or false. Rewrite any false statement so that it is true. (There may be more than one correct new statement.)

6. If the remainder is 0, then both the divisor and the quotient are factors of the dividend.

7. The remainder can be equal to the divisor.

8. Any nonzero number divided by itself is equal to 1.

9. Any nonzero number divided by 1 is equal to 1.

Name: Date:

Describe the mistake made while performing long division on the expression and then correctly divide the expression to find the actual result.

10. $16\overline{)3280}$ with quotient 25:

```
     25
16)3280
   32
    08
     0
    80
    80
     0
```

11. $12\overline{)1680}$ with quotient 140:

```
     140
12)1680
   12
    48
    48
     0
```

Skill Check

Go to Software Work through Practice in Lesson 1.4 of the software before attempting the following exercises.

Find each quotient and remainder.

12. $0\overline{)15}$

13. $145 \div 6$

14. $\frac{8338}{22}$

15. $202\overline{)65656}$

Apply Skills

Work through the problems in this section to apply the skills you have learned related to dividing whole numbers.

Since these problems are in a section on division, division will be used to find the answer. However, if they were not in this section, it may not be obvious that division should be used. Think about how you would know which operation to use for each problem. Are there keywords to indicate division is to be used? Does the context of the problem indicate division? After solving each problem, give a reason for using division. Circle any key words or phrases that suggest division.

Quick Tip

Key words that indicate division are *divide*, *quotient*, *ratio*, and *per*.

16. Sophia decides to use a store credit deal where she won't have to pay any interest on her purchase if she pays off the entire purchase amount within a year. She bought $1920 worth of furniture and plans to make equal-sized monthly payments for six months. How much will Sophia pay per month?

17. Xander has $150 to spend on computer games at an online summer sale.

a. If the average price of a computer game in the sale is $16 (including tax), how many games can he buy?

b. How much money will he have left over?

18. Barbara's Bombtastic Bakery sells their signature cupcake in boxes of four. If they make 312 of these signature cupcakes in the morning, how many boxes of four can be sold during the day?

19. A box of cereal has a net weight of 429 grams. If a serving of cereal is 33 grams, how many servings are there per box of cereal?

20. A lantern manufacturer is preparing their newly created lanterns for storage at the warehouse. They have 7685 lanterns packaged and ready to go.

a. They can fit 8 packaged lanterns into a box. How many boxes do they need?

b. How many lanterns are left over after packing them into boxes?

c. They can safely stack 20 boxes on a wooden pallet. How many pallets do they need? (Only use the completely full boxes.)

d. If 8 pallets can fit on a single shelving system in the warehouse, how many shelving systems will these lanterns take up?

Name: Date:

1.5 Rounding and Estimating with Whole Numbers

Objectives

Know how to round whole numbers.

Estimate sums and differences by using rounded numbers.

Estimate products by using rounded numbers.

Estimate quotients by using rounded numbers.

Success Strategy

Rounding and estimating come in handy when checking your work. It can give you a rough idea if your answer is reasonable or not. This is especially useful when you don't have a calculator on hand!

Understand Concepts

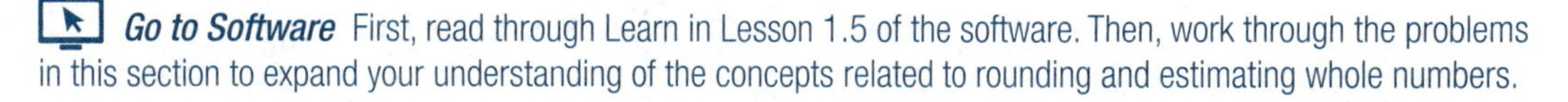

Go to Software First, read through Learn in Lesson 1.5 of the software. Then, work through the problems in this section to expand your understanding of the concepts related to rounding and estimating whole numbers.

Rounding Rules

1. Look at the single digit just to the right of the digit that is in the place of desired accuracy.
2. **If this digit is 5 or larger,** increase the digit in the desired place of accuracy by one and replace all digits to the right with zeros.
3. **If this digit is less than 5,** leave the digit in the desired place of accuracy the same and replace all digits to the right with zeros.

1. When rounding whole numbers, the reason that we round 5 up instead of down may not be clear. Think back to what you know of the place value system. We use a base ten system which means we have ten digits to work with in each place.

Quick Tip

Remember that **digits** are the symbols that make up a number just as letters are the symbols that make up a word.

a. What are the digits in the base ten number system? Write them in increasing order.

b. What are the first five digits in the base ten number system from part **a.**?

c. What are the last five digits in the base ten number system from part **a.**?

d. Circle the correct words: The first five digits are rounded (up/down) and the last five digits are rounded (up/down).

Read through the following information on estimation and work through the problems.

Estimating

1. Round each number to the place of the **leftmost** digit.
2. Perform the indicated operation with these rounded numbers.

Lesson Link

Estimation will be covered again in multiple sections of Chapter 3. Chapter 3 deals with decimal numbers, but the concepts for rounding and estimating remain the same.

The method of estimation taught in this course (to round everything to the leftmost digit) is the quickest and easiest method of estimation, but it may not always give the most useful estimate. While estimation is never completely accurate, you can develop a sense of judgment to obtain an estimate that is useful for the situation. (For homework in the Hawkes Learning software, however, use the leftmost digit rule!)

Listed in the box are some alternative methods of rounding that can be used when finding an estimate. The rounding method you use while estimating will depend on the situation and how accurate you need the estimate to be.

Alternate Methods for Rounding While Finding an Estimate

a. Round to the nearest hundred.

b. Round to the nearest ten.

c. Round to the nearest multiple of fifty.

d. Round **up** to the nearest ten, hundred, etc.

2. Think of two additional ways to round numbers while estimating that we can add to the list.

Quick Tip

In some situations, such as estimating a purchase price, overestimating is preferred so a person doesn't spend more money than they have. In other situations, such as purchasing inventory, underestimating is preferred so there isn't too much inventory in stock that doesn't sell.

3. Overestimating or underestimating can both have negative consequences. These need to be taken into consideration before determining which rounding method to use. This table shows some situations where estimation can be used along with some negative consequences of underestimating or overestimating. Fill in the missing consequences.

Situation	Underestimate	Overestimate
Ordering bakery ingredients	Not enough ingredients for the week	
Estimating purchase price		Think purchase price is more money than you have to spend
Purchasing inventory items		Too much in stock, doesn't sell

The next problems describe situations where estimation may be used. Answer the questions while keeping the previous two problems in mind.

4. The recipe for a signature treat at a bakery requires 11 eggs. Each week they make this recipe 28 times. The owner needs to approximate how many eggs are used to order ingredients for the next week.

a. Which rounding method would you use to find the estimate? (Pick a method from the list or create a new method.)

b. Find the estimated value and the actual value.

c. Does this overestimate or underestimate the actual value needed? What are possible consequences of this?

d. Do you think the rounding method you chose provides a good estimate? Give a reason to support your answer.

5. Samuel is moving his used book store to a larger location and needs to buy medium-sized boxes to move all of the books. Each box can hold 36 novels. If he currently has 782 novels in stock, approximately how many boxes will he need just for the novels?

   a. Which rounding method would you use to find the estimate?

   b. Find the estimated value and the actual value.

   c. Does this overestimate or underestimate the actual value needed? What are possible consequences of this?

   d. Do you think the rounding method you chose provides a good estimate? Give a reason to support your answer.

6. A trucking company wants to estimate how many miles will be traveled before scheduling a new delivery. The delivery requires three stops. The truck must travel 573 miles to the first stop, 845 miles to the second stop, and 639 miles to the third stop.

   a. Which rounding method would you use to find the estimate?

   b. Find the estimated value and the actual value.

   c. Does this overestimate or underestimate the actual value needed? What are possible consequences of this?

d. Do you think the rounding method you chose provides a good estimate? Give a reason to support your answer.

7. Michelle is starting her own business and has $44,750 in start-up money. It will cost her $17,990 to obtain a business license, register her company name, and rent a furnished office space for six months. Michelle wants to approximate how much money she will have left.

a. Which rounding method would you use to find the estimate?

b. Find the estimated value and the actual value.

c. Does this overestimate or underestimate the actual value needed? What are possible consequences of this?

d. Do you think the rounding method you chose provides a good estimate? Give a reason to support your answer.

Skill Check

Go to Software Work through Practice in Lesson 1.5 of the software before attempting the following exercises.

Round as indicated.

8. 768, to the nearest ten

9. 6495, to the nearest thousand

Estimate the given expression. Use the leftmost digit rounding rule.

10. $475 + 715 + 852$

11. $11{,}593 - 849$

12. $315 \cdot 72$

13. $4972 \div 23$

Name: Date:

Apply Skills

Work through the problems in this section to apply the skills you have learned related to rounding and estimating whole numbers.

Comparing an estimate to its actual value is the easiest way to determine if the estimate is reasonable or not. While this isn't always reasonable to do in everyday life (why estimate when you are going to determine the actual value anyway?), it is useful in practice to develop your sense of judgment for which rounding method to use when estimating.

14. Emilia is sending her three children to different week long daytime summer camps. The cost of the summer camp is $239 for the youngest child, $487 for the middle child, and $350 for the oldest child.

a. Estimate how much it will cost to send the three kids to summer camp by using the left-most digit rounding rule.

b. Is this estimate reasonable? If not, explain what is unreasonable about the estimate and find another way to round the numbers.

15. Oliver has $160 and is going to buy a ticket to visit an amusement park for $44. He wants to know how much money he'll have left to spend on food and souvenirs.

a. Estimate how much money Oliver will have left by using the leftmost digit rounding rule.

b. Is this estimate reasonable? If not, explain what is unreasonable about the estimate and find another way to round the numbers.

16. Omar drives a semi-trailer truck for 11 hours each day. His average speed is 68 miles per hour.

a. Estimate how many miles Omar drives per day by using the leftmost digit rounding rule.

b. Is this estimate reasonable? If not, explain what is unreasonable about the estimate and find another way to round the numbers.

17. A statewide food drive plans to share the donated food equally between food kitchens across the state. People donated a total of 1488 cans of green beans. There are 48 food kitchens that will receive donations.

a. Estimate how many cans of green beans each food kitchen will receive by using the left-most digit rounding rule.

b. Is this estimate reasonable? If not, explain what is unreasonable about the estimate and find another way to round the numbers.

Name: Date:

1.6 Exponents and Order of Operations

Objectives

Identify the base and exponent in an exponential expression.

Know how to evaluate expressions containing exponents.

Evaluate expressions with 1 and 0 as exponents.

Know the rules for order of operations.

Success Strategy

When learning the order of operations, doing a lot of practice in the software can really pay off. The order of operations is not intuitive which means it is a topic which can be difficult for many students.

Understand Concepts

Go to Software First, read through Learn in Lesson 1.6 of the software. Then, work through the problems in this section to expand your understanding of the concepts related to exponents and the order of operations.

Exponential Expressions

Repeated multiplication of the same number by itself can be shortened into an **exponential expression**.

For example, $5 \cdot 5 \cdot 5 = 5^3$. In this expression, the 5 is called the **base** and the 3 is called the **exponent**, or **power.**

A number with an exponent of 1 is equal to itself. For example, $2^1 = 2$.

A number with an exponent of 0 is equal to 1. For example, $7^0 = 1$.

When working with exponents, a^0 is defined to be 1 for any value of a. Let's explore why this is true for two different numbers.

Quick Tip

An exponential expression such as 2^4 is read as "two to the fourth power" or "two to the power of four."

1. First, we'll choose $a = 3$. The first two powers of 3 have been completed in the table.

a. Fill in the values for 3^2 and 3^1.

Variable Notation	$a = 3$	Pattern
a^4	$3^4 = 81$	
a^3	$3^3 = 27$	
a^2	$3^2 =$ __	
a^1	$3^1 =$ __	
a^0	$3^0 =$ __	

b. Looking at the values of column two of the table, what is the pattern when moving from one row to the row below it? Write this pattern in the table.

c. If you continue this pattern, going from 3^1 to 3^0, what algebraic expression would you need to evaluate?

d. Evaluate the expression from part **b.** and place the answer in the table. Does it match the rule $a^0 = 1$?

2. Now, you will follow the same method as Problem 1 for a different value of a. Choose from the counting numbers, which are the values 1, 2, 3, ... (Remember that 0 is not a counting number.)

   a. Pick any counting number you want (besides 3) for the base a and fill in the first four rows of the second column of the table.

Variable Notation	Your Value: a = __	Pattern
a^4		
a^3		
a^2		
a^1		
a^0		

   b. Looking at the values of column two of the table, what is the pattern when moving from one row to the row below it? Write this pattern in the table.

   c. If you continue this pattern, going from your value of a^1 to a^0, what algebraic expression would you need to evaluate?

   d. Evaluate this expression and place the answer in the table. Does it match the $a^0 = 1$ rule?

3. Do you now understand why a^0 is defined to be 1 for any value of a? If your answer is no, try to figure out what doesn't make sense or ask your instructor for help.

Name: Date:

Lesson Link

Knowing the perfect squares of numbers will be helpful in Section 5.7 when simplifying square roots is covered. Knowing the perfect squares and perfect cubes will be helpful in Chapter 14 when simplifying radicals is covered.

4. Learning the perfect squares and perfect cubes of numbers can help you perform calculations more quickly. Fill in the table with the squared and cubed values of each number.

Number	1	2	3	4	5	6	7	8
Square								
Cube								

Number	9	10	11	12	13	14	15	16
Square								
Cube								

The order of operations is an important concept in mathematics with an interesting history. Read through the following information about the order of operations and then work through the following problems.

Order of Operations

1. Simplify within grouping symbols such as parentheses ( ), brackets [ ], or braces { }. If there are multiple grouping symbols, start with the innermost grouping.
2. Evaluate any numbers or expressions indicated by exponents.
3. From left to right, perform any multiplication or division in the order they appear.
4. From left to right, perform any addition or subtraction in the order they appear.

5. Why do you think the operations need to be performed from left to right? Try simplifying the expression $2 \cdot 4 + 15 - 9 - 6 \div 2$ in several different ways, once following the order of operations and at least once following a different ordering.

Did you know that the order of operations you learn in math courses is a relatively new set of rules? They were created in the early 1900s and solidified into the current form along with the creation of computers and computer languages. Before the 1600s, mathematical notation was not commonly used and mathematical expressions and equations were written out in words. Any phrasing that was ambiguous (that is, could be understood in more than one way) was avoided.

Quick Tip

While searching on the Internet, refer to the *Doing Research on the Web* information at the beginning of the chapter.

6. Using the keywords "history of math symbols", see if you can find when different math symbols were first used. Write the symbols and dates here.

Read the following paragraph about expressions and work through the problems

Expression

An **expression** is a string of mathematical symbols that makes sense according to the rules of math. These symbols can include numbers, variables, and any operation symbol.

For now we will only be working with numerical expressions. This just means that the pieces in the expressions will be limited to numbers and the operation symbols you have learned so far. A numerical expression will always represent a specific number and can be simplified to that number.

Quick Tip TECH

One way to verify that the order of operations was followed correctly while simplifying an expression is to check it by using a scientific calculator that allows the entire expression to be typed in before calculating. If the expression is entered exactly as it is written, the calculator will follow the order of operations and give the correct result.

7. When following the order of operations to simplify expressions, grouping symbols are used to change the order in which the operations are evaluated. The following two expressions contain the same numbers and operations in the same order, but the second expression uses parentheses. Simplify each expression and compare the results.

a. $50 \div 5 \cdot 2 + 8 \cdot 4^2$

b. $50 \div (5 \cdot 2) + (8 \cdot 4)^2$

c. Was there a difference between which operation you evaluated first in part **a.** and the operation you evaluated first in part **b.**? Explain why or why not.

Determine if any mistakes were made while simplifying each expression. If any mistakes were made, describe the mistake and then correctly simplify the expression to find the actual result.

8. $2 + 3 \cdot 7 - 10 \div 2$
$= 5 \cdot 7 - 5$
$= 35 - 5$
$= 30$

9. $3^2 - 5 + 4^2 \div 8$
$= 9 - 5 + 16 \div 8$
$= 9 - 5 + 2$
$= 9 - 7$
$= 2$

Skill Check

Go to Software Work through Practice in Lesson 1.6 of the software before attempting the following exercises.

Evaluate each exponential expression.

10. 2^5

11. 3^4

Evaluate each numerical expression by using the order of operations.

12. $8 \cdot 2 + 5 \cdot 6 \div 3$

13. $2 \cdot 52 - 4 \cdot 22$

Name: Date:

Quick Tip

Grouping symbols located within grouping symbols should be evaluated from the innermost set to the outermost set.

14. $3+(8\cdot5+4)\div11-7$

15. $(4+1)^2\left[(12-9)^2-14\div2\right]$

Apply Skills

Work through the problems in this section to apply the skills you have learned related to exponents and the order of operations.

Before working through the following problems, review the *Basic Steps for Problem Solving* found in the First Day of Class Resources.

Lesson Link

The method for problem solving will be discussed more in depth in Section 1.7.

16. Neville bought 15 boxes of trading cards. Each box has 10 packs of trading cards. Each pack of trading cards contains 20 cards. He adds 132 cards that he already owns to the newly purchased cards. Then, Neville evenly distributes all of the cards to 6 of his friends. How many trading cards would each person get?

a. If you simplify the expression $15 \cdot 10 \cdot 20 + 132 \div 6$ using the order of operations, will you get the correct answer? If not, explain what is wrong with the expression.

b. What is the correct answer? If necessary, rewrite the expression from part **a.** to get the correct result by following the order of operations.

17. Robert is purchasing shirts for his weekend soccer team. The shirts he wants to buy are normally \$25 each but are on sale for \$10 off. His team has a total of 11 players. How much will he spend to buy the shirts?

a. If you simplify the expression $11 \cdot \$25 - \$10$ using the order of operations, will you get the correct answer? If not, explain what is wrong with the expression.

b. What is the correct answer? If necessary, write the expression from part **a.** to get the correct result by following the order of operations.

18. Camila is a seamstress and is creating wedding dresses. She has 126 yards of silk fabric. For each dress, the skirt requires 4 yards of silk and the bodice requires 2 yards of silk. How many dresses can she make with the amount of silk she has?

a. If you simplify the expression $126 \div 4 + 2$ using the order of operations, will you get the correct answer? If not, explain what is wrong with the expression.

b. What is the correct answer? If necessary, rewrite the expression from part **a.** to get the correct result when following the order of operations.

Name: Date:

1.7 Problem Solving with Whole Numbers

Objectives

Learn the basic strategy for solving word problems.

Analyze and solve word problems involving numbers.

Analyze and solve word problems involving consumer items.

Analyze and solve word problems involving checking accounts.

Analyze and solve word problems involving average.

Success Strategy

If you haven't read through The Basic Steps of Problem Solving located in the First Day of Class Resources at the beginning of the workbook, you may want to do that now. Problem solving is a skill that takes practice and time to master.

Understand Concepts

Go to Software First, read through Learn in Lesson 1.7 of the software. Then, work through the problems in this section to expand your understanding of the concepts related to problem solving.

The **average** of a list of numbers is a sort of "middle" or "center" value of the list of numbers. Let's explore what this really means.

Lesson Link

Average will be covered again, along with two other types of "center," in Section 6.1.

Finding the Average of a Set of Numbers

1. Find the sum of the given list of numbers.
2. Divide the sum from Step 1 by the number of numbers in the list.

1. Let's look at the average of two numbers.

a. Take the numbers 2 and 12 and plot them on a number line.

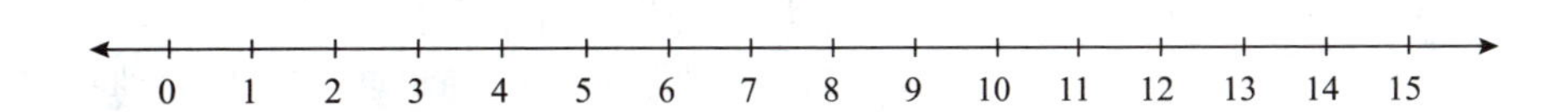

Lesson Link

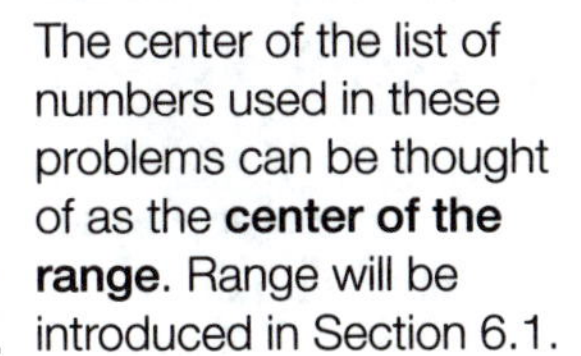

The center of the list of numbers used in these problems can be thought of as the **center of the range**. Range will be introduced in Section 6.1.

b. To find the center of a list of numbers, find the middle number between the smallest and largest number. Plot this number on the number line in part **a.** and label the number as the halfway point.

c. Find the average of 2 and 12. Compare this to the middle number you just found. Plot this number on the number line in part **a.** and label this as the average.

d. Is the following statement true or false? If it is false, rewrite the statement so that it is true.

The average of two numbers is always equal to the middle number between those two numbers.

2. Now, we will look at the average of four numbers.

   a. Take the numbers 2, 4, 6, 12 and plot them on the number line.

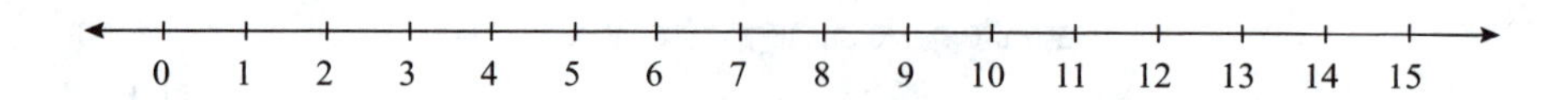

   b. To find the center of a list of numbers, find the middle number between the smallest and largest number. Plot this number on the number line from part **a.** and label the number as the halfway point. (**Note:** The middle number might not be a number in the list.)

   c. Find the average of the four numbers. Compare this to the middle number you found. Plot this number on the number line from part **a.** and label this as the average.

   d. Is the following statement true or false? If it is false, rewrite the statement so that it is true.

   The average of a set of numbers is always equal to the middle number of the set.

3. When you are told that average is a type of "middle number," what do you think this means?

Basic Problem-Solving Strategy

1. Read the problem carefully.
2. Draw any type of figure or diagram that might be helpful and decide which operations are needed.
3. Perform the operations to solve the problem.
4. Check your work.

4. In previous sections you solved applications problems and pointed out key words or phrases that indicated which operation was to be performed. Fill in the following table with those key words and phrases.

Addition	Subtraction	Multiplication	Division

Name: Date:

Determine whether each statement is true or false. Rewrite any false statement so that it is true. (There may be more than one correct new statement.)

5. Key words will always be obvious in word problems.

6. After determining an answer, you should check your work.

7. Your answer should always make sense in the context of the problem.

Skill Check

Go to Software Work through Practice in Lesson 1.7 of the software before attempting the following exercises.

Solve the following number problems.

8. The sum of five and fourteen is multiplied by the quotient of twenty-seven and nine. What is the product?

9. Twice the difference of nineteen and eleven is increased by five. What is the sum?

Apply Skills

Work through the problems in this section to apply the skills you have learned related to problem solving.

Life typically won't present you with nicely packaged word problems like the kind you see in a math course. However, there are some common situations in which you will need to apply problem solving skills that can be generalized in the form of a word problem. Such situations range from every day finances and purchases to understanding and interpreting medical data. The following problems will cover a few of these situations.

10. Maxwell is redecorating his living room. He needs to buy curtains and curtain rods for the four windows in the room. Each window needs one curtain rod and two curtain panels. He finds curtain panels that he likes for $34 each. The curtain rods that he likes are $22 each. He also has a coupon for $40 off of his total purchase. How much will Maxwell spend on curtains and curtain rods (before tax)?

Quick Tip

A **budget** is a plan for how to spend or save money over a certain period of time.

11. Lorenzo recently graduated from college and is overwhelmed by his monthly expenses. He has decided to keep a simple budget to make sure he doesn't spend too much money and can pay his bills on time. He lives in an apartment with a friend and pays $375 per month for rent. His monthly share of the utilities is $18 for water, $45 for gas and electric, $43 for cable TV and Internet, and his cell phone bill is $55 per month. He decides to budget $250 per month for food and $150 for entertainment. He has two credit cards that he wants to pay $75 and $150 per month towards. Finally, his student loan payments are $322 per month.

Quick Tip

Utilities are services such as electric or water which are supplied to the public by a company.

a. Fill in the table with the indicated information.

Category	Expenses
Rent & Utilities	
Food & Entertainment	
Credit Card Bills	
Student Loan	

b. How much money does Lorenzo budget for these expenses each month?

c. If Lorenzo makes $1758 per month (after taxes), how much money does he have left over for savings and other expenses?

Name: Date:

Quick Tip

If you are uncomfortable sharing real information for your expenses, make up numbers that seem reasonable to you.

12. Creating a personal budget can be very beneficial, but it can be confusing at first if you don't know how much money you place in each category. Think of all the things you spend money on each month. Would you add any categories to the simple budget presented in the previous problem? Would you remove any categories? Make a simple monthly budget for your expenses.

Category	Expenses
Rent & Utilities	
Food & Entertainment	
Credit Card Bills	
Student Loan	

Lesson Link

Bar graphs are often used to compare data from different categories or taken at different times. Bar graphs will be discussed more in depth in Section 6.2.

Some bar graphs use abbreviated numbers, such as the one in Problem 11. The number of "New Cases of Diagnosed Diabetes" is presented in thousands, which means that to get the actual number you need to add "000" to the end of each value. For example, in 1985 there were 564,000 new diabetes cases diagnosed.

13. The bar graph shown here describes the number of new diabetes cases diagnosed from 1985 to 2010 in the United States. Use this information to answer the questions.
Source: http://www.cdc.gov/diabetes/statistics/incidence/fig1.htm

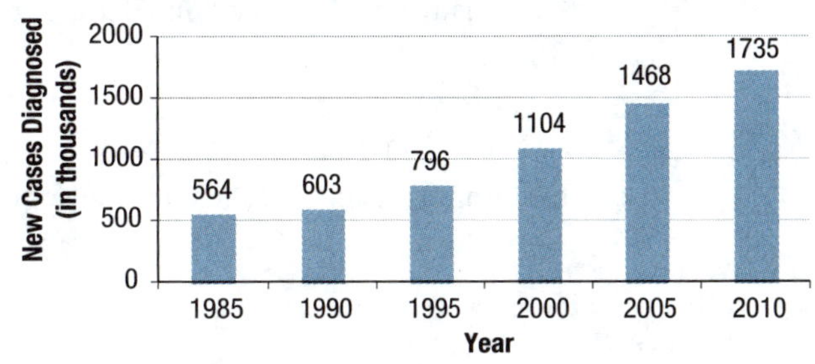

a. How many new cases of diabetes were diagnosed in 2000?

b. How many more cases of diabetes were diagnosed in 2010 than in 1985?

c. What is the average number of cases diagnosed during the years reported in the bar graph?

d. Do you notice a trend in the number of new diabetes cases diagnosed from 1985 to 2010? If yes, describe the trend.

Sometimes there will be more data presented in a bar graph than is needed to answer a question. Being able to ignore extra information is a useful skill in problem solving.

Quick Tip

More information may be given in a problem than is needed. Knowing when to ignore extra information is a skill that comes with practice.

14. The graph shown here describes the leading causes of death in 2010 in the United States.
Source: http://www.cdc.gov/nchs/fastats/deaths.htm

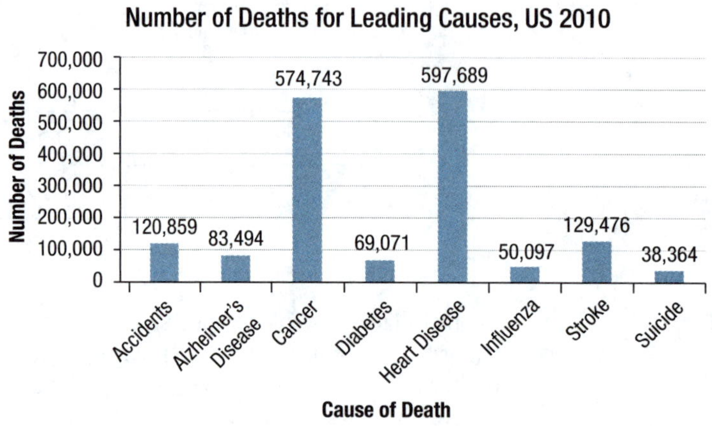

Number of Deaths for Leading Causes, US 2010

Number of Deaths
700,000
600,000
500,000
400,000
300,000
200,000
100,000
0
120,859
83,494
574,743
69,071
597,689
50,097
129,476
38,364
Accidents
Alzheimer's Disease
Cancer
Diabetes
Heart Disease
Influenza
Stroke
Suicide
Cause of Death

a. What was the leading cause of death in the United States in 2010?

b. How many people in the United States died of diabetes in 2010?

c. Determine the top three leading causes of death in the United States in 2010.

d. How many people in the United States died of the top three leading causes of death in 2010?

Name: Date:

1.8 Tests for Divisibility (2, 3, 4, 5, 6, 9, and 10)

Objectives

Know the tests for checking divisibility by 2, 3, 4, 5, 6, 9, and 10.

Apply the concept of divisibility to products of whole numbers.

Success Strategy

If you have an exam at the end of the chapter, now would be a good time to start reviewing concepts and skills from the previous sections. You can also consider forming a study group with your classmates to prepare for the exam.

Understand Concepts

Go to Software First, read through Learn in Lesson 1.8 of the software. Then, work through the problems in this section to expand your understanding of the concepts related to the tests for divisibility.

When a number can be divided by another number so that the remainder is 0, we say that the dividend is **divisible by** the divisor. Another way to say this is that the divisor **divides** evenly into the dividend. This section will explore divisibility and discuss tests for determining divisibility by certain numbers.

Quick Tip

Recall that whole numbers are the counting numbers along with 0.

Divisibility Rules For Whole Numbers

Divisible By	Rule
2	If the last digit is an even number (0, 2, 4, 6, 8)
3	If the sum of the digits of the number is divisible by 3
4	If the last two digits of the number form a number divisible by 4
5	If the last digit is a 0 or 5
6	If the number is divisible by both 2 and 3
9	If the sum of the digits of the number is divisible by 9
10	If the last digit is a 0

True or False: Determine if each statement is true or false. If false, provide an example to support your answer and rewrite the false statement so that it is true.

1. If a number is divisible by 4, then it is also divisible by 2.

2. If a number is divisible by 2 and 5, then it is also divisible by 10.

3. If a number is divisible by 3 and 4, then it is divisible by 12.

4. If a number is divisible by 2 and 4, then it is divisible by 8.

Looking at Problems 2, 3, and 4, you may think that you can create a new divisibility rule by combining two of the divisibility rules you already know. There are, however, several restrictions on creating new divisibility rules. The following problems will help you find one of these restrictions and then apply it to make new divisibility rules.

5. First, look at the rule for divisibility by 6. This states that a number is divisible by 6 if it is also divisible by 2 and 3.

a. The number 6 is the product of two factors: 6 = _____ · _____

b. Is either one of these factors divisible by the other?

6. Next, look at Problem 4 which claims that if a number is divisible by 2 and 4, then it is divisible by 8.

a. Similar to the number 6, we can write 8 as a product of two factors: 8 = _____ · _____

b. Is either of these factors divisible by the other?

c. Let's look at the number 12. Is this number divisible by both 2 and 4?

d. Is 12 divisible by 8? Give a reason for your answer or show your work.

Quick Tip

The answer to part **d.** is known as a **counterexample**, which is just a way to say "an example that breaks the rule." Only one counterexample is needed to prove that something is false.

e. What does this tell us about the rule stated in Problem 4?

7. What do you think the restriction is when creating new divisibility rules from the rules you already know?

8. Create two more divisibility rules from the rules introduced in this section.

Name: Date:

Skill Check

Go to Software Work through Practice in Lesson 1.8 of the software before attempting the following exercises.

Use the tests for divisibility to determine which of the numbers 2, 3, 4, 5, 6, 9, and 10 (if any) will divide exactly into the given number.

9. 816

10. 2231

11. 605

12. 4050

Apply Skills

Work through the problems in this section to apply the skills you have learned related to the tests for divisibility.

The divisibility rules can be used to quickly determine if a certain number of objects can be evenly divided between a specific number of people. The only limitation is that you need a divisibility rule for that specific number. In some situations you may be better off just performing the division and checking for a remainder. However, the following problems illustrate situations where the divisibility rules covered in this section can be used.

13. A city raises $15,255 in scholarship money for local students. The mayor wants to evenly split the money between multiple students.

a. Given the options to split the money between 3, 4, 6, or 9 students, in which ways can the money be evenly split among the students?

b. How much scholarship money would each student receive for the options in part **a.** that result in an even split among the students?

14. Jack can purchase business flyers in amounts of 270, 375, or 522. The amount of flyers purchased needs to be split evenly between 9 locations to distribute to potential customers.

a. Which test for divisibility would you use to figure out the amount of flyers to buy?

b. Will any of these amounts split evenly between 9 locations? If so, which option(s) and how do you know?

15. Bethany is participating in a Walk to End Lupus Now charity event. She is on a team that wants to raise $12,400 for the event. Each team member agrees to raise the same amount of money to reach their goal.

a. The possible team sizes are 5, 6, 9, or 10 members. Which of the team sizes allows the goal amount of money to be raised to be evenly split among the team members?

b. For each team size in part **a.** that results in an even split, how much would each team member have to raise in donations?

16. Cameron is raising money for charity by selling candy bars. He buys 5 large cases of candy bars. Each case of candy bars contains smaller boxes of candy bars. Each case contains 3 rows with 3 boxes in each row. Each box of candy bars contains individual candy bars. Each box has 5 rows with 7 candy bars in each row.

a. Write a product to describe the total amount of candy bars Cameron bought. Do not simplify.

b. Cameron has 35 volunteers to sell the candy bars. Can this amount of candy bars be evenly divided among the 35 volunteers? How do you know?

c. Cameron has permission to sell candy bars outside of 6 different stores. Can this amount of candy bars be evenly divided among the 6 stores? How do you know?

Name: Date:

1.9 Prime Numbers and Prime Factorizations

Objectives

Understand the difference between prime numbers and composite numbers.

Use the Sieve of Eratosthenes. (Software only)

Determine whether a number is prime.

Find the prime factorization of a composite number.

Find all of the factors of a composite number.

Success Strategy

If you are having difficulty learning a concept or skill, try searching for it on the Internet. There are a variety of free resources available and many of them include videos. Keep a list of websites that you find useful to refer back to in the future.

Understand Concepts

Go to Software First, read through Learn in Lesson 1.9 of the software. Then, work through the problems in this section to expand your understanding of the concepts related to prime numbers and prime factorization.

Prime Numbers and Composite Numbers

A **prime number** is a counting number which has exactly two factors, itself and 1.
A **composite** number has more than two factors. The number 1 is neither prime nor composite.

Let's explore prime numbers and their relationship to composite numbers.

1. Every composite number can be written as a product of primes, which is known as its **prime factorization**. Consider only the first four prime numbers, which are 2, 3, 5, and 7.

a. Try to write the numbers 2 through 20 as a product of prime numbers by using only these four prime numbers.

b. Were there any numbers that you couldn't create as a product of primes by using the first four prime numbers? If so, list them here.

c. Why do you think these numbers could not be created from the first four prime numbers?

d. If we include the numbers from part **b.** with our list of prime numbers, would we be able to create all numbers between 20 and 30 as a product of prime numbers? If not, which numbers would we not be able to create?

e. If we continue to add each new smallest number that we couldn't create to the list of primes, would we continue to have the same problem?

f. What does this suggest about the amount of prime numbers?

Quick Tip

The method of finding the prime factorization of a number taught in the software is commonly called a **factor tree**. The lines connecting a factor with its smaller factors are called **branches**. Each branch should end in a prime number.

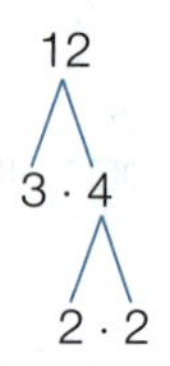

While factor trees are very useful for finding the prime factorization of numbers, there is an alternate method. This other method is a form of repeated division by prime numbers until the resulting quotient is also a prime number.

2. The setup is similar to long division, but the division bar is turned upside-down. $\underline{|140}$

Since 140 is even, it is divisible by 2.

$$\begin{array}{r} 2\underline{|140} \\ 70 \end{array}$$

Now, find a prime number which divides into 70. 7 is an option.

$$\begin{array}{r} 2\underline{|140} \\ 7\underline{|70} \\ 10 \end{array}$$

Next, find a prime number which divides into 10. 5 is an option.

$$\begin{array}{r} 2\underline{|140} \\ 7\underline{|70} \\ 5\underline{|10} \\ 2 \end{array}$$

The quotient is 2, a prime number, so we are done.

The prime factorization of this number is the product of all the prime divisors and the last quotient. What is the prime factorization of 140?

3. Try this alternate method to find the prime factorization of 78.

Lesson Link

Finding the prime factorization of numbers is important when finding least common multiples in Section 2.3 and simplifying radicals in Section 14.2.

4. Either method can be used when finding the prime factorization of a number. Which method do you prefer? Write an explanation for your choice.

Name: Date:

True or False: Determine whether each statement is true or false. Rewrite any false statement so that it is true. (There may be more than one correct new statement.)

5. A prime number has exactly two divisors.

6. A prime number has more than two divisors.

7. Both 1 and 0 are prime numbers.

8. A prime number cannot be even.

Skill Check

Go to Software Work through Practice in Lesson 1.9 of the software before attempting the following exercises.

Determine if the given number is prime or composite. If it is composite, find at least 3 factors.

9. 51

10. 109

Find the prime factorization.

Lesson Link

The rules for divisibility covered in Section 1.8 are useful for finding factors of numbers.

11. 630

12. 114

13. 128

14. 525

Apply Skills

Knowing all of the factors of a number is very useful when splitting a number of objects into equal sized groups. It can help you determine how many groups of each size you can have and pick out which sizes are suitable for your needs. Work through the problems in this section to apply the skills you have learned related to prime numbers and prime factorization.

15. Professor Delaney has 42 students in his lecture class. He needs to split the students into groups for an assignment. Each group needs to have the same number of students.

a. What are the possible sizes for the groups?

b. Are any of the answers from part **a.** unreasonable for the group size? If so, what is unreasonable about these answers?

c. Professor Delaney decides that he wants each group to have no more than 6 students. How many groups would there be for each possible group size?

16. A radio station has 120 concert tickets that they plan to give away during a week-long contest. The packages of concert tickets need to each have the same number of tickets.

a. What are the possible sizes for the packages of tickets?

b. The radio station wants to have between 10 and 25 packages of tickets. Which package sizes meet all of the criteria and how many tickets will they each have?

Name: Date:

Chapter 1 Projects

Project A: Can You Still Get Your Kicks on Route 66?

An activity to demonstrate the use of whole numbers in real life

US Route 66 has been called many names: the Will Rogers Highway, the Main Street of America, and even the Mother Road. This highway, which became one of the most famous roads in America, ran east to west from Chicago, Illinois, to Los Angeles, California, passing through 8 states and 3 time zones. This road made such an impact on American life that a movie and a television series were made about it, and a song was written about it—which led to the title of this project.

1. Using the table of lengths found under the History subheading at the Internet link in the table, determine the eight states crossed by Route 66 and the miles traveled through each state (as of 1926) and write your results below.

Information from: http://en.wikipedia.org/wiki/U.S._Route_66	
State	**Miles Traveled**
1.	
2.	
3.	
4.	
5.	
6.	
7.	
8.	
Total	

2. In 1926, Route 66 covered a total of 2448 miles. Add up the individual mileages that you recorded for each state and report the total in the bottom right hand cell of the table.

a. Does the total you obtained from adding the individual mileages through each state differ from the reported number of 2448 miles?

b. By how much?

c. Why do you think there is a difference between the two values?

3. Use the data in the table to answer the following questions.

a. Which state has the fewest number of miles of Route 66 running through it?

b. Which state has the most number of miles of Route 66 running through it?

c. What is the difference between these two values?

4. Assuming that each state that Route 66 crossed was approximately the same width, calculate the average number of miles traveled in each of the 8 states. (Use 2448 miles for the total.)

5. Refer to the table for the following problems.

   a. Round the total miles calculated to the nearest ten.

   b. Round the total miles calculated to the nearest hundred.

   c. Which one of these numbers do you think is the better estimate of miles traveled on Route 66? Be sure to explain your reasoning.

6. Suppose your car gets an average of 25 mpg (miles per gallon).

   a. Using the rounded value from part **a.** of Problem 5, calculate the total number of gallons needed to travel the entire length of Route 66.

   b. Round the number of gallons obtained in part **a.** to the nearest ten.

   c. Round the number of gallons obtained in part **a.** to the nearest hundred.

   d. Compare the two values obtained in parts **b.** and **c.** Are they the same or different?

   e. Why do you think this happens?

7. If you were going to travel Route 66 through Oklahoma and make two rest stops along the way so that you would travel the same number of miles between rest stops, how many miles would you travel between stops?

8. Can you determine that the total miles across Oklahoma is divisible by 3 without actually doing the division? Explain how.

9. Assume that you drive 52 mph (miles per hour) on average across the entire route. Will you be able to drive through either California or Illinois in 6 hours?

Name: ______________________ Date: ______________________

Project B: Aspiring to New Heights!

An activity to demonstrate the use of whole numbers in real life

You may have never heard of the Willis Tower, but it once was the tallest building in the United States. This structure was originally named the Sears Tower when it was built in 1973 and it held the title of the tallest building in the world for almost 25 years. The name was changed in 2009 when Willis Group Holdings obtained the right to rename the building as part of their lease for a large portion of the office space in the building.

The Willis Tower, which is 1451 feet tall and located in Chicago, Illinois, was the tallest building in the Western Hemisphere until May 2, 2013. On this date a 408 foot spire was placed on the top of One World Trade Center in New York to bring its total height to a patriotic 1776 feet. One World Trade Center now claims the designation of being the tallest building in the United States and the Western Hemisphere.

1. The Willis Tower has an unusual construction. It is comprised of 9 square tubes of equal size, which are really separate buildings, and the tubes rise to different heights. The footprint of the building is a 225 foot by 225 foot square.

a. Since the footprint of the Willis Tower is a square measuring 225 feet on each side and it is comprised of 9 square tubes of equal size, what is the dimension of each tube? (It might help to draw a diagram.)

b. What is the perimeter around the footprint, or base, of the Willis Tower?

c. What is the area of the base of the Willis Tower?

d. What is the area of the base for one square tube of the Willis Tower?

2. Suppose there are plans to alter the landscape around the Willis Tower. The city engineers have proposed adding a concrete sidewalk 6 feet wide around the base of the building. A drawing of the proposal is shown. (**Note:** This drawing is not to scale.)

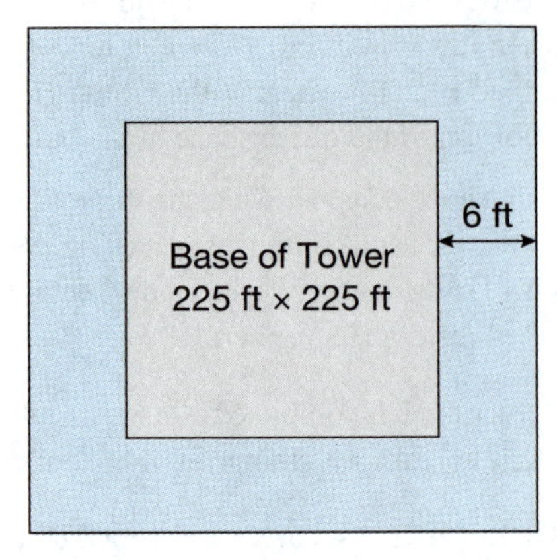

a. Determine the total area of the base of the tower including the new sidewalk.

b. Write down the area of just the base of the tower that you determined in part **c.** of Problem 1.

c. Determine the area covered by the concrete sidewalk around the building. (**Hint:** You only want the area between the two squares.)

d. If a border were to be placed around the outside edge of the concrete, how many feet of border would be needed?

e. If the border is only sold by the yard, how many yards of border will be needed? (**Note:** 1 yard = 3 feet.)

Name: Date:

Math@Work

Basic Inventory Management

As an business manager you will need to evaluate the company's inventory several times per year. While evaluating the inventory, you will need to ensure that enough of each product will be in stock for future sales based on current inventory count, predicted sales, and product cost. Let's say that you check the inventory four times a year, or quarterly. You will be working with several people to get all of the information you need to make the proper decisions. You need the sales team to give you accurate predictions of how much product they expect to sell. You need the warehouse manager to keep an accurate count of how much of each product is currently in stock and how much of that stock has already been sold. You will also have to work with the product manufacturer to determine the cost to produce and ship the product to your company's warehouse. It's your job to look at this information, compare it, and decide what steps to take to make sure you have enough of each product in stock for sales needs. A wrong decision can potentially cost your company a lot of money.

Suppose you get the following reports: an inventory report of unsold products from the warehouse manager and the report on predicted sales for the next quarter (three months) from the sales team.

Unsold Products	
Item	**Number in Stock**
A	5025
B	150
C	975
D	2150

Predicted Sales	
Item	**Expected Sales**
A	4500
B	1625
C	1775
D	2000

Suppose the manufacturer gives you the following cost list for the production and shipment of different amounts of each inventory item.

Item	Amount	Cost	Amount	Cost	Amount	Cost
A	500	$875	1000	$1500	1500	$1875
B	500	$1500	1000	$2500	1500	$3375
C	500	$250	1000	$400	1500	$525
D	500	$2500	1000	$4250	1500	$5575

1. Which items and how much of each item do you need to purchase to make sure the inventory will cover the predicted sales?

2. If you purchase the amounts from Problem 1, how much will this cost the company?

3. By ordering the quantities you just calculated, you are ordering the minimum of each item to cover the expected sales. If the actual sales during the quarter are higher than expected, what might happen? How would you handle this situation?

4. Which skills covered in this chapter were necessary to help you make your decisions?

Foundations Skill Check for Chapter 2

The foundation of your mathematical knowledge is made of previously learned skills and concepts. Many of these skills and concepts are needed when learning new topics in math. Building new skills and concepts on a solid foundation is easier than building on a foundation that is weak or missing pieces. This page lists several skills covered previously in the book and software that are needed to learn new skills in Chapter 2. To make sure you are prepared to learn these new skills, take the self-test below and determine if any specific skills need to be reviewed.

Each skill includes an easy (**e.**), medium (**m.**), and hard (**h.**) version. You should be able to complete each exercise type at each skill level. If you are unable to complete the exercises at the easy or medium level, go back to the given lesson and review until you feel confident in your ability. If you are unable to complete the hard exercise for a skill, or are able to complete it but with minor errors, a review of the skill may not be necessary. You can wait until the skill is needed in the chapter to decide whether or not you should work through a quick review.

1.2 Find the sum.

e. $\begin{array}{r} 42 \\ +37 \\ \hline \end{array}$

m. $\begin{array}{r} 896 \\ +247 \\ \hline \end{array}$

h. $\begin{array}{r} 75{,}324 \\ 119{,}283 \\ +\ 66{,}787 \\ \hline \end{array}$

1.2 Find the difference.

e. $\begin{array}{r} 65 \\ -24 \\ \hline \end{array}$

m. $\begin{array}{r} 334 \\ -194 \\ \hline \end{array}$

h. $\begin{array}{r} 2845 \\ -1876 \\ \hline \end{array}$

1.3 Find the product.

e. $\begin{array}{r} 36 \\ \times 5 \\ \hline \end{array}$

m. $\begin{array}{r} 299 \\ \times 12 \\ \hline \end{array}$

h. $\begin{array}{r} 3554 \\ \times\ 213 \\ \hline \end{array}$

1.4 Divide using long division. Indicate any remainder.

e. $6\overline{)96}$

m. $380 \div 15$

h. $1134 \div 27$

1.9 Find the prime factorization.

e. 30

m. 126

h. 792

Chapter 2: Fractions and Mixed Numbers

Study Skills

2.1 Introduction to Fractions and Mixed Numbers

2.2 Multiplication and Division with Fractions and Mixed Numbers

2.3 Least Common Multiple (LCM)

2.4 Addition and Subtraction with Fractions

2.5 Addition and Subtraction with Mixed Numbers

2.6 Order of Operations with Fractions and Mixed Numbers

Chapter 2 Projects

Math@Work

Foundations Skill Check for Chapter 3

Math@Work

Introduction

If you plan on going into hospitality management, you have a variety of career areas to choose from. You can work at a hotel, a restaurant, a cruise ship, or even run your own business. Clear and accurate communication involving numbers is a large aspect of the job, from the management of inventory to the cost of services. Regardless of which area of hospitality management you choose to pursue, understanding and using fractions will likely be a large part of your job.

Suppose you choose restaurant management for a career. How do you determine how much of each item needs to be prepared for a dinner service? How do you know how much of each ingredient is required to prepare all of the food? Do you have enough of these ingredients in stock? Finding the answers to these questions (and many more) require several of the skills covered in this chapter and the previous chapter. At the end of the chapter, we'll come back to this topic and explore how math is used when managing a restaurant.

Study Skills

The Frayer Model

If you are a visual learner, you can try using a graphic organizer to help you understand more complex mathematical concepts or terms. A simple graphic organizer to use is called the Frayer Model (Frayer, 1969). You can use the Frayer Model on any size piece of paper or index card. You can fold the paper into four quadrants, or quarters, or just mark the paper with a vertical and horizontal line to split the paper into 4 quadrants.

Steps for Using the Frayer Model

1. First, fold a piece of paper into four quadrants by folding in half twice, once vertically and once horizontally.
2. In the center of the paper, write the topic. You might want to circle or highlight the topic.
3. In the top left quadrant, write the definition, formula, or concept.
4. In the top right quadrant, list any facts that you know (use your text or notes to fill in any gaps later).
5. In the bottom left quadrant, list or draw any examples of the concept or term
6. In the bottom right quadrant, list or draw any nonexamples.

Don't worry about making a complete list of examples and nonexamples because there may be an infinite number of them. Just include enough to help you understand the concept or term.

Here is an example of how this graphic organizer would work for the concept of perimeter:

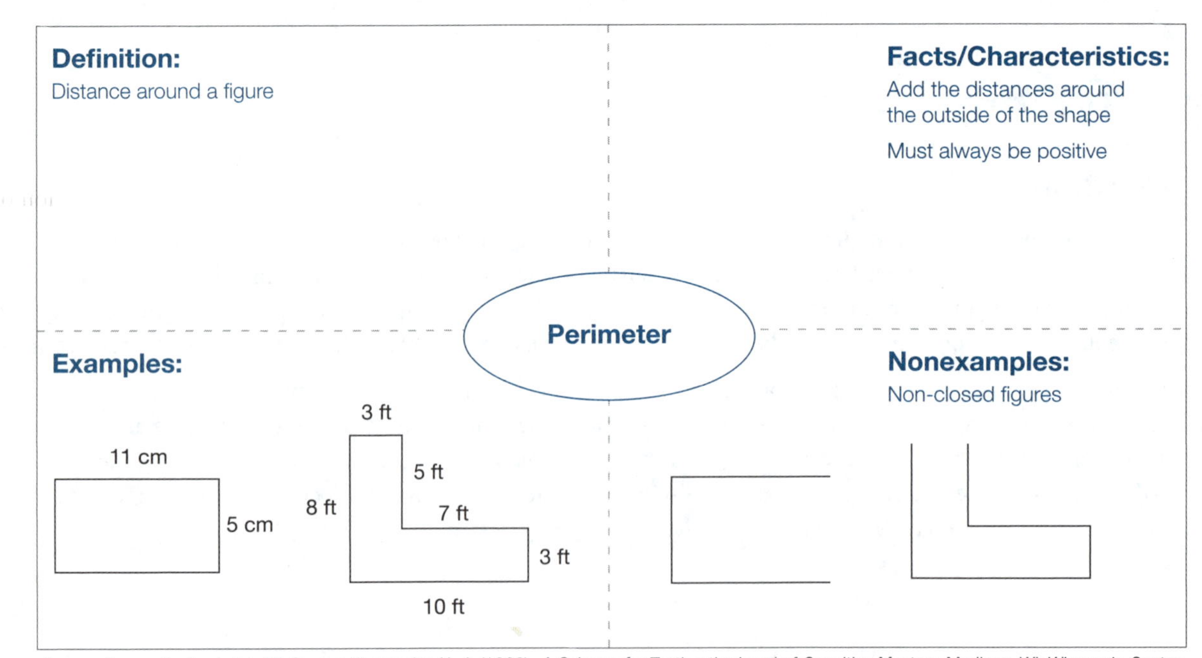

Source: Frayer, D., Frederick, W. C., and Klausmeier, H. J. (1969). *A Schema for Testing the Level of Cognitive Mastery.* Madison, WI: Wisconsin Center for Education Research.

Name: Date:

2.1 Introduction to Fractions and Mixed Numbers

Objectives

Understand the basic concepts of fractions.

Multiply fractions.

Find equivalent fractions.

Reduce fractions to lowest terms.

Change mixed numbers to improper fractions.

Change improper fractions to mixed numbers.

Success Strategy

The topic of fractions is a difficult one for many students. Patience, practice, and perseverance are the keys to working with fractions.

Understand Concepts

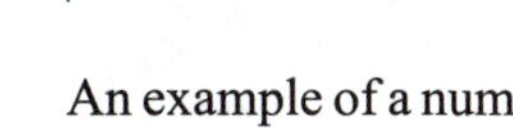

Go to Software First, read through Learn in Lesson 2.1 of the software. Then, work through the problems in this section to expand your understanding of the concepts related to fractions and mixed numbers.

An example of a number in **fraction form** is $\frac{1}{5}$, where 1 is the **numerator** and 5 is the **denominator**.

Fractions can be used to represent parts of a whole item. For the fraction $\frac{1}{5}$, the number of parts that make up a whole item is indicated by the denominator and the number of parts present is indicated by the numerator.

1. Label the parts of a fraction. $\frac{15}{24}$ ← ______________ ← ______________

2. Fractions are a part of our daily lives, so understanding what a fraction represents is important. Some common situations that use fractions involve weights and volume. For instance, we can buy $\frac{1}{2}$ of a pound of lunch meat from a deli and a leaf blower can hold $\frac{3}{4}$ of a gallon of fuel. List three examples of situations in your life where fractions are used.

Multiplication with Fractions

To multiply two fractions, multiply the numerators together and multiply the denominators together. In mathematical notation, if a, b, c, and d are whole numbers where $b \neq 0$ and $d \neq 0$, then $\frac{a}{b} \cdot \frac{c}{d} = \frac{a \cdot c}{b \cdot d}$.

Lesson Link

Multiplication with whole numbers was covered in Section 1.3.

3. When multiplying whole numbers, you can think of the process as repeated addition. This can extend to multiplying a whole number by a fraction, for example, $4 \cdot \frac{1}{2} = \frac{1}{2} + \frac{1}{2} + \frac{1}{2} + \frac{1}{2} = 2$, but it doesn't extend easily to multiplying a fraction by a fraction. As a result, it can be difficult to understand what is happening when you multiply two fractions. Using a visual description of the multiplication of two small fractions may help with understanding.

a. What fraction is represented by the shaded part of this figure?

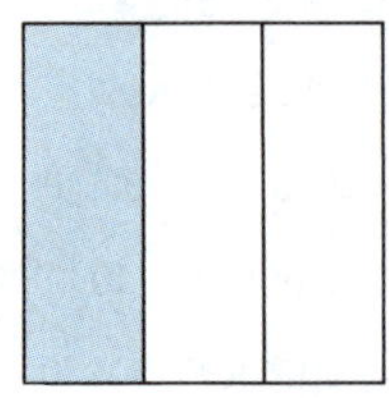

b. What fraction is represented by the shaded part of this figure?

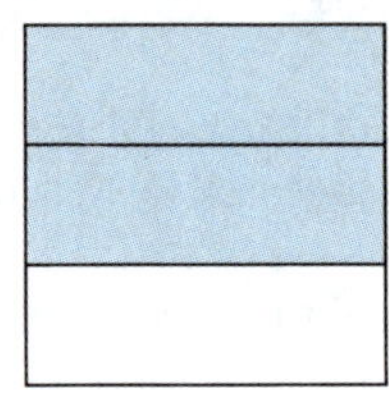

c. Multiplying these two fractions can be visualized by placing the figure from part **a.** on top of the figure from part **b.** What fraction is represented by the overlapping shaded part of this figure?

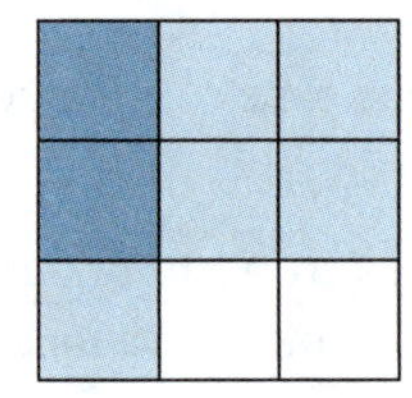

d. Write an equation to describe what happened in part **c.**

Types of Fractions

A **proper fraction** is a fraction in which the numerator is less than the denominator.

An **improper fraction** is a fraction in which the numerator is larger than the denominator.

A **mixed number** is the sum of a whole number and a proper fraction.

Depending on the situation, you may need to write fractions as either improper fractions or as mixed numbers. The next two problems will guide you through the process of changing an improper fraction to a mixed number or a mixed number to an improper fraction.

Quick Tip

In a fraction, the **numerator** can be thought of as the parts of a whole item and the **denominator** can be thought of as how many equal parts make one whole.

4. Suppose you need to write the improper fraction $\frac{8}{3}$ as a mixed number.

a. How many parts, or pieces, represent a whole item in the fraction $\frac{8}{3}$?

Name: Date:

b. How many parts, or pieces, are represented by the fraction $\frac{8}{3}$?

Quick Tip

The division in part **c.** is the same as dividing the numerator by the denominator.

c. To find how many wholes are in the fraction $\frac{8}{3}$, divide your answer from part **b.** by your answer from part **a.** Be sure to note any remainder.

d. The remainder from part **c.** tells you how many parts are left over which do not make up a whole. This value becomes the numerator of a proper fraction with the same denominator as the original improper fraction. Write this proper fraction here.

e. Combining the number of wholes with the proper fraction from part **d.** gives the mixed number equivalent of $\frac{8}{3}$. Write the mixed number equivalent of $\frac{8}{3}$.

5. Suppose you have the mixed number $2\frac{3}{4}$ and you need to write it as an improper fraction.

a. How many parts, or pieces, represent a whole item in the fraction $\frac{3}{4}$?

b. To determine how many parts are represented by the mixed number, multiply the whole number of the mixed number by the answer from part **a.** and add that value to the numerator of the fraction part. Calculate how many parts of a whole item are in $2\frac{3}{4}$.

c. The value from part **b.** becomes the numerator of the improper fraction that is equivalent to the mixed number. Write the improper fraction equivalent of $2\frac{3}{4}$. (**Note:** The denominator in the improper fraction is the same as the denominator in the mixed number.)

Equivalent Fractions

To find an equivalent fraction, multiply the numerator and denominator of the fraction by the same number. In mathematical notation, if a, b, and k are whole numbers where $b \neq 0$ and $k \neq 0$, then $\frac{a}{b} \cdot \frac{k}{k} = \frac{a \cdot k}{b \cdot k}$.

Lesson Link

Recall from Section 1.4 that a number cannot be divided by 0. This means that fractions cannot have 0 as a denominator.

6. Equivalent fractions can be used to plot fractions with different denominators on the same number line.

a. For each fraction, find an equivalent fraction with a denominator of 16: $\frac{1}{2}, \frac{1}{4}, \frac{3}{4}, \frac{1}{8}, \frac{6}{8}, \frac{7}{8}$.

b. Plot the equivalent fractions on the number line and label them with their original form.

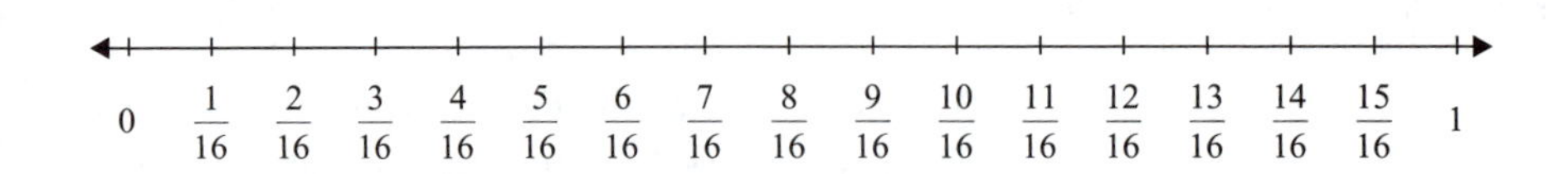

c. Were any of the given fractions from part **a.** equivalent fractions? If so, which fractions?

d. Are equivalent fractions plotted at the same point on the number line or at different points?

Read through the following information on rulers and work through the following problem.

Lesson Link

The US measurement system and the Metric measurement system are both covered in the Appendix.

Rulers are commonly used to determine the length of an object. In the US measurement system, a ruler can have inches divided into fractions of an inch, as shown. A ruler is a representation of the number line with the distance between each whole number representing the length of an inch.

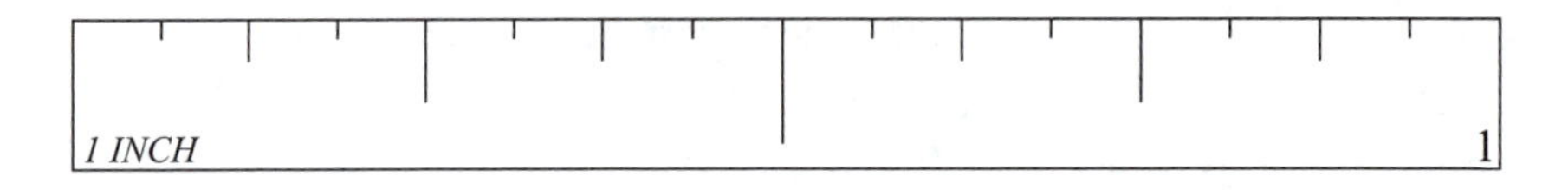

7. Use the enlarged image of the ruler above to complete the following problems.

a. Draw an arrow pointing to $\frac{1}{4}$ of an inch on the ruler given above and label it with the measurement.

b. What is $\frac{1}{2}$ of $\frac{1}{4}$ of an inch? Draw an arrow labeling this value on the ruler given above.

c. Draw an arrow pointing to $\frac{3}{8}$ of an inch on the ruler given above and label it with the measurement.

d. What is $\frac{1}{2}$ of $\frac{3}{8}$ of an inch? Draw an arrow labeling this value on the ruler given above.

Name: Date:

Quick Tip

To find out the value of different coins, visit http://en.wikipedia.org/wiki/Coins_of_the_United_States_dollar

8. In the US currency system, each coin represents a fraction of a dollar.

a. A dollar is worth 100 cents. Where would we place the 100 if we were to write the value of a coin as a fraction of a dollar—the numerator or denominator?

b. Where would we place the value of the coin if we were to write the value of a coin as a fraction of a dollar—the numerator or denominator?

c. Fill in the table with the nonreduced fraction equivalent of each coin value in column two and the reduced fraction equivalent in column three.

Coin Name	Nonreduced	Reduced
Penny		
Nickel		
Dime		
Quarter		
Half-dollar		

Skill Check

Go to Software Work through Practice in Lesson 2.1 of the software before attempting the following exercises.

Find the indicated products.

9. $\frac{3}{5}\cdot\frac{2}{5}$

10. $\frac{1}{3}\cdot\frac{2}{5}\cdot\frac{4}{7}$

Find the missing numerator that will make the fractions equivalent.

11. $\frac{2}{3}=\frac{?}{15}$

12. $\frac{1}{4}=\frac{?}{48}$

Change the number from a mixed number to an improper fraction or an improper fraction to a mixed fraction.

13. $3\frac{2}{7}$

14. $\frac{25}{8}$

Apply Skills

Work through the problems in this section to apply the skills you have learned related to fractions and mixed numbers.

Quick Tip

When the answer is a fraction, reduce it to lowest terms unless otherwise instructed.

15. A computer stores data on a hard drive in the form of bits, bytes, and sectors.

a. Each byte is made up of eight bits. What fraction of a byte is a bit?

b. A sector on a hard drive is traditionally five hundred twelve bytes. A byte is what fraction of a sector?

c. A bit is what fraction of a sector? (**Hint:** The key word "of" means multiply.)

d. If a computer stores 192 bytes of data, what fraction of a sector does that amount of data take up? Reduce to lowest terms.

16. There is $\frac{2}{3}$ of a cake left over from a party. You and four friends want to evenly divide the cake to eat as dessert.

a. What fraction of the *leftover* cake will each person get?

b. What portion of the *entire* cake will each person get?

17. Omar is the inventory manager at a warehouse. He determines that $\frac{7}{12}$ of the current inventory has not been sold. Of this unsold inventory, the sales team predicts that $\frac{2}{3}$ will be sold during the next two weeks. What fraction of the current inventory does the sales team predict will be sold during the next two weeks?

18. The gas tank on Zamira's car holds 14 gallons of gas. She puts 8 gallons of gas in the car. What fraction of the gas tank does the 8 gallons take up?

Name: Date:

2.2 Multiplication and Division with Fractions and Mixed Numbers

Objectives

Multiply mixed numbers.

Multiply and reduce with fractions and mixed numbers.

Understand the term reciprocal.

Learn to divide with fractions and mixed numbers.

Success Strategy

Drawing a visual model of what a fraction represents in a problem can be helpful.

Understand Concepts

Go to Software First, read through Learn in Lesson 2.2 of the software. Then, work through the problems in this section to expand your understanding of the concepts related to multiplication and division with fractions and mixed numbers.

Multiplication with fractions was introduced in Section 2.1. This section will discuss working with more complicated multiplication and also introduce division with fractions.

1. A common mistake that occurs when multiplying mixed numbers together is shown below.

$$2\frac{1}{3}\cdot 4\frac{1}{2} \neq (2\cdot 4)\left(\frac{1}{3}\cdot\frac{1}{2}\right) \neq 8\frac{1}{6}$$

a. The value $2\frac{1}{3}$ does not mean that 2 is being multiplied by $\frac{1}{3}$. What operation is understood to be between 2 and $\frac{1}{3}$? (**Hint:** Think about how you say $2\frac{1}{3}$.)

b. What is the mistake that was made in the simplification above?

Lesson Link 

Changing mixed numbers to improper fractions was covered in Section 2.1

2. To avoid complicated expressions when multiplying the mixed numbers from Problem 1, we can rewrite the initial expression to make it easier to work with. The mixed numbers can be rewritten as improper fractions before multiplying.

a. Rewrite the expression $2\frac{1}{3}\cdot 4\frac{1}{2}$ as a product of improper fractions.

Quick Tip

When simplifying expressions with mixed numbers, the answer should be written as a mixed number unless otherwise instructed.

b. Perform the multiplication with improper fractions from part **a.** to verify that the correct simplification of $2\frac{1}{3}\cdot 4\frac{1}{2}$ is not $8\frac{1}{6}$. What is the correct answer?

Reciprocals

The **reciprocal** of $\frac{a}{b}$ is $\frac{b}{a}$ where $a \neq 0$ and $b \neq 0$.

For example, the reciprocal of $\frac{5}{7}$ is $\frac{7}{5}$.

3. The definition of a reciprocal may seem simple enough, but there are some additional facts to be aware of. This problem will help you discover these facts.

Quick Tip

Any whole number can be written as a fraction by making the whole number the numerator and using 1 as the denominator.

a. Write 0 as a fraction. (There may be more than one correct answer.)

b. Read the definition of reciprocal and relate it to the fraction from part **a.** Does 0 have a reciprocal? If yes, what is it? If no, why not?

c. What is the reciprocal of $\frac{3}{4}$?

d. What is the product of $\frac{3}{4}$ and its reciprocal? Show your work.

e. The product of the reciprocals from part **d.** is true for the product of every number and its reciprocal. Write a general rule for this using the notation from the definition of a reciprocal.

Dividing by a Fraction

To **divide a number by a fraction**, multiply the number by the reciprocal of the fraction. For example, you can rewrite the expression $\frac{1}{5} \div \frac{2}{3}$ as $\frac{1}{5} \cdot \frac{3}{2}$. In mathematical notation, if a, b, c, and d are whole numbers and $b \neq 0$, $c \neq 0$, and $d \neq 0$, then $\frac{a}{b} \div \frac{c}{d} = \frac{a}{b} \cdot \frac{d}{c}$.

This rule can be thought of as a shortcut of the full process for dividing by a fraction. The following problems explore this process of how division by a fraction works.

4. The fraction bar can be thought of as which operation?

Name: Date:

Lesson Link

This type of fraction will be covered in Section 2.6.

Equivalent fractions were covered in Section 2.1.

5. As a result of Problem 4, we can write $\frac{1}{5} \div \frac{2}{3}$ as $\frac{\frac{1}{5}}{\frac{2}{3}}$. This is a complicated-looking fraction.

Our goal now is to rewrite this fraction as an equivalent fraction with a denominator equivalent to 1.

a. To find an equivalent fraction, what steps must we take?

b. To get a denominator equivalent to 1, what number do we need to multiply the $\frac{2}{3}$ by? What is the mathematical name of this number?

Lesson Link

Recall from Section 2.1 that equivalent fractions are found by multiplying the fraction by a value equivalent to 1.

c. Fill in the missing fractions to find an equivalent fraction that has a denominator equivalent to 1.

$$\frac{\frac{1}{5} \cdot \boxed{\frac{}{}}}{\frac{2}{3} \cdot \boxed{\frac{}{}}}$$

d. The denominator should simplify to 1. Verify that this is true.

e. What is the simplified value of the numerator? Reduce if possible.

f. The quotient of any number divided by 1 is equal to what value?

g. The denominator is now equivalent to 1. The final value of $\frac{1}{5} \div \frac{2}{3}$ is equal to what value?

6. Compare the result from Problem 5 part **g.** to the result you get when you follow the method of multiplying by the reciprocal as shown in the box on the bottom of the previous page. Show your work.

7. Every whole number can be written as a fraction by writing the whole number in the numerator and writing 1 in the denominator. When a whole number is written as a fraction, the 1 in the denominator is sometimes referred to as an "invisible denominator." Use the key words "invisible denominator" to find out what this means. Write a brief summary of your findings here.

Skill Check

Go to Software Work through Practice in Lesson 2.2 of the software before attempting the following exercises.

Quick Tip

When multiplying or dividing with mixed numbers, change the mixed numbers to improper fractions first.

Perform the indicated operation. Be sure to reduce your answer.

8. $2\frac{2}{5}\cdot 3\frac{1}{4}$

9. $1\frac{6}{8}\cdot 1\frac{1}{4}$

10. $\frac{5}{6}\cdot\frac{3}{5}$

11. $\frac{3}{4}\cdot\frac{5}{6}\cdot\frac{4}{15}$

12. $\frac{10}{13}\div\frac{5}{26}$

13. $\frac{2}{3}\div 4$

14. $2\frac{1}{17}\div 1\frac{1}{4}$

15. $1\frac{1}{32}\div 2\frac{3}{4}$

Name: ______________________ Date: ______________________

Apply Skills

Work through the problems in this section to apply the skills you have learned related to multiplication and division with fractions and mixed numbers.

16. A cookie recipe makes 8 dozen cookies and calls for $2\frac{1}{2}$ cups of flour. You only want to make 2 dozen cookies for yourself.

a. What fraction of an entire recipe will you be making?

b. How many cups of flour will you need to make 2 dozen cookies?

Quick Tip

Word problems can contain **extraneous** information, that is, information that is not used to solve the problem. Be sure to pick out what is necessary to answer the questions.

17. Elizabeth runs a day care and is responsible for planning the daily snacks each week for 14 toddlers. She plans on serving apple slices as a snack twice this week. The serving size for a toddler is $\frac{1}{4}$ of a cup of apple. She plans on giving each toddler 2 servings of apple during each snack time, which she determines to be $\frac{1}{3}$ of an apple per toddler.

a. What portion of an apple will each toddler eat each week?

b. How many apples does Elizabeth need to buy for the week to make snacks for the 14 toddlers? Round your answer to the next whole number, if necessary.

c. Will there be any apple left over? If yes, what fraction of an apple?

18. Part of Jayden's job is to clean oriental rugs and he uses a cleaning solution that is sold in 5-ounce containers. On average, he uses 2 ounces of solution per week.

a. What fraction of a container does Jayden use per week?

b. Jayden cleans 6 rugs each week. If he uses the same amount of solution to clean each rug, what fraction of the container does he use per rug?

19. Ms. Bambic is a high school drama coach. To retain the funding for the schools' drama productions, she has to keep track of the amount of students who are interested in participating. She determines that $\frac{1}{24}$ of the entire student body auditioned for the upcoming drama production.

a. Only $\frac{2}{5}$ of all of the students who audition are able to be cast in the upcoming drama production. What fraction of the entire student body will be cast in the production?

b. If there are 840 students attending the high school, how many will be a part of the cast of the upcoming drama production?

20. After assembling the bodice and skirt of a wedding dress, Camila needs to add a lace border to the bottom of the skirt. She has $34\frac{1}{2}$ yards of lace. Each skirt requires $2\frac{1}{4}$ yards of lace.

a. How many wedding dress skirts can Camila trim with lace? (Write as a mixed number.)

b. How many complete dresses can be made with the amount of lace she has?

c. How much lace will she have left over after the complete dresses are trimmed with lace? (**Hint:** How many yards of lace are needed to trim the amount of skirts from part **b.**?)

Name: Date:

2.3 Least Common Multiple (LCM)

Objectives

Understand the meaning of the term Least Common Multiple.

Use prime factorizations to find the LCM of a set of numbers.

Use Venn diagrams to find the LCM of two numbers.

Recognize the application of the LCM concept in a word problem.

Success Strategy

There is a lot of new terminology in this chapter. You should write down key terms and definitions in a notebook or use index cards based on the Frayer Model discussed at the beginning of this chapter.

Understand Concepts

Go to Software First, read through Learn in Lesson 2.3 of the software. Then, work through the problems in this section to expand your understanding of the concepts related to least common multiples.

The **least common multiple** (LCM) of a set of numbers is the smallest number that is a multiple of each number in the set. This number is very important when adding and subtracting fractions, as you will see in Section 2.4. The LCM is also useful for a variety of other reasons, as you will see in the Apply Skills section.

Quick Tip

The **multiples of a number** are the products of that number with the counting numbers.

For example,

$12 \cdot 1 = 12$, $12 \cdot 2 = 24$, etc.

1. One way to find the LCM of a set of counting numbers is to list the multiples of each number in the set until you find the smallest multiple that each number has in common. Use this method to find the LCM of 12 and 15.

Multiples of 12:

Multiples of 15:

The LCM of 12 and 15 is

The method of listing multiples can be time consuming. The following method, which uses prime factorization, is more efficient, although it may not always be the easiest method to use.

To Find the LCM of a Set of Counting Numbers

1. Find the prime factorization of each number.
2. Identify the prime factors that appear in any one of the prime factorizations.
3. Form the product of these primes using each prime the most number of times it appears in any one of the prime factorizations.

Quick Tip

The LCM of a set of numbers which includes 0 would be 0. For this reason, we do not find the LCM of sets of numbers which include 0.

2. Use the method described in the box above to find the LCM of 12 and 15.

12 =

15 =

The LCM of 12 and 15 is

Read through the following paragraph on Venn diagrams and work through the problems.

Quick Tip

Elements in a Venn diagram can be numbers, letters, dates, or any type of information. This workbook will only use numbers.

A more visual way to find the LCM of two numbers is to use a **Venn diagram** to organize the prime factors. Venn diagrams are made of two overlapping circles and show the relationships between the elements that are contained in the circles. Venn diagrams are great for visually organizing any information. In the Venn diagrams used in this section, each circle will represent a number and the elements inside of each circle will be the prime factors of that number.

3. Let's construct a Venn diagram to determine the LCM of 60 and 42.

a. Find the prime factorizations for 60 and 42.

60 =

42 =

Quick Tip

Only place one copy of the common prime factors in the overlapping area of the Venn Diagram.

b. A Venn diagram for the two numbers is shown in the figure. For our problem, the left circle represents 60 and the right circle represents 42. The overlapping section represents the prime factors that 60 and 42 have in common. Place the common primed factors in the overlapping section.

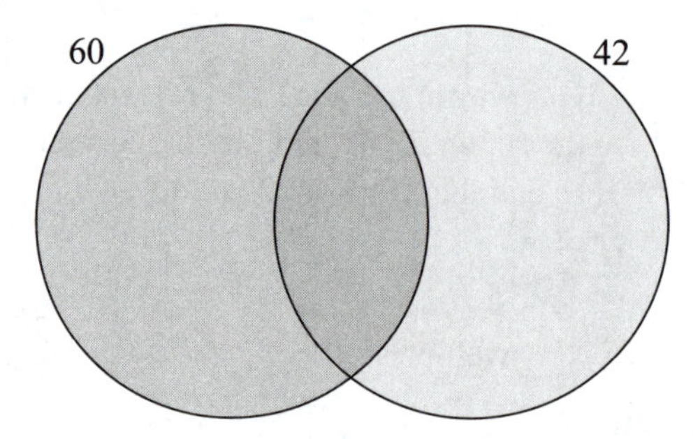

c. Place any prime factors unique to 60 in the circle that represents 60, but not in the overlapping part of the circle. That is, place the prime factors of 60 that are not prime factors of 42 in the outer part of the left circle of the Venn diagram.

d. Place any factors unique to 42 in the circle that represents 42, but not in the overlapping part of the circle.

e. The product of all the prime numbers from the three regions in the Venn diagram is the LCM. What is the LCM of 60 and 42?

Quick Tip

The method of finding the LCM of two numbers with a Venn diagram can be extending to finding the LCM of three numbers.

4. Use the Venn diagram method to find the following LCMs.

a. Find the LCM of 12 and 15.

b. Find the LCM of 140 and 150.

5. You now are aware of three methods for finding the LCM of a set of numbers: listing multiples, finding common prime factors, and using Venn diagrams. You have also used each method to find the LCM of 12 and 15. Which method of finding the LCM of two numbers do you prefer? Write an explanation for your choice.

6. Math is full of patterns and rules, so looking for patterns is a good habit to pick up. Sometimes you may think you have found a pattern or rule, but later you find out that it is incorrect. This problem explores how one rule (or pattern) can be deceiving and not as straight-forward as you may originally think.

a. To find the LCM of 3 and 5, you can simply multiply 3 times 5. Verify this is true by using your preferred method of finding the LCM.

b. Why do you think this works for 3 and 5? (**Hint:** What type of numbers are 3 and 5?)

c. Can you find the LCM of 8 and 15 using the same method as in part **a.**? Show work to support your answer.

d. Can you find the LCM of 4 and 14 using the same method as in part **a.**? Show work to support your answer.

e. Does your reasoning from part **b.** work for part **c.** or part **d.**? Why or why not?

f. Try to create a general rule for when you can multiply two numbers together to find their LCM.

7. Fractions have an interesting history and they haven't always been used in the form that we use them now. Perform an Internet search with the key words "history of fractions" to search for the answers to the following questions.

   a. Which civilization was the first to use fractions?

   b. Who was the mathematician that first mentioned the use of the fraction bar?

   c. Write two additional interesting facts you found about the history of fractions during your search.

Determine if any mistakes were made while simplifying each expression. If any mistakes were made, describe the mistake and then correctly find the LCM.

8. Find the LCM of 6 and 14.

   $6 = 2 \cdot 3$

   $14 = 2 \cdot 7$

   $\text{LCM} = 2 \cdot 2 \cdot 3 \cdot 7 = 84$

9. Find the LCM of 12 and 14.

   $12 = 2 \cdot 2 \cdot 3$

   $14 = 2 \cdot 7$

   $\text{LCM} = 2 \cdot 3 \cdot 7 = 42$

Skill Check

Go to Software Work through Practice in Lesson 2.3 of the software before attempting the following exercises.

Find the LCM for each of the following sets of counting numbers.

10. 8, 12

11. 2, 5, 8

12. 22, 33, 121

13. 15, 25, 30, 45

Name: Date:

Apply Skills

Work through the problems in this section to apply the skills you have learned related to LCM. Keep in mind that problems that involve finding the LCM typically do not state that the LCM is needed to find the solution. Circle any key words of phrases that suggest the LCM needs to be found.

14. Angela is the property manager of an apartment complex. For this month's community activity, she is planning an ice cream social with hot fudge sundaes. She plans to buy vanilla ice cream and fudge topping. Each container of ice cream has 12 servings. Each jar of fudge topping has enough fudge for 18 single serving sundaes.

a. What is the least amount of ice cream sundaes Angela can make if she doesn't want any leftover ingredients?

b. How many containers of ice cream and jars of fudge topping should Angela buy if she wants to make as many sundaes as the answer to part **a.**?

15. To make marshmallow crispy treats you need 6 cups of puffed rice cereal and 10 ounces of marshmallows. A box of puffed rice cereal has approximately 18 cups. A large bag of marshmallows weighs 40 ounces.

a. How many recipes of marshmallow crispy treats can you make from one box of cereal?

b. How many recipes of marshmallow crispy treats can you make from one bag of marshmallows?

c. How many recipes of marshmallow crispy treats must you make so there is no leftover puffed rice cereal in a box or marshmallows in a bag?

d. How many boxes of cereal and how many bags of marshmallows would you need to purchase to have no leftover puffed rice cereal or marshmallows?

Quick Tip

Try to draw a figure or diagram to help visualize the situations in Problems 16 and 17.

16. The maintenance crew of a building is given the task to lay tile in the reception room. They are to lay small black and white tiles in rows with three different repeating patterns across the width of the room. In the first row, every eighth tile is a black tile. In the second row, every twelfth tile is a black tile. In the third row, every twenty-fourth tile is a black tile. The width of the reception room takes 48 tiles to span. The length of the reception room takes 60 tiles to span.

a. Across the width of the room, how often will the black tiles align in the three patterns?

b. How many times will the black tiles align across the width of the reception room?

c. Will the black tiles in the first two rows align more often than the black tiles in all three rows align? (**Hint:** Compare the LCMs.)

d. How many times will the 3 rows of repeating patterns appear across the length of the reception room?

e. If a special tile that is three rows long is used in place of the three black tiles whenever the black tiles align, how many special tiles will be needed for the entire reception room?

17. Vincent is a jeweler and is creating a new line of beaded jewelry. For one necklace design, he has three rows of beads. In the first row, every sixth bead is blue. In the second row, every fifteenth bead is blue. In the third row, every tenth bead is blue. Vincent would like to place a charm in place of the blue beads every time the blue beads align.

a. How often will the blue beads align?

b. Vincent wants to make the necklace 330 beads long. How many charms will be on the necklace?

Name: Date:

2.4 Addition and Subtraction with Fractions

Objectives

Add fractions with the same denominator.

Subtract fractions with the same denominator.

Add fractions with different denominators.

Subtract fractions with different denominators.

Success Strategy

For many students, adding and subtracting fractions can be more difficult than multiplying or dividing fractions, so be sure to do extra practice in the software.

Understand Concepts

Go to Software First, read through Learn in Lesson 2.4 of the software. Then, work through the problems in this section to expand your understanding of the concepts related to addition and subtraction with fractions.

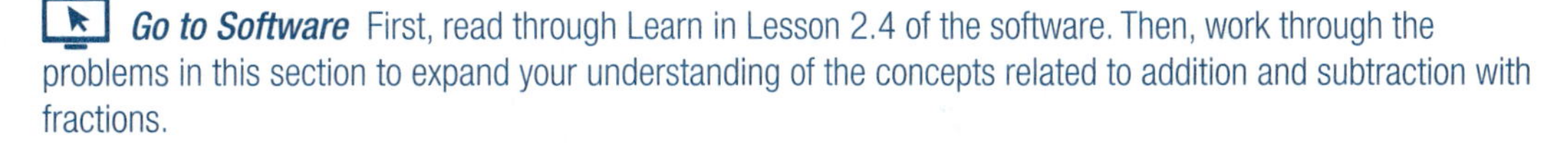

Adding and Subtracting Fractions with the Same Denominator

1. Add the numerators.
2. Keep the common denominator the same.
3. Reduce, if possible.

When adding and subtracting fractions, the denominators of the fractions need to be the same before you can perform the operation. The next problem will explore why this is necessary.

1. Suppose you have two rectangles. The first rectangle is divided into two equally sized pieces. The second rectangle is divided into three equally sized pieces. Each rectangle has one piece shaded. We want to combine these two shaded pieces together to see what fraction of a whole rectangle they represent.

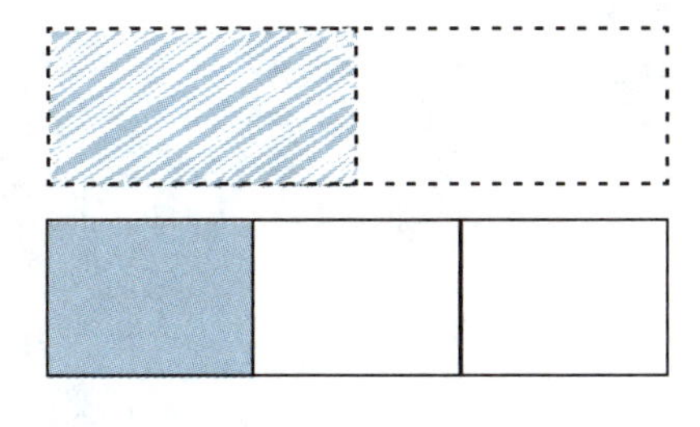

a. Suppose we combine both of the shaded pieces of both rectangles inside of one of the rectangles with their current divisions. Can we look at the shaded area and know exactly what fraction of the whole it takes up without any further manipulations?

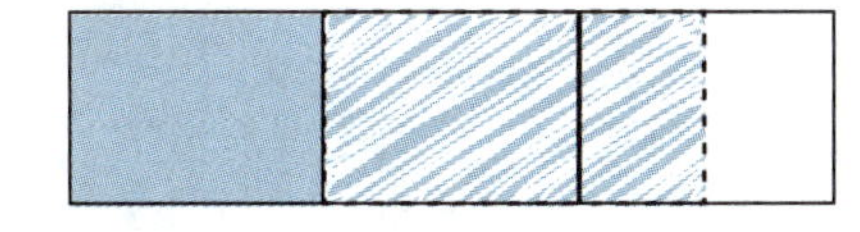

Quick Tip

Fractions with different denominators can be multiplied and divided, but when added or subtracted, they must have the same denominator.

b. If the answer to part **a.** was yes, tell what fraction of the whole is shaded. If the answer was no, what is the problem?

c. Find the LCM of 2 and 3.

d. Find equivalent fractions of $\frac{1}{2}$ and $\frac{1}{3}$ that have a denominator equal to the LCM from part **c.**

e. Draw two rectangles to represent the fractions you found in part **d.**

f. If we combine the two shaded areas from part **e.** into one of these two rectangles, can we easily determine what fraction of the whole rectangle is shaded? If so, draw a rectangle that represents the combined shaded area and tell what fraction of the whole rectangle is shaded.

Quick Tip

The **LCD** is the LCM of a set of denominators.

When first learning how to find the least common denominator (LCD) of two fractions, it may seem like a time consuming process. There are a few situations where you don't need to fully work out the prime factorizations to find the LCD of two fractions, which can speed up the process. The next problem shows three different situations for finding the LCD of a pair of fractions.

Quick Tip

The methods described in Problem 2 can also be applied to whole numbers when finding their LCM.

2. Find another pair of fractions that fits each situation and find the LCD of each pair.

Situation	Example	Your Example
A. If both denominators are prime numbers, the LCD is their product.	$\frac{1}{5}; \frac{2}{7}$ LCD: 35	
B. If one denominator divides into the other, the LCD is the larger number.	$\frac{1}{2}; \frac{5}{8}$ LCD: 8	
C. If you know that the two denominators are not prime numbers but have no common factors besides 1, then the LCD is their product.	$\frac{3}{4}; \frac{1}{9}$ LCD: 36	

Quick Tip

It might not be clear initially which situation applies to a pair of numbers. You will gain more experience in determining which to use with repeated practice.

3. Match the situation from Problem 2 to each pair of fractions. Find the LCD of each pair.

a. $\frac{5}{6}; \frac{2}{24}$ **b.** $\frac{7}{15}; \frac{3}{4}$ **c.** $\frac{10}{13}; \frac{1}{11}$

Name: Date:

4. Numerator and denominator are important words when dealing with fractions. Knowing what the names mean is helpful in understanding the role each term plays in mathematics. Perform an Internet search using the key words "numerator" and "denominator" to answer the following questions.

a. Where did the term "numerator" come from and what does it mean?

b. Where did the term "denominator" come from and what does it mean?

Adding and Subtracting Fractions with Different Denominators

1. Find the LCD.
2. Change each fraction into an equivalent fraction with the LCD as the denominator.
3. Add or subtract the fractions from left to right.
4. Reduce, if possible.

Determine if any mistakes were made while simplifying each expression. If any mistakes were made, describe the mistake and then correctly simplify the expression to find the actual result.

5. $\frac{3}{4}-\frac{2}{8}+\frac{1}{8}=\frac{3}{4}-\frac{3}{8}$
$=\frac{6}{8}-\frac{3}{8}$
$=\frac{3}{8}$

6. $\frac{5}{6}+\frac{2}{5}=\frac{7}{11}$

7. $\frac{5}{2}+\frac{3}{10}=\frac{1}{2}+\frac{3}{2}$
$=\frac{4}{2}$
$=2$

8. $\frac{1}{3}+\frac{3}{4}=\frac{4}{12}+\frac{9}{12}$
$=\frac{13}{24}$

Skill Check

Go to Software Work through Practice in Lesson 2.4 of the software before attempting the following exercises.

Perform the indicated operation. Reduce if possible.

9. $\frac{4}{15}+\frac{1}{15}$

10. $\frac{1}{10}+\frac{3}{100}$

11. $\frac{16}{25}-\frac{11}{25}$

12. $\frac{5}{6}-\frac{4}{15}$

13. $\frac{7}{10}-\frac{41}{100}$

14. $\frac{2}{5}+\frac{4}{35}+\frac{7}{15}$

Apply Skills

The following problems use the skills you have gained relating to addition and subtraction with fractions. Be sure to pay attention to whether you should add or subtract in each problem. Circle any key words or phrases that indicate which operations to use.

15. The stockroom of an office has three open boxes of pens. An intern has a task to consolidate the open boxes of pens, that is, put all of the pens together into one container. One box is $\frac{1}{8}$ full, another box is $\frac{1}{4}$ full, and the final box is $\frac{1}{2}$ full.

a. Will all of the pens fit into a single box? Show work to support your answer.

b. If so, how much room will be left in the box? If not, what fraction of a box of pens will not fit?

Name: Date:

16. Violeta created a simple budget for herself. She knows how much of her monthly income she wants to put towards housing, food, utilities, transportation, and savings. The circle graph shows the percentages of her income that she wants to put towards each category.

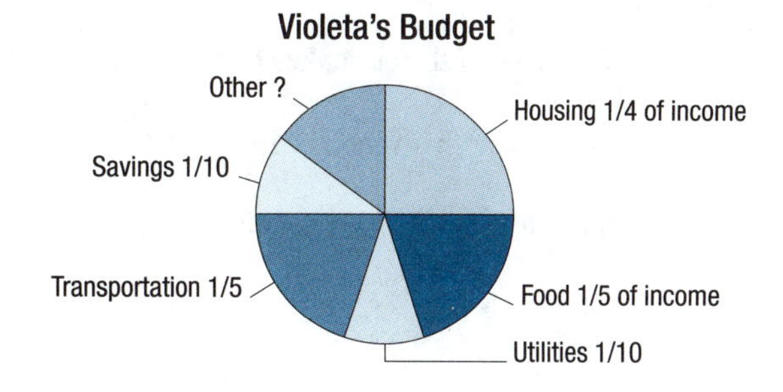

a. What fraction of her monthly income is spent on other expenses?

b. What fraction of her monthly income does she spend on housing, food, and utilities?

c. If she makes \$2400 per month, how much does she spend on housing, food, and utilities?

17. Charles has two types of health insurance: a primary insurance and a supplemental insurance. With his primary insurance, he has to pay $\frac{1}{4}$ of the cost of a hospital stay. His supplemental insurance will cover $\frac{3}{5}$ of any "out of pocket" costs for a hospital stay (that is, they will cover $\frac{3}{5}$ of any amount the primary insurance does not pay).

a. What fraction of the out of pocket costs does the supplemental insurance not cover?

b. What fraction of the amount of the hospital stay will be covered by the supplemental insurance?

c. What fraction of the costs of the hospital stay does Charles have to pay?

18. Barbara's Bombtastic Bakery wants at least $\frac{1}{2}$ of the total cupcakes each day to contain some type of chocolate cake. For tomorrow's menu, they want $\frac{1}{12}$ of the cupcakes to be chocolate fudge, $\frac{1}{6}$ of the cupcakes to be chocolate marble, and $\frac{1}{4}$ of the cupcakes to be Dutch chocolate. If they use this menu, will $\frac{1}{2}$ of the total cupcakes have chocolate cake? If not, will the total amount of chocolate cupcakes be over or under $\frac{1}{2}$.

19. Marten, a jeweler, is creating a new ring. He currently has two stones that he is using for the ring. One is a sapphire that is $\frac{3}{8}$ of a carat in size. The other stone is an alexandrite that is also $\frac{3}{8}$ of a carat in size.

Quick Tip

A **carat** is a unit of measurement for the weight of precious gem stones. One carat is equal to 100 milligrams.

a. What is the total weight of the gemstones for the ring?

b. Marten decides to add a garnet to the ring. If he wants the total carat weight of the ring to be $\frac{15}{16}$ of a carat, what should the weight of the garnet be?

Name: Date:

2.5 Addition and Subtraction with Mixed Numbers

Objectives

Add mixed numbers.

Subtract mixed numbers.

Subtract mixed numbers with borrowing.

Success Strategy

There are two methods shown in this lesson for adding and subtracting mixed numbers. Be sure to try solving a few problems each way to determine which method you prefer.

Understand Concepts

Go to Software First, read through Learn in Lesson 2.5 of the software. Then, work through the problems in this section to expand your understanding of the concepts related to addition and subtraction with mixed numbers.

Adding and Subtracting Mixed Numbers

1. Add or subtract the fraction parts.
2. Add or subtract the whole number parts.
3. Write the answer as a mixed number with the fraction part less than 1 (as a proper fraction).

Lesson Link

Subtrahend and *minuend* are proper terms used with the operation of subtraction, which was covered in Section 1.2.

When subtracting mixed numbers, the fraction part of the subtrahend may be larger than the fraction part of the minuend. This means you have to "borrow a 1" from the whole number part of the minuend. When subtracting with mixed numbers, borrowing a 1 is a slightly different idea than borrowing a 1 when subtracting whole numbers.

1. When simplifying the expression $5\frac{1}{5}-2\frac{3}{5}$ you will need to borrow a 1 since $\frac{1}{5}<\frac{3}{5}$. The following steps will guide you through the process.

a. Rewrite the mixed number $5\frac{1}{5}$ as an expression using addition.

b. The borrowed 1 will always come from the whole number. Rewrite the expression from part **a.**, with the whole number part as the sum of a number plus 1.

Quick Tip

Every whole number can be written as a fraction.

c. Rewrite the expression from part **b.** with the 1 written as a fraction with the same denominator as the fraction in part **a.**

d. Simplify the expression in part **c.** and write the result as a mixed number (with an improper fraction part).

e. Replace $5\frac{1}{5}$ in the original expression with the mixed number from part **d.** Simplify the expression by subtracting the whole numbers and subtracting the fractions.

2. When adding or subtracting mixed numbers, one option is to work with the whole numbers and fraction parts separately, as we did in Problem 1. An alternative method is to change the mixed numbers to improper fractions before performing any operations.

Quick Tip

Make sure the answer is in the same format as the problem. For example, if the problem has mixed numbers, the answer should be a mixed number.

a. Use both methods to simplify $5\frac{1}{2}+2\frac{7}{8}$. Show your work.

b. What extra step did this improper fraction method avoid when adding?

c. Use both methods to simplify $5\frac{1}{5}-2\frac{3}{5}$. Show your work.

d. What step does this improper fraction method remove when subtracting?

3. Which method for adding and subtracting mixed numbers do you like better? Write an explanation for your choice.

Name: Date:

Skill Check

Go to Software Work through Practice in Lesson 2.5 of the software before attempting the following exercises.

Perform the indicated operation. Reduce if possible.

4. $6\frac{1}{7}+2\frac{3}{7}$

5. $3\frac{4}{5}+\frac{7}{20}$

6. $6\frac{1}{12}+2\frac{3}{8}+1\frac{2}{3}$

7. $9\frac{7}{12}-2\frac{5}{18}$

8. $4\frac{3}{8}-2\frac{7}{8}$

9. $12\frac{1}{2}-5\frac{5}{6}$

Quick Tip

When there are multiple methods to solve a problem, use the method you feel most comfortable with unless instructed to use a specific method.

Apply Skills

The following problems use the skills you gained relating to adding and subtracting mixed numbers. Be sure to pay attention to whether you should add or subtract in each problem. Circle any key words or phrases that indicate which operations to use.

10. The dry ingredients in a cookie recipe are $3\frac{1}{4}$ cups of flour, $\frac{3}{4}$ of a cup of brown sugar, $\frac{3}{4}$ of a cup of granulated sugar, $\frac{1}{16}$ of a cup of baking soda, and $1\frac{1}{2}$ cups of chocolate chips.

a. How many total cups of dry ingredients does the recipe require?

b. Will all of the dry ingredients fit into a $6\frac{1}{2}$ cup container? Why or why not?

11. David is an independent contractor and specializes in making decorative walkways. For his current job he is placing large decorative stones in concrete. To make the concrete for the job, David mixes $8\frac{1}{4}$ quarts of gravel, $7\frac{1}{2}$ quarts of sand, 3 quarts of cement, and $1\frac{1}{2}$ quart of water. How many quarts of concrete will David have?

12. Ellen needs $7\frac{5}{16}$ pounds of fertilizer to cover her lawn. She currently has $5\frac{5}{8}$ pounds of fertilizer.

a. How many pounds of fertilizer does she need to purchase?

b. There is a sale on half-pound bags of fertilizer. How many half-pound bags will Ellen need to buy?

13. Daniel makes his own taco seasoning by combining $2\frac{1}{8}$ tablespoons (T) of chili powder, $1\frac{1}{2}$ T ground cumin, $1\frac{3}{4}$ T paprika, $1\frac{1}{3}$ T dried oregano, $\frac{3}{4}$ T garlic powder, $\frac{3}{4}$ T onion powder, and $1\frac{3}{4}$ T ground cayenne pepper. He then makes tacos with a recipe that calls for $2\frac{3}{4}$ T taco seasoning.

a. How many tablespoons of taco seasoning does Daniel make?

b. How many tablespoons of taco seasoning does Daniel have left after making the tacos?

Name: Date:

2.6 Order of Operations with Fractions and Mixed Numbers

Objectives

Compare fractions by finding a common denominator and comparing the numerators.

Evaluate expressions with fractions.

Simplify complex fractions.

Evaluate expressions with mixed numbers.

Find the average of a group of fractions or mixed numbers.

Success Strategy

Be sure to review the order of operations with whole numbers from Section 1.6 before using it to simplify operations with fractions.

Understand Concepts

Lesson Link

Equivalent fractions, which were covered in Section 2.1, are commonly used when comparing fractions.

Quick Tip

The approach used in Situation **D.** works for all comparisons. It may be more time consuming than the three other situations.

Go to Software First, read through Learn in Lesson 2.6 of the software. Then, work through the problems in this section to expand your understanding of the concepts related to order of operations with fractions and mixed numbers.

There are four different situations that you may run into when comparing two fractions. Each situation requires a different approach.

Situation	Rule	Example	Why the Rule Works
A. Common denominators	Compare the numerators. The larger fraction will have the larger numerator.	$\frac{3}{8}$ vs. $\frac{5}{8}$	Since the denominators are the same, we only compare the numerators. We can determine that $\frac{5}{8}$ is larger than $\frac{3}{8}$ because 5 is larger than 3.
B. Numerators are 1, different denominators	Compare the denominators. The larger fraction will have the smaller denominator.	$\frac{1}{2}$ vs. $\frac{1}{3}$	Dividing a whole object into 2 equal parts gives you larger pieces than if you divide a whole object into 3 equal parts. So, $\frac{1}{2}$ is larger than $\frac{1}{3}$.
C. Common numerators	Compare the denominators. The larger fraction will have the smaller denominator.	$\frac{3}{7}$ vs. $\frac{3}{8}$	According to Situation B, we can determine that $\frac{1}{7}$ is larger than $\frac{1}{8}$ because 7 is smaller than 8. Multiplying both $\frac{1}{8}$ and $\frac{1}{7}$ by 3 will give us $\frac{3}{8} < \frac{3}{7}$.
D. Different numerators and different denominators	Find equivalent fractions and compare as in Situation A.	$\frac{5}{6}$ vs. $\frac{7}{8}$	The equivalent fractions with the same denominators are $\frac{20}{24}$ vs. $\frac{21}{24}$. By Situation A, $\frac{21}{24}$ is the larger fraction because 21 > 20. Therefore, $\frac{7}{8}$ is larger.

1. Use the information in the table to determine which situation the comparison falls into and then determine which fraction is larger.

a. $\frac{5}{14}$ vs. $\frac{9}{14}$

b. $\frac{9}{10}$ vs. $\frac{4}{5}$

c. $\frac{4}{5}$ vs. $\frac{4}{9}$

d. $\frac{1}{13}$ vs. $\frac{1}{20}$

Complex Fractions

A **complex fraction** is a fraction which contains one or more fractions in the numerator, the denominator, or both.

To Simplify a Complex Fraction

1. Simplify the numerator so that it is a single fraction.
2. Simplify the denominator so that it is a single fraction.
3. Divide the numerator by the denominator and reduce if possible.

2. Think back to Problem 5 in Section 2.2 where we discussed dividing fractions.

a. We determined that $\frac{1}{5} \div \frac{2}{3} = \frac{\frac{1}{5}}{\frac{2}{3}}$. Is the expression $\frac{\frac{1}{5}}{\frac{2}{3}}$ a complex fraction? If it is, how do you know?

Lesson Link

The rules for dividing by a fraction were introduced in Section 2.2.

b. In the rules for simplifying complex fractions, one step is to "divide the numerator by the denominator." When both the numerator and denominator are fractions, this is the same as dividing two fractions. If we use the rule that is taught when dividing a fraction by a fraction, how can we rewrite this step?

Determine if any mistakes were made while simplifying each complex fraction. If any mistakes were made, describe the mistake and then correctly simplify the complex fraction to find the actual result.

3. $$\frac{\frac{3}{4}}{\frac{3}{2}} = \frac{3}{4} \cdot \frac{3}{2} = \frac{9}{8}$$

4. $$\frac{1\frac{1}{4} + 2\frac{1}{4}}{3\frac{1}{4}} = \frac{3\frac{1}{2}}{3\frac{1}{4}} = 3\frac{1}{2} \cdot 3\frac{4}{1} = \frac{7}{2} \cdot \frac{7}{1} = \frac{49}{2}$$

Lesson Link

The order of operations for whole numbers was introduced in Section 1.6. Notice that the order of operations is the same for fractions.

Order of Operations

1. Simplify within grouping symbols. If there are multiple grouping symbols, start with the innermost grouping.
2. Evaluate any numbers or expressions indicated by exponents.
3. From left to right, perform any multiplication or division in the order they appear.
4. From left to right, perform any addition or subtraction in the order they appear.

Name: Date:

5. A common error when simplifying a long expression that contains fractions has to do with not following the order of operations. Here is an example of this common error:

$$\frac{1}{2}+\frac{1}{4}\div\frac{1}{3}\cdot\frac{1}{5}\neq\frac{1}{2}+\frac{1}{4}\div\frac{1}{15}$$

What error happened in the above simplification? That is, what part of the order of operations was not followed?

6. To avoid the error from Problem 5, let your first step when simplifying be to change the division symbol to a multiplication symbol and change the fraction following the symbol to its reciprocal. Using this initial step, the expression above changes as follows: $\frac{1}{2}+\frac{1}{4}\div\frac{1}{3}\cdot\frac{1}{5}=\frac{1}{2}+\frac{1}{4}\cdot\frac{3}{1}\cdot\frac{1}{5}$.

a. Simplify the new expression by following the order of operations.

b. If you multiply $\frac{3}{1}\cdot\frac{1}{5}$ together first, does this result in a different simplification than if you correctly followed the order of operations?

c. If the answer to part **b.** is yes, what are the different solutions? If the answer is no, what property allows you to perform multiplication in a different order than left to right?

Lesson Link

The associative and commutative properties for multiplication were introduced in Section 1.3.

True of False: Determine whether each statement is true or false. Rewrite any false statement so that it is true. (There may be more than one correct new statement.)

7. The sum of two fractions that are between 0 and 1 can never be greater than 1.

8. The product of two fractions between 0 and 1 can never be greater than 1.

Skill Check

 ***Go to Software*** Work through Practice in Lesson 2.6 of the software before attempting the following exercises.

Arrange the numbers in order from smallest to largest.

9. $\frac{5}{6}, \frac{3}{4}, \frac{9}{10}$

10. $\frac{3}{8}, \frac{5}{16}, \frac{17}{32}$

Evaluate each expression using the order of operations.

11. $\frac{3}{20} \cdot \frac{5}{7} \div \frac{1}{4} - \frac{3}{14} \cdot \frac{2}{3}$

12. $4\frac{1}{3} - 1\frac{1}{2}\left(2 - \frac{2}{3}\right)^2$

Simplify the following complex fractions.

13. $\dfrac{\frac{3}{10}}{\frac{9}{14}}$

14. $\dfrac{\frac{1}{2} + \frac{1}{4}}{2\frac{1}{8}}$

Apply Skills

The following problems combine the skills and concepts you've learned so far in Chapter 2. As with many of the application problems in this workbook, the key words or phrases for operations aren't always obvious. Knowing how to solve the problem depends on thinking about the situation and the questions asked. Circle any key words or phrases that indicate which operation is to be used to solve the problem.

15. Mason owns a small bakery and is trying to keep track of how much of each ingredient is used per day. Each day for their breakfast service, the bakery makes 3 batches of muffins that take $4\frac{1}{4}$ cups of flour per batch, 6 batches of bagels that take $3\frac{3}{4}$ cups of flour per batch, and 4 batches of scones that take $5\frac{1}{4}$ cups of flour per batch.

a. Write an expression to determine how much flour is used for the breakfast food each day.

b. Simplify the expression from part **a.** to determine how much flour the bakery uses for their breakfast service.

Name: Date:

16. A veterinarian's assistant is doing an inventory of supplies from three different closets in the office. In the first closet, she finds $2\frac{1}{2}$ boxes of syringes. In the second closet, she finds three boxes of syringes, each $\frac{3}{4}$ full. In the third closet, she finds 3 full boxes and two boxes that are each $\frac{1}{3}$ full. How many total boxes of syringes will she write on the inventory sheet?

17. The college's undergraduate student council is voting on a bill to allow the council room to be used for free student tutoring twice a week. For the vote to pass in the student council, $\frac{3}{4}$ of the student council must vote in favor. The student council has an equal number of freshman, sophomores, juniors, and seniors.

a. What fraction of the student council is in each year level?

b. Suppose the following amount of students on the council vote in favor of the bill: $\frac{3}{4}$ of the freshman, $\frac{1}{2}$ of the sophomores, $\frac{3}{5}$ of the juniors, and $\frac{2}{5}$ of the seniors. Write an expression to determine what fraction of all of the student council members voted in favor of the bill.

c. Simplify the expression from part **b.** to determine what fraction of all the students in the student council voted in favor of the bill.

d. Did the bill pass? Why or why not?

18. Tiffany orders the specialty Italian sub sandwich from a local deli. The deli advertises that the sub has at least $1\frac{1}{4}$ pounds of meat. The person assembling the sandwich adds $\frac{1}{4}$ pound of ham, $\frac{2}{3}$ pound of salami, $\frac{1}{8}$ pound of hot capicola, and $\frac{1}{4}$ pound of mortadella. Does this sandwich have at least $1\frac{1}{4}$ pounds of meat?

19. Owen has a side business where he sells jars of dry cookie mix that you just have to mix in an egg, $\frac{1}{2}$ cup oil, and 1 teaspoon of vanilla before baking. To make the sugar cookie mix, he combines 3 cups granulated sugar, 3 cups powdered sugar, $\frac{1}{16}$ cups kosher salt, $\frac{1}{16}$ cups baking soda, $\frac{1}{16}$ cups cream of tartar, and 12 cups flour. This mixture is then divided into glass jars which hold $3\frac{1}{32}$ cups of dry ingredients each.

Quick Tip

Be sure to properly use parentheses when writing the expression for part **a.** of Problem 19.

a. Write an expression to determine how many jars of cookie mix Owen made.

b. Simplify the expression in part **a.** to determine how many jars of cookie mix Owen made.

20. Jennifer works as a Technical Support Specialist for a software company and the number of hours she works each day varies based on the number of calls they receive. Last week she worked $8\frac{1}{2}$ hours on Monday, $7\frac{3}{5}$ hours on Tuesday, $9\frac{1}{4}$ hours on Wednesday, $8\frac{9}{10}$ hours on Thursday, and 8 hours on Friday.

a. How many total hours did Jennifer work last week?

b. What was the average number of hours Jennifer worked per day?

Quick Tip

1 hour is equal to 60 minutes. To convert hours to minutes, multiply the fractional part of the hour by 60.

c. On Thursday, the fractional part of an hour that she worked is equivalent to how many minutes?

Name: Date:

Chapter 2 Projects

Project A: If You Can't Take the Heat, Get Out of the Kitchen!

An activity to demonstrate the use of fractions in real life

No, this famous saying did not come from a famous TV chef, but from a former US President Harry S. Truman. President Truman was a plainspoken politician, and he used this phrase in 1949 when explaining to his staff that they should not worry about the criticism they were receiving about their appointments. It has become a popular phrase implying that if you can't take the pressure of the job you have been given to do, then find another job and let someone else who can handle the pressure take over.

If you have watched any episodes of the Iron Chef, you may have noted some similarities in working for a president and working for a famous chef. It's a tough job and definitely not a good career choice for the timid. In this activity you will be working in the kitchen baking cookies. Things may get a little heated with all the math work you are about to do, so be careful—don't get burned!

The following is a list of ingredients for making 3 dozen (medium-sized) chocolate chip cookies. For baking it is generally easier to purchase butter or margarine in the form of sticks. There are 4 sticks of butter in a box that weighs 16 ounces or 1 pound.

$2\frac{1}{4}$ cups all-purpose flour	$1\frac{1}{2}$ teaspoon vanilla extract
1 teaspoon baking soda	2 large eggs
1 teaspoon salt	2 cups chocolate chips
1 cup (2 sticks) butter, softened	1 cup chopped nuts
$\frac{3}{4}$ cup granulated sugar	$\frac{3}{4}$ cup packed brown sugar

1. What fraction of a pound of butter does 2 sticks of butter represent?

2. If you need to make 6 dozen cookies for the Scout meeting next week, how will you need to scale the recipe?

3. How much of the following ingredients will you need to make 6 dozen cookies?

____ cups all-purpose flour	____ teaspoon vanilla extract
____ teaspoon baking soda	____ large eggs
____ teaspoon salt	____ cups chocolate chips
____ cup butter, softened	____ cup chopped nuts
____ cup granulated sugar	____ cup packed brown sugar

4. What fraction of a dozen eggs does the value in Problem 3 represent?

5. If you only want to make 1 dozen cookies (just for yourself), how will you need to scale the recipe?

6. How much of the following ingredients will you need to make 1 dozen cookies?

____ cups all-purpose flour	____ teaspoon vanilla extract
____ teaspoon baking soda	____ large eggs
____ teaspoon salt	____ cups chocolate chips
____ cup butter, softened	____ cup chopped nuts
____ cup granulated sugar	____ cup packed brown sugar

7. Look at the amount of eggs required for your scaled recipe from Problem 6.

a. Why doesn't the amount of eggs in your answer to Problem 6 make sense?

b. How would you handle this in making the cookie dough?

Name: Date:

Project B: A Cookout Dilemma!

An activity to demonstrate the use of the least common multiple in real life

Have you ever watched the movie *Father of the Bride*? Steve Martin plays the role of the bride's father and he is somewhat anxious about the wedding. Go to this website to watch a video clip where he goes to the supermarket to buy hot dogs and hot dog buns for the cookout: http://www.youtube.com/watch?v=j0A-DeOYOJ0

Steve's character claims that the hot dog company and the bun company are working together to swindle the American public by selling hot dogs in packages of 8 and buns in packages of 12. Do you recognize this as a problem involving finding the least common multiple of 8 and 12?

1. Let's assume that there are going to be 20 people at the cookout and everyone will eat only 1 hot dog. What is the **least** number of packages of hot dogs with 8 in a package and the **least** number of packages of buns with 12 buns in a package needed so that there is an equal amount of hot dogs and buns.

a. First find the prime factorization of both 8 and 12.

b. Now find the LCM using the results from part **a.**

c. How many packages of hot dogs will be needed?

d. How many packages of buns will be needed?

e. Will there be enough hot dogs and buns so that each person at the cookout can have one hot dog on a bun?

f. If there are enough hot dogs and buns,will any be left over? If so how many? If there are not enough hot dogs and buns, how many more are still needed?

2. We all know how good hot dogs cooked on the grill are—no one could eat just one! What if everyone eats 2 hotdogs each?

   a. What is the **least** number of packages of hot dogs with 8 in a package and the **least** number of packages of buns with 12 buns in a package that will be needed to feed everyone 2 hot dogs and buns. Remember, there needs to be an equal amount of hot dogs and buns.

   b. Will there be any hot dogs and buns left over? If so how many?

3. Suppose that you did not require the number of hot dogs and buns to be equal. Answer the following questions.

   a. Would there be a solution for Problem 2 that costs less?

   b. How many packages of hot dogs would you buy?

   c. How many packages of buns would you buy?

   d. How many hot dogs would be left over and how many buns would be left over?

4. Does it bother you to always end up with buns left over after a cookout because of the uneven packaging of hot dogs and buns by their manufacturers? Explain why or why not?

Name: Date:

Math@Work

Hospitality Management: Preparing for a Dinner Service

As the manager of a restaurant you will need to make sure everything is in place for each meal service. This means that you need to predict and prepare for busy times, such as a Friday night dinner rush. To do this, you will need to obtain and analyze information to determine how much of each meal is typically ordered. After you estimate the number of meals that will be sold, you need to communicate to the chefs how much of each item they need to expect to prepare. An additional aspect of the job is to work with the kitchen staff to make sure you have enough ingredients in stock to last throughout the meal service.

You are given the following data, which is the sales records for the signature dishes during the previous four Friday night dinner services.

Week	Meal A	Meal B	Meal C	Meal D
1	30	42	28	20
2	35	38	30	26
3	32	34	26	26
4	30	32	28	22

Meal C is served with a risotto, a type of creamy rice. The chefs use the following recipe, which makes 6 servings of risotto, when they prepare Meal C. (**Note:** The abbreviation for tablespoon is T and the abbreviation for cup is c.)

$5\frac{1}{2}$ c chicken stock

$2\frac{1}{3}$ T chopped shallots

$\frac{1}{2}$ c red wine

$1\frac{1}{2}$ c rice

2 T chopped parsley

$4\frac{3}{4}$ c thinly sliced mushrooms

2 T butter

2 T olive oil

$\frac{1}{2}$ c Parmesan cheese

1. For the past four Friday night dinner services, what was the average number of each signature meal served? If the average isn't a whole number, explain why you would round this number either up or down.

2. Based on the average you obtained for Meal C, calculate how much of each ingredient your chefs will need to make the predicted amount of risotto.

3. The head chef reports the following partial inventory: $10\frac{3}{4}$ c rice, $15\frac{3}{4}$ c mushrooms, and 10 T shallots. Do you have enough of these three items in stock to prepare the predicted number of servings of risotto?

4. Which skills covered in this and the previous chapter helped you make your decisions?

Foundations Skill Check for Chapter 3

This page lists several skills covered previously in the book and software that are needed to learn new skills in Chapter 3. To make sure you are prepared to learn these new skills, take the self-test below and determine if any specific skills need to be reviewed.

Each skill includes an easy (**e.**), medium (**m.**), and hard (**h.**) version. You should be able to complete each exercise type at each skill level. If you are unable to complete the exercises at the easy or medium level, go back to the given lesson in the software and review until you feel confident in your ability. If you are unable to complete the hard problem for a skill, or are able to complete it but with minor errors, a review of the skill may not be necessary. You can wait until the skill is needed in the chapter to decide whether or not you should work through a quick review.

1.3 Find the product.

e. $\begin{array}{r} 47 \\ \times\ 8 \\ \hline \end{array}$

m. $\begin{array}{r} 389 \\ \times\ 15 \\ \hline \end{array}$

h. $\begin{array}{r} 1357 \\ \times\ 312 \\ \hline \end{array}$

1.4 Divide using long division. Indicate any remainder.

e. $7\overline{)83}$

m. $250 \div 24$

h. $\dfrac{1768}{34}$

2.5 Find the sum and reduce if possible.

e. $1\frac{1}{5}+3\frac{4}{5}$

m. $4\frac{1}{2}+6\frac{7}{16}$

h. $3\frac{5}{6}+5\frac{7}{15}$

2.5 Find the difference and reduce if possible.

e. $8\frac{9}{16}-4\frac{5}{16}$

m. $12\frac{1}{4}-7\frac{3}{4}$

h. $22\frac{1}{4}-3\frac{5}{6}$

2.6 Simplify the expression using the order of operations.

e. $\frac{1}{4}\div\frac{1}{4}+\frac{2}{3}\cdot\frac{1}{5}$

m. $\frac{4}{10}+\frac{7}{10}\div\frac{1}{4}\cdot\frac{1}{2}-\frac{4}{5}$

h. $\left(\frac{1}{3}\right)^2+2\frac{5}{18}-\frac{7}{36}\div\frac{1}{9}$

Chapter 3: Decimal Numbers

Study Skills

3.1 Introduction to Decimal Numbers

3.2 Addition and Subtraction with Decimal Numbers

3.3 Multiplication with Decimal Numbers

3.4 Division with Decimal Numbers

3.5 Decimal Numbers and Fractions

Chapter 3 Projects

Math@Work

Foundations Skill Check for Chapter 4

Math@Work

Introduction

If you plan on going into accounting, there are several careers that you can pursue ranging from bookkeeper to certified public accountant. There are also a variety of settings you can work in. You can work for a business keeping track of their financial dealings or you can run your own accounting service and consult with individuals or businesses to help them manage their finances. In any of these jobs, you can expect to deal with money and decimal numbers in your day to day work. As a part of your job, you will need to quickly and accurately work with decimals and compare results. After performing any necessary calculations, you will need to effectively communicate your findings to the customers or your supervisors who need this information.

Suppose you start a career as a bookkeeper at a company with a traveling sales team. How do you know how much to reimburse an employee for mileage traveled? How do you split up an expense report according to different expense codes? How do you determine if the company is making a profit? Finding the answers to these questions (and many more) require several of the skills covered in this chapter and the previous chapters. At the end of the chapter, we'll come back to this topic and explore how math is used in bookkeeping.

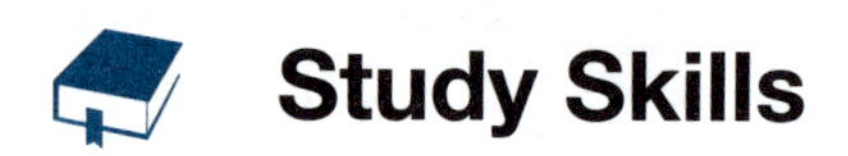

Study Skills

How to Read a Math Textbook

Reading a textbook is very different than reading a book for fun. You have to concentrate more on what you are reading because you will likely be tested on the content. Reading a math textbook requires a different approach than reading literature or history textbooks, because it contains a lot of symbols and formulas in addition to words. Here are some tips to help you successfully read a math textbook.

1. Don't Skim When reading math textbooks you need to look at everything: titles, learning objectives, definitions, formulas, text in the margins, and any text that is highlighted, outlined, or in bold font. Also pay close attention to any tables, figures, charts, and graphs.

2. Minimize Distractions Reading a math textbook requires much more concentration than a novel by your favorite author, so pick a study environment with few distractions and a time when you are most attentive.

3. Start at the Beginning Don't start in the middle of an assigned section. Math tends to build on previously learned concepts and you may miss an important concept or formula that is crucial to understanding the rest of the material in the section.

4. Highlight and Annotate Put your book to good use and don't be afraid to add comments and highlighting. If you don't understand something in the text, reread it a couple of times. If it is still not clear, note the text with a question mark or some other notation, so that you can ask your instructor about it.

5. Go Through Each Step of the Examples Make sure you understand each step of an example and if you don't, mark it so you can ask about it in class. Sometimes math textbooks leave out intermediate steps to save space. Try filling these in yourself in the spaces or margins of the book. Also, try working the examples on your own.

6. Take Notes Write down important definitions, symbols or notation, properties, formulas, theorems, and procedures. Review these daily as you do your homework and before taking quizzes and tests. Practice rewriting definitions in your own words so that you understand them better.

7. Use Available Resources Many textbooks come with CDs or have companion websites to help you with understanding the content. These resources may contain videos that help explain more complex steps or concepts. Do some searching on the Internet for topics you don't understand.

8. Read the Material Before Class Try to read the material from your book before the instructor lectures on it. After the lecture, reread the section again to help you retain the information as you look over your class notes.

9. Understand the Mathematical Definitions Many terms used in everyday English have a different meaning when used in mathematics. Some examples include equivalent, similar, average, median, prime, and product. Two equations can be equivalent to one another without being equal. Similar triangles can be different sizes as long as their sides are in the same proportion. An average can be computed mathematically in several ways. It is important to note these differences in meaning in your notebook where you keep important definitions and formulas.

10. Try Reading the Material Aloud Reading aloud makes you focus on every word in the sentence. Leaving out a word in a sentence or math problem could give it a totally different meaning, so be sure to read the text carefully and reread if necessary.

Name: Date:

3.1 Introduction to Decimal Numbers

Objectives

Read and write decimal numbers.

Compare decimal numbers.

Round decimal numbers.

Success Strategy

Pay close attention to the placement of the decimal point when working with decimal numbers as it will change your results by a power of ten for each incorrect place.

Understand Concepts

Go to Software First, read through Learn in Lesson 3.1 of the software. Then, work through the problems in this section to expand your understanding of the concepts related to decimal numbers.

1. The place value system for **decimal numbers** is an extension of the place value system for whole numbers. Whole numbers are written to the left of the **decimal point** and fractions are written to the right.

a. Fill in the second row with the place value name for each position.

b. Fill in the third row with the whole number power of ten that each place value represents.

Lesson Link

This place value table is an extension of the place value table introduced in Section 1.1.

Ones Period			Decimal Point						
		Ones (or units)							
		1	•						

Reading and Writing Decimal Numbers

1. Read (or write) the whole number.
2. Read (or write) the word "and" in place of the decimal number.
3. Read (or write) the fraction part as a whole number. Then name the fraction with the name of the place of the last digit to the right.

Quick Tip

The part of the decimal number to the right of the decimal point is called the **fraction part** of the decimal number.

2. As we discussed in Section 1.1, using the word "and" is not considered to be proper in math when writing the English word equivalent of whole numbers. The proper mathematical use occurs when writing the English word equivalent of decimal numbers. When is the word "and" properly used in math?

The next three problems will discuss that, while people don't always use "and" properly in conversation and the media, you can usually figure out what people mean by how the number is said or the context of the number.

3. How does the title *One Hundred and One Dalmatians* make it clear that the "and" does not refer to a decimal point?

4. How does the number in "five and four tenths miles" make it clear that the "and" refers to a decimal point?

5. List three improper ways that people say decimal numbers, such as 1.95, in conversation and the media. Try to include at least one improper way that people say monetary values.

Lesson Link

We will show that these two formats are equivalent in Section 3.4.

6. In Section 1.1, we learned how to write whole numbers in expanded form. Decimal numbers can also be written in expanded form. There are two options for writing decimal numbers in expanded form.

Method 1 is with fractions: 13.65 can be written as $1(10)+3(1)+6\left(\frac{1}{10}\right)+5\left(\frac{1}{100}\right)$

Method 2 is with decimal numbers: 13.65 can be written as $1(10)+3(1)+6(0.1)+5(0.01)$

a. Write the number 4.587 in expanded form with both fractions and decimal numbers.

b. Write the number 956.24 in expanded form with both fractions and decimal numbers.

c. Either method can be used to write the expanded form of decimal numbers. Which method do you prefer? Explain why.

Name: Date:

Comparing Decimal Numbers

1. Moving left to right, compare digits with the same place value. (Insert zeros to the right to continue the comparison, if necessary.)
2. When one compared digit is larger, then the corresponding number is larger.

Read the following information about number lines and work through the next problem.

When comparing decimal numbers, a number line can help you visualize the comparison. The increments on the number line can appear in many different ways and represent any distance. For whole numbers, we can mark every whole number from 0 to 10. For fractions, we can use a number line similar to a ruler with every $\frac{1}{16}$ of an inch marked off. For decimal numbers, we can show whole numbers and every 0.5 increment. The increments indicated on a number line depend on the numbers to be plotted and the distance between the increments.

7. Plot each data set on the given number line. First determine the length of the intervals needed and mark them on the number line. Be sure to label each increment marker below the number line.

a. \$4.00, \$1.75, \$1.25, \$3.50

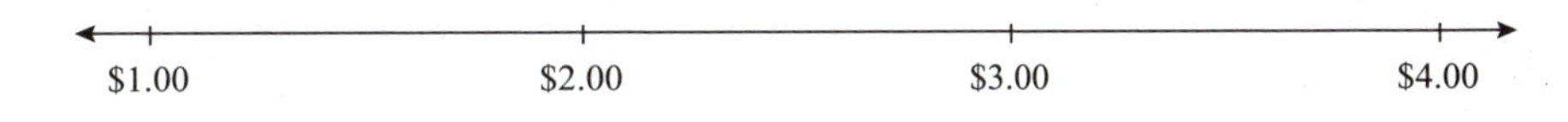

b. 1.4, 1.8, 1.5, 1.2

While working with decimal numbers, it is useful to know what the word decimal means and some of the history of the decimal number system. Use the internet to research and answer the following questions.

8. Go to www.m-w.com and look up the definition of the word "decimal".

a. Write the definition of "decimal" here.

b. According to the definition of decimal, are whole numbers considered to be decimal numbers even though they don't use decimal points? Explain your answer.

9. Perform an internet search using the key words "history of decimals".

a. Which civilization first developed and used decimal numbers or decimal fractions?

b. How long ago was the modern notation for decimal numbers developed?

Quick Tip

A **decimal fraction** is a fraction which has a denominator equal to a power of ten.

Rounding Rules

1. Look at the single digit just to the right of the digit that is in the place of desired accuracy.
2. If this digit is 5 or larger, increase the digit in the desired place of accuracy by one and replace all digits to the right with zeros.
3. If this digit is less than 5, leave the digit in the desired place of accuracy the same and replace all digits to the right with zeros.
4. If the place of accuracy is in the fraction part of the decimal number, drop any zeros to the right of the place of accuracy. If the place of accuracy is in the whole number part of the decimal number, drop any zeros to the right of the decimal point.

Skill Check

Go to Software Work through Practice in Lesson 3.1 of the software before attempting the following exercises.

Write the following decimal numbers in their English word equivalent.

10. 125.7

11. 4.758

12. Write the given number in standard notation.

a. Two hundred and fifteen thousandths

b. Two hundred fifteen thousandths

13. Round 15.2796 to the place value indicated.

a. tenths

b. hundredths

14. Round 7195.5917 to the place value indicated.

a. tens

b. tenths

Name: Date:

Apply Skills

Work through the problems in this section to apply the skills you have learned related to decimal numbers.

Quick Tip

The given place value of a number is determined by the rightmost digit.

15. For each measurement, determine to which place value the measurement is given.

a. A pile of sodium hydroxide is placed on a scale and the mass is determined to be 30.0483 grams.

b. The length of an amoeba is measured to be 0.27 millimeters.

c. The radius of the Earth is measured to be 6378.1 kilometers.

16. In the 2012 Olympics, Veronica Campbell-Brown ran the 100-meter dash in ten and eighty-one hundredths of a second. Shelly-Ann Fraser-Pryce ran the 100-meter dash in ten and seventy-five hundredths of a second. Carmelita Jeter ran the same race in ten and seventy-eight hundredths of a second. Source: espn.go.com

a. Write each finish time as a decimal number. Be sure to label the finish times with the racer's name.

b. Order the finish times from least to greatest.

Quick Tip

The runner who finishes a race in the least amount of time is the winner.

c. Who won the race?

17. Sarah and Richard are partners in a chemistry lab. They are working together to record measurements for an experiment in their lab notebook. Sarah is measuring and Richard is recording. Sarah gives the following measurements.

"We have two point oh four grams of salicylic acid, five and one tenth milliliters of acetic anhydride, and forty-nine point eight milliliters of water."

a. How would Richard fill out the record table?

Chemical	Amount
Salicylic Acid	
Acetic Anhydride	
Water	

b. Which numbers did Sarah say in a nonstandard English word equivalent? How would these be written in their correct English word equivalents?

18. The meat department at the grocery store has packages of ground beef in the cooler case. You see packages with the following weights: 1.15 pounds, 1.39 pounds, 1.28 pounds, 1.21 pounds, and 1.35 pounds.

a. Plot these values on a number line.

b. You have a recipe which calls for 1.25 pounds of ground meat. Which package has the closest weight?

19. You go to a deli counter at the grocery store and order 1.5 pounds of ham, 1.25 pounds of mozzarella cheese, and 0.75 pounds of salami. The deli clerk gives you 1.57 pounds of ham, 1.19 pounds of mozzarella cheese, and 0.75 pounds of salami.

a. Did the clerk give you more or less ham than you asked for?

b. Did the clerk give you more or less mozzarella cheese than you asked for?

c. Did the clerk give you more or less salami than you asked for?

d. Why do you think the clerk might not give you the exact amount you ask for?

Name: Date:

3.2 Addition and Subtraction with Decimal Numbers

Objectives

Add decimal numbers.

Subtract decimal numbers.

Estimate sums and differences.

Success Strategy

To keep decimal numbers lined up correctly while adding and subtracting, try turning a piece of lined notebook paper so the lines run vertically. Enter one number per column and give the decimal point its own column.

Understand Concepts

Go to Software First, read through Learn in Lesson 3.2 of the software. Then, work through the problems in this section to expand your understanding of the concepts related to addition and subtraction with decimal numbers.

Adding and Subtracting Decimal Numbers

1. Write the addition or subtraction problem vertically, with the numbers aligned at the decimal point.
2. Add or subtract as with whole numbers.
3. Check that the decimal point is in the correct position in the sum or difference.

1. When adding or subtracting decimal numbers, it is important to line up the numbers at the decimal point. The following example will show why this is necessary by looking at the sum of 3.248 and 24.25.

Lesson Link

Two different methods of writing decimals in expanded form were introduced in Section 3.1.

a. Write both addends in expanded form in the format of your choice.

b. Which digits have the same place value, from greatest to least?

c. Suppose we performed a vertical addition with the problem set up as follows.

$$\begin{array}{r} 3.248 \\ +\,24.25 \\ \hline \end{array}$$

Why would adding vertically give you the wrong answer? (**Hint:** Try to use your answer from part **b.**)

Quick Tip

To help answer these problems, come up with an example of each statement and pay close attention to the wording for each place value.

True or False: Determine whether each statement is true or false. Rewrite any false statement so that it is true. (There may be more than one correct new statement.)

2. When you carry a 1 from the hundredths place and add it to the tenths place, you can think of this as adding one to the entire sum.

3. When you borrow a 1 from the hundredths place, you can think of this as borrowing ten thousands.

4. When you carry a 2 from the thousandths place and add it to the hundredths place, you can think of this as adding 0.02 to the sum.

Read through the following paragraph on estimation and work through the problems.

Lesson Link

The method of estimation for decimal numbers is the same as the method for whole numbers introduced in Section 1.5.

As we covered in Chapter 1, the method of estimation taught in this course (to round everything to the leftmost nonzero digit) is the quickest and easiest method of estimation, but it may not always give the most useful estimate. While estimation is never completely accurate, you can develop a sense of judgment to obtain an estimate that is useful for the situation. (For homework in the Hawkes Learning software, be sure to use the leftmost nonzero digit rule!)

Listed below are some alternative methods of rounding decimal numbers that can be used when finding an estimate. The rounding method you use while estimating will depend on the situation and how accurate an estimate you need.

Alternate Methods for Rounding Decimal Numbers

a. Round to the nearest half (0.5).

b. Round to the nearest whole number.

c. Round **up** to the nearest tenth, hundredth, thousandth, etc.

5. Hannah has 0.149 liters of sulfuric acid in her lab. She uses 0.062 liters to perform an experiment. Approximately how much sulfuric acid does Hannah have left?

   a. Which rounding method would you use to find the estimate?

   b. Find the estimated value and the actual value.

   c. Does the estimate from part **b.** overestimate or underestimate the actual value left?

   d. Do you think the rounding method you chose provides a good estimate? Give a reason to support your answer.

Name: Date:

6. Julian's doctor wants him to keep track of his weight loss. He lost 2.45 pounds the first week, 3.14 pounds the second week, 4.49 pounds the third week, and 2.38 pounds the fourth week. Approximately how much weight did Julian lose during those four weeks?

   a. Which rounding method would you use to find the estimate?

   b. Find the estimated value and the actual value.

   c. Does the estimate from part **b.** overestimate or underestimate the actual value lost?

   d. Do you think the rounding method you chose provides a good estimate? Give a reason to support your answer.

Skill Check

 ***Go to Software*** Work through Practice in Lesson 3.2 of the software before attempting the following exercises.

Perform the indicated operation.

7. $2.5 + 7.3 + 0.4$

8. $6.91 + 0.06 + 0.25$

9. $14.52 + 1.972 + 0.2005$

10. $15.885 - 12.274$

11. $7.2 - 4.61$

12. $4 - 1.648$

Apply Skills

Work through the problems in this section to apply the skills you have learned related to addition and subtraction with decimal numbers.

13. Abigail works three nights a week as a waitress. During the week, she made the following amounts in tips: \$38.49, \$56.72, \$43.70. How much did Abigail make in tips during the week?

14. During a chemistry lab experiment, Irene took 50.000 grams of a chemical compound and heated it over a burner. Several minutes later, she determined that the mass of the chemical compound was 42.809 grams after heating. What is the difference between the starting mass and the final mass?

15. Macy has \$8500 to spend on redecorating her home. She spends \$6493.74 on furniture. Approximately how much money does Macy have left after buying the furniture?

Quick Tip

A **credit** adds money to an account. A **debit** removes money from an account.

16. Gareth runs a small business and keeps track of expenses and income. At the beginning of the week, there was \$9529.54 in the business checking account.

a. During the week, Gareth spent \$383.72 on office supplies, paid his employees a total of \$2149.22, and spent \$234.63 on business lunches. How much did Gareth spend during the week?

b. How much will be left after the account is debited the amount from part **a.**?

c. During the week, the business made \$5286.81. How much will be in the business account after the earnings are credited?

Name: ______________________ Date: ______________________

3.3 Multiplication with Decimal Numbers

Objectives

Multiply decimal numbers.

Multiply by powers of 10.

Estimate products.

Work applications using decimal numbers.

Success Strategy

The tip from the previous section about using a lined sheet of notebook paper to keep decimal numbers aligned works for multiplication as well. The only difference is that you ignore the decimal point until you get the final result.

Understand Concepts

Go to Software First, read through Learn in Lesson 3.3 of the software. Then, work through the problems in this section to expand your understanding of the concepts related to multiplication with decimal numbers.

Multiplication with decimal numbers is similar to multiplication with whole numbers. The main difference is figuring out where to place the decimal point in the product.

Multiplying Decimal Numbers

1. Multiply the two numbers as if they were whole numbers.
2. Count the number of places to the right of the decimal points in both numbers being multiplied and add them together.
3. Place a decimal point in the product so that the number of places to the right of the decimal point is the same as that found in Step 2.

Quick Tip

When performing multiplication with decimal numbers, the digits do not need to be lined up according to place value.

1. Decimal numbers are often multiplied by a power of ten. It's important to understand the relationship between the exponential form of the power of ten and the number of places the decimal point is moved during multiplication by the power of ten. A better understanding of this relationship can result in quicker mental calculations.

Whole Number	Exponential Form	0.12345 · (Whole Number)	Number of Places the Decimal Point Moved to the Right
1	10^0	0.12345 · 1 = 0.12345	0
10	10^1		
100			
1000			
10,000			

Lesson Link 

Exponents were introduced in Section 1.6.

a. Fill in column two of the table with the exponential forms of the corresponding whole numbers in column one. The first two have been filled in for you.

Lesson Link

Scientific notation, covered in Section 11.2, uses powers of 10.

b. What is the relationship between the zeros in the whole number in column one and the exponent of the power of ten in column two?

c. Complete column three by multiplying 0.12345 by each whole number from column one.

d. Fill in column four with the number of places that multiplication by the whole number moved the decimal point to the right.

e. What is the relationship between the exponential form of the power of ten and the number of places it moves the decimal point during multiplication?

2. Let's look at an expanded process of multiplying $(2.45)(3.1)$ to see why the method described in the box works.

a. What number do you need to multiply 2.45 by to make it a whole number? (**Hint:** It's a power of 10.)

b. What number do you need to multiply 3.1 by to make it a whole number?

c. Rewrite $2.45 \cdot 3.1$ as a product of whole numbers by multiplying each number as suggested by parts **a.** and **b.** Then, simplify the expression.

d. What is the product of the answers from parts **a.** and **b.**?

e. Divide the product from part **c.** by the product from part **d.**

f. Simplify $(2.45)(3.1)$ using the method given in the box at the beginning of the section. Do both methods result in the same product? If so, why do you think the products are the same?

Lesson Link

Section 3.2 introduced several different rounding options to use while estimating.

The method for estimating products with decimal numbers is the same as the method for whole numbers that was introduced in Section 1.5. Work through the following problems involving estimating with decimal numbers.

3. Dylan purchased 24.6 cubic yards of soil. Each cubic yard of soil costs $14.58. Approximately how much money did Dylan spend on the soil?

a. Which rounding method would you use to find the estimate?

Name: Date:

Quick Tip

When working with monetary amounts, round the answer to the nearest cent unless otherwise instructed.

b. Find the estimated value and the actual value. (Round to the nearest hundredth, if necessary.)

c. Does this overestimate or underestimate the actual cost?

d. Do you think the rounding method you chose provides a good estimate? Give a reason to support your answer.

4. The rear driver's side tire on Regina's car has a leak that causes it to lose 0.143 pounds per square inch (PSI) of air pressure per hour. Approximately how much air pressure will the tire lose after 8.5 hours?

a. Which rounding method would you use to find the estimate?

b. Find the estimated value and the actual value.

c. Does this overestimate or underestimate the actual amount of pressure lost?

d. Do you think the rounding method you chose provides a good estimate? Give a reason to support your answer.

Skill Check

Go to Software Work through Practice in Lesson 3.3 of the software before attempting the following exercises.

Find each product.

5. $0.2 \cdot 12$

6. $0.4 \cdot 0.7$

7. $0.5 \cdot 3.7$

8. $100 \cdot 24.973$

Apply Skills

Work through the problems in this section to apply the skills you have learned related to multiplication with decimal numbers.

9. Julian is a car mechanic and charges $70.75 per hour of labor. It takes him 3.5 hours to repair a car. How much will he charge for the labor? (Round to the nearest cent, if necessary.)

10. Kevin needs 28.4 gallons of white interior paint to paint several rooms in a house. The cost per 5-gallon bucket of paint is $62.87.

a. Kevin can only buy the type of paint he needs in the 5-gallon bucket size. How many buckets of paint should he buy?

b. Approximately how much will Kevin spend on paint?

c. How much will Kevin spend on the paint (not including tax)? Round your answer to the nearest cent.

Quick Tip

The key word *of* is commonly used in multiplication problems. For example, two tenths of fifteen would be written as $0.2 \cdot 15$.

11. Water makes up eighty-two hundredths of the mass and weight of an apple.

a. A large apple has a mass of 205 grams. What is the mass of the water in the apple?

b. The same apple has a weight of 7.23 ounces. What is the weight of the water in the apple?

12. A landscaper buys small rectangular stones to create a walkway. Each stone has a length of 4.75 inches and a width of 3.5 inches. The stones are sold in bundles of 100.

a. What is the area of each stone?

b. What is the total area that one bundle of stones can cover?

c. If the landscaper buys 5 bundles of stones, will there be enough to create a walkway with an area of 8500 square inches? Why or why not?

Name: Date:

3.4 Division with Decimal Numbers

Objectives

Divide decimal numbers.

Divide by powers of 10.

Use the rules for order of operations with decimal numbers.

Estimate quotients.

Work applications using decimal numbers.

Success Strategy

Remember to check your work whenever possible. Division of decimal numbers can always be checked by verifying that the product of the quotient and the divisor is equal to the dividend.

Understand Concepts

Go to Software First, read through Learn in Lesson 3.4 of the software. Then, work through the problems in this section to expand your understanding of the concepts related to division with decimal numbers.

Dividing with Decimal Numbers

1. Move the decimal point in the divisor to the right so that the divisor is a whole number.
2. Move the decimal point in the dividend the same number of places to the right.
3. Place the decimal point in the quotient directly above the new decimal point in the dividend.
4. Divide as with whole numbers.

Quick Tip

In division problems, the **divisor** divides into the **dividend** and the result is the **quotient**.

1. When working with whole numbers, we sometimes get a remainder when performing division. Now that we are using decimal numbers, we will handle the remainder differently.

Division Problem	Remainder	Fraction Equivalent of Remainder	Decimal Equivalent of Remainder	Quotient as a Decimal Number
$7 \div 4$	3	$\frac{3}{4}$		
$13 \div 3$				
$25 \div 8$				

a. Find the remainder of each division problem and place it in column two.

b. To find the fractional equivalent of the remainder, write the remainder as the numerator of the fraction and the divisor as the denominator of the fraction. Find the fraction equivalent of the remainders and place them in column three.

c. Find the decimal equivalent of each remainder and place these values in column four. Round your answer to the nearest thousandth if necessary.

d. Perform the long division in column one and write the quotient as a decimal number and place in column five. Round your answer to the nearest thousandth if necessary.

e. Explain what happens to the remainder when we use decimal numbers.

Lesson Link

Multiplying by a power of 10 was discussed in Section 3.3.

2. Decimal numbers are often divided by powers of ten. It is important to understand the relationship between the exponential form of the power of ten and the number of places the decimal point is moved during division by the power of ten. A better understanding of this relationship can result in quicker mental calculations.

Whole Number Form	Exponential Form	54321.6 ÷ (Whole Number Form)	Number of Places the Decimal Point Moved to the Left
1	10^0	54321.6 ÷ 1 = 54321.6	0
10	10^1		
100			
1000			
10,000			

a. Fill in column two of the table with the exponential forms of each corresponding whole number in column one. The first two have been filled in for you.

b. Complete column three by dividing 54321.6 by each whole number from column one.

c. Fill in column four with the number of places each division by the whole number moved the decimal point to the left.

d. What is the relationship between the exponential form of the power of ten and the number of places it moves the decimal point during division?

Lesson Link

Section 3.2 introduced several different rounding options to use while estimating.

The method for estimating quotients with decimal numbers is the same as the method for whole numbers that was introduced in Section 1.5. Work through the next problem involving estimating with decimal numbers.

3. The International Space Station (ISS) orbits the earth 16 times each day. How many hours does it take the ISS to orbit the earth one time?

a. Which rounding method would you use to find the estimate?

b. Find the estimated value and the actual value.

c. Does this overestimate or underestimate the orbit time of the ISS?

d. Do you think the rounding method you chose provides a good estimate? Give a reason to support your answer.

Name: Date:

Lesson Link

The order of operations that was introduced in Section 1.6 for whole numbers and in Section 2.6 for fractions is the same for decimal numbers.

Order of Operations

1. Simplify within grouping symbols. If there are multiple grouping symbols, start with the innermost grouping.
2. Evaluate any numbers or expressions indicated by exponents.
3. From left to right, perform any multiplication or division in the order they appear.
4. From left to right, perform any addition or subtraction in the order they appear.

Determine if any mistakes were made while simplifying each expression. If any mistakes were made, fix them and simplify to find the actual result.

4. $12.98+(1.3+1.4)^2=12.98+(1.69+1.96)$
$=12.98+3.65$
$=16.63$

5. $3(2.25+2\cdot 1.75)=6.75+6\cdot 5.25$
$=6.75+31.5$
$=38.25$

Skill Check

Go to Software Work through Practice in Lesson 3.4 of the software before attempting the following exercises.

Find the quotient. Round to the nearest thousandth if necessary.

6. $7.28 \div 2$

7. $0.12 \div 0.3$

8. $4.5 \div 0.25$

9. $3.4 \div 0.9$

Apply Skills

Work through the problems in this section to apply the skills you have learned related to division with decimal numbers.

10. In 2012, Usain Bolt ran the 100-meter dash in 9.63 seconds. In 2008, he ran the 200-meter dash in 19.30 seconds.

a. Calculate Usain Bolt's speed for the 100-meter dash in meters per second. Round your answer to the nearest thousandth.

Quick Tip

To find the **speed** of an object, divide the distance traveled by the time it takes the object to travel that distance.

b. Calculate Usain Bolt's speed for the 200-meter dash in meters per second. Round your answer to the nearest thousandth.

c. In which race did Usain Bolt have the fastest speed?

11. The average rainfall for Seattle, WA, is shown in the table. Use this data to answer the following questions. Source: National Climatic Data Center, NOAA, data for 1961–1990

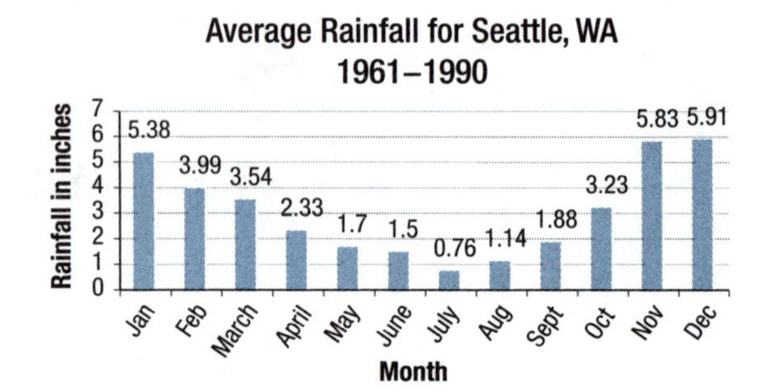

Lesson Link

Finding the average of a list of numbers was introduced in Section 1.7.

a. Which month had the highest average rainfall?

b. What is the difference in rainfall between the month with the highest average rainfall and the month with the lowest average rainfall?

c. What is the average of the monthly rainfall averages from January through June? Round your answer to the nearest hundredth.

12. Barbara's Bombtastic Bakery sells their custom decorated cakes in several different sizes. The sizes, number of servings, and prices are shown in the table.

Size	Servings	Price
6-inch Round	8	$9.95
8-inch Round	16	$19.50
Half Sheet	30	$35.75
Full Sheet	60	$69.95

a. During one day, the bakery sold ten 6-inch round cakes, three 8-inch round cakes, and five half sheet cakes. How much did customers pay in total for cakes during that day?

b. A group of five friends are throwing an end-of-the-semester party and they order two full sheet cakes. They split the cost of the cake evenly. How much will each friend pay for the cake?

Name: ______________________ Date: ______________________

3.5 Decimal Numbers and Fractions

Objectives

Change decimal numbers to fractions.

Change fractions to decimal numbers.

Operate with both fractions and decimal numbers.

Success Strategy

Most of the problems in this section involve converting fractions to decimal numbers or vice versa. Be sure to spend some extra time practicing these skills in the software.

Understand Concepts

Go to Software First, read through Learn in Lesson 3.5 of the software. Then, work through the problems in this section to expand your understanding of the concepts related to decimal numbers and fractions.

Changing Decimal Numbers to Fractions

1. For the numerator, write the fraction part of the decimal number as a whole number with no decimal point.
2. For the denominator, write the power of ten that names the position of the rightmost digit of the numerator.
3. Reduce if possible.

Determine if any mistakes were made while converting each decimal number to a reduced fraction. If any mistakes were made, describe the mistake and then correctly convert the decimal number to a reduced fraction.

1. $0.75 = \frac{1}{75}$

2. $0.15 = \frac{15}{100} = \frac{3}{20}$

Changing Fractions to Decimal Numbers

To change a fraction to a decimal number, divide the numerator by the denominator. Round the decimal number to the desired place of accuracy.

Determine if any mistakes were made while converting each fraction to a decimal number. If any mistakes were made, describe the mistake and then correctly convert the fraction to a decimal number.

3. $\frac{1}{4} = 0.4$

4. $\frac{1}{3} = 0.33$

Types of Decimal Numbers

If the remainder is eventually 0, the decimal number is said to be **terminating**. If the remainder is never 0, the decimal number is said to be **nonterminating**.

Nonterminating decimal numbers can be repeating or nonrepeating. A nonterminating repeating decimal number has a repeating pattern to its digits. If a number is either terminating or nonterminating with repeating digits, then the number is called a rational number.

5. In Section 2.1, we found the reduced fractional form of different coins in the US currency system. The values of coins represent fractions and decimals that are commonly used in math.

Quick Tip

Thinking of these common fractions in terms of money and coins may be very useful when performing mental calculations.

Coin Name	Reduced Fraction	Decimal Number Equivalent
Penny		
Nickel		
Dime		
Quarter		
Half-dollar		

a. Fill in the reduced fraction value for each coin in column two of the table. This should match the values in column three of the table in Section 2.1.

b. Fill in the decimal number equivalent for each coin.

c. When working with money, do you find it easier to think of coins in terms of fractions or decimal numbers?

Read through the following paragraph on simplifying expressions with fractions and decimals and then work through the problems.

When simplifying expressions that have both fractions and decimal numbers, it can be useful to change all fractions to decimal numbers or all decimal numbers to fractions before simplifying. Some fractions turn into repeating decimal numbers, such as $\frac{1}{6}$ which is equal to $0.1\overline{6}$. Since we would need to round these numbers to write them in decimal form, the simplification wouldn't be as accurate as if we left these numbers in fraction form. Sometimes you need to make a decision between accuracy of the answer or ease of simplification. For instance, $\frac{1}{3}+0.739$ would involve either rounding $\frac{1}{3}$ to decimal form or writing 0.739 in fraction form as $\frac{739}{1000}$ which results in a large denominator.

6. Simplify $0.25+\frac{1}{3}+1.5$.

a. Would rounding be involved if you change all fractions to decimals?

b. Change the numbers to all fractions or all decimal numbers based on your answer to part **a.**

c. Simplify your expression from part **b.**

7. Simplify $2-\frac{3}{4}+0.967$.

a. Would rounding be involved if you change all fractions to decimals?

b. Change the numbers to all fractions or all decimal numbers based on your answer to part **a.**

c. Simplify your expression from part **b.**

Read through the following paragraph on the number line and work through the problems.

One way to order numbers from least to greatest is to plot them on a number line. This is a visual way to compare numbers which are fractions, decimals, and whole numbers. Some sets of numbers will be easier to plot on a number line with fractional increments and others will be easier to plot on a number line with decimal increments.

8. Consider the list of numbers $\frac{1}{2}$, 0.75, 0, $\frac{2}{3}$.

a. Are any of the numbers in the list equivalent to a repeating decimal?

b. Change all of the numbers in the list to fractions.

c. Plot each value on the number line.

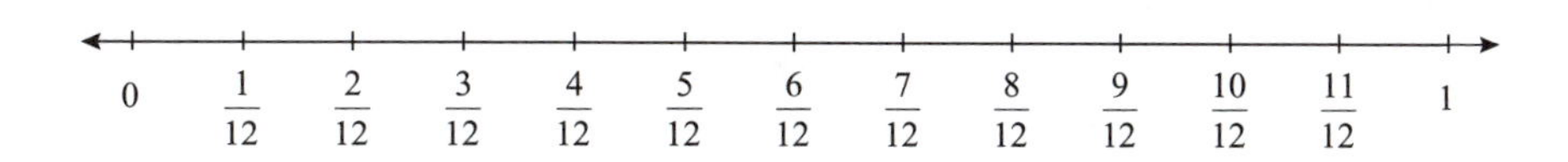

d. Why do you think we changed the numbers to fractions before plotting them on the number line?

9. Consider the list of numbers $\frac{6}{5}$, 1.5, 2, $\frac{11}{5}$.

   a. Are any of the numbers in the list equivalent to a repeating decimal?

   b. Change all of the numbers in the list to decimal numbers.

   c. Plot each value on the number line.

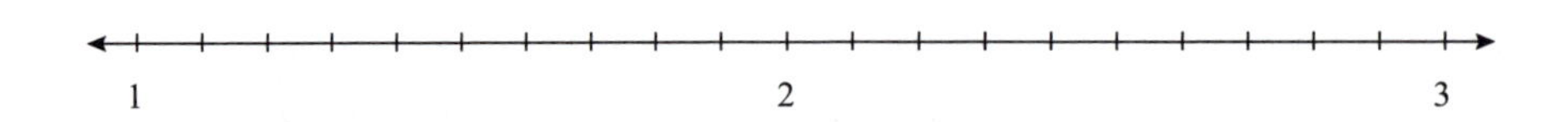

   d. Why do you think we changed the numbers to decimal numbers before plotting them on the number line?

Skill Check

Go to Software Work through Practice in Lesson 3.5 of the software before attempting the following exercises.

Change each decimal number to fractional form and reduce if possible.

10. 0.625

11. 0.84

Change each fraction to decimal form. If the decimal is nonterminating, write the answer using the bar notation over the repeating digits.

12. $\frac{3}{16}$

13. $\frac{1}{12}$

Determine which number is larger by changing both numbers to fractions or both to decimal numbers.

14. $\frac{7}{8}$; 0.878

15. $\frac{22}{7}$; 3.3

Name: Date:

Apply Skills

Work through the problems in this section to apply the skills you have learned related to decimal numbers and fractions.

16. During a physics lab, two lab partners separately performed the experiment three times and recorded how long it took to complete. The first lab partner obtained measurements of 2.55 seconds, 2.25 seconds, and 2.60 seconds. The second lab partner obtained measurements of $2\frac{1}{2}$ seconds, $2\frac{1}{3}$ seconds, and $2\frac{2}{5}$ seconds. They need to find the average time from their measurements.

a. Would the lab partners find a more accurate average by changing all numbers to decimals or all numbers to fractions? Explain why.

b. What is the average time based on the answer from part **a.**?

17. During the month of June, Artem earned two paychecks worth $1634.53 and $1620.50 each. He plans on putting $\frac{1}{3}$ of his June income into his savings account. How much will Artem put into his savings account?

18. Lotte is redecorating her living room. The back wall is 8 feet high and 14.75 feet wide. She wants to put wood paneling on the bottom $\frac{2}{5}$ of the back wall. The price of the paneling per square foot is $2.45.

a. How many square feet of wooden paneling must Lotte buy?

b. Approximately how much will Lotte spend on the wooden paneling?

c. If the paneling can only be sold by the whole square foot, how much will Lotte pay (before taxes) for the amount of paneling she needs?

19. Tuition and fees for a local college are given in the table. The amounts given are per credit hour except the lab fee which is per lab course. Use this information to answer the following questions.

Reason	Amount
1 Credit Hour	$353.10
Facility Fee	$17.65
Library Fee	$3.00
Lab Fee	$15.50

a. Brad registers for 12 credit hours, which includes a chemistry lab. How much are his tuition and fees for the semester?

b. Marta registers for 14 credit hours, which includes two biology labs. How much are her tuition and fees for the semester?

20. Find the tuition and fee schedule on your college's website. Determine how the tuition you paid this semester was calculated.

Reason	Amount
1 Credit Hour	
Facility Fee	
Library Fee	
Lab Fee	
Other:	
Other:	
Other:	

Name: Date:

Chapter 3 Projects

Project A: Is It Really More Expensive to Eat Healthier?

An activity to demonstrate the use of decimal numbers in real life

People commonly think that it costs more to eat healthy. But is it really more expensive? In this project you are going to do a mini research study to determine if it costs more to eat fresh foods, which are generally healthier, than it does to eat prepackaged foods, which are usually assumed to be less healthy due to the preservatives and additives they contain. Keep in mind that your results may vary from others in the class depending upon the food choices made.

1. Using your recent grocery bill or an online resource such as a weekly store advertisement, pick four fresh vegetables or fruits that are sold by the pound and record each item's price per pound in the table below.

Food Item	Price per Pound

2. Now compute the average price per pound of these four items by summing up the prices in the table and dividing by 4.

3. Now pick four prepackaged items and list the prices and the number of ounces in the package in the table below.

Food Item	Cost of Food Item	Number of Ounces

4. Use the information from the table in Problem 3 to calculate the following values.

a. Compute the total cost of all four prepackaged food items.

b. Compute the total number of ounces of all four prepackaged food items.

c. Convert ounces to pounds by dividing the number in part **b.** by 16 (16 ounces = 1 pound). Round your decimal value to the nearest thousandth.

5. Now compute the price per pound of the prepackaged food items by dividing the total price of the four items by the total number of pounds. Round your answer to the nearest cent.

6. Compare the average price per pound of the fresh items from Problem 2 to the price per pound of the prepackaged items determined in Problem 5. Which costs more—fresh foods or prepackaged items? Why do you think this is so?

7. How likely are you to change your future food purchases based on this analysis?

Name: Date:

Project B: What Would You Weigh on the Moon?

An activity to demonstrate the use of decimal numbers in real life

The table below contains the surface gravity of each of the planets in the same solar system as the Earth, as well as Earth's moon, and the Sun. The acceleration due to gravity g at the surface of a planet is given by the formula

$$g = \frac{GM}{R^2}$$

where M is the mass of the planet, R is the planet's radius, and G is the gravitational constant. From the formula, you can see that a planet with a larger mass M will have a greater value for surface gravity. Also the larger the radius of the planet, the smaller the surface gravity.

If you look at different sources, you may find that the estimated surface gravity varies slightly from one source to another due to different values for the radius of some planets, especially the gas giants: Jupiter, Saturn, Uranus, and Neptune.

Celestial Body	Surface Gravity (m/s^2)	Relative Surface Gravity	Fractional Equivalent
Earth	9.78	1.00	
Jupiter	23.1	2.36	
Mars	3.72		
Mercury	3.78		
Moon	1.62		
Neptune	11.15		
Saturn	9.05		
Sun	274.00		
Uranus	8.69		
Venus	9.07		

1. Compare the surface gravity of each planet or celestial body to the surface gravity of the Earth by forming a fraction with each planet's surface gravity in the numerator and the Earth's gravity in the denominator. (This is referred to as relative surface gravity.) Round your answer to the nearest hundredth and place your results in the third column of the table. The values for Earth and Jupiter have been calculated for you. (**Note:** Comparing Earth to itself results in a value of 1.)

2. For Jupiter, the relative surface gravity value of 2.36 means that the gravity on Jupiter is 2.36 times that of Earth, therefore your weight on Jupiter would be approximately 2.36 times your weight on Earth. (Although mass is a constant and doesn't change regardless of what planet you are on, your weight depends on the pull of gravity.) Explain what the relative surface gravity value means for Mars.

3. Calculate your weight on the Moon by taking your present weight (in kilograms or pounds) and multiplying it by the Moon's relative surface gravity.

4. Approximately how many times larger is the surface gravity of the Sun compared to that of Mars? Round to the nearest whole number.

5. Convert each value in column three to a mixed number and place the result in column four. Be sure to reduce all fractions to lowest terms.

Name: Date:

Math@Work

Bookkeeper

As a bookkeeper, you will often receive bills and receipts for various purchases or expenses from employees of the company you work for. You will need to split the bill by expense code, assign costs according to customer, and reimburse an employee for their out-of-pocket spending. To do this you will need to know the company's reimbursement policies, the expense codes for different spending categories, and which costs fall into a particular expense category.

Suppose two employees from the sales department recently completed sales trips. Employee 1 flew out of state and visited two customers, Customer A and Customer B. This employee had a preapproved business meal with Customer B and was traveling for three days. Employee 2 drove out of state to visit Customer C. This employee stayed at a hotel for the night and then drove back the next day. The expenses for the two employees are as follows.

Employee 1	
Flight and Rental Car	$470.50
Hotel	$278.88
Meals	$110.56
Business Meal	$102.73
Presentation Materials	$54.86

Employee 2	
Miles Driven	578.5 miles
Fuel	$61.35
Hotel	$79.60
Meals	$53.23
Presentation Materials	$67.84

The expense categories used by your company to track spending are: Travel (includes hotel, flights, mileage, etc.), Meals (business), Meals (travel), and Supplies. Traveling employees are reimbursed up to $35 per day for meals while traveling and for all preapproved business meals. They also receive $0.565 per mile driven with their own car.

1. How much will you reimburse each employee for travel meals? Did either employee go over their allowed meal reimbursement amount?

2. What were the total expenses for each employee?

3. The company you work for keeps track of how much is spent on each customer. When a sales person visits multiple customers during one trip, the tracked costs are split between the customers. Fill in this table according to how much was spent on each customer for the different expense categories. (**Note:** For meals, only include the amount the employee was reimbursed.)

Expense	Customer A	Customer B	Customer C
Travel			
Meals (business)			
Meals (travel)			
Supplies			
Total			

Foundations Skill Check for Chapter 4

This page lists several skills covered previously in the book and software that are needed to learn new skills in Chapter 4. To make sure you are prepared to learn these new skills, take the self-test below and determine if any specific skills need to be reviewed.

Each skill includes an easy (**e.**), medium (**m.**), and hard (**h.**) version. You should be able to complete each exercise type at each skill level. If you are unable to complete the exercises at the easy or medium level, go back to the given lesson in the software and review until you feel confident in your ability. If you are unable to complete the hard problem for a skill, or are able to complete it but with minor errors, a review of the skill may not be necessary. You can wait until the skill is needed in the chapter to decide whether or not you should work through a quick review.

2.1 Reduce the fraction to lowest terms.

e. $\frac{5}{20}$ **m.** $\frac{32}{48}$ **h.** $\frac{130}{182}$

3.3, 3.4 Find the product or quotient.

e. $14 \div 10$ **m.** $0.0013 \cdot 1000$ **h.** $158.35 \div 100{,}000$

3.3 Find the product.

e. $0.7 \cdot 0.5$ **m.** $(1.5)(0.6)$ **h.** $(24.5)(1.3)$

3.4 Find the quotient and round your answer to the nearest hundredth.

e. $36 \div 1.2$ **m.** $8\overline{)375}$ **h.** $5 \div 2.75$

3.5 Change the fraction to a decimal number.

e. $\frac{3}{4}$ **m.** $\frac{4}{25}$ **h.** $\frac{7}{11}$

3.5 Simplify the expression using the order of operations.

e. $\frac{1}{2}+0.75-\frac{1}{5}$ **m.** $\frac{3}{4}\left(1.90-\frac{4}{5}\right)$ **h.** $(1.25+3.75+4.5)\cdot\frac{1}{3}$

Chapter 4: Ratios and Proportions, Percent, and Applications

Study Skills

4.1 Ratios and Proportions

4.2 Solving Proportions

4.3 Decimal Numbers and Percents

4.4 Fractions and Percents

4.5 Solving Percent Problems Using the Proportion $P/100 = A/B$

4.6 Solving Percent Problems Using the Equation $R \cdot B = A$

4.7 Applications: Discount, Sales Tax, Commission, and Percent Increase/Decrease

4.8 Applications: Profit, Simple Interest, and Compound Interest

Chapter 4 Projects

Math@Work

Foundations Skill Check for Chapter 5

Math@Work

Introduction

If you plan on going into nursing, there are a wide variety of areas to specialize in. Pediatric nursing, trauma nursing, medical-surgical nursing, and forensic nursing are just a few of the options you can choose from. Nurses not only need a strong ability to communicate with doctors and other coworkers, but they should also have a solid understanding of basic mathematics, which is important in performing tasks such as administering medication and determining if a patient has a healthy blood pressure. No matter which field of nursing you choose to make your career in, math will be a part of your daily job to ensure that your patients receive proper medical care.

Suppose you choose to become a pediatric nurse in a hospital. Throughout your work day, you'll be faced with many questions that require math to find the answer. How often does an IV need to be replaced according to the doctor's prescription? How much medication should be given to the patient to meet the prescribed dosage? Is the patient's blood pressure and heart rate normal? Finding the answers to these questions (and many more) require several of the skills covered in this chapter and previous chapters. At the end of the chapter, we'll come back to this topic and explore how math is used as a pediatric nurse.

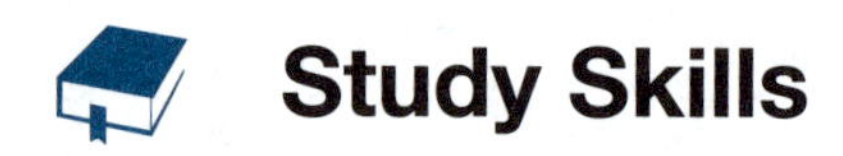

Study Skills

Tips for Success in a Math Course

1. Reading Your Textbook/Workbook One of the most important skills when taking a math class is knowing how to read a math textbook. This skill was explained in the study skills for Chapter 4. Reading a section before the instructor teaches the content and then reading it again afterwards are important strategies for success in a math course. Even if you don't have time to read the entire assigned section, you should get an overview by reading the introduction and summary, and looking at section objectives, headings, and vocabulary terms.

2. Taking Notes Take notes in class using a method that works for you. There are many different note-taking strategies, such as the Cornell Method and Concept Mapping. You can try researching these and other methods on the Internet to see if they might work better than your current note-taking system. Be sure to date your class notes and write the topic or section heading at the top of the page so that you can organize your notes later.

3. Review Always go back and read through your notes as soon as possible after class to make sure they are readable, write down any questions you have, and fill in any gaps that you missed during class. Mark any information that is incomplete so that you can get it from the textbook or your instructor later. It's important for you to review your notes as soon as possible after class so that you can make any changes while the information is fresh in your mind.

4. Organize As you review your notes each day, be sure to label them using categories such as definitions, theorems, formulas, examples, and procedures. You could also highlight each category with a different colored highlighter as long as you are consistent with the highlighting throughout your notes.

5. Study Aids Use index or note cards to help you remember important definitions, theorems, formulas, or procedures. Use the front of the card for the vocabulary term, theorem name, formula name, or procedure description. Write the definition, the theorem, the formula, or the procedure on the back of the index card, along with a description in your own words. You might also try using the Frayer Model presented in Chapter 2.

6. Practice, Practice, Practice! Math is like playing a sport. You don't get good at basketball if you don't practice—the same is true of math. Math can't be learned by just listening and watching your instructor work through problems. You have to be actively involved in doing the math yourself. Work through the examples in the book, do some practice exercises at the end of the section or chapter, and keep up with homework assignments on a daily basis.

7. Homework When doing homework, always allow plenty of time to get it done before it is due. Work some practice problems before starting the assigned problems to make sure you know what you are doing and to build up your confidence. Check your answers when possible to make sure they are correct. With word or application problems, always review your answer to see if it appears reasonable. Use the estimation techniques that you have learned to determine if your answer makes sense. Try working the problem a different way to see if you come up with the same answer.

8. Understand, Don't Memorize Don't try to memorize formulas or theorems without understanding them. Try describing or explaining them in your own words or look for patterns in formulas so that you don't have to memorize them. In this chapter, you will learn several formulas to find the perimeter of different shapes. You don't need to memorize every perimeter formula if you understand that perimeter is equal to the sum of the lengths of the sides of the figure.

9. Study Plan to study 2 to 3 hours outside of class for every hour spent in math class. If your math class meets for one hour four times a week, then you should spend 8 to 12 hours outside of class, reviewing, studying, and practicing. If math is your most difficult subject then study it while you are alert and fresh. Also, pick a study time when you will have the least interruptions or distractions so that you can concentrate.

10. Manage Your Time Don't spend more than 10 to 15 minutes working on a single problem. If you can't figure out the answer, put it aside and go on to another one. You may learn something from the next problem that will help you with the one you couldn't do. Mark the problems that you skip so that you can ask your instructor about it at the next class. It may also help to work a similar, but perhaps easier, problem that appears near that problem in the exercises. Most textbooks include the answers to the even- or odd-numbered exercises, so if you are assigned an odd-numbered problem for homework, work the even-numbered problem right before or after it for practice.

Name: Date:

4.1 Ratios and Proportions

Objectives

Use ratios to compare two quantities.

Verify proportions.

Success Strategy

You may want to review and practice finding the least common denominator of two fractions in Section 2.4 before working through this section.

Understand Concepts

Go to Software First, read through Learn in Lesson 4.1 of the software. Then, work through the problems in this section to expand your understanding of the concepts related to ratios and proportions.

Ratios, Rates, and Proportions

A **ratio** is a comparison of two quantities by division. If the units are the same in the numerator and denominator or there are no units, we can write the ratio without them.

A **rate** is a ratio with different units in the numerator and denominator. The units should always be written when working with a rate.

A **proportion** is a statement that two ratios are equal. In math symbols, if a, b, c, and d are real numbers, with $b \neq 0$, and $d \neq 0$, then $\frac{a}{b} = \frac{c}{d}$.

Quick Tip

Keeping the units in mind when working through a problem is a good idea. The units are often an important part of the final answer and can help you make sense of the solution.

1. Determine if the following are ratios, rates, or proportions.

a. 5 to 9

b. $\frac{24 \text{ miles}}{1 \text{ hour}}$

c. $\frac{4 \text{ gallons}}{\$14} = \frac{10 \text{ gallons}}{\$35}$

d. $\frac{18 \text{ feet}}{24 \text{ feet}}$

Read the following paragraph about proportions and work through the problems.

A proportion is considered to be **true** if the ratios on both sides are equal or equivalent fractions. Otherwise, the proportion is considered to be **false**. Suppose you have the proportion $\frac{10}{15} = \frac{18}{25}$ and you want to compare the two ratios to determine if the proportion is true. Two methods that can be used to do this are explained in Problems 2 and 3.

Quick Tip

Before finding the LCD, determine if the fraction can be reduced. If so, this can reduce the work needed to find the LCD. If both sides of the proportion reduce to the same fraction, then the proportion is true.

2. First method: Since ratios can be thought of as fractions, one way to compare the two ratios is to find equivalent fractions for each by using the least common denominator (LCD) and then compare the numerators.

a. What is the LCD for $\frac{10}{15}$ and $\frac{18}{25}$?

b. Find equivalent fractions in the proportion by using the LCD from part **a.**

Lesson Link

Methods for comparing fractions were introduced in Section 2.6.

c. Is the proportion $\frac{10}{15} = \frac{18}{25}$ true or false?

3. **Second method:** Another way to determine if a proportion is true or false is to "clear the denominators." To do this we need to multiply both ratios by a whole number that reduces both denominators to 1. While this number does not need to be the LCD, using the LCD will keep the numbers smaller and easier to work with.

a. Find the LCD of $\frac{10}{15}$ and $\frac{18}{25}$.

Quick Tip

An important rule when working with any equation is that any operation that is done to one side of the equation must also be done to the other side of the equation.

b. If we multiply the ratio $\frac{10}{15}$ by the answer to part **a.**, we must multiply $\frac{18}{25}$ by that same number or the resulting equation will not be equivalent. Rewrite the proportion $\frac{10}{15} = \frac{18}{25}$ by multiplying each ratio by the LCD found in part **a.** Simplify your answer.

c. Is the proportion true or false? How does this answer compare to part **b.** of Problem 2?

4. The comparison values you get from both methods may be different (compare the values from Problem 2 part **b.** with Problem 3 part **b.**), but both approaches are valid to use.

a. What do the two methods have in common?

b. Which method do you prefer? Explain why you prefer this method.

Quick Tip

Finding the cross products is also known as **cross multiplying**.

Cross Products

The **cross products** of the two ratios in the proportion $\frac{a}{b} = \frac{c}{d}$ is found by multiplying a by d and multiplying b by c.

$$\frac{a}{b} = \frac{c}{d} \longrightarrow a \cdot d = b \cdot c$$

If the cross products are equal, then the proportion is true.

Name: Date:

Quick Tip

Remember that any whole number can be written as a fraction by placing it over a denominator of 1.

5. The cross products of a proportion are used to compare the two ratios in the proportion to see if they are equivalent. Perform cross multiplication to verify the proportion $\frac{10}{15} = \frac{18}{25}$ is true. Do you get the same results as you did for part **b.** of Problem 3?

Skill Check

Go to Software Work through Practice in Lesson 4.1 of the software before attempting the following exercises.

Write each comparison as a ratio or rate reduced to lowest terms.

6. 75 miles to 3 gallons of gas

7. \$1.25 per 25 ounces

Determine if each proportion is true or false.

8. $\frac{25}{6} = \frac{50}{8}$

9. $\frac{3}{4} = \frac{15}{20}$

10. $\frac{4}{5} = \frac{2.4}{3}$

11. $\frac{1\frac{1}{2}}{1\frac{1}{3}} = \frac{\frac{1}{2}}{\frac{1}{3}}$

Apply Skills

Work through the problems in this section to apply the skills you have learned related to ratios and proportions.

12. Lachlan is comparing nutrition labels to see if two products have equivalent ratios of calories per serving size. The first product has 110 calories per 80-gram serving. The second product has 82.5 calories per 60-gram serving.

a. Set up a proportion using the given information.

b. Do the two products have equivalent calories per serving ratios?

Quick Tip

Usually the phrase "per serving" means "per 1 serving."

Quick Tip

A **unit rate** is a rate that has a denominator of 1, such as 60 miles in 1 hour, which is often abbreviated 60 miles per hour.

c. Another way to compare two ratios is to convert them to unit rates, which is an equivalent ratio with a denominator of 1. You can convert a ratio to a unit rate by dividing the denominator into the numerator. Convert the ratios on both sides of the proportion in part **a.** to a unit rate.

d. Do the two products have the same number of calories per 1-gram serving based on your results from part **c.**?

13. When making concrete, the ratio of sand to cement should be 5 parts sand to 2 parts cement. A batch of concrete is made with the ratio of 30 parts sand per 14 parts cement.

a. Set up a proportion using the given information.

b. Was the batch of concrete made with the correct ratio of sand to cement? Explain how you know.

Quick Tip

Intravenous therapy is when medication in liquid form is injected directly into the bloodstream through a vein. This is commonly abbreviated as IV and often referred to as a "drip."

14. A doctor orders an IV for a patient to run at 125 mL per hour. A nurse sets the IV to run at a rate of 1000 mL per 8 hours.

a. Set up a proportion using the given information.

b. Did the nurse set the IV to the correct rate? Explain how you know.

15. A student teacher is given a guideline that a test should have an average of 6 points per problem. The student teacher makes a test that is worth 78 points and has 13 problems.

a. Set up a proportion using the given information.

b. Did the student teacher make the test according to the guideline? Explain how you know.

Name: Date:

4.2 Solving Proportions

Objectives

Solve proportions.

Use proportions to solve problems

Success Strategy

Be sure to pay attention to the units when setting up proportions. The units of the numerator should match and the units of the denominator should match.

Understand Concepts

Go to Software First, read through Learn in Lesson 4.2 of the software. Then, work through the problems in this section to expand your understanding of the concepts related to solving proportions.

Solving a Proportion

1. Find the cross products.
2. Divide both sides of the equation by the coefficient of the variable.
3. Simplify.

Quick Tip

The word *scale* is often used when working with proportions. A *scaled value* has been adjusted to fit a certain need. Things can be scaled in relation to size or amount. For instance, residential plans are commonly drawn on a quarter-scale, which means that one inch on the scale drawing is equivalent to four feet of the actual version.

The first step in solving proportions in word problems is to set up the proportion. When setting up proportions to solve for an unknown variable, there are multiple setups to choose from. Consider the situation where a baker is scaling a cookie recipe. The recipe uses 4 cups of flour and makes 36 cookies. The baker wants to determine how many cups of flour are needed to make 60 cookies.

1. We first need to analyze the problem to determine how to set up the proportion.

a. What is the unknown value in this problem?

b. What are the units used in this problem?

2. One way to set up a proportion is to have matching units in the numerators, matching units in the denominators, and the variable in a numerator. In this form, it is important to have one ratio be the original quantities and the other ratio to be the scaled quantities. For our baking problem, the proportion in this form would be

$$\frac{4 \text{ cups}}{36 \text{ cookies}} = \frac{x \text{ cups}}{60 \text{ cookies}}$$

a. Are the original values on the right or left side of the equation?

Quick Tip

The answer remains the same regardless of whether the scaled values are placed on the right or left side of the equation.

b. Are the scaled values on the right or left side of the equation?

c. How many cups of flour are needed to make 60 cookies?

Lesson Link

Reciprocal was covered in Section 2.2. Recall that to find the reciprocal of a number, the numerator and the denominator are switched.

3. Another way to set up this problem is to take the **reciprocal** of both sides of the proportion from Problem 2. This will result in the variable being in the denominator.

a. Write the reciprocal of each side of the proportion from Problem 2.

b. The units have switched from the numerator to the denominator and vice versa. Does this change the solution to the proportion?

4. Yet another way to set up a proportion is to have the matching units in the same ratio with the variable in the numerator. In this form, it is important to have the original quantities together in either the numerator or the denominator. For our baking problem, the proportion in this form would be

$$\frac{x \text{ cups}}{4 \text{ cups}} = \frac{60 \text{ cookies}}{36 \text{ cookies}}$$

a. Are the original values in the numerators or the denominators?

b. Are the scaled values in the numerators or the denominators?

c. Do you still get the same solution for x?

d. Can you think of another way to set up this proportion that gives you the same answer? Describe your method and solve the proportion to make sure you get the same answer. (**Hint:** See Problem 3.)

Name: Date:

Nursing Notation

In nursing, the following two notations are commonly used to set up and solve proportions when preparing medications for patients.

$$a : b = c : d \quad \text{and} \quad a : b :: c : d$$

The values b and c are known as the **means** and a and d are known as **extremes**.

In both notations, the $a : b$ represents a ratio where a is the numerator and b is the denominator. Similarly, $c : d$ is a ratio where c is the numerator and d is the denominator.

Quick Tip

To help you remember which values are the means and which are the extremes, remember that **mean** refers to "the middle value," so the numbers in the middle are the means.

5. The two notations represent the same proportion. What does the symbol "::" stand for?

6. One way to read "$a : b :: c : d$" is "a is to b as c is to d." Write in words how would you read "10 mg : 1 mL :: 12.5 mg : 1.25 mL."

Quick Tip

If you are a nursing student, you may want to practice solving proportions using both notations (the standard notation and the medical notation).

Solving Proportions Using Nursing Notation

Method 1: Change the notation into standard proportion notation and then solve.

For example, $1 : 2 :: 3 : x$ would be $\frac{1}{2} = \frac{3}{x}$. Solve for x.

Method 2: Keep the notation the same, and multiply the means together and the extremes together to get the result of $a \cdot d = b \cdot c$.

For example, $1 : 2 :: 3 : x$ would cross multiply to become $1 \cdot x = 2 \cdot 3$. Solve for x.

Quick Tip

Multiplying together the means and the extremes is the same process as finding the cross products of a proportion.

7. Verify that both methods in the instruction box give the same results for $8 : 15 :: 12 : x$.

Skill Check

Go to Software Work through Practice in Lesson 4.2 of the software before attempting the following exercises.

Solve each proportion.

8. $\frac{3}{5}=\frac{x}{25}$

9. $\frac{y}{7}=\frac{21}{49}$

10. $\frac{10}{x}=\frac{25}{40}$

11. $\frac{15}{25}=\frac{72}{y}$

Apply Skills

Work through the problems in this section to apply the skills you have learned related to solving proportions. Try using different formats for setting up the proportion from Problems 2, 3, and 4 to determine which setup you like best.

12. A nurse is told that a patient's IV is to run at a rate of 75 mL per hour. If the nurse uses a 250 mL bag of IV solution, how long will it be until the nurse needs to replace it?

a. Set up a proportion with the given information.

b. Solve the proportion.

c. Interpret your answer from part **b.** Write a complete sentence.

13. A civil engineer is making a scale model of a water pump to test before building the full-sized version. One inch on the model represents ten inches on the actual water pump. If the full-sized water pump is designed to be 82 inches long, what is the length of the scale model?

a. Set up a proportion with the given information.

b. Solve the proportion.

c. Interpret your answer from part **b.** Write a complete sentence.

14. In the United States, one new person is infected with HIV every ten and a half minutes. How long will it take for fifty new people to become infected with HIV? Source: http://www.cdc.gov/hiv/statistics/basics/

a. Set up a proportion with the given information.

b. Solve the proportion.

c. Interpret your answer from part **b.** Write a complete sentence.

Quick Tip

Many 2-cycle engines run on a mixture of gasoline and oil. This ratio varies depending on the engine.

15. A certain type of chain saw runs on a mixture of 128 ounces of gasoline for every 4 ounces of oil. If you want to use 5 ounces of oil to make the fuel mixture, how many ounces of gasoline are required?

a. Set up a proportion with the given information.

b. Solve the proportion.

c. Interpret your answer from part **b.** Write a complete sentence.

16. A machinist runs a machine that can make 6 parts every 5 minutes. How long will it take to make 100 parts for a customer? Round your answer to the nearest tenth.

a. Set up a proportion with the given information.

b. Solve the proportion.

c. Interpret your answer from part **b.** Write a complete sentence.

Name: Date:

4.3 Decimal Numbers and Percents

Objectives

Understand percents.

Change decimal numbers to percents.

Change percents to decimal numbers.

Success Strategy

Being able to convert between decimals and percents is an important skill in real life situations. Extra practice in the software would be beneficial in improving this skill.

Understand Concepts

Go to Software First, read through Learn in Lesson 4.3 of the software. Then, work through the problems in this section to expand your understanding of the concepts related to decimal numbers and percents.

1. The % symbol is a recent form to represent $\frac{1}{100}$ when talking about percents in writing. The % symbol slowly developed over several hundred years from its original form. Use the key words "percent symbol history" to answer the following questions.

a. When was a symbol for percent first used?

b. Before a symbol was created, what words were used when writing about percents? What do these words mean?

Writing a Decimal Number as a Percent

1. Move the decimal two places to the right.
2. Attach the percent sign. Add zeros when necessary.

0.25 ⟶ 0.25. ⟶ 25%

Writing a Percent as a Decimal Number

1. Remove the percent sign.
2. Move the decimal two places to the left. Add zeros when necessary.

56% ⟶ .56.0 ⟶ 0.56

2. Multiplying and dividing by powers of 10 and how those actions relate to the placement of the decimal point in a number was discussed in Sections 3.3 and 3.4. Use this information for the next problem.

a. Instead of moving the decimal point two places to the right, what is another way to write the rule for changing a decimal number into a percent?

b. Instead of moving the decimal point two places to the left, what is another way to write the rule for changing a percent into a decimal number?

Lesson Link

In Section 3.5, we discussed that coins represent a fraction of a dollar where 100 cents is equal to one dollar.

3. Money is useful for connecting the fraction form, decimal form, and percent form of some common values since it is something we are familiar with and use often. Fill in this table with each form of the coin value.

Coin Name	Unreduced Fraction	Decimal Number Equivalent	Percent Form
Penny	$\frac{1}{100}$	0.01	1%
Nickel			
Dime			
Quarter			
Half-dollar			

True or False: Determine whether each statement is true or false. Rewrite any false statement so that it is true. (There may be more than one correct new statement.)

4. The coins listed in the table are worth less than 100% of a dollar.

5. The coins listed in the table have a decimal form less than 0.01.

6. A dollar coin is worth 100% of a dollar.

Name: ____________________ Date: ____________

Skill Check

Go to Software Work through Practice in Lesson 4.3 of the software before attempting the following exercises.

Change each fraction to a percent.

7. $\frac{26}{100}$

8. $\frac{13.9}{100}$

Change each decimal number to a percent.

9. 0.06

10. 0.153

Change each percent to a decimal number.

11. 85%

12. 1.4%

Quick Tip

This problem involves some common fractions that are used in both math and real life.

13. Fill in the missing values in the table. Reduce all fractions to lowest terms.

Fraction	Decimal	Percent
	0.12	
		$\frac{3}{8}\%$
$1\frac{9}{16}$		
	2.34	
		30%

Apply Skills

Work through the problems in this section to apply the skills you have learned related to decimal numbers and percents.

14. The following chart shows the distribution of employed citizens in the United States in 2007 based on the age ranges given along the horizontal axis. Source: http://www.cdc.gov/

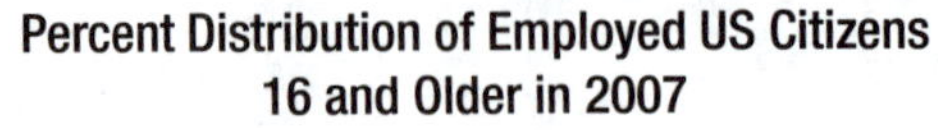

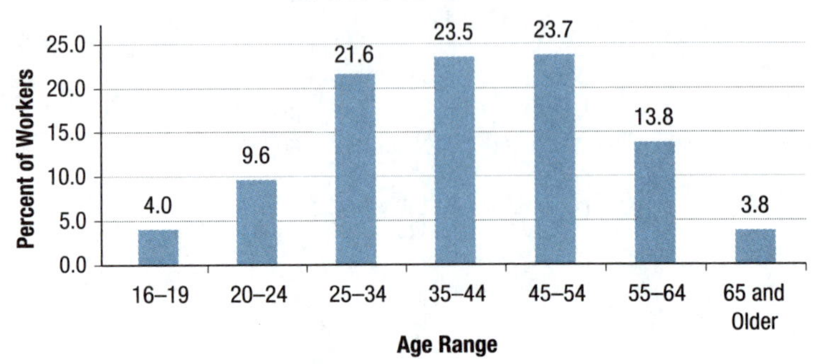

a. What is the sum (as a percent) of all the age groups?

b. What percent of workers are below the age of 45?

c. What percent of workers are age 25 or older?

d. What trend do you notice in the data with relation to age?

e. What is the difference between the largest percent and the smallest?

f. How does the lowest age category compare to the highest?

Name: Date:

4.4 Fractions and Percents

Objectives

Change fractions and mixed numbers to percents.

Change percents to fractions and mixed numbers.

Success Strategy

Changing fractions to percents requires the use of long division. You may find it helpful to review long division in Section 1.4 of the software before working through this section.

Understand Concepts

Go to Software First, read through Learn in Lesson 4.4 of the software. Then, work through the problems in this section to expand your understanding of the concepts related to fractions and percents.

Writing a Fraction as a Percent

1. Find the decimal number equivalent of the fraction.
2. Move the decimal two places to the right.
3. Attach the percent sign.

Writing a Percent as a Fraction

1. Write the percent as a fraction over 100 and remove the percent symbol.
2. Reduce if possible.

Lesson Link

Reducing a fraction to lowest terms was introduced in Section 2.1.

Quick Tip

You may not see percents written as fractions in real world situations very often, but there are times when this format is useful. In Section 4.5 you will learn how to use the fractional form of a percent to solve percent problems by using a proportion.

Read the following paragraph about solving word problems, then work through the problems.

When solving word problems where the solution is a percent, finding the answers will typically involve writing the percent as a fraction and using quantitative reasoning skills. You need to determine which value in the question represents the whole, or total, amount (denominator) and which value represents a part of the whole (numerator). The following problems will guide you through the process.

1. Tim needs to earn 120 credit hours to receive a Bachelor's degree, and he currently has 96 credit hours. What percent of the credit hour requirement has Tim completed?

a. What is the total amount of credit hours needed?

b. How many credit hours has Tim earned?

c. Write the portion of total credit hours that Tim has earned as a fraction.

d. Answer the question asked by changing the fraction from part **c.** into a percent.

2. Christine's doctor recommends that she takes 1000 milligrams of calcium per day. She takes a vitamin which contains 400 milligrams of calcium and she knows that today she has taken 350 milligrams from her food. What percent of the recommended value of calcium has Christine taken?

a. What is the total amount of calcium Christine needs to take per day?

b. How many milligrams of calcium has Christine taken so far today?

c. Write the portion of the recommended amount of calcium that Christine has taken today as a fraction.

d. Answer the question asked by changing the fraction from part **c.** into a percent.

Read through the following paragraph about writing percents as fractions then solve the problems.

When writing a percent as a fraction, you will often need to reduce the fraction. When reducing fractions where both the numerator and denominator end in a zero, the zero can be removed. This is the result of dividing a 10 out of the numerator and denominator and the fact that $\frac{10}{10} = 1$.

For example, $\frac{50}{70} = \frac{5 \cdot 10}{7 \cdot 10} = \frac{5}{7}$.

3. State how many zeros can be removed when reducing each fraction. Then, reduce the fraction.

a. $\frac{40}{70}$

b. $\frac{40}{700}$

c. $\frac{100}{70}$

d. $\frac{25,000}{50,000}$

True or False: Determine whether each statement is true or false. Rewrite any false statement so that it is true. (There may be more than one correct new statement.)

4. $\frac{1}{4}$ is equivalent to $\frac{1}{4}\%$.

Name: Date:

5. 0.005 is equivalent to $\frac{1}{2}$%.

6. 25% and $\frac{1}{4}$ have the same decimal number equivalent.

Lesson Link 

You will learn more about the likelihood of events in relation to probability in Section 6.4.

Read the following paragraph about percents and probability and then work through the following problems.

In media advertising, the **likelihood**, or **chance**, that something happens is commonly given as a ratio. This ratio can be written as a fraction or a percent. When commercials report that "4 out of 5 dentists" recommend a certain type of toothpaste, what they mean is $\frac{4}{5}$ of all dentists (or 80% of all dentists) recommend this toothpaste.

7. Interpret the following statements using complete sentences by converting the given ratio into a fraction and a percent. (You will have two sentences for each statement.)

a. 3 out of 4 choosy moms prefer Jif brand peanut butter.

b. 81 out of 1000 babies are born with a low birth weight. Source: http://www.cdc.gov

c. 5 out of 100 adults with a post graduate degree are smokers. Source: http://www.cdc.gov/

Skill Check

Go to Software Work through Practice in Lesson 4.4 of the software before attempting the following exercises.

Change each fraction or mixed number to a percent. Round to the nearest hundredth of a percent if necessary.

8. $\frac{1}{25}$

9. $\frac{5}{6}$

10. $2\frac{3}{8}$

Change each percent to a fraction or mixed number. Reduce if possible.

11. 72%

12. 125%

13. 62.5%

Apply Skills

Work through the problems in this section to apply the skills you have learned related to fractions and percents.

14. Write the percent in each statement as a fraction and reduce when possible.

a. A Nutri-Grain cereal bar gives you 20% of your daily value of calcium.

b. Thirty-two percent of the student population at a college are first-year students.

c. In June of 2013, 7.6% of US citizens that were able to work were unemployed. Source: http://data.bls.gov/timeseries/LNS14000000

15. Margot made a purchase that costs \$50 and was charged \$3.50 in sales tax. What is the sales tax rate as a percent of the cost?

Quick Tip

To learn more about the daily recommended intake of sodium, visit http://www.cdc.gov/features/dssodium/

16. A Big Mac from McDonald's is estimated to have 970 mg of sodium. The maximum recommended daily value of sodium for adults in the United States is 2300 mg. What percent of the maximum daily value of sodium does the Big Mac provide? Round to the nearest hundredth of a percent.

17. Americans consume an average of 3436 mg of sodium per day. The recommended adequate intake level of sodium is 1500 mg. What percent of the adequate sodium intake level does the average American consume per day? Round to the nearest hundredth of a percent.

18. In 2011, 15 out of every 100 people in the United States were living in poverty. In the same year, 48 out of every 100 people in the US, above the age of 15, owned a smartphone. Source: US Census Bureau

a. What percent of the US population was living in poverty in 2011?

b. What percent of the US population over the age of 15 owned a smartphone in 2011?

Name: Date:

4.5 Solving Percent Problems Using the Proportion *P*/100 = *A*/*B*

Objectives

Understand the proportion $\frac{P}{100} = \frac{A}{B}$.

Use the proportion $\frac{P}{100} = \frac{A}{B}$ to solve percent problems.

Success Strategy

There are two different methods for solving percent problems. One way is to use a proportion, which is taught in this section. The next section will show a different method. Be sure to practice both methods to determine which one you prefer.

Understand Concepts

Go to Software First, read through Learn in Lesson 4.5 of the software. Then, work through the problems in this section to expand your understanding of the concepts related to solving percent problems by using the proportion $P/100 = A/B$.

The Percent Proportion

The proportion $\frac{\mathbf{P}}{\mathbf{100}} = \frac{\mathbf{A}}{\mathbf{B}}$ can be used when solving percent problems.

$P =$ Percent

$A =$ Amount

$B =$ Base

Quick Tip

The amount may sometimes be larger than the base. This happens when the percent is greater than 100.

1. Write a short paragraph to compare the parts of the ratio $\frac{A}{B}$ in the percent proportion to the parts of a fraction. Be sure to include how they are similar and how they are different.

When setting up a proportion to solve a percent problem, it is important to read the problem statement carefully. Before setting up the proportion, you should determine which values in the proportion are known and which value is unknown.

Quick Tip

Remember that the unknown value in the proportion will be represented by a variable.

2. Review Pólya's problem-solving steps, which are located in the *First Day of Class Resources* at the beginning of the workbook. Write your own version of the problem-solving steps to solve a percent problem using the percent proportion.

Step 1 is:

Step 2 is:

Step 3 is:

Step 4 is:

Being able to identify the important information in a word problem is a valuable skill that takes time to develop. There are three pieces to a percent proportion: the amount A, the base B, and the percent P. Two of the pieces are typically given to you in the problem statement. The other piece will be the unknown, which is the value that needs to be solved for.

For the following problems, determine the pieces of the percent proportion. If a value is unknown based on the problem statement, write "unknown" as your answer. You do not need to solve the problem.

Quick Tip

The percent in a word problem will typically be given with the % symbol or the word *percent*.

3. 45% of what number is 36?

a. What is the percent?

b. What is the amount?

c. What is the base?

d. Write the percent proportion.

4. During the past year, Franklin spent 32% of his income on his mortgage and home repairs. If he earned $38,750 during the year, how much money did he spend on his mortgage and home repairs?

a. What is the percent?

b. What is the amount?

c. What is the base?

d. Write the percent proportion.

5. Tilda purchased a pair of jeans that cost $48. The cost of tax on her purchase was $4.08. What percent of the purchase price was the tax rate?

a. What is the percent?

b. What is the amount?

c. What is the base?

d. Write the percent proportion.

Skill Check

Go to Software Work through Practice in Lesson 4.5 of the software before attempting the following exercises.

Use the proportion $P/100 = A/B$ to solve for the unknown quantity.

6. Find 60% of 25.

7. What is 12% of 20?

Name: Date:

8. What percent of 75 is 15?

9. 27 is 30% of what number?

Apply Skills

The following problems focus on two important parts of solving percent problems using the proportion $P/100 = A/B$. The first part is setting up the proportion based on the information provided. The second part is understanding what the answer means and putting it into words. Outside of a math course, problems which need math to solve them will often involve this type of translation into mathematical notation and back into words.

10. In 2011, 70.7% of babies in the state of Washington were vaccinated against Hepatitis B within 3 days of birth. If 86,929 babies were born in Washington in 2011, how many babies were vaccinated against Hepatitis B within 3 days of their birth? Sources: http://www.doh.wa.gov/Portals/1/Documents/Pubs/422-099-VitalStatistics2011Highlights.pdf and http://www.cdc.gov/vaccines/stats-surv/nis/data/tables_2011.htm

a. Set up the proportion.

b. Solve for the unknown variable. Round your answer to the nearest whole number.

c. What does this answer mean? Write a complete sentence.

11. A family has a monthly income of $3700 and they spend $925 per month on housing costs. What percent of their monthly income goes towards housing costs?

a. Set up the proportion.

b. Solve for the unknown variable.

c. What does this answer mean? Write a complete sentence.

12. The average semester tuition at a state college is \$5650. The college board decides to raise tuition by 7%. How much will tuition increase?

a. Set up the proportion.

b. Solve for the unknown variable.

c. What does this answer mean? Write a complete sentence.

d. What will the tuition per semester be after the tuition increase?

e. Suppose a student starts college the semester after the tuition increase. How much will a 4-year degree cost if there are no more tuition increases during those 4 years? (**Hint:** There are two semesters per school year.)

13. In the United States, approximately 7% of the population are military veterans. If there are approximately 22,000,000 veterans in the United States, what is the approximate population of the United States (rounded to the nearest whole number)? Source: US Census Bureau

a. Set up the proportion.

b. Solve for the unknown variable. Round your answer to the nearest whole number.

c. What does this answer mean? Write a complete sentence.

Name: Date:

4.6 Solving Percent Problems Using the Equation $R \cdot B = A$

Objectives

Understand the equation $R \cdot B = A$.

Use the equation $R \cdot B = A$ to solve percent problems.

Success Strategy

The previous section introduced a method for solving percent problems using a proportion. In this section, you will learn another method that uses an equation. Be sure to practice both methods to determine which one you prefer.

Understand Concepts

Go to Software First, read through Learn in Lesson 4.6 of the software. Then, work through the problems in this section to expand your understanding of the concepts related to solving percent problems by using the equation $R \cdot B = A$.

The Percent Equation

The equation $R \cdot B = A$ can be used when solving percent problems.

$R =$ Rate, or percent, written as a decimal number

$A =$ Amount

$B =$ Base

Quick Tip

Remember, a **proportion** is an equation involving two ratios.

1. The previous section introduced the proportion $\frac{P}{100} = \frac{A}{B}$ for solving percent problems. This problem will investigate how this proportion relates to the equation $R \cdot B = A$.

   a. What are the similarities and differences between the variables used in the proportion and those used in the equation?

   b. The fraction form of the percent is $\frac{P}{100}$ and the decimal form of the percent is R. This means that $\frac{P}{100} = R$. Substitute (or replace) $\frac{P}{100}$ with R in the proportion $\frac{P}{100} = \frac{A}{B}$.

   c. Rewrite the equation from part **b.** for A by writing R as $\frac{R}{1}$ and cross multiplying.

   d. Which property of multiplication says that $R \cdot B = A$ and $B \cdot R = A$ are equivalent equations?

   e. Since we can rearrange the proportion $\frac{P}{100} = \frac{A}{B}$ to look like the equation $R \cdot B = A$, what does this tell you about the two equations?

Lesson Link

Solving formulas for different variables will be covered in Section 8.5.

Lesson Link

The properties of multiplication for whole numbers were covered in Section 1.3.

2. Use the information from Problem 1 to answer the following exercises.

a. Rewrite the proportion $\frac{65}{100} = \frac{A}{40}$ as an equation.

b. Use both the proportion and the equation to solve for A. Do you get the same value after using both methods?

c. Which method do you prefer for solving percent problems, the proportion or the equation? Explain your answer.

For the following problems, determine the pieces of the percent equation. If a value is unknown based on the problem statement, write "unknown" as your answer. You do not need to solve the problem.

Quick Tip

The percent will typically be given with the % symbol or the word *percent*.

3. What percent of 42 is 8?

a. What is the rate?

b. What is the amount?

c. What is the base?

d. Write the percent equation.

4. At a hospital, 8% of emergency room visits are for non-urgent care needs. One day, there were 24 non-urgent care patients. How many patients did the emergency room have that day?

a. What is the rate?

b. What is the amount?

c. What is the base?

d. Write the percent equation.

5. Watermelons are approximately 91% water by weight. A medium-size seedless watermelon weighs 14 pounds. What is the weight of the water content in the watermelon?

a. What is the rate?

b. What is the amount?

c. What is the base?

d. Write the percent equation.

Name: Date:

Skill Check

Go to Software Work through Practice in Lesson 4.6 of the software before attempting the following exercises.

Use the equation $R \cdot B = A$ to solve for the unknown quantity.

6. 20% of 80 is what number?

Quick Tip

The word *of* is often used to indicate multiplication.

7. What percent of 24 is 6?

8. 46 is what percent of 115?

9. Find 175% of 48.

Apply Skills

The following problems focus on two important parts of solving percent problems by using the equation $R \cdot B = A$. The first part is setting up the equation based on the information provided. The second part is understanding what the answer means and putting it into words. Outside of a math course, problems which need math to solve them will often involve this type of translation into mathematical notation and back into words.

10. At the beginning of the year, Nevaeh invests $1400 in a fund that is expected to grow at a rate of 7.3% per year. How much interest should she expect to earn by the end of the year?

a. Set up the equation.

b. Solve for the unknown variable.

c. What does the answer to part **b.** mean? Write a complete sentence.

11. In a survey of 150,000 adults, 6300 people admitted to falling asleep while driving during the previous 30 days. What percent of the people surveyed fell asleep while driving during the previous 30 days? Source: http://www.cdc.gov/mmwr/pdf/wk/mm6151.pdf

a. Set up the equation.

b. Solve for the unknown variable.

c. What does the answer to part **b.** mean? Write a complete sentence.

12. A salaried worker makes $35,000 per year. During the yearly review, he is given a 4.5% increase in pay. How much of a pay increase did he receive?

a. Set up the equation.

b. Solve for the unknown variable.

c. What does the answer to part **b.** mean? Write a complete sentence.

d. If the worker is paid two times each month, how much will each paycheck increase after the raise?

Name: Date:

4.7 Applications: Discount, Sales Tax, Commission, and Percent Increase/Decrease

Objectives

Calculate a discount.

Calculate sales tax.

Calculate percent increase and percent decrease.

Calculate commission.

Use and understand reference values.

Success Strategy

This section introduces an expanded version of Pólya's problem-solving steps. This approach works for all types of problems, not only math problems. It would benefit you to make a copy of these steps and keep it handy when you need to solve a problem.

Understand Concepts

Go to Software First, read through Learn in Lesson 4.7 of the software. Then, work through the problems in this section to expand your understanding of the concepts related to applications of percent problems.

The application problems presented in this section can be solved using either the proportion or the equation covered in the previous two sections. Whichever method you use, be sure to identify the pieces of the equation that are given to you, determine the unknown value, write the equation, and then solve for the unknown.

Quick Tip

The **discount** and the **discounted price** are not the same thing. The discount is the amount taken off of the original price. The discounted price is the price paid after the discount is taken off.

Finding the Discounted Price

There are two methods for finding the **discounted price** of an item.

1. Find the amount of the discount and then subtract that amount from the original price.
2. Subtract the discount percent from 100% to find the percent of the sale price you will pay. Then, multiply that percent by the original price to find the discounted price.

For the next two problems, use both methods of finding the discounted price of an item to find the answer.

1. You receive a coupon for 35% off of any item at a local store. You decide to buy something that originally costs \$26. What is the discounted price of the item after you use the coupon?

2. You have a coupon code for an online store which gives you 15% off of any order over \$50. You spend \$65 and use the coupon. How much will your order cost after applying the coupon?

3. Which method do you prefer to find the discounted price? Explain your reasoning.

Quick Tip

To find the amount of sales tax based on purchase price, multiply the purchase price by the sales tax rate.

4. The sales tax percent is a combination of both state tax and county tax in most states. Use the key words "sales tax" along with your county's name or state's name to answer the following questions.

   a. What is the combined county and state sales tax rate where you live?

   b. Which county in your state has the highest sales tax rate?

   c. If you make a purchase for $250 at a local store, how much will you pay in sales tax?

Quick Tip

Calculating the amount of income tax paid is similar to calculating sales tax. Multiply the amount of pay by the income tax rate.

5. Income tax is an amount of money deducted from your paycheck and given to the federal and state governments. It is calculated in a similar way as sales tax. The main difference is that the tax amount is deducted from your paycheck and is based on a percentage of your earnings.

   Go to www.forbes.com and enter "federal tax bracket" into the search bar. Find the current year's federal tax bracket information and use that to answer the following questions. (For the following questions, "single" means "unmarried.")

   a. A single person makes $45,000 per year. What percent of their income goes towards federal income tax? How much will they pay?

   b. A married couple earns $85,000 per year. What percent of their income goes towards the federal income tax? How much will they pay?

   c. A single person earns $85,000 per year. What percentage of their income goes towards federal income tax? How does this compare to the taxes paid by the married couple earning the same amount per year from part **b.**?

   d. What percent of your income goes toward federal income tax?

Name: ______________________ Date: ______________________

Quick Tip

Percent increase is sometimes called **appreciation**. Percent decrease is sometimes called **depreciation**.

Percent Increase and Percent Decrease

Percent increase or **percent decrease** is the percent of the original value that the value of an item changes over time. To find the percent increase or percent decrease:

1. Find the difference in values. For percent increase, subtract the original value from the new value. For percent decrease, subtract the new value from the original value.
2. Determine what percent the difference is of the original value.

Use the steps given in the blue box to solve the following problems about percent increase and percent decrease.

6. After Maria's annual review, her pay was increased to \$9.25 per hour. Before her raise she made \$8.75 per hour. What was the percent increase in her pay? Round your answer to the nearest tenth, if necessary.

7. Kobe got a new job and it now takes him 15 minutes to drive to work. The commute time to his previous job was 25 minutes. What was the percent decrease in Kobe's commute time? Round your answer to the nearest tenth, if necessary.

Skill Check

Work through Practice in Lesson 4.7 of the software before attempting the following exercises.

Solve each problem using either the percent proportion or equation.

8. 50% of what number is \$12.35?

9. What percent of 270 is 94.5?

10. 25% of 338 is what number?

11. $\frac{8}{100} = \frac{x}{12{,}000}$

Apply Skills

Work through the problems in this section to apply the skills you have learned related to applications of percent problems.

Quick Tip

A **break** can be inserted into an axis of the graph if the data set doesn't begin at zero. The values listed after the break will be related to the data set that is graphed.

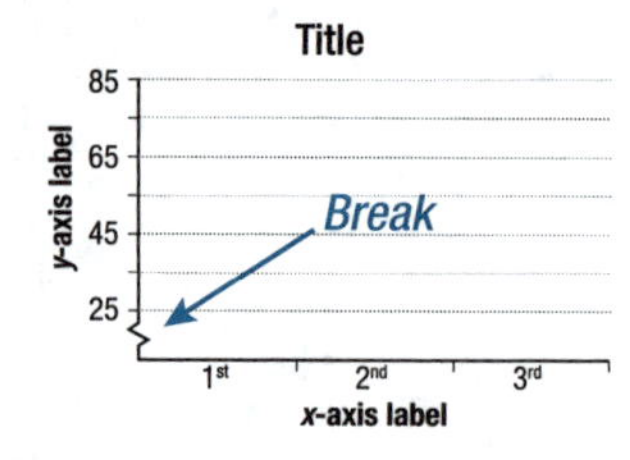

12. People who run a business often have to review quarterly profits for their business to make adjustments to sales plans or item production. The line graph shows the profit per quarter at a small business. Round your answers to the nearest tenth when necessary.

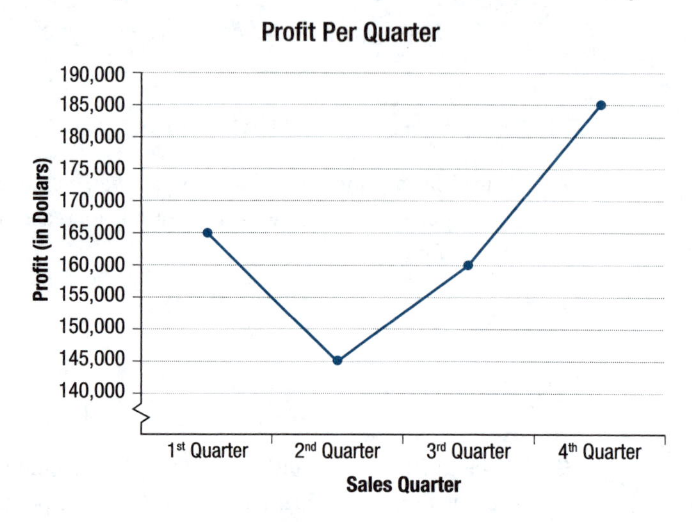

a. Which quarter had the most profit?

b. Between which two quarters did sales decrease the most?

c. What was the percent decrease from part **b.**?

d. Between which two quarters did sales increase the most?

e. What was the percent increase from part **d.**?

Name: Date:

Quick Tip

Commission is a fee paid to an agent or sales person for a service. It is commonly a percent of the sales price.

13. Levi is a full-time sales associate at a computer store. He earns a weekly salary of \$220 and earns 15% commission on all of his sales.

a. During one week, Levi sold \$8500 in merchandise. Sales tax in the county where Levi works is 8.5%. What was the total sales tax paid on all of the sales Levi made?

b. What would Levi's paycheck for the week be before taxes? (**Hint:** Levi's total earnings is equal to salary plus commission.)

c. Levi's combined federal and state income tax is 23%. How much will his paycheck be after taxes?

d. During the next week, one of Levi's customers returns \$1250 in merchandise. When commission-based purchases are returned, the amount of commission based on the value of the returned merchandise is deducted from the salesperson's next paycheck. How much will be deducted from Levi's next paycheck (before taxes)?

14. At an electronics superstore, a 46-inch flat screen TV has a retail price of \$575. Round your answers to the nearest cent when necessary.

a. During a holiday sale, management decides to offer a 25% discount on the TV. What is the discounted price of the TV?

b. The superstore also offers a credit card where the customer can save an additional 5% on their entire purchase (after all other discounts and before sales tax) by charging the sale to the store card. If a customer uses the store credit card to buy the discounted TV, what would the final discounted price of the TV be?

c. The superstore is located in a county with a 6.5% sales tax. What would be the final sales price of the TV for a customer who uses the store credit card?

Reference Values and Percent Change

In determining the percent change of a quantity, it is important to identify the **reference value**. The reference value is usually the starting value or original amount of a quantity before the change occurred. Many times the data is in chronological or time order, so the reference value is easy to determine. Look at the following example.

> The batting average of Freddie Freeman, the Atlanta Braves first baseman, was .259 for the 2012 season. In 2013 his batting average was .319. Calculate the percent increase in his batting average from 2012 to 2013.
> Source: http://espn.go.com/mlb/player/stats/_/id/30193/freddie-freeman

In this problem we are trying to determine the percent increase of Freddie Freeman's batting average from 2012 to 2013, so the reference value will be his batting average from the earlier time period of 2012, or .259. Also, the key word increase and the fact that $.259 < .319$ tells you that the starting value or reference point is his 2012 batting average.

In calculating percent changes with regard to sales and company profits, the reference point may not be so obvious and may depend on what information is desired. Let's look at an example.

> Last year, a local company that manufactures denim clothing had annual costs of \$235,000 and sales of \$567,000. Calculate the percent profit made by the company.

For this problem, it is easy to determine the amount of profit. You take *sales – cost*, which gives you \$332,000. The difficult part is in determining what the denominator needs to be for calculating percent profit. Should the denominator be the amount of cost or the amount of sales? Most companies actually calculate both as a measure of company profitability, but in a typical problem it will most likely be specified as to which one you should use as the reference value. You can see that, depending on the reference value, the percent profit can be vastly different:

As a percent of sales: $(\text{amount of profit} \div \text{sales}) \cdot 100 = (\$332{,}000 \div \$567{,}000) \cdot 100 = 58.5\%$

As a percent of cost: $(\text{amount of profit} \div \text{cost}) \cdot 100 = (\$332{,}000 \div \$235{,}000) \cdot 100 = 141.3\%$

Another good illustration of the use of reference values can be found by looking at the data in the following table on the preferred news source of a group of 990 people based on their age.

Age (years)	Internet	Other (radio, TV, newspapers)
40 and under	350	360
Over 40	30	250

If you ask the following questions, would you get the same results for both? They sound like they are the same thing, but are they really?

1. What percentage of the people surveyed who are age 40 or older get their news from the Internet?
2. What percentage of the people surveyed who get their news from the Internet are 40 or older?

For both questions, the numerator of the ratios will be 30 since both involve the number of people 40 or older who get their news from the Internet. For Question 1, the denominator of the ratio will be the number of people who are 40 or older in the table, which is $30 + 250 = 280$. So the percentage would be $100 \cdot \frac{30}{280} \approx 10.7\%$. For Question 2, the denominator of the ratio will be the number of people who get their news from the Internet, which is $350 + 30 = 380$. So the percentage would be $100 \cdot \frac{30}{380} \approx 7.9\%$. This shows that the answers to the two questions are different because the reference values in the denominators are different.

In determining the reference value when calculating a percent, be sure to read the problem very carefully. If the reference value isn't explicitly stated in the problem, look for key words and phrases like "increase," "decrease," "changed from," or "changed to," to help you identify the starting value or original amount.

Name: Date:

4.8 Applications: Profit, Simple Interest, and Compound Interest

Objectives

Calculate percent of profit.

Calculate simple interest.

Calculate compound interest.

Success Strategy

Keep your list of problem-solving steps handy when working through this section. Also, read each interest problem carefully so that you are determining the correct type of interest—simple vs. compound interest.

Understand Concepts

Go to Software First, read through Learn in Lesson 4.8 of the software. Then, work through the problems in this section to expand your understanding of the concepts related to applications of percent problems.

Percent of Profit

The amount of profit earned when an item is sold is the difference between the selling price and the cost of the item.

Percent of profit can be calculated in two ways:

1. **Based on cost:** Divide the amount of profit by the cost of the item.

$$\frac{\text{profit}}{\text{cost}} = \%\text{ of profit based on cost}$$

2. **Based on selling price:** Divide the amount of profit by the selling price of the item.

$$\frac{\text{profit}}{\text{selling price}} = \%\text{ of profit based on selling price}$$

1. In the proportion $\frac{P}{100} = \frac{A}{B}$, which variable would be used for

a. the profit?

b. the cost or selling price?

c. the percent of profit?

2. In the equation $R \cdot B = A$, which variable would be used for

a. the profit?

b. the cost or selling price?

c. the percent of profit?

Quick Tip

In 2007, the Credit Card Accountability Responsibility and Disclosure Act was passed to help ensure that consumers are informed of important details related to any credit accounts they open. There are many sources online to learn more about this law.

Quick Tip

A **certified deposit** (CD) is a low-risk investment that has a fixed time period during which the money cannot be accessed.

Quick Tip

The interest rate period and time period must be in the same units.

Quick Tip

Most interest-bearing accounts today use compound interest instead of simple interest.

3. Interest is a part of most people's financial life. Interest can be earned in our favor, which means that we gain money from the interest. Interest can also be earned from us, which means that we owe money due to interest. Identify if interest is earned in your favor (paid to you) or earned from you (paid by you) for the following types of accounts.

a. Savings accounts

b. Credit cards

c. Mortgages

d. Certified Deposits

e. Student loans

Interest Formulas

Simple interest is calculated with the formula $I = P \cdot r \cdot t$, where

I = interest earned or paid

P = principal (or starting amount)

t = time, in years

r = interest rate

Compound interest uses the same formula but $t = \frac{1}{n}$, where n is equal to the number of times per year the account is compounded. To find compound interest, follow these steps:

1. Use the simple interest formula with $t = \frac{1}{n}$,
2. Add the interest earned to the principle amount,
3. Repeat steps 1 and 2 as many times as the interest is to be compounded.

4. Use the above information in the Interest Formulas box to answer the following questions.

a. How many times do you use the equation $I = P \cdot r \cdot t$ when calculating simple interest?

b. How many times do you use the equation $I = P \cdot r \cdot t$ when calculating compound interest?

Name: Date:

5. An account is opened with an initial deposit of $1000 and earns 9% annual interest.

a. Calculate the simple interest on the account for a year.

b. Suppose it is a compound interest account that is compounded quarterly (4 times per year). Calculate the interest earned on the account after a year.

c. Which account earned more interest during the year, the simple interest account or the compound interest account?

d. Why do you think the account from part **c.** earned more interest?

6. When calculating interest with compound interest accounts, it is important to know the key words for different values of n in $t = \frac{1}{n}$. Fill in the missing information in the following table.

Key Word	Times Per Year (Value of n)	Value of $t = \frac{1}{n}$
Annually		
Semi-annually		
Quarterly		
Monthly		
Weekly		

Skill Check

Go to Software Work through Practice in Lesson 4.8 of the software before attempting the following exercises.

Substitute the given values into $I = P \cdot r \cdot t$ and then simplify to find the simple interest earned.

7. $P = \$1000$, $r = 5\%$, $t = 3$ years

8. $P = \$2000$, $r = 4\%$, $t = 5$ years

9. $P = \$5000$, $t = 2$ years, $r = 2\%$

10. $P = \$10{,}000$, $r = 6\%$, $t = 4$ years

Apply Skills

Work through the problems in this section to apply the skills you have learned related to applications of percent problems.

11. Barbara's Bombtastic Bakery sells custom decorated 8-inch round cakes for \$18.50 each. The cost of materials to make and package the cake (not including labor costs for baking and decorating) is \$4.50. Round your answers to the nearest hundredth.

a. What is the amount of profit earned on each cake?

b. What is the percent of profit based on cost?

c. What is the percent of profit based on sale price?

Name: Date:

12. A furniture store has an in-store credit deal where you can have 0% interest on your purchase if you pay it off within 6 months. If the balance is not paid off within 6 months, you must pay for 6 months of simple interest on the original purchase price at an annual rate of 15%.

a. If you purchased $1400 in furniture and did not pay off the balance within 6 months, how much interest will be added to your account?

b. If you made equal monthly payments, how much would you need to pay each month to pay off the furniture in 6 months so you will not have to pay interest?

13. You have a savings account that earns 4% annual interest compounded quarterly. You make an initial one-time deposit of $1000 into the savings account.

a. Fill in the table to calculate the interest after each compounding period for a year. Round your answer to the nearest hundredth when necessary.

Quarter	Starting Principle	Interest Earned
1		
2		
3		
4		

b. How much interest was earned on the savings account during that 1 year?

c. What percent of the initial deposit is the amount of interest earned? This value is called the **effective interest rate** for the account.

14. According to the Federal Reserve, in July 2013, Americans owed $849.8 billion in credit card debt. If the average annual credit card interest rate is 15%, how much interest is earned on this debt in one month? Round your answer to the nearest hundredth.

Name: Date:

Chapter 4 Projects

Project A: Take Me Out to the Ball Game!

An activity to demonstrate the use of percents and percent increase/decrease in real life

The Atlanta Braves baseball team has been one of the most popular baseball teams for fans not only from Georgia, but throughout the Carolinas and the southeastern United States. The Braves franchise started playing at the Atlanta-Fulton County Stadium in 1966 and this continued to be their home field for 30 years. In 1996, the Centennial Olympic Stadium that was built for the 1996 Summer Olympics was converted to a new ballpark for the Atlanta Braves. The ballpark was named Turner Field and was opened for play in 1997.

Round all percentages to the nearest whole percent.

1. The Atlanta-Fulton County Stadium had a seating capacity of 52,769 fans. Turner Field has a seating capacity of 50,096 people. Source: http://atlanta.braves.mlb.com/atl/ballpark/history.jsp

a. Determine the amount of decrease in seating capacity between Turner Field and the original Braves stadium.

b. Determine the percent decrease in seating capacity at Turner Field (based on the original stadium).

2. The Centennial Olympic Stadium had approximately 85,000 seats. Some of the seating was removed in order to convert it to the Turner Field ballpark. Rounding the number of seats in Turner Field to the nearest thousand, what is the approximate percent decrease in seating capacity from the original Olympic stadium?

3. When Turner Field opened in 1997, the average attendance at a Braves game was 42,771 people. In 2012, the average attendance was 29,878 people. What is the percent decrease in attendance from 1997 to 2012? Source: espn.go.com

4. In 2013 the average attendance at a Braves game was 31,465 people. What is the percent increase in average attendance per game from 2012 to 2013?

5. In July of 2013, Chipper Jones, a popular Braves third baseman, retired. He started his career with the Braves in 1993 at the age of 21. Source: espn.go.com

a. In 2001, Chipper had 189 hits in 572 at-bats. Calculate Chipper's batting average for the season by dividing the number of hits by the number of at-bats. Round to 3 decimal places.

b. In 2008, Chipper had 160 hits in 439 at-bats. Calculate Chipper's batting average for the season by dividing the number of hits by the number of at-bats. Round to 3 decimal places.

c. Calculate the percent change in Chipper's batting average from 2001 to 2008.

d. Does this represent a percent increase or decrease?

6. In 2001, Chipper had 102 RBIs (runs batted in). In 2008, Chipper had only 75 RBIs.

a. Calculate the percent change in RBIs from 2001 to 2008.

b. Does this represent a percent increase or decrease?

Name: Date:

Project B: Getting a Different Perspective on Things!

An activity to demonstrate the use of ratios and rates in real life

The table below contains population estimates taken from the World Atlas.

Country/Region	Population Estimate	Area in km^2	Area in mi^2	Density/mi^2 (to nearest tenth)
Bangladesh	164,425,000	144,000	55,599	
Brazil	193,364,000	8,511,965		
China	1,339,190,000	9,596,960		
Germany	81,757,600	357,021		
India	1,184,639,000	3,287,590		
Japan	127,380,000	377,835		
Russia	141,927,297	17,075,200		
Thailand	63,525,062	514,000		
United Kingdom	62,041,708	244,820		
United States	309,975,000	9,629,091		
Source: http://www.worldatlas.com/aatlas/populations/ctypopls.htm				

It is extremely hard to put these population estimates into perspective since they are so large. One way to do this is to look at the ratio of the number of people per unit of area which is called **population density**. Since the units of the numerator and denominator of this expression are different, this ratio is called a **rate**. The units should always be specified on values representing rates for clarity.

In the table above, the population estimates are in the second column and the area in km^2 of the corresponding country or region is in column three.

1. Since the use of metric equivalents is not common in the United States, let's first convert the area in square kilometers (ki^2) in column three to area in square miles (mi^2) by using the relationship 1 km = 0.6213712 mi and noting that the units are squared. For example, to convert the area in km^2 to mi^2 for Bangladesh, you could set up the proportion

$$\frac{x \text{ mi}^2}{144,000 \text{ km}^2} = \frac{(0.6213712 \text{ mi})^2}{(1 \text{ km})^2}.$$

Solving this proportion for x using cross products yields $x = 55,599$ when rounded to the nearest whole number. Using a similar proportion to the one above, convert the measurements in column three to mi^2 and place the values in column four, rounding to the nearest whole number. (The reason for the large number of decimal places in the conversion factor is to increase the accuracy of our calculations because our area values are so large.)

2. Now compute the population density per square mile for each country by dividing the population estimate in column two by the corresponding area in column four. Round your answer to the nearest tenth and place the result in the last column.

3. Which country in the table has the largest number of people per square mile?

4. Which country in the table has the smallest number of people per square mile?

5. Look at the population density of the United States.

   a. How does the population density of the United States compare to the population density of other countries?

   b. Does this surprise you? Explain why or why not.

6. Was it easier to put the population numbers in perspective once they were converted to people per square mile, a population density? Why or why not?

7. Look at the numbers for countries like Bangladesh, India, and Japan. Using the Internet, do some research to determine two consequences resulting from overpopulation in an area.

Name: Date:

Math@Work

Pediatric Nurse

As a pediatric nurse working in a hospital setting, you will be responsible for taking care of several patients during your work day. You will need to administer medications, set IVs, and check each patient's vital signs (such as temperature and blood pressure). While doctors prescribe the medications that nurses need to administer, it is important for nurses to double check the dosage amounts. Administering the incorrect amount of medication can be detrimental to the patient's health.

During your morning nursing round, you check in on three new male patients and obtain the following information.

	Patient A	Patient B	Patient C
Age	10	9	12
Weight (pounds)	81	68.5	112
Blood Pressure	97/58	100/59	116/73
Temperature (°F)	99.7	97.3	101.4
Medication	A	B	A

The following table shows the bottom of the range for abnormal blood pressure (BP) for boys. If either the numerator or the denominator of the blood pressure ratio is greater than or equal to the values in the chart, this can indicate a stage of hypertension.

Abnormal Blood Pressure for Boys by Age	
	Systolic BP / Diastolic BP
Age 9	109/72
Age 10	111/73
Age 11	113/74
Age 12	115/74
Source: http://www.nhlbi.nih.gov/health/public/heart/hbp/bp_child_pocket/bp_child_pocket.pdf	

Medication Directions	
Medication	**Dosage Rate**
A	40 mg per 10 pounds
B	55 mg per 10 pounds

1. Do any of the patients have a blood pressure which may indicate they have hypertension? If yes, which patient(s)?

2. Use proportions to determine the amount of medication that should be administered to each patient based on weight.

3. The average body temperature is 98.6 degrees Fahrenheit. You are supposed to alert the doctor on duty if any of the patients have a temperature 2.5 degrees higher than average. For which patients would you alert a doctor?

4. Which skills covered in this chapter and the previous chapters were necessary to help you make your decisions?

Foundations Skill Check for Chapter 5

This page lists several skills covered previously in the book and software that are needed to learn new skills in Chapter 5. To make sure you are prepared to learn these new skills, take the self-test below and determine if any specific skills need to be reviewed.

Each skill includes an easy (**e.**), medium (**m.**), and hard (**h.**) version. You should be able to complete each problem type at each skill level. If you are unable to complete the problems at the easy or medium level, go back to the given lesson in the software and review until you feel confident in your ability. If you are unable to complete the hard problem for a skill, or are able to complete it but with minor errors, a review of the skill may not be necessary. You can wait until the skill is needed in the chapter to decide whether or not you should work through a quick review.

1.6 Simplify the expression using the order of operations.

e. $6 + 14 \div 2$ **m.** $3 \cdot (14 - 2 \cdot 6)^2$ **h.** $3 + 15 \div 3 \cdot 5 - 16 \cdot 2 \div 4$

2.6 Simplify the expression using the order of operations.

e. $\frac{3}{4} + \frac{7}{8} \cdot \frac{2}{3}$ **m.** $\frac{12}{25} \div \frac{1}{5} \cdot \frac{1}{3} + 2 \cdot \frac{1}{5}$ **h.** $3 \cdot \left(2\frac{1}{5} - 1\frac{4}{5}\right)^2 - \frac{4}{25} \cdot 2\frac{1}{2}$

3.4 Simplify the expression using the order of operations.

e. $12.5 + 3 \div 2$ **m.** $6.25 + 1.5 \div 2 \cdot 4$ **h.** $(2.5 - 1)^2 + 3 - 1.1^2$

3.5 Simplify the expression using the order of operations.

e. $\frac{4}{5} \cdot 0.5 + \frac{1}{2} \cdot 7$ **m.** $3.14 \cdot \left(\frac{8}{2}\right)^2$ **h.** $\frac{1}{2} \cdot 1.5 \cdot (2.5 + 1)$

4.2 Solve the proportion for the variable.

e. $\frac{3}{4} = \frac{9}{x}$ **m.** $\frac{5}{6} = \frac{y}{9}$ **h.** $\frac{18}{x} = \frac{42}{56}$

Chapter 5: Geometry

Study Skills

5.1 Angles

5.2 Perimeter

5.3 Area

5.4 Circles

5.5 Volume and Surface Area

5.6 Triangles

5.7 Square Roots and the Pythagorean Theorem

Chapter 5 Projects

Math@Work

Foundations Skill Check for Chapter 6

Math@Work

Introduction

If you plan to study architecture, there are several different careers you can choose from that have a varying range of daily tasks. You can create the broad sketches to present to clients, you could design the detailed construction documents, or you could be on site during construction to ensure everything goes smoothly according to the plans. Every career path in architecture requires many math skills, from measuring walls and door frames to working with geometric structures and converting units of measurement. No matter which area of architecture you choose, you will also need to effectively communicate with other members of your team.

Suppose you decide to pursue a career as a project architect at a large architecture firm. While creating detailed construction drawings, you will need to know how to answer several questions that will be asked during the creation of the project. What is the final square footage of the building and individual rooms? Is the cost of materials needed to construct the project within the budget? Determining the answers to these questions (and many more) require several of the skills covered in this chapter and the previous chapter. At the end of the chapter, we'll come back to this topic and explore how math is used as an architect.

Study Skills

Tips for Improving Math Test Scores

Preparing for a Math Test

1. Don't try to cram right before the test or wait until the night before to study. You should be reviewing your notes and note cards every day in preparation for quizzes and tests.
2. If the textbook has a chapter review or practice test in it after each chapter, work through the problems as practice for the test.
3. If the textbook has accompanying software with review problems or practice tests, use it for review.
4. Review and rework through any homework problems, especially those that you found difficult.
5. If you are having trouble understanding certain concepts or working any types of problems, schedule a meeting with your instructor or arrange for a tutoring session (if your college offers this service) well in advance of the next test.

Test-Taking Strategies

1. Scan the test as soon as you get it to determine the number of questions, their levels of difficulty, and their point values so you can adequately gauge how much time you will have to spend on each question.
2. Start with the questions that seem easiest or that you know how to work immediately. If there are problems with large point values, work them next since they count for a larger portion of your grade.
3. Show all steps in your math work so that you may receive partial credit. This will also make it quicker to check your answers later once you are finished since you will not have to work through all of the steps again.
4. If you are having difficulty remembering how to work a problem, skip over it and come back to it later so that you don't use up all your time.

After the Test

The material learned in most math courses is cumulative, which means the concepts you missed on the current test may be needed to understand concepts in future chapters. That's why it is extremely important to review your returned tests and correct any misunderstandings that may hinder your performance on future tests.

1. Be sure to correct any work you did wrong on the test so that you know the correct way to do the problem in the future. If you are not sure what you did wrong, get help from a peer who scored well on the test or schedule time with your instructor to go over the test.
2. Analyze the test questions to determine if the majority came from your class notes, homework problems, or the textbook. This will give you a better idea of where to spend the majority of your time studying before the next test.
3. Analyze the errors you made on the test. Were they careless mistakes? Did you run out of time? Did you not understand the material well enough? Were you unsure of which method to use?
4. Based on your analysis, what should you do differently before the next test? Where do you need to focus your time?

Name: Date:

5.1 Angles

Objectives

Recognize points, lines, and planes.

Know the definition of an angle and how to measure an angle.

Be able to classify an angle by its measure.

Recognize complementary angles and supplementary angles.

Recognize congruent angles, vertical angles, and adjacent angles.

Know when lines are parallel and perpendicular.

Success Strategy

There are a lot of terms in this section, so be sure to devote a section in your notebook to writing down all of the terms and their definitions. You could also use index cards and the Frayer Model from Chapter 2.

Understand Concepts

Quick Tip

Pay close attention to the notation used to denote each geometric form. There may be more than one way to refer to some of these.

Go to Software First, read through Learn in Lesson 5.1 of the software. Then, work through the problems in this section to expand your understanding of the concepts related to angles.

1. Geometry has a lot of important terminology. Knowing what these terms mean and being able to give an example of each is important when learning geometry. Fill in the following table by drawing an example and writing a definition or description of each term.

Term	Example	Definition or Description
Point		
Line		
Plane		
Line Segment		
Ray		
Angle		

Quick Tip

It is important to use the correct mathematical notation when talking about angles or degrees in math. Using the wrong symbol can lead to confusion when someone else reads your work.

2. This problem explores the mathematical notation related to angles and how to translate the symbols into English words.

a. The symbol for the word "angle" is $\angle$. Translate the symbols "$\angle A$" into words.

b. The symbol for the phrase "measure of" is m. Translate the symbols "$m\angle A$" into words.

c. The symbol for the word "degree(s)" is °. Translate the symbols "72°" into words.

d. Putting these all together, translate the symbols "$m\angle A = 85°$" into words.

3. There are several different types of angles and relationships between angles. Knowing what these types of angles are and being able to give an example of each is important when learning geometry. Fill in the following table by drawing an example and writing a definition or description of each term.

Term	Example	Definition or Description
Acute		
Right		
Obtuse		
Straight		
Adjacent Angles		
Congruent Angles		

Name: Date:

4. You need to be careful when naming adjacent angles. If you don't properly name the angles, it will be unclear which angle you are referring to.

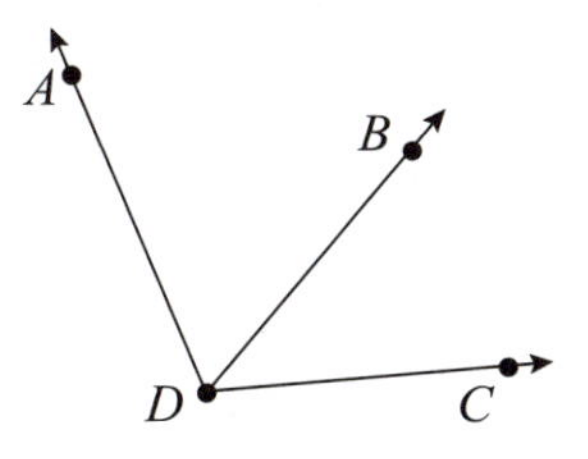

a. How many angles are in the figure?

b. If someone writes "$\angle D$", is it clear which angle they are referring to? Explain why or why not.

c. Name all of the angles by referring to the three points associated with each angle. Remember that the vertex point needs to be the center point listed in the angle name.

Quick Tip

An easy way to remember or distinguish between the terms **complementary** and **supplementary** is that complementary comes before supplementary when written in alphabetical order and 90° is less than 180°, so complementary goes with 90° and supplementary goes with 180°.

Angle Relationships

Complementary angles are angles whose measures add to 90°.

Supplementary angles are angles whose measures add to 180°. When two lines intersect, they form two pairs of **vertical angles.** The vertical angles are opposite of each other. Vertical angles are congruent, which means they have the same measure.

Intuition and observation, along with logic, can be used to show that mathematical properties or theorems are true. If you understand why a property or statement is true, then remembering and using it properly will be easier. The following problem will help you understand the above statement that vertical angles are congruent.

5. According to the vertical angles property, since $\angle 1$ and $\angle 3$ in the figure are congruent, then $m\angle 1 = m\angle 3$.

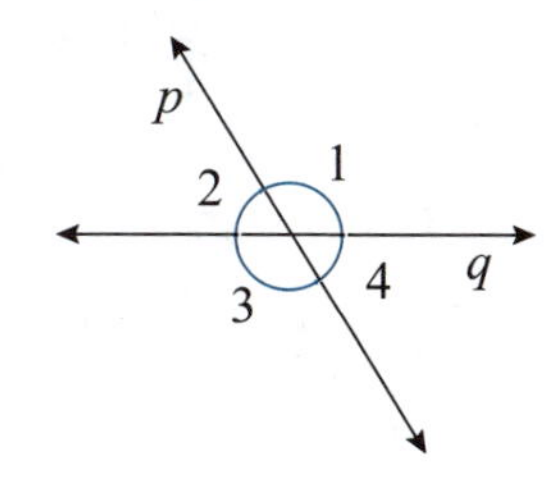

a. Which angles are supplementary with $\angle 1$?

Quick Tip

In the figure for Problem 5, $\angle 2$ and $\angle 4$ are also vertical angles.

b. Write an equation of the form $m\angle X + m\angle Y = 180°$ for each pair of supplementary angles from part **a.** by replacing X and Y with the angle names for each supplementary pair.

c. Which angles are supplementary with $\angle 3$?

d. Write an equation of the form $m\angle X + m\angle Y = 180°$ for each pair of supplementary angles from part **c.** by replacing X and Y with the angle names for each supplementary pair.

e. What must be true about $m\angle 1$ and $m\angle 3$ for both pairs of equations from parts **b.** and **d.** to be true?

Line Relationships

Two lines **intersect** if they cross at any point.

Two lines are **parallel** if they never cross.

Two lines are **perpendicular** if they intersect at a 90° angle.

A **transversal** is a line in a plane that intersects two or more lines at different points.

6. Consider the figure to the right to answer the following questions.

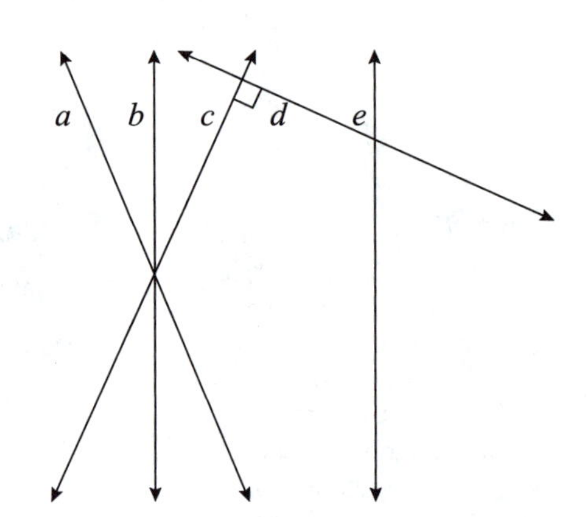

a. Are any lines parallel? If yes, list them.

b. Are any lines perpendicular? If yes, list them.

Quick Tip

In geometric figures, a 90° angle is often represented by a small square.

c. Is line c a transversal of lines a and b? If no, why not?

d. Is line d a transversal of lines c and e? If no, why not?

Angles Created by Transversals

When two parallel lines are intersected by a transversal, both of the following statements are true.

1. Four angles are created on each parallel line. The angles in matching corners are called **corresponding angles**. Corresponding angles are congruent.
2. The pairs of angles on opposite sides of the transversal, but inside the two parallel lines, are called **alternate interior angles**. Alternate interior angles are congruent.

Name: Date:

7. Consider the figure to the right and answer the following questions.

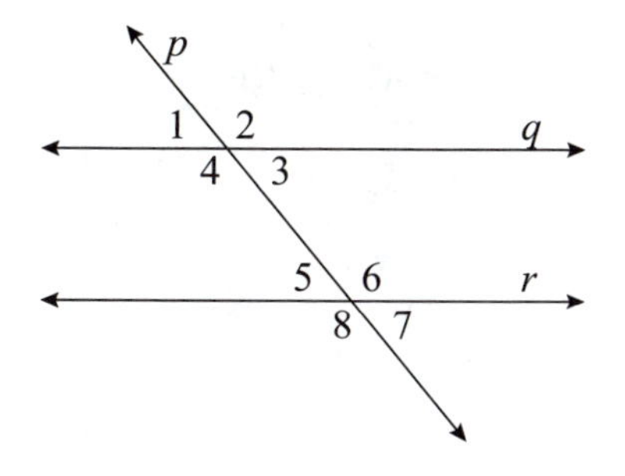

   a. In the figure, $\angle 1$ and $\angle 5$ are corresponding angles. List any other pair(s) of corresponding angles.

   b. In the figure, $\angle 4$ and $\angle 6$ are alternate interior angles. List any other pair(s) of alternate interior angles.

   c. List any pair(s) of vertical angles.

   d. List any pair(s) of supplementary angles.

   e. If $m\angle 1 = 45°$, use the properties of angles to find the measures of the other angles in the figure.

Skill Check

Go to Software Work through Practice in Lesson 5.1 of the software before attempting the following exercises.

8. Assume $\angle 1$ and $\angle 2$ are complimentary.

   a. If $m\angle 1 = 15°$, what is $m\angle 2$?

   b. If $m\angle 2 = 43°$, what is $m\angle 1$?

9. Assume $\angle 3$ and $\angle 4$ are supplementary.

   a. If $m\angle 3 = 115°$, what is $m\angle 4$?

   b. If $m\angle 4 = 74°$, what is $m\angle 3$?

Apply Skills

Work through the problems in this section to apply the skills you have learned related to angles.

10. Consider the proposed road plan shown here.

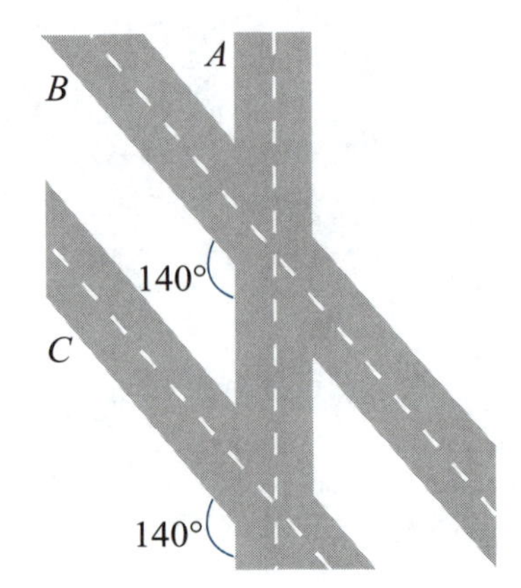

a. Are the right-hand lane and left-hand lane of a roadway parallel or perpendicular?

b. A city building code prohibits the construction of roadway intersections that result in an angle of less than 45°. Does the proposed road plan violate this building code? Why or why not?

c. Roads B and C are parallel roads. Label the angle measures formed by the roads on the figure.

d. What properties allowed you to determine the answers to part **c.**?

11. The navigator of a submarine sees that there are two unknown ships located at points A and B.

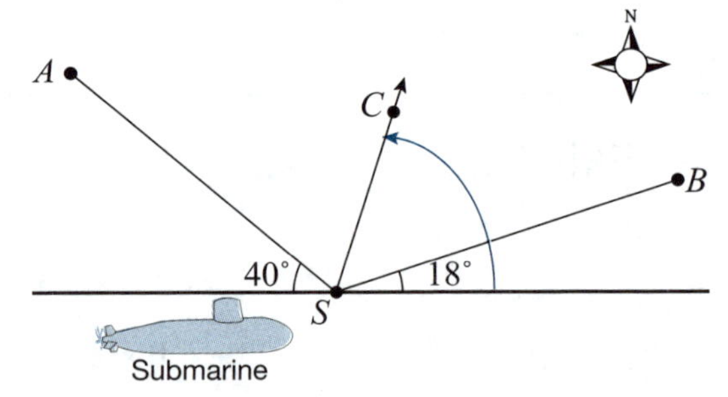

a. Is the angle formed by the two unknown ships and the submarine acute, obtuse, right, or straight?

b. What is the measure of the angle formed by the two unknown ships and the submarine?

Quick Tip

To **bisect** an angle means to divide the angle into two angles of equal measure.

c. In order to remain undetected, the navigator wants to keep as much distance as possible between the submarine located at point S and the two unknown ships. In order to do this, he sets a course at an angle which bisects the angle between the unknown ships. The submarines northeasterly course is towards point C in the figure. At what angle from horizontal, indicated by the arrow, is the submarine traveling?

Name: Date:

5.2 Perimeter

Objectives

Know what types of geometric figures are polygons.

Find the perimeters of polygons.

Success Strategy

You don't need to memorize all of the formulas in this section. To find the perimeter of a figure, just add the lengths of all sides of the figure.

Understand Concepts

Go to Software First, read through Learn in Lesson 5.2 of the software. Then, work through the problems in this section to expand your understanding of the concepts related to perimeter.

In this section, we will discuss the perimeter of **polygons**. A polygon is a closed plane figure with three or more sides. Each side of a polygon is a line segment and two sides meet at a **vertex**. While having formulas to find the perimeter of a shape isn't necessary, they can be used to practice substituting values into equations and simplifying. You can confirm that you substituted the correct values into the perimeter formula by finding the sum of all of the side lengths of the figure and then comparing the solutions.

Quick Tip

Substitute is a verb which means "to put or use in place of another." So, when substituting a value for a variable in a formula, put that value in place of the variable.

Quick Tip

There are polygons that have more than 4 sides. A **pentagon** has 5 sides, a **hexagon** has 6 sides, a **heptagon** has 7 sides, an **octagon** has 8 sides, a **nonagon** has 9 sides, and a **decagon** has 10 sides.

1. Fill in the table with the name of each shape and the formulas to find the perimeter. The variables in the figure represent the lengths of the sides of the figure.

Formulas for Perimeter

Shape	Shape Name	Perimeter Formula
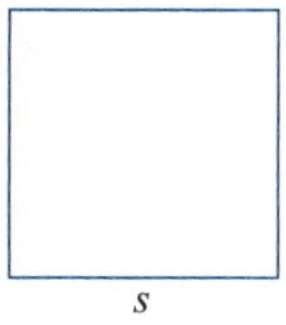		
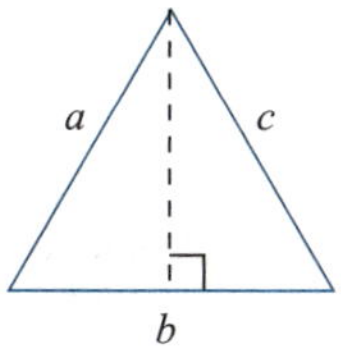		
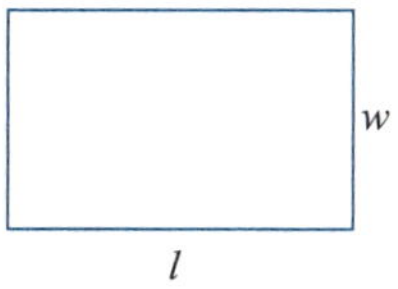		
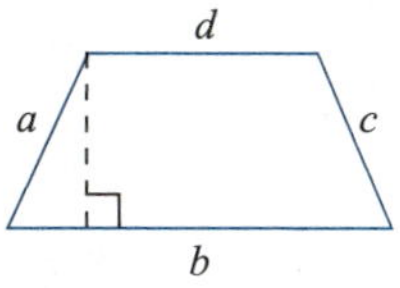		
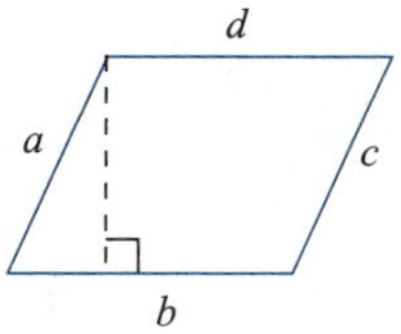		

Lesson Link

Perimeter was first introduced in relation to whole numbers in Section 1.2.

2. Look at the completed table from Problem 1 and answer the following questions.

a. Are any of the formulas the same? If so, for which shapes?

b. Why do you think that the shapes from part **a.** have the same perimeter formula?

c. Instead of using a formula, what do you need to remember to find the perimeter of any geometric shape?

When using a formula to solve a problem, it is important to recognize what the variables in the formula represent. Some are easy to identify, such as the variable for the side length of a square, which is usually referred to as s. Other formulas are flexible with which letter can be used for the variables. For example, a formula involving the side lengths of a triangle could be expressed in multiple ways.

3. For each description, determine which formula from the table should be used to find the perimeter and which measurement will be substituted for each variable in the formula. Then, find the perimeter.

a. A square has a side length of 5 inches.

b. A rectangle has width 4 inches and length 7 inches.

c. A parallelogram has side lengths 5 inches and 4 inches.

d. A triangle has sides 3 inches, 4 inches, and 5 inches.

e. A trapezoid has top length 5 inches, bottom length 15 inches, and side lengths 9 inches and 12 inches.

Name: Date:

Skill Check

Quick Tip

Some common measurements of length are inches (in.), feet (ft), yards (yd), miles (mi), centimeters (cm), and meters (m).

Go to Software Work through Practice in Lesson 5.2 of the software before attempting the following exercises.

Find the perimeter of each figure.

4.

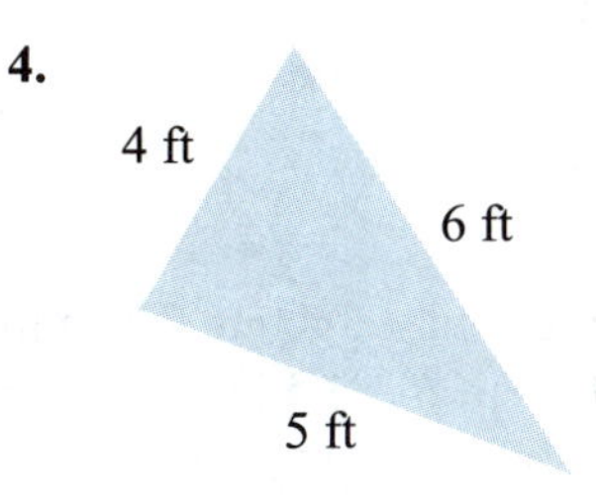

5.

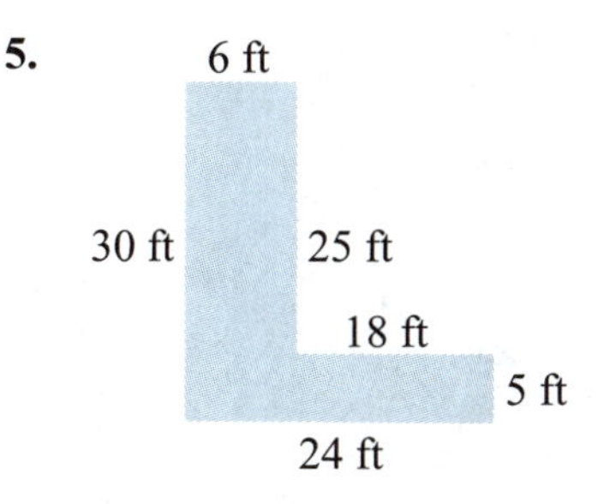

6.

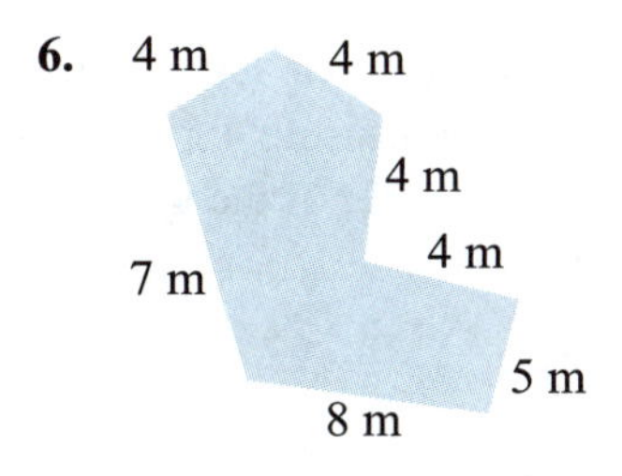

7.

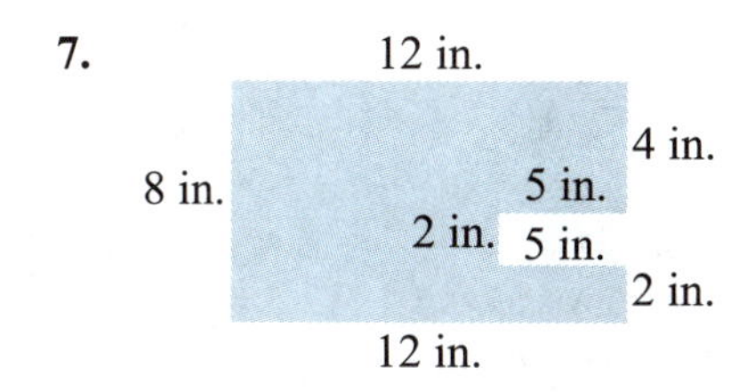

Apply Skills

Work through the problems in this section to apply the skills you have learned related to perimeter. Circle or underline any key words that indicate which perimeter formula should be used.

8. A police officer needs to tape off a crime scene with caution tape. The smallest area he can tape off is outlined by trees and road signs, which he can wrap the tape around. The trees and road signs mark the vertices of the figure.

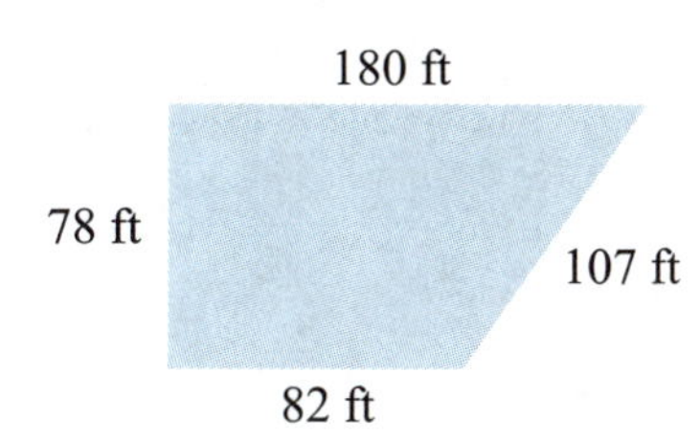

a. What is the perimeter of the crime scene?

b. The officer needs 6 feet of caution tape in addition to the perimeter to properly tape off the crime scene. What is the total amount of caution tape needed?

Quick Tip

Remember that drawing a figure can be helpful when solving a word problem.

9. Jessica wants to redecorate her living room by updating items she already owns.

a. Jessica wants to add a decorative fringe to a throw rug. The rug is a rectangle with length 8 feet and width 5 feet. If Jessica wants to buy 1 foot more than the perimeter of the rug, how many feet of fringe must she buy?

Quick Tip

In a **regular polygon**, all sides have the same length.

b. Jessica wants to outline her mirror with tube lighting. The mirror is in the shape of a regular octagon. The length of one side of the mirror is 5 inches. How many inches of tube lighting must Jessica buy?

Quick Tip

Trim is a type of material that is used for decorating something, especially around its edges. Window trim can be made of wood, vinyl, or other materials.

c. Jessica wants to put new trim around the windows in her living room. She has two windows of the same size. The windows measure 4.5 feet tall and 6 feet wide. She needs an additional 0.25 feet of trim for each corner of the window. How many feet of trim will she need to buy?

Quick Tip

Neoprene is a synthetic rubber made for use in variety of applications such as laptop sleeves, wet suits, and automotive fan belts.

10. An engineer designing a new smartphone decides to add a soft neoprene edging to the phone. The phone itself is $4\frac{1}{2}$ inches tall and $2\frac{2}{5}$ inches wide.

a. How much neoprene edging is needed to go along the outside edge of each smartphone?

b. The neoprene edging will cost $0.12 per inch. How much will the edging cost per phone?

Name: Date:

5.3 Area

Objectives

Understand the concept of area.

Know the formulas for finding the area of five polygons.

Success Strategy

It is important to understand the difference between the perimeter and the area of a figure. Also note that perimeter is measured in standard units and area is measured in square units.

Understand Concepts

Go to Software First, read through Learn in Lesson 5.3 of the software. Then, work through the problems in this section to expand your understanding of the concepts related to area.

Quick Tip

Figure examples and related formulas can be found in the Learn portion of Lesson 5.3 of the software.

1. Fill in the table by sketching a figure of each shape with the variables labeled and the formulas to find the area.

Formulas for Area		
Shape Name	**Figure**	**Area Formula**
Square		
Triangle		
Rectangle		
Trapezoid		
Parallelogram		

The concept of area was introduced in Section 1.3.

2. In Problem 4 of Section 1.3, you wrote down some area formulas that you remembered from previous courses. Compare those formulas to the formulas presented in this section.

a. Does this section cover any formulas you did not write down? If so, what are they?

b. Did you write down any formulas not covered in this section? If so, what are they?

c. Many formulas are presented in different situations with different notation or variables. Did any of the formulas you remembered use different notation than the notation used in this section?

Finding the Area of a Figure with a Section Cut Out

1. Find the area of the full figure (ignoring the cut out section).
2. Find the area of the cut out section.
3. Subtract the area of the cut out section from the area of the full figure.

Quick Tip

Height is the perpendicular distance from the base of the figure to the highest point. The height of a figure is not always equal to a side length.

3. This problem will guide you through the steps to determine the area of a figure with sections cut out.

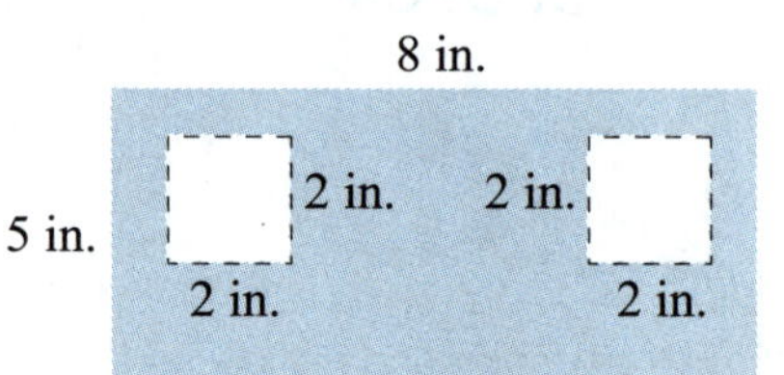

a. List the shapes in the figure. Indicate whether the shape is a cut out.

b. What formula is needed to find the area of each shape in the figure?

c. What is the area of the full figure (that is, the largest shape)?

d. What is the total area of the cut out sections?

e. Find the area of the figure by subtracting the area in part **d.** from the area in part **c.**

4. Learning how to break complicated geometric shapes into simpler shapes that are easier to work with is a skill that you can develop with practice.

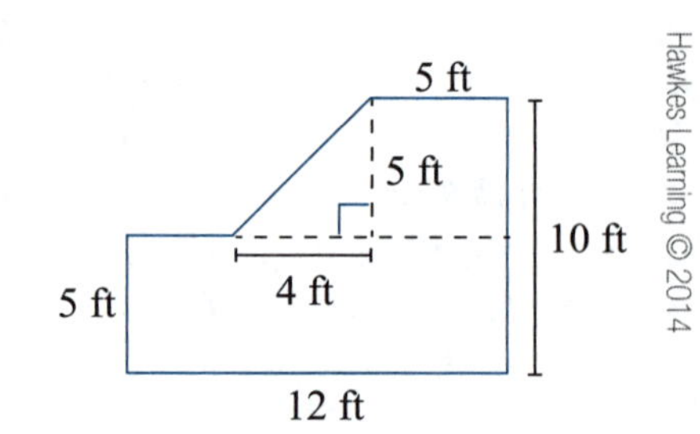

a. What shapes are formed by the dashed lines in the figure?

b. Which area formulas do you need to determine the area of each of the simpler shapes?

Name: Date:

c. Find the area of each of the simpler shapes in the diagram.

d. Find the sum of the areas from part **c.** to determine the area of the complicated shape.

Skill Check

Go to Software Work through Practice in Lesson 5.3 of the software before attempting the following exercises.

Find the area of each figure.

5.

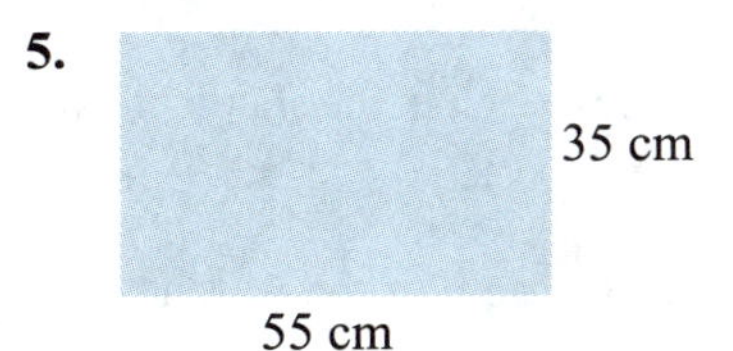

6.

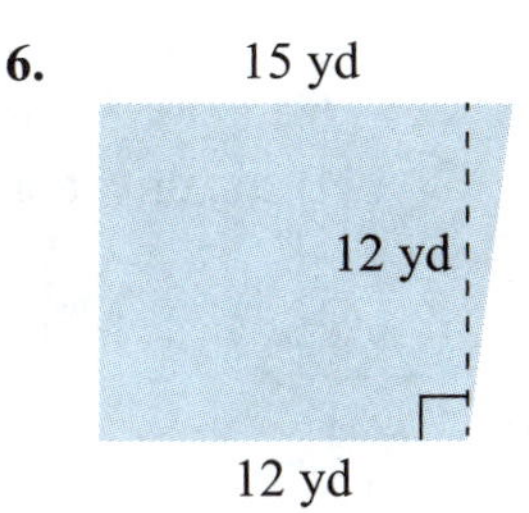

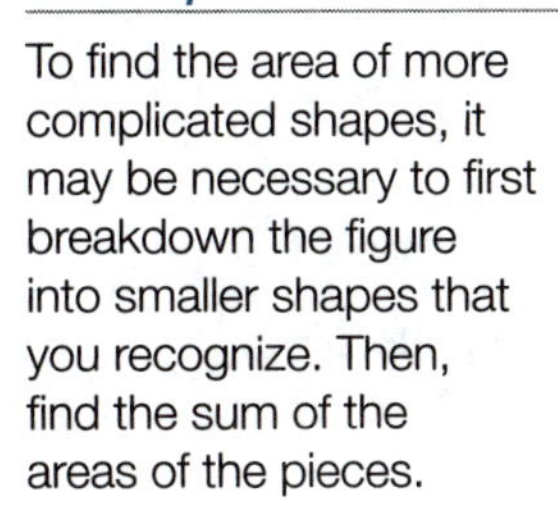

Quick Tip

To find the area of more complicated shapes, it may be necessary to first breakdown the figure into smaller shapes that you recognize. Then, find the sum of the areas of the pieces.

7.

4 in.
7 in.
10 in. 3 in.
3 in.
12 in.

8.

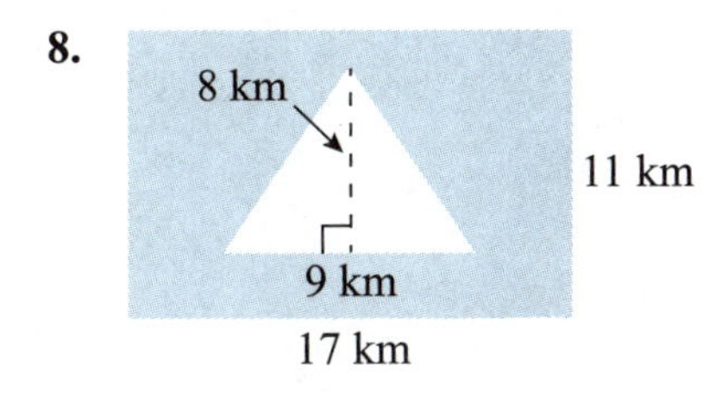

Apply Skills

Work through the problems in this section to apply the skills you have learned related to area. Circle or underline any key words that indicate which area formula should be used.

9. A parking space is in the shape of a rectangle that is $2\frac{1}{2}$ meters wide and 5 meters long. What is the area of the parking space?

10. The main stage at a theater is in the shape of a trapezoid. The owner of the theater is planning to install a new specially designed flooring system on the stage. The stage is 12 feet wide in the front and 15 feet wide in the back. The stage is 10 feet deep.

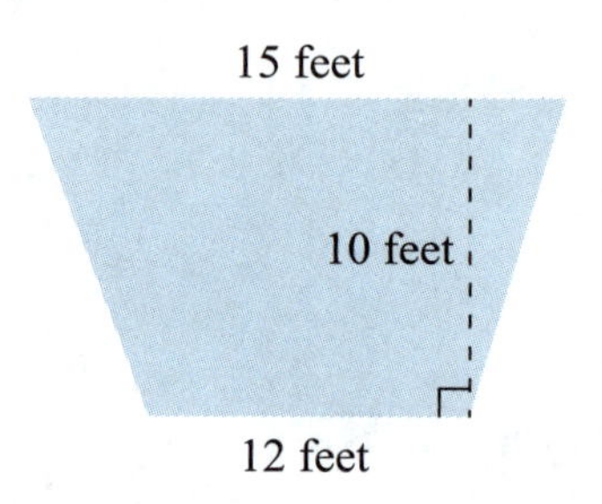

a. What is the area of the stage?

b. The wooden flooring system costs $35.50 per square foot for purchase and installation. How much will it cost to purchase and install the new stage floor?

11. A warehouse has several different rooms, each in the shape of a rectangle. The floor of one room in the warehouse is 25 feet by 40 feet.

a. What is the area of the floor for this room of the warehouse?

b. A pallet for storage measures 4 feet by 3.5 feet. What is the area of a pallet?

c. The warehouse room is empty except for 38 pallets laying flat on the floor. What is the area of the empty floor space in the room?

12. Lee is making a box. He starts with a piece of cardboard that is 14 inches by 20 inches.

a. What is the area of the piece of cardboard?

b. Lee cuts a square with a side length of 3 inches from each corner of the cardboard. What is the area of the cardboard with the corners removed?

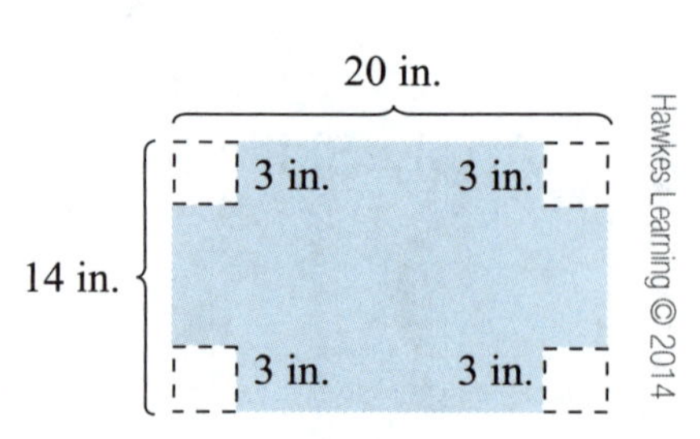

c. When the sides are folded up, what will be the area of the bottom of the box? (**Hint:** Find the length and width of the base first.)

Name: Date:

5.4 Circles

Objectives

Know the definition of a circle and its related terms.

Be able to find the circumference (perimeter) and area of a circle.

Success Strategy

Most calculators have a button for π, a special constant that you will be working with in this section. You should find where this value is on your calculator and learn how to use it.

Understand Concepts

Quick Tip

Pi Day is celebrated in the United States on March 14 every year. Pi Approximation Day is celebrated on July 22 (22/7 in the day/month format) since 22/7 is a common fractional approximation of π.

Go to Software First, read through Learn in Lesson 5.4 of the software. Then, work through the problems in this section to expand your understanding of the concepts related to circles.

1. When working with circles, the value π is something you should be familiar with. This mathematical constant has a long and interesting history. Use the key words "history of pi" to find answers to the following questions.

a. What ratio does π represent?

b. What was the first civilization to approximate the value of π to find the area of a circle?

c. Who was the first mathematician to approximate the value of π?

d. Who was the first person to use the symbol π to stand for this value?

2. Reference the figure to the right and fill in the definition of each term in the table.

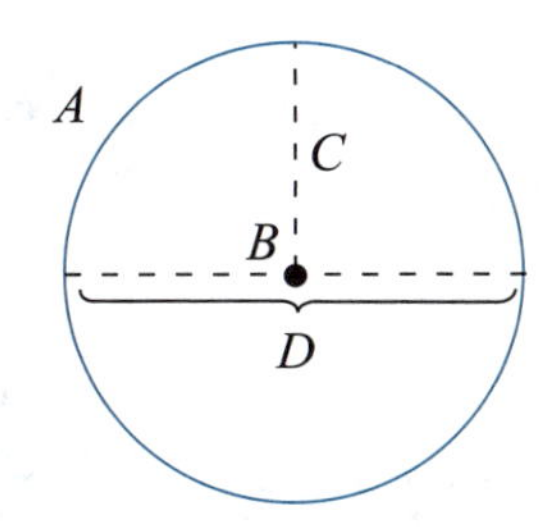

Quick Tip

"N/A" means "not applicable" or "no answer." This notation is used when a question doesn't apply to a certain case or the answer is not available.

Term	Part of Figure	Definition
Circle	N/A	
Circumference	*A*	
Center	*B*	
Radius	*C*	
Diameter	*D*	

Circle Formulas

Circumference of a Circle $C = 2\pi r$ or $C = \pi d$

Area Enclosed by a Circle $A = \pi r^2$

When working with circles, you will need to determine the values of the variables r and d. Occasionally you need to determine the value of one of these variables given the value of the other. It is important to know if a problem statement is giving you the value of the radius or of the diameter. For the next two problems, answer the questions based on the information given.

Quick Tip

A common approximation of π is 3.14. If a problem in this workbook requires the use of π, use 3.14 unless otherwise directed.

3. Suppose you need to find the circumference of this circle.

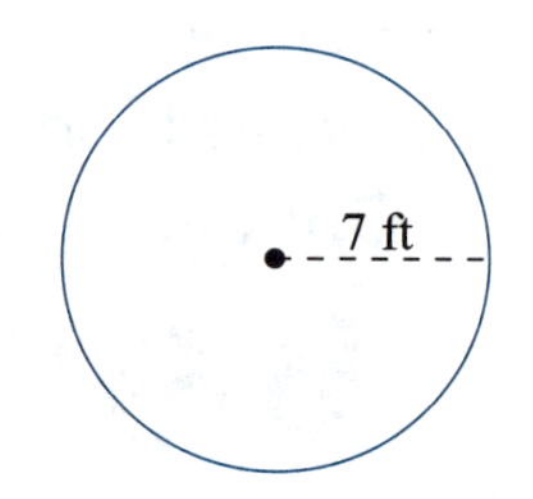

a. The value of which measurement is given?

b. Which equation should be used to find the circumference?

c. Find the circumference of the circle.

4. Suppose you need to find the area of the circle.

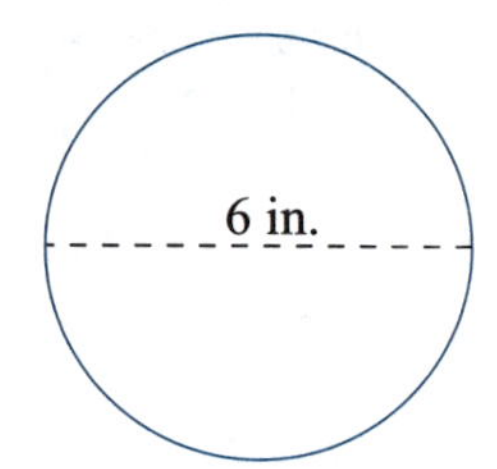

a. The value of which measurement is given?

b. Do you need r or d to find the area of the circle?

c. Find the area of the circle.

Quick Tip

Half of a circle is called a **semicircle**. The area of a semicircle is one half of the area of a circle. The perimeter of a semicircle is half the circumference of a circle plus the diameter.

5. Some complicated geometric figures contain circles or parts of a circle. Knowing how to identify the shapes that make up these figures is a skill that you can develop with practice.

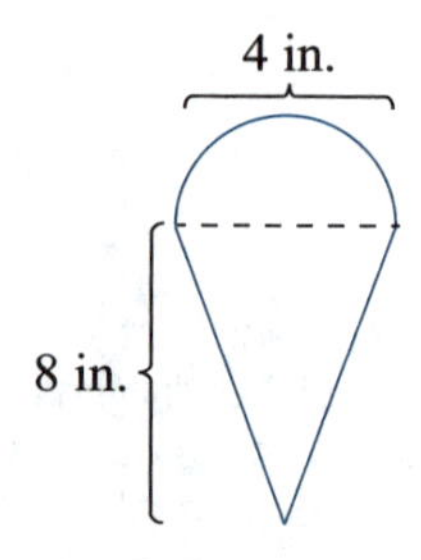

a. What two shapes can you identify in the figure?

b. Which area formulas do you need to determine the area of the shapes in the figure?

c. Find the area of each shape in the figure.

Name: Date:

d. Find the sum of the areas from part **c.** to determine the area of the entire figure.

Skill Check

Quick Tip

Remember that perimeter is the distance around a figure. For a circle, the perimeter is also known as the **circumference**.

Go to Software Work through Practice in Lesson 5.4 of the software before attempting the following exercises.

Find **a.** the perimeter and **b.** the area of each figure. Use $\pi \approx 3.14$ and round to the nearest hundredth.

6.

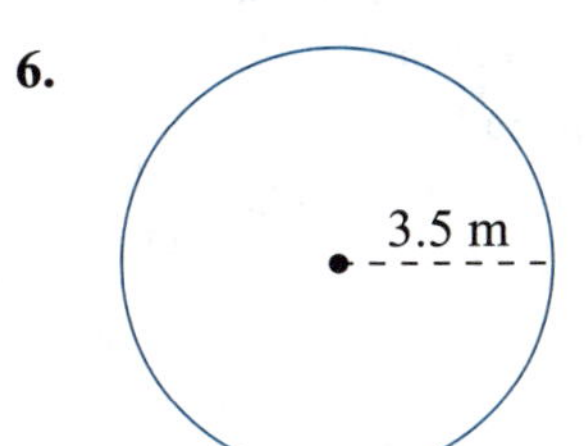

7.

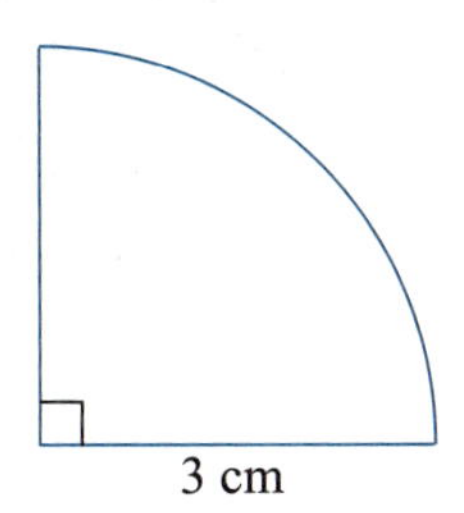

8.

12 yd

12 yd

9.

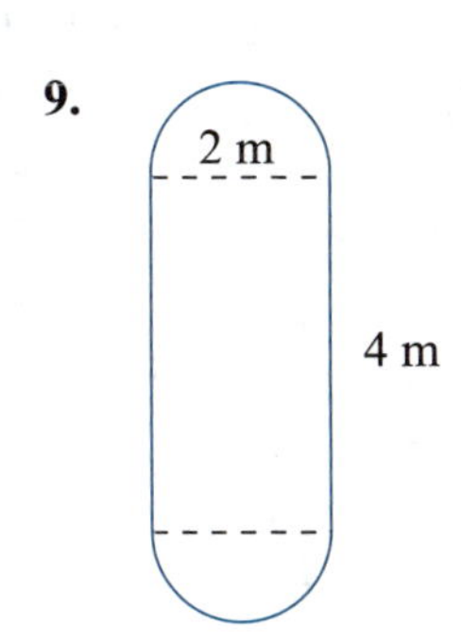

Apply Skills

Quick Tip

The size of the pizza indicates the diameter of the pizza.

Work through the problems in this section to apply the skills you have learned related to circles. Use $\pi = 3.14$ and round your answers to the nearest hundredth.

10. The price for three different sizes of a 1-topping pizza at Romito's pizza are shown in the table.

Price for a 1-Topping Pizza		
9-inch	**12-inch**	**16-inch**
$7.25	$10.25	$13.50

a. Find the area of each pizza. (**Note:** Units will be in square inches.)

b. Use ratios to find the price per square inch for each pizza size.

c. Based on your answer to part **b.**, which pizza is the best value?

11. The city is planning to put a fountain in the middle of the public park. The park is a rectangle with a length of 70 feet and a width of 45 feet. The base of the fountain will be a circle with diameter 10 feet. What area of the public park will not be taken up by the fountain? (**Hint:** Draw a picture to help you solve this problem.)

12. The parking lot of the emergency room at a hospital is in the shape of a rectangle with length 100 feet and width 85 feet. There is also a semicircle with radius 14 feet near the entrance for the ambulance drop off. What is the area of the parking lot, including the drop off area?

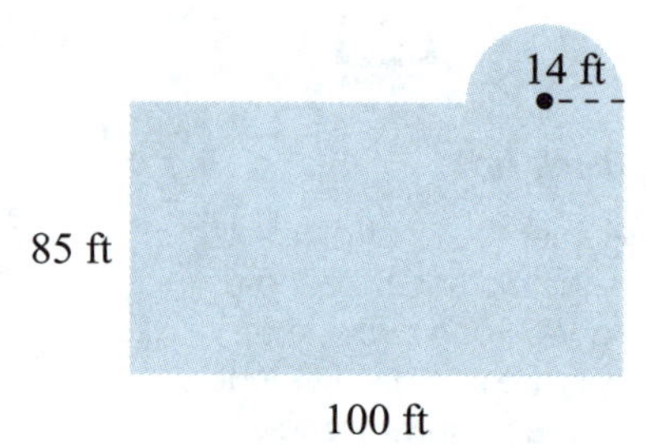

Lesson Link

Finding the area of a washer is similar to finding the area of a shape with cut outs. This was covered in Section 5.3.

13. A machine shop receives an order for 80-millimeter (mm) wide washers with an area of 2198 mm^2. They have machines set up to make the following washers.

Machine	Inner Radius	Outer Radius	Area of Washer
A	10 mm	40 mm	
B	20 mm	40 mm	
C	30 mm	40 mm	

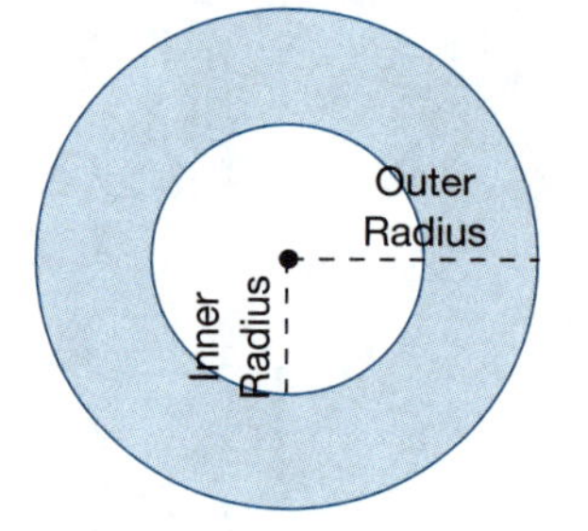

a. Are all of the machines set up to produce a washer with the correct outer radius?

b. Without calculating the areas, which machine will produce the washer with the largest area?

c. Calculate the area of each washer and place the areas in column four of the table.

d. Are any of the machines set up to create the washer size that was ordered? If yes, which machine?

Name: Date:

5.5 Volume and Surface Area

Objectives

Understand the concept of volume.

Know the formulas for finding the volume of five geometric solids.

Understand the concept of surface area.

Know the formulas for finding the surface area of three geometric solids.

Success Strategy

To help you determine when to use volume formulas and when to use surface area formulas, remember that volume refers to how much an object can hold and surface area is a measurement of the outside area of the object.

Understand Concepts

Quick Tip

The shape names and related formulas can be found in the Learn portion of Lesson 5.5 of the software.

Go to Software First, read through Learn in Lesson 5.5 of the software. Then, work through the problems in this section to expand your understanding of the concepts related to volume and surface area.

1. Fill in the table with the name of each shape and the formulas for volume and surface area. If the surface area formula is not given for that shape, write "N/A" in the box.

Formulas for Volume and Surface Area			
Shape	**Shape Name**	**Volume Formula**	**Surface Area Formula**
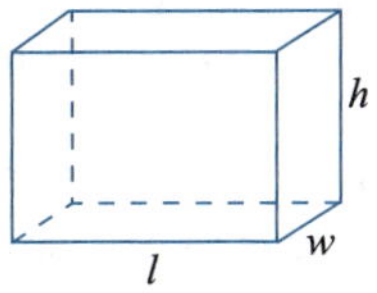			
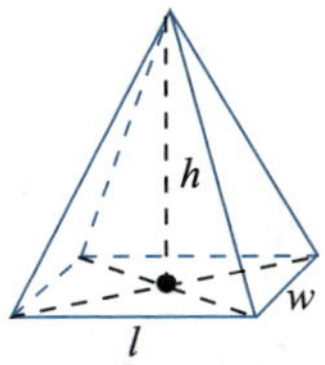			
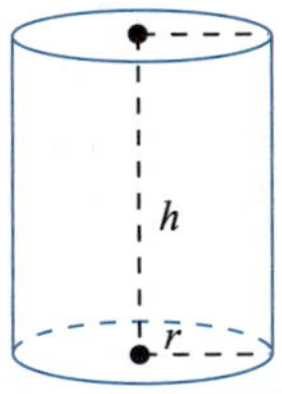			
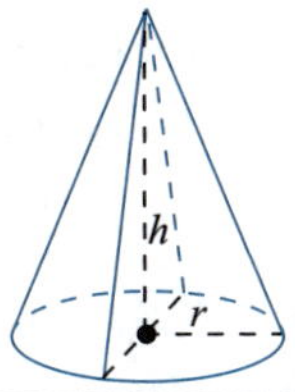			
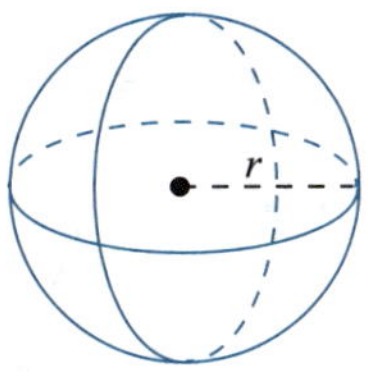			

2. Perimeter, area, and volume are all measurements involving standard units of length that have different meanings and uses. Each of these corresponds with a dimension of space and there are units that go along with each dimension. Fill in the table with the missing information if the unit of measurement is inches.

Measurement	Dimension	Units of Measurement	Shape Example
Perimeter	1-Dimensional		
Area	2-Dimensional		
Volume	3-Dimensional		

Knowing how formulas were developed can help you understand them and use them correctly. The next two problems will guide you through the logic behind the volume formula and the surface area formula for circular cylinders.

Quick Tip

It is helpful to visualize a soup can when working with circular cylinders.

3. The volume of a right circular cylinder is given by the formula $v = \pi r^2 h$.

   a. One way to think of a right circular cylinder is as a lot of equal sized circles stacked on top of each other. What is the equation for the area of a circle?

   b. Suppose a circle has a radius of 2 inches. What is the area of the circle? (Use $\pi = 3.14$.)

   c. Circles are 2-dimensional shapes, which mean they have a width and a length. A right circular cylinder is a 3-dimensional object. Which additional dimension does the circular cylinder have that the circle does not have?

   d. If the formula for the area of a circle is multiplied by this missing measurement, will we obtain the formula for the volume of a circular cylinder?

   e. What is the volume of the circular cylinder if it has a radius of 2 inches and a height of 5 inches?

Name: Date:

4. The surface area of a right circular cylinder is given by the formula $SA = 2\pi r^2 + 2\pi rh$.

a. A right circular cylinder can be divided into three pieces. The top and bottom are circles. The "tube" piece can be cut down one side and flattened into a rectangle. Draw the pieces of a disassembled right circular cylinder.

b. What are the area formulas for a circle and a rectangle? Label the variables from these formulas on your drawing from part **a.**

c. For the rectangle from part **a.**, two of the side lengths are the same as the circumference of the circles which form the top and bottom of the cylinder. What is the formula for the circumference of a circle that uses radius?

d. The circumference of a circle is equal to which variable on your rectangle from part **a.**?

e. Which measurement of the rectangle represents the height of the cylinder?

f. Rewrite the area formula for the rectangle from part **b.** by using the information from parts **c.**, **d.**, and **e.**

g. What do you need to do with the area formulas for the rectangle and the circles to create the surface area formula of a right circular cylinder?

Skill Check

Quick Tip

Remember that 3.14 is a common approximation for π.

Go to Software Work through Practice in Lesson 5.5 of the software before attempting the following exercises.

Find the volume of each figure. Round your answers to the nearest hundredth when necessary.

5.

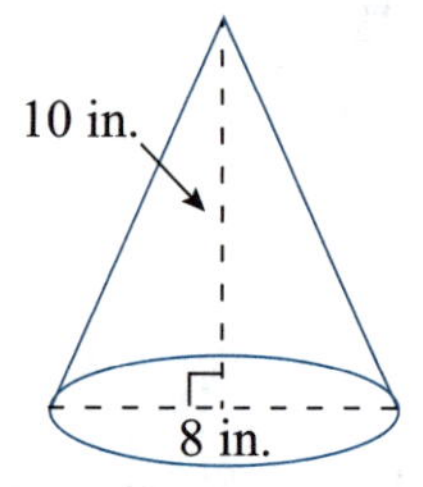

6.

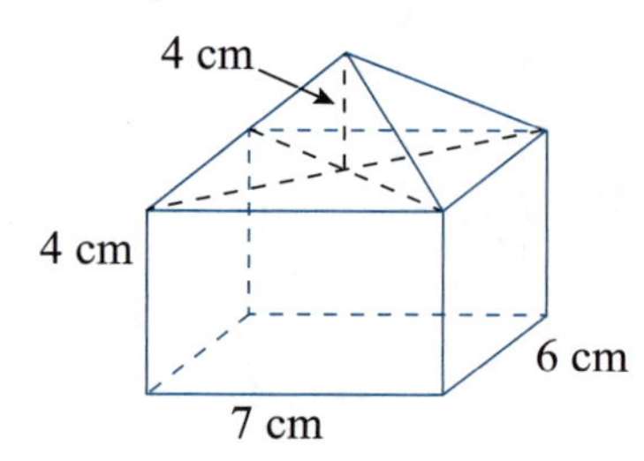

Find the surface area of each figure. Round your answers to the nearest hundredth when necessary.

7.

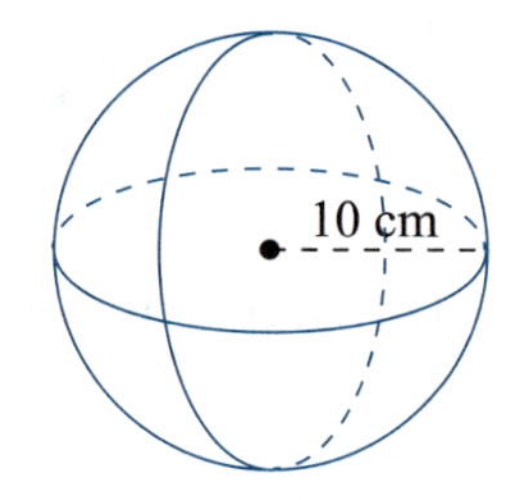

8.

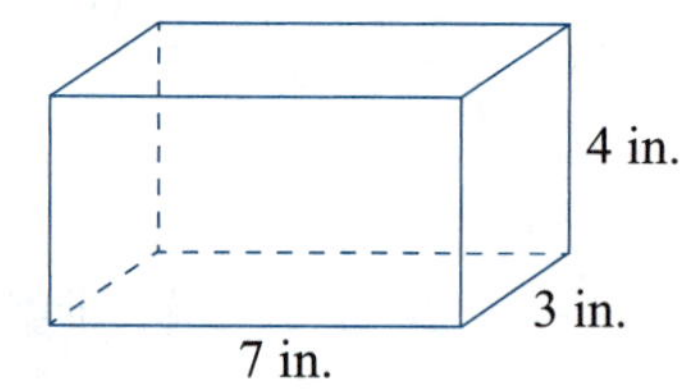

Apply Skills

Work through the problems in this section to apply the skills you have learned related to volume and surface area. Use $\pi = 3.14$ and round your answers to the nearest hundredth if necessary.

9. A can of soup is 4 inches high and has a diameter of 2.6 inches.

a. Find the surface area of the soup can.

b. Find the volume of the soup can.

Name: Date:

10. Barbara's Bombtastic Bakery sells wedding cakes and sets a price based on the number of servings. The volume of a serving of wedding cake is equivalent to the volume of a rectangular piece of cake with measurements 1 inch by 2 inches by 4 inches. Each serving of wedding cake costs \$1.25.

a. What is the volume of one slice of wedding cake?

b. The bottom tier of a round wedding cake has a diameter of 16 inches. If the cake tier is 4 inches high, what is the volume of this tier of the cake? Round to the nearest cubic inch.

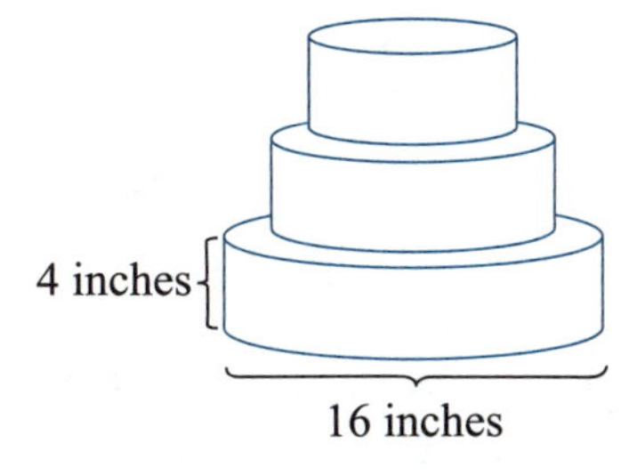

c. How many equivalent slices of wedding cake are in the 16-inch-diameter wedding cake? Round to the nearest whole slice.

d. How much should Barbara's Bombtastic Bakery charge for this tier of the wedding cake?

Quick Tip

A **cube** is a rectangular solid where the length, width, and height have equal measures.

11. A glass ornament in the shape of a sphere is to be packaged in a box with soft foam pellets for protection. The ornament has a diameter of 4 inches. The box is a cube whose side length is 5 inches.

a. What is the volume of the glass ornament?

b. What is the volume of the box?

c. The volume of the box which is not taken up by the glass ornament will be filled with the foam pellets. What volume of the box will be filled with foam pellets?

12. Jerry is a tool and die maker, and is creating a specialized solid steel cone for a customer. The cone needs to be 12.125 cm tall, with a radius of 4.4 cm. How much steel will be used to create the solid steel cone?

13. The Louvre Pyramid is a rectangular pyramid made of glass and metal which is located in the courtyard of the Louvre Palace in Paris, France. The pyramid has an approximate height of 21.6 meters and each side of the base has a length of 35 meters. Source: www.aviewoncities.com

a. What is the volume of the Louvre Pyramid?

b. Each triangular piece that makes up a side of the pyramid has a height of approximately 27 meters. What is the surface area of each triangular piece? (**Note:** The height of the triangular side is a different measurement than the height of the pyramid.)

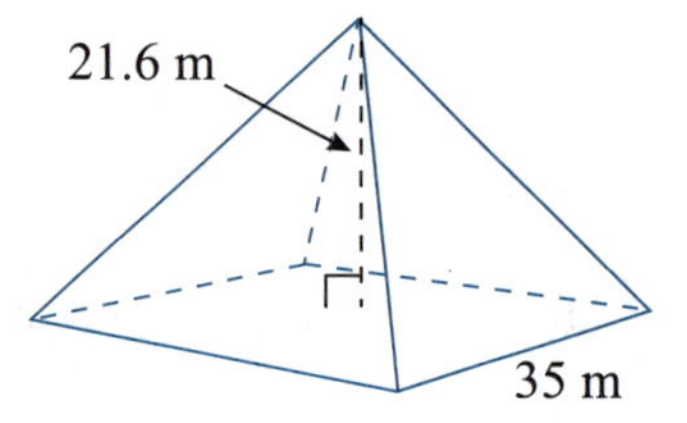

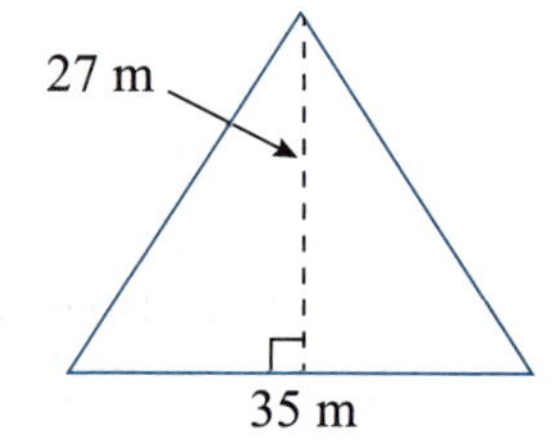

c. The surface area of the pyramid is equal to the area of the base of the pyramid plus the total area of the four triangular faces. What is the surface area of the pyramid?

Name: Date:

5.6 Triangles

Objectives

Be able to classify triangles by sides.

Be able to classify triangles by angles.

Know the properties of similar triangles.

Know the properties of congruent triangles.

Success Strategy

There are a lot of terms in this section, so be sure to devote a section in your notebook to writing down all of the terms and their definitions. You could also use index cards and the Frayer Model from Chapter 2.

Understand Concepts

Quick Tip

Triangle properties and examples can be found in Learn of Lesson 5.6 of the software.

Quick Tip

Did you know that in construction the shape that has the most structural strength is the triangle? This is why you see the triangle shape in bridge designs. Go to www.teachengineering.org for more information.

Go to Software First, read through Learn in Lesson 5.6 of the software. Then, work through the problems in this section to expand your understanding of the concepts related to triangles.

1. For each type of triangle, describe the properties of the triangle and draw an example.

Classification by Sides		
Name	**Properties**	**Example**
Scalene		
Isosceles		
Equilateral		

Classification by Angles		
Name	**Properties**	**Example**
Acute		
Right		
Obtuse		

2. Mathematical notation for angles was introduced in Section 5.1. Geometry also has a specific notation for triangles. This problem explores the notation for triangles and how to translate the notation into English words.

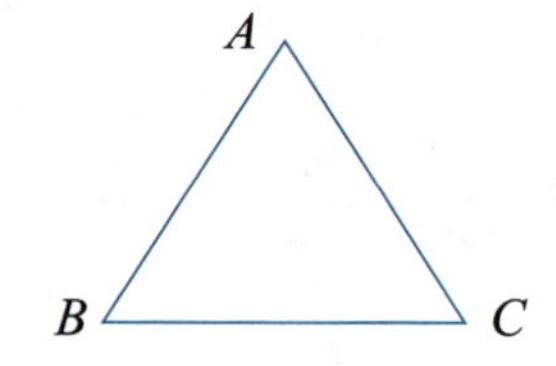

   a. The symbol for the word "triangle" is Δ. Translate the symbols "ΔABC" into English words.

Quick Tip

To move in a **counterclockwise** direction means to move in a direction that is the opposite direction in which the hands of a clock rotate.

   b. A triangle is named by listing the angles in order as you move clockwise or counterclockwise around the figure. How many different ways can you name this triangle?

Three Properties of Triangles

1. The sum of the measures of the angles is 180°.
2. The sum of the lengths of any two sides must be greater than the length of the third side.
3. Longer sides are opposite angles with larger measures.

3. The following problem will help you understand why the sum of the measures of the angles of a triangle is equal to 180°.

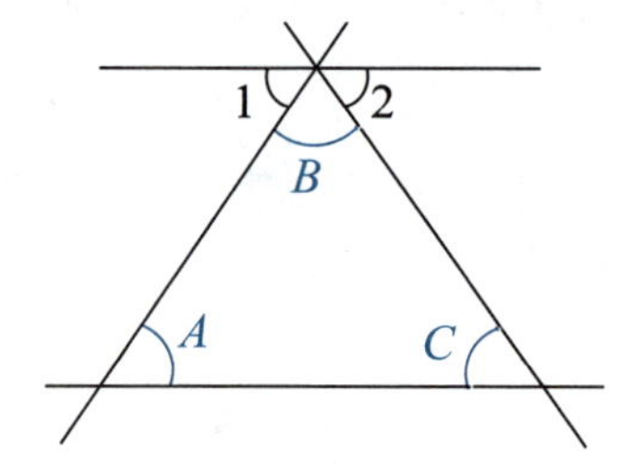

   a. What is the angle measure of a straight line?

   b. What do you know about the sum of $m\angle 1$, $m\angle 2$, and $m\angle B$?

Lesson Link

Alternate interior angles were defined in Section 5.1.

   c. Which of the labeled angles in this figure are alternate interior angles?

   d. What do you know about alternate interior angles?

   e. What does this tell you about the sum of the angle measures of a triangle?

Name: Date:

Similar Triangles

Similar triangles have two properties:

1. Corresponding angles have the same measure.
2. Lengths of corresponding sides are proportional.

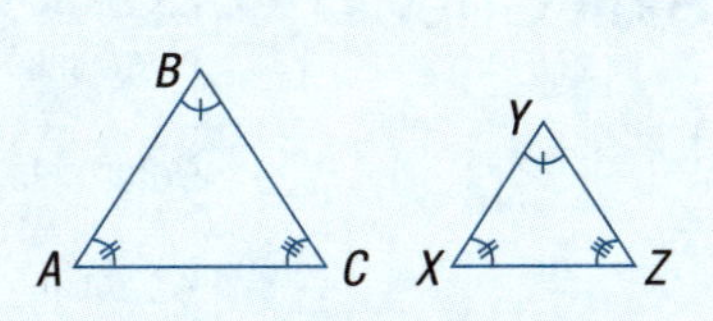

The notation to indicate two triangles are similar is $\triangle ABC \sim \triangle XYZ$.

Similar triangles have sides that are proportional according to the relationship:

$$\frac{AB}{XY} = \frac{BC}{YZ} = \frac{AC}{XZ}$$

Quick Tip

The order that the *pairs* of congruent angles are listed in the notation $\triangle ABC \sim \triangle XYZ$ can vary. The important part is to correctly pair the corresponding angles of the two similar triangles. For example, $\triangle BCA \sim \triangle YZX$ represents the same similar triangles as $\triangle ABC \sim \triangle XYZ$.

When writing the names of two similar triangles or two congruent triangles, it is important to write the corresponding vertices in the same order for both triangles. This means that for $\triangle ABC \sim \triangle XYZ$, $\angle A$ corresponds with $\angle X$, $\angle B$ corresponds with $\angle Y$, and $\angle C$ corresponds with $\angle Z$.

4. A common mistake when writing the notation for similar triangles is to incorrectly match up the corresponding angles. Determine if any mistakes were made in the notation for each pair of similar triangles. If any mistakes were made, describe the mistake and then write the notation correctly.

a. $\triangle ABC \sim \triangle XYZ$

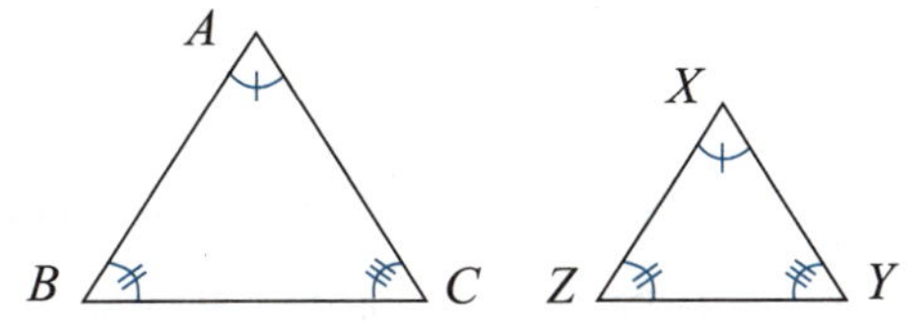

b. $\triangle DEF \sim \triangle LNM$

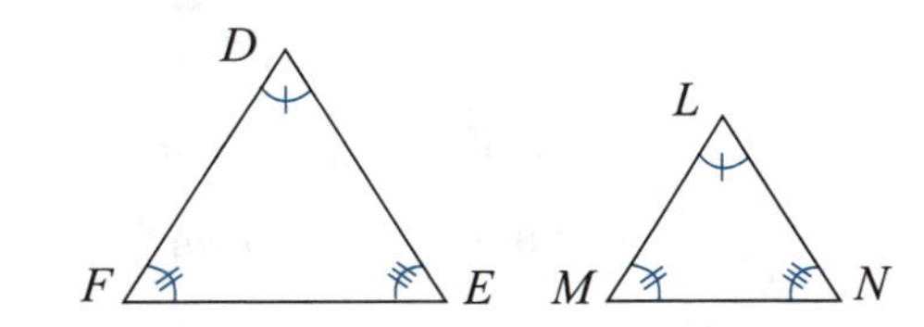

Congruent Triangles

Congruent triangles have two properties: Notation: $\triangle ABC \cong \triangle XYZ$

1. Corresponding angles have the same measure.
2. Lengths of corresponding sides are equal.

There are three ways to determine if triangles are congruent:

Side-Side-Side (SSS): If the three sides of one triangle are equal in length to the corresponding sides of another triangle, then the two triangles are congruent.

Side-Angle-Side (SAS): If two sides of one triangle are equal in length to the corresponding sides of another triangle and the angle between the two sides are congruent, then the two triangles are congruent.

Angle-Side-Angle (ASA): If two angles of one triangle are congruent to corresponding angles of another triangle and the corresponding sides between the two angles are equal in length, then the two triangles are congruent.

Skill Check

Go to Software Work through Practice in Lesson 5.6 of the software before attempting the following exercises.

Determine whether or not a triangle with the given dimensions exists using the second property of triangles.

5. 4 in., 5 in., 7 in.

6. 9 ft, 32 ft, 41 ft

Determine whether or not the pairs of triangles are similar. If they are similar, use the proper notation to indicate the similarity.

Quick Tip

You may need to visually rotate the triangles to line up the corresponding angles or sides.

7.

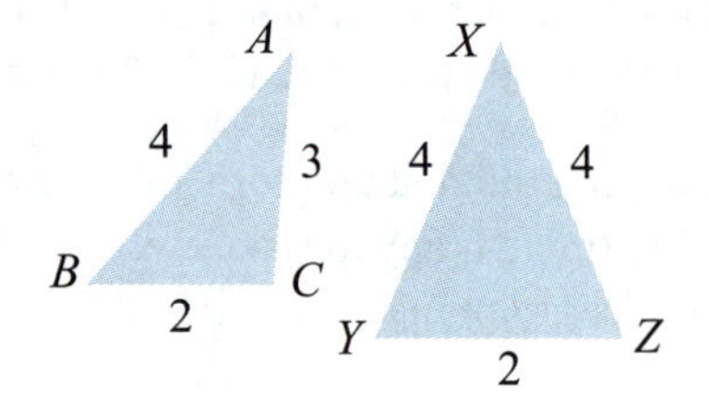

8.

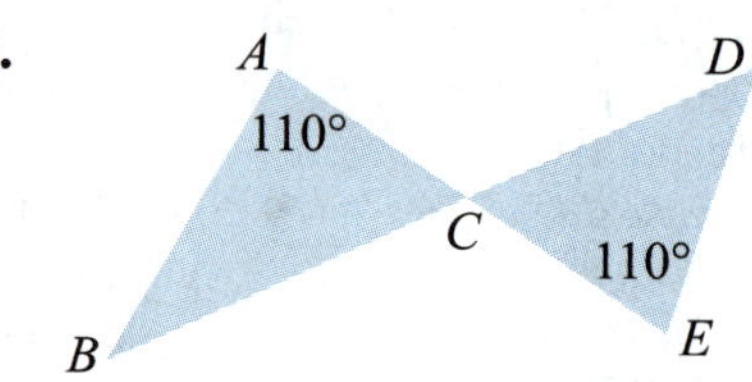

Determine whether or not the pairs of triangles are congruent. If they are congruent, state the property that makes them congruent: SSS, SAS, or ASA.

9.

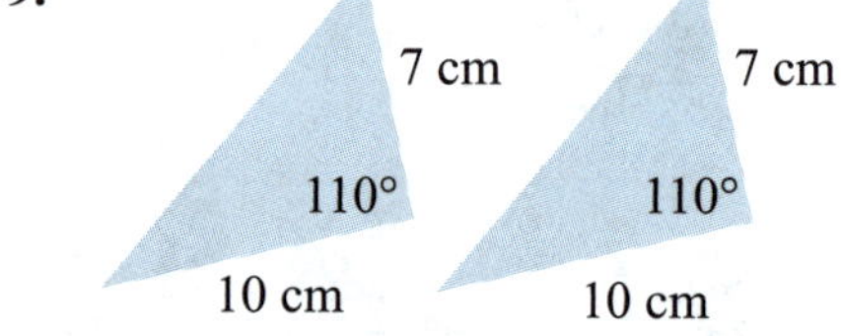

10.

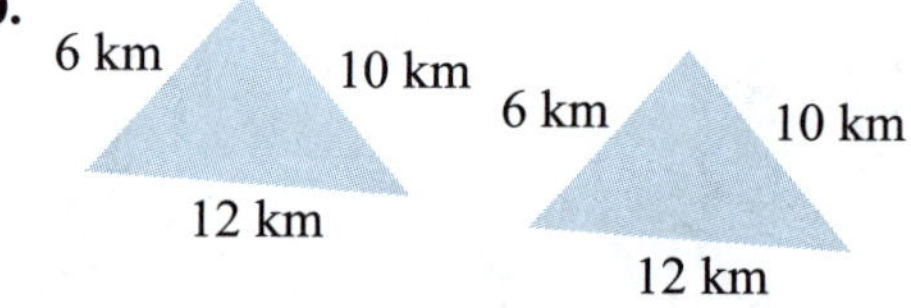

11.

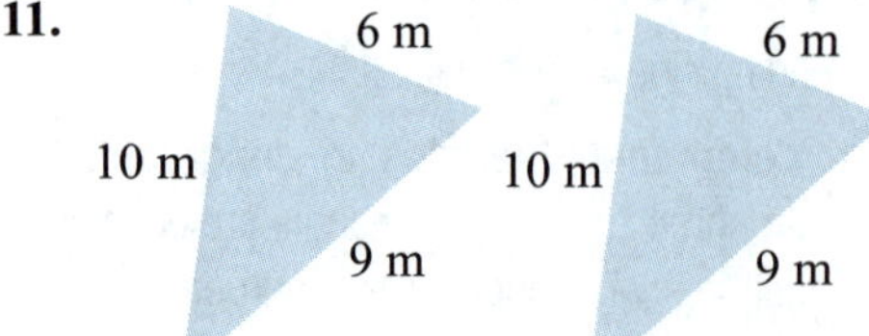

12.

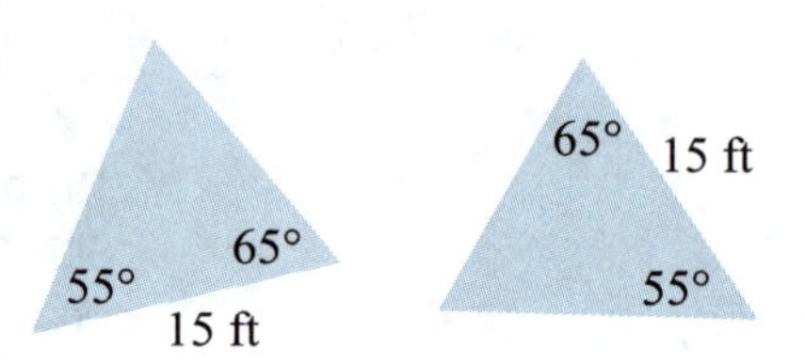

Name: Date:

Apply Skills

Work through the problems in this section to apply the skills you have learned related to triangles. Round answers to the nearest hundredth if necessary.

13. A building has two ramps going up to the entrances located on opposite sides of the building. Both ramps have an incline of 4.5° and form a right angle with the building. The first ramp has a base length of 8 feet. The second ramp has a base of length of 15.25 feet and a height of 1.2 feet.

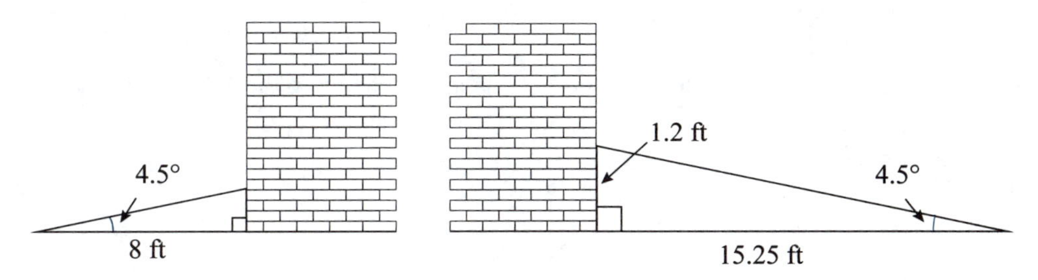

a. Do these ramps form similar triangles, congruent triangles, or neither?

b. What is the height of the 8-foot-long ramp?

14. The pieces to assemble a spice rack include two congruent triangles. The triangles need to have corresponding angles lined up.

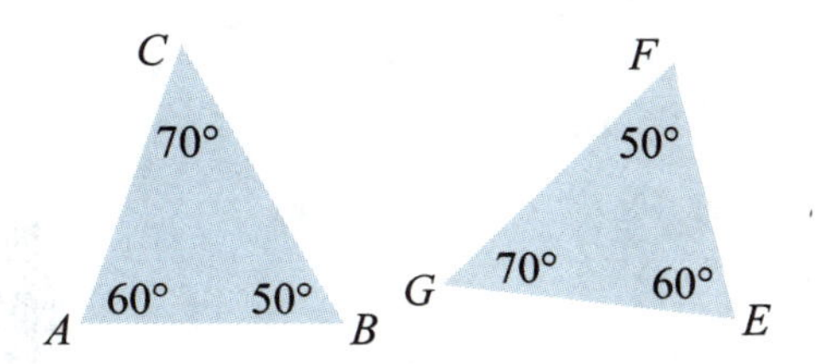

a. Match up the corresponding angles on the triangles. Write that the two triangles are congruent using proper mathematical notation.

b. Would it be enough for the manufacturer to label just one corresponding angle on each triangle? That is, can you determine how the other angles correspond just by knowing how one angle corresponds?

15. A billboard advertisement has a right triangle as part of its design. In the scaled version that the graphic designers made, the base of the triangle is 8 inches and the height of the triangle is 5 inches. On the full sized billboard, the base of the triangle is 96 inches. What is the height of the triangle on the billboard?

Quick Tip

In geometric figures, a 90° angle is often represented by a small square.

16. Handicap ramps must be at an angle no greater than 4.76° from horizontal.

a. What is the measure of $\angle x$?

Lesson Link

Relationships between angles were covered in Section 5.1.

b. What is the relationship between the 4.76° angle and $\angle x$?

Quick Tip

Indirect measurement can be used to find the height of an object when measuring the object directly is not possible.

17. While performing field research, a historian needs to determine the height of an abandoned lighthouse. Since he is unable to directly measure the height of the lighthouse, he determines the height indirectly. He places a 2-foot-long stick in the ground and measures the length of the shadow it casts. He then measures the length of the shadow cast by the lighthouse. What is the height of the abandoned lighthouse? (**Note:** The light house and the stick are both at right angles from the ground.)

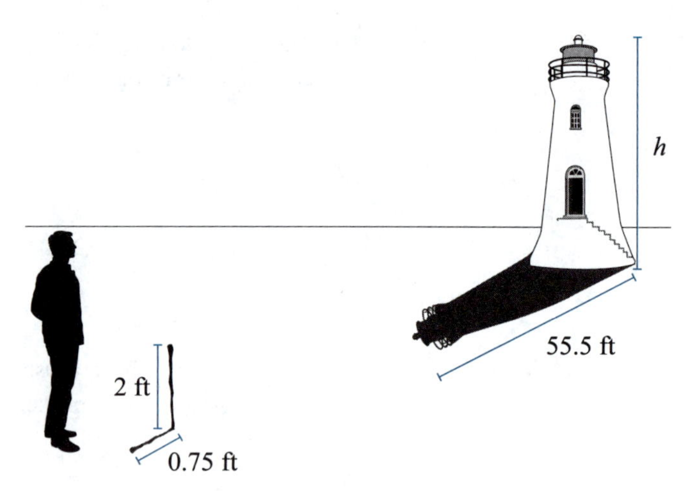

Name: ______________________ Date: ______________________

5.7 Square Roots and the Pythagorean Theorem

Objectives

Find the square roots of a perfect square.

Approximate square roots using a calculator.

Use the Pythagorean Theorem.

Success Strategy

Locate the square root button on your calculator and practice using it so that you can correctly work the problems in this section.

Understand Concepts

Go to Software First, read through Learn in Lesson 5.7 of the software. Then, work through the problems in this section to expand your understanding of the concepts related to square roots and the Pythagorean Theorem.

1. Label the parts of the radical expression.

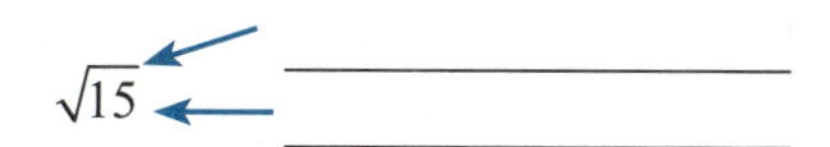

Square Roots and Irrational Numbers

A **perfect square** is the square of a counting number. The square root of a perfect square is a counting number. The square root of a number which is not a perfect square is an **irrational number**. An **irrational number** is an infinite nonrepeating decimal.

Lesson Link

Perfect squares were first introduced in Section 1.6 when learning about exponents.

2. Fill in the table with the first 20 perfect squares and their square roots.

Perfect Square										
Square Root										
Perfect Square										
Square Root										

3. Before calculators were commonly used in classrooms, people had to calculate the square roots of numbers that were not perfect squares by hand. Use the keywords "square roots without calculator" to find at least two different methods of calculating a square root by hand.

a. Describe one method of finding square roots without a calculator.

b. What benefit do you think there is in learning how to calculate a square root by hand?

Quick Tip

The **legs** of a right triangle are represented by the variables a and b. The **hypotenuse**, which is opposite the right angle and is always the longest side, is represented by the variable c.

The Pythagorean Theorem

In a right triangle, the square of the length of the hypotenuse is equal to the sum of the squares of the lengths of the two legs.

$$c^2 = a^2 + b^2$$

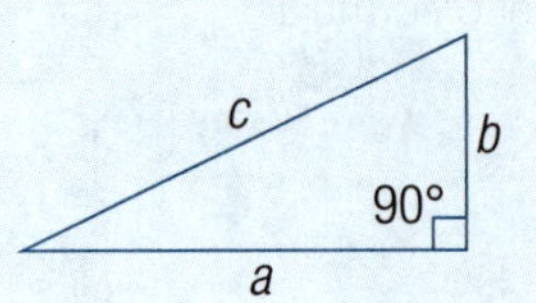

4. A visual verification of the Pythagorean Theorem uses three squares to make a right triangle. We will use squares with side lengths 3, 4, and 5.

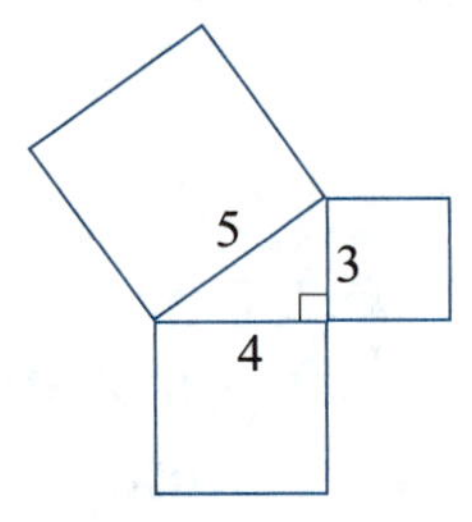

a. Find the area of each square.

b. Write the area of the largest square as the sum of the other two squares.

c. Use the equation from the Pythagorean Theorem to verify that the triangle made by the squares is a right triangle. Show your work.

d. What is the similarity between the work from part **b.** and part **c.**?

Read the following paragraph about Pythagorean triples and work through the problems.

A **Pythagorean triple** is a set of three counting numbers that satisfy the Pythagorean Theorem. One example of a Pythagorean triple, (3, 4, 5), was used in the proof of the Pythagorean Theorem presented in Problem 4. There are several formulas that can be used to create these triples. The next problems explore how to create Pythagorean triples.

5. The easiest method to create more Pythagorean triples is to start with a known Pythagorean triple. In this case, you multiply each number in the triple by the same whole number. For example, since we know (3, 4, 5) is a Pythagorean triple, then $2 \cdot 3$, $2 \cdot 4$, and $2 \cdot 5$, which is (6, 8, 10), is also a Pythagorean Triple.

a. Use the Pythagorean Theorem formula to verify that (6, 8, 10) forms a Pythagorean triple.

b. Use this method to create two more Pythagorean triples.

Name: Date:

Quick Tip

This set of formulas does not produce all of the Pythagorean triples. Different formulas to produce Pythagorean triples can be found on the Internet.

6. One of many sets of formulas to create Pythagorean triples is

$$a = n^2 - m^2$$
$$b = 2nm$$
$$c = n^2 + m^2$$

where n and m are integers and $n > m$.

For example, if we use $n = 2$ and $m = 1$ we get the triple (3, 4, 5). We have already verified that this is a triple.

a. Substitute two integers n and m (that are not 2 and 1), where $n > m$, into the formulas to create a Pythagorean triple.

b. Use the formula of the Pythagorean Theorem to verify that the triple you created in part **a.** is a Pythagorean triple.

Skill Check

Go to Software Work through Practice in Lesson 5.7 of the software before attempting the following exercises.

To calculate a square root using a calculator, press the [√] button and then enter the number you are finding the square root of, followed by the [=] button. Determine the square root of each number to the nearest thousandth.

Quick Tip TECH

Some basic calculators require the number to be entered before the [√] button is pressed. Care must be taken when using calculators so the desired answer is given.

7. 14

8. 72

9. 24

10. 6724

Apply Skills

Quick Tip

The Pythagorean Theorem is a useful tool in mathematics as shown by the application problems in this section.

Work through the problems in this section to apply the skills you have learned related to square roots and the Pythagorean Theorem.

11. A police officer needs to tape off a crime scene. The crime took place in a park that has a fence along one side and a shed near the fence.

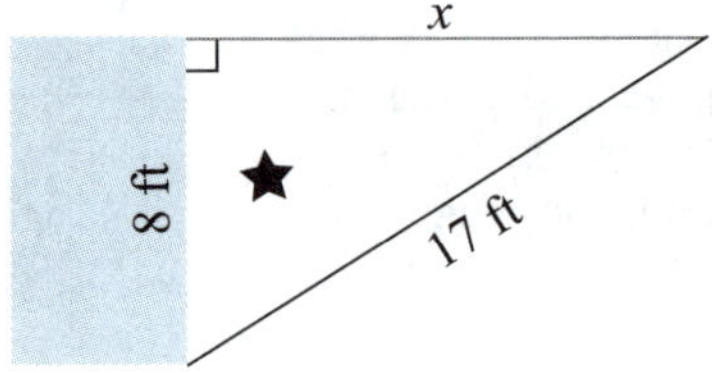

a. The shed is 8 feet long and the fence is 17 feet long. The officer wants to attach the caution tape from the edge of the shed to the end of the fence, labeled with an x in the diagram. How much caution tape does he need?

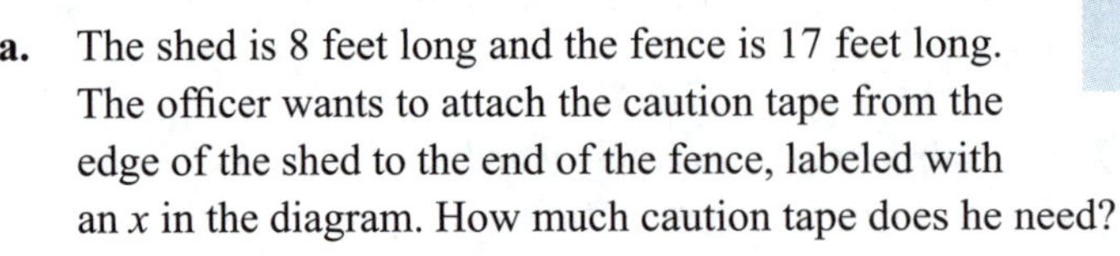

b. What is the area of the taped off crime scene?

12. Aya has a triangular wooden porch attached to the side of her house. The "legs" of the triangular porch measure 12 feet and 16 feet. She is decorating for a party and has 18 feet of party lights.

a. Does she have enough lighting to put along the entire railing of the longest side of the porch?

b. If yes, how many feet of party lights are left over? If no, how many additional feet of party lights are needed?

13. The maximum walking speed of an animal depends on the length of their legs. To calculate the maximum speed an animal can walk (in feet per second), multiply the square root of the animal's leg length, in feet, by 5.66.

a. A giraffe has legs that measure 6 feet in length. What is this giraffe's maximum walking speed to the nearest hundredth?

b. A man has legs that measure 3 feet in length. What is this man's maximum walking speed to the nearest hundredth?

c. Since the giraffe's legs are twice as long as the man's legs, does this mean that the giraffe could walk twice as fast? If not, how did the speeds compare?

Name: Date:

Chapter 5 Projects

Project A: Before and After

An activity to demonstrate the use of geometric concepts in real life

Suppose HGTV came to your home one day and said, "Congratulations, you have just won a FREE makeover for any room in your home! The only catch is that you have to determine the amount of materials needed to do the renovations and keep the budget under $2000." Could you pass up a deal like that? Would you be able to calculate the amount of flooring and paint needed to remodel the room? Remember it's a FREE makeover if you can!

Let's take a room that is rectangular in shape and measures 16 feet 3 inches in width by 18 feet 9 inches in length. The height of the ceiling is 8 feet. The plan is to repaint all the walls and the ceiling and to replace the carpet on the floor with hardwood flooring. For a more sophisticated look, you are also going to install crown molding, which is a decorative type of trim used along the top of a wall where the ceiling and the wall meet.

1. Take the length and width measurements that are in feet and inches and convert them to a fractional number of feet and reduce to lowest terms. (Remember that there are 12 inches in a foot. For example, 12 feet 1 inch is $12\frac{1}{12}$ feet.)

2. Now convert the measurements from Problem 1 into to decimal numbers.

3. Determine the number of square feet of flooring needed to redo the floor. (Express your answer in terms of a decimal and do not round the number.)

4. If the flooring comes in boxes that contain 24 square feet, how many boxes of flooring will be needed? (Remember that the store only sells whole boxes of flooring.)

5. If the flooring you have chosen costs $74.50 per box, how much will the hardwood flooring for the room cost (before sales tax)?

6. Figure out the surface area of the four walls and the ceiling that need to be painted, based on the room's dimensions. (We will ignore any windows, doors, or closets since this is an estimate.)

7. Assume that a gallon of paint covers 350 square feet and you are going to have to paint the walls and the ceilings twice to cover the current paint color. Determine how many gallons of paint you need to paint the room. (Assume that you can only buy whole gallons of paint. Any leftover paint can be used for touch-ups.)

8. If the paint you have chosen costs $18.95 per gallon, calculate the cost of the paint (before sales tax).

9. Determine how many feet of crown molding will be needed to go around the top of the room.

10. The molding comes in 12-foot lengths only. How many 12-foot lengths will you need to buy?

11. If the molding costs $2.49 per linear foot, determine the cost of the molding (before sales tax).

12. Calculate the cost of all the materials for the room makeover (before sales tax).

    a. Were you able to stay within budget for the project?

    b. If so, then what extras could you add? If not, what could you adjust in this renovation to stay within budget?

    c. Using sales tax in your area, calculate the final price of the room makeover with sales tax included.

Name: Date:

Project B: Building a Circular Patio

An activity to demonstrate the use of geometric concepts in real life

Bob has just finished converting the back bedroom of his house to a sunroom, which has dimensions of 11 feet by 11 feet. He has decided that his next project will be to build a concrete patio in a circular shape outside the sunroom to use for his grill.

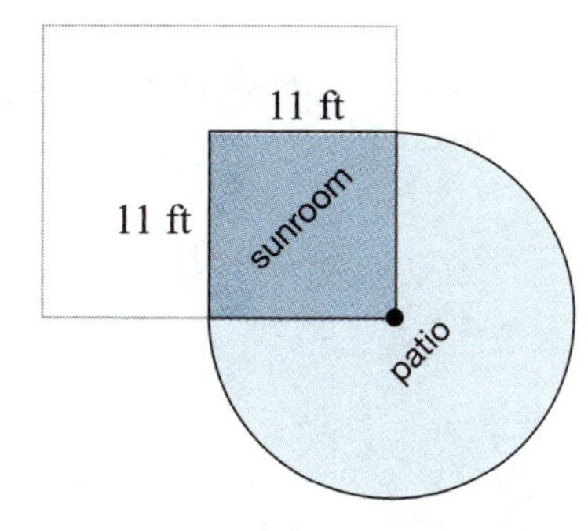

1. What will the radius of the circular patio be?

2. Calculate the area of the patio. Use $\pi = 3.14$ as an approximation and round your answer to the nearest hundredth. (**Hint:** One-fourth of the area of the entire circle is taken up by the sunroom.)

3. If the concrete for the patio is to be 6 inches thick, calculate the volume of concrete needed in **cubic feet**. Use $\pi = 3.14$ as an approximation and round your answer to the nearest hundredth. (**Hint:** The patio is a cylinder with one-fourth of it missing.)

4. Convert the volume result from Problem 3 to cubic yards. Round your answer to the nearest hundredth. (**Hint:** Use the ratio $\left(\frac{1 \text{ yard}}{3 \text{ feet}}\right)^3$ and multiply this by the volume of concrete needed.)

5. The concrete that Bob wants to use costs $75 per cubic yard. Determine the cost of the concrete for the patio (before sales tax). Round your answer to the nearest cent.

6. Bob's wife, Brenda, wants to put a short decorative fence around the patio. How many feet of fencing will be needed. Round your answer to the nearest hundredth.

7. The fencing that Brenda wants to use costs $6.50 per foot. Determine the cost of the fencing (before sales tax). (**Hint:** Brenda can only buy the fencing in one foot segments.)

Name: Date:

Math@Work

Architecture

As a project architect, you will be part of a team that creates detailed drawings of the project that will be used during the construction phase. It will be your job to ensure that the project will meet guidelines given to you by your company, such as square footage requirements and budget constraints. You will also need to meet the design requirements requested by the client.

Suppose you are part of a team that is designing an apartment building. You are given the task to create the floor plan for an apartment unit with two bedrooms and one bathroom. The apartment management company that has contracted your company to do the project has several requirements for this specific apartment unit.

1. One bedroom is the "master bedroom" and must have at least 60 square feet more than the other bedroom.
2. All walls must intersect or touch at 90 degree angles.
3. The kitchen must have an area of no more than 110 square feet.
4. The apartment must be between 1000 square feet and 1050 square feet.

A preliminary sketch of the apartment is shown here.

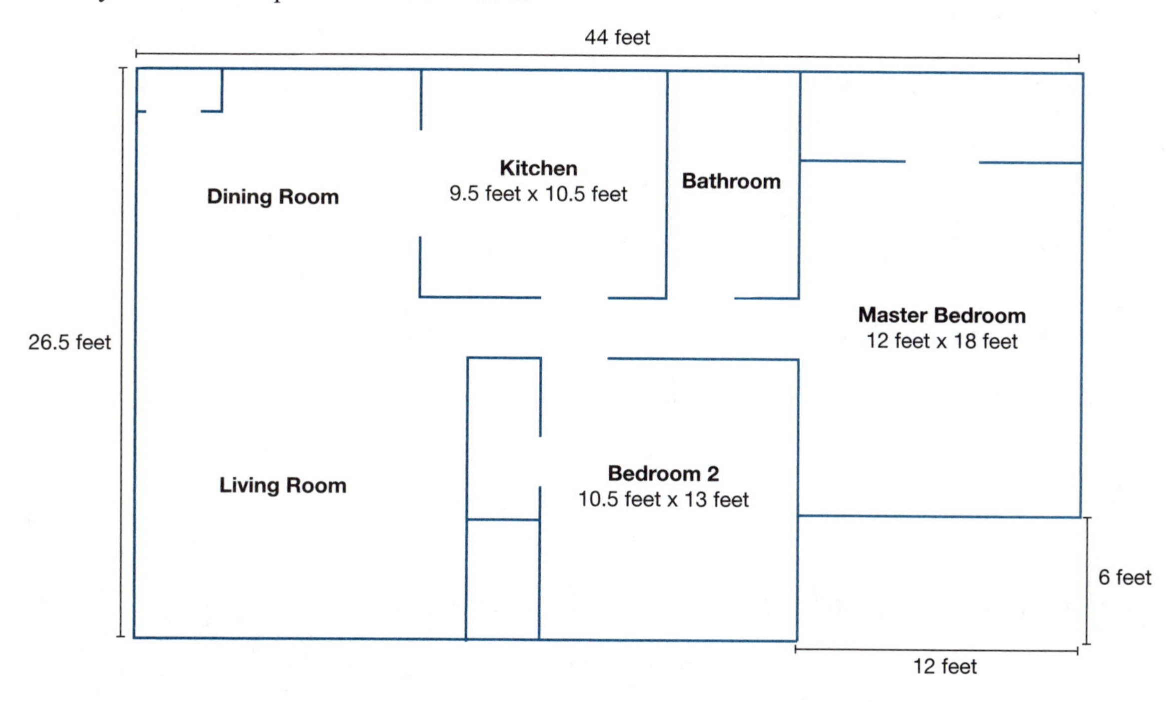

1. Does the apartment have the required total square footage that was requested? Is it over or under the total required?

2. Does the apartment blueprint meet the other requirements given by the client? If not, what does not meet the requirements?

3. For this specific apartment unit, the total construction cost per square foot is estimated to be $75.75. Approximately how much will it cost to construct each two-bedroom apartment based on the floor plan?

Foundations Skill Check for Chapter 6

This page lists several skills covered previously in the book and software that are needed to learn new skills in Chapter 6. To make sure you are prepared to learn these new skills, take the self-test below and determine if any specific skills need to be reviewed.

Each skill includes an easy (**e.**), medium (**m.**), and hard (**h.**) version. You should be able to complete each problem type at each skill level. If you are unable to complete the problems at the easy or medium level, go back to the given lesson in the software and review until you feel confident in your ability. If you are unable to complete the hard problem for a skill, or are able to complete it but with minor errors, a review of the skill may not be necessary. You can wait until the skill is needed in the chapter to decide whether or not you should work through a quick review.

1.4 Find the quotient. Indicate any remainder.

e. $51 \div 3$ **m.** $8844 \div 33$ **h.** $9936 \div 27$

2.1 Reduce the fraction to lowest terms.

e. $\frac{25}{100}$ **m.** $\frac{36}{48}$ **h.** $\frac{51}{85}$

3.4 Find the quotient. Round to the nearest hundredth.

e. $14.5 \div 1000$ **m.** $72 \div 1.2$ **h.** $267 \div 1.2$

3.5 Simplify the expression using the order of operations. Round to the nearest hundredth.

e. $\frac{1}{2} + 1.75$ **m.** $\left(2.35 + 1\frac{2}{5} + 1\frac{1}{2}\right) \div 3$ **h.** $\left(\frac{3}{10}\right)^2 + 2.75\left(\frac{2}{5}\right)$

3.5 Arrange the numbers in order from least to greatest.

e. $0.5, \frac{5}{1}, \frac{1}{5}$ **m.** $\frac{7}{3}, 2.3, 2.33$ **h.** $\frac{4}{25}, 0.15, \frac{9}{50}$

Chapter 6: Statistics, Graphs, and Probability

Study Skills

6.1 Statistics: Mean, Median, Mode, and Range

6.2 Reading Graphs

6.3 Constructing Graphs from Data Sets

6.4 Probability

Chapter 6 Projects

Math@Work

Foundations Skill Check for Chapter 7

Math@Work

Introduction

If you are considering a job as a statistician, there are several areas that you can specialize in, depending on what type of business you desire to work for. Some of the many areas where you may find statisticians employed include biology, government, insurance, manufacturing, and medicine. In a manufacturing company, you could work in research, designing experiments or analyzing data from statistical studies to improve products or make processes more efficient. You could also work in quality control, sampling raw materials to ensure quality or monitoring manufacturing processes to make sure they remain on target and within an allowed range of variation. An actuarial statistician in the insurance industry collects and analyzes large amounts of accident and mortality data to determine insurance premium costs and benefits. A biostatistician designs experiments to answer questions about public health matters such as the relationship of obesity and heart disease. Regardless of the area you choose, accurate analysis of data and clear communication of results are extremely important as a statistician.

Suppose you are a statistician employed in the quality control department of a manufacturing company. There are several questions you'll need to answer on a regular basis. Do the parts produced by a certain machine process meet specifications? Is a particular machine producing parts that are on target? Is the variation of the parts produced within the accepted range? Does the machine process need adjustment to keep from producing unacceptable parts? Finding the answers to these questions requires several of the skills introduced in this chapter and previous chapters. At the end of this chapter, we'll come back to this topic and explain some basic analysis techniques used by statisticians to summarize and interpret data.

Study Skills

Reading, Analyzing, and Interpreting Graphs

An important skill for today's society and workplace is the ability to analyze and process information. With the invention of the Internet, people are bombarded with large amounts of information on a daily basis. How do you make sense of all the data that you come in contact with every day? One way to quickly summarize information is to use a chart or graph. Many savvy advertisers use graphics to convey their message and get people's attention. Graphs can also be used to misrepresent information, so being able to analyze and correctly interpret data in a graph is an important skill.

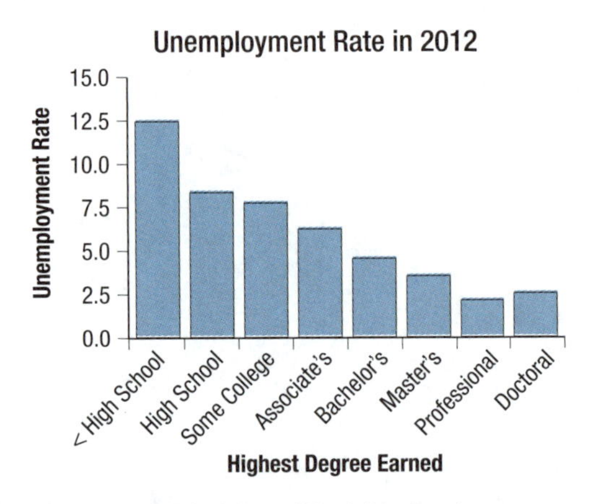

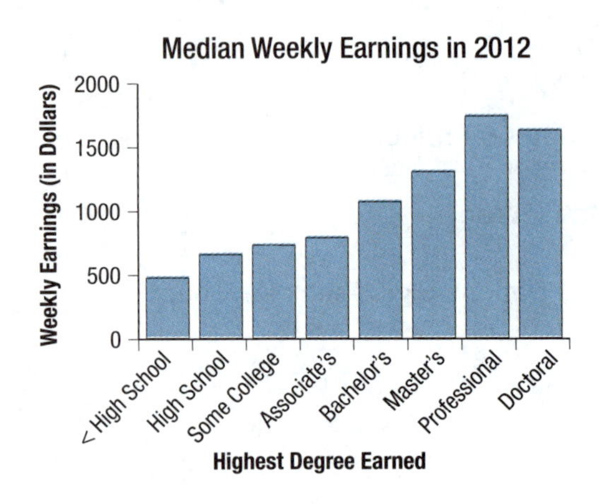

Source: US Bureau of Labor Statistics and The College Board.

Graphs that are not properly created can be misleading, so here are some important points to keep in mind when reading information from a graph. As you read through the list below, see if you can answer the questions as they relate to the bar graphs above on earnings and employment status based on education level.

1. Read the title of the graph to be sure you understand the data being displayed (and what is not displayed).
2. Be sure to read the axis labels carefully. Sometimes the units of the graph are scaled in thousands or percent.
3. Review the legend or key (if there is one) which describes the different types or categories of data that are displayed.
4. Is there a trend or pattern in the data? Does the trend or pattern make sense? Sometimes the scaling of a graph can exaggerate or minimize a trend depending on the author's purpose.
5. If more than one set of data is displayed, how do they compare? Graphs such as double bar graphs or line graphs can display more than one data set on the same graph. If the data appear on separate graphs and a comparison of the data is being made, make sure the scales on the axes are the same.
6. Don't forget to look at the source of the information which is generally displayed below the graph in small print. Evaluate the reliability of the source and whether the data is current.
7. There is often other information displayed at the bottom of the graph that indicates how the data was collected, sources of error, data that may have been omitted, etc.

Based on the graphs above, how would you answer the following questions?

1. As education level increases, what trend do you observe in the unemployment rate? (Is unemployment increasing or decreasing?)
2. As education level increases, what trend do you observe in median weekly earnings? (Are median weekly earnings increasing or decreasing?)
3. What do you think is the most important information these graphs are trying to get across to readers?

Name: Date:

6.1 Statistics: Mean, Median, Mode, and Range

Objectives

Recognize common statistical terms.

Calculate mean, median, mode, and range.

Success Strategy

Understanding statistics is important in today's society because data is collected on almost any imaginable topic. For extra practice, try to find some videos on the statistics mentioned in this section by searching the Internet.

Understand Concepts

Quick Tip

Quality control is a process used to maintain standards in services and manufacturing of products. Quality control involves random inspections of goods and services to make sure they meet requirements.

Go to Software First, read through Learn in Lesson 6.1 of the software. Then, work through the exercises in this section to expand your understanding of the concepts related to statistics.

Statistics is the study of gathering, organizing, and interpreting numerical information. This information comes in the form of data, which are values measuring some characteristic of interest, such as weight, cost, time, etc. This collection of data can represent the entire population of items or just a sample, or part, of the population. Data is typically presented as a **set**, which is a list or collection of values or numbers that can contain repeated values. Statistics is important in the quality control process for businesses, in the analysis of data in the sciences, and the communication of research results.

Range

The **range** of a data set is the difference between the largest and smallest data item.

1. The range of the data set shows how close together or how spread apart the data points in a data set are. The smaller the range, the closer the data points are grouped together. The larger the range, the more the data points are spread apart.

a. Find the range of the data set: 1, 9, 7, 5, 12, 4.

b. Find the range of the data set: 3, 5, 3, 4, 5, 2, 4, 2.

c. Find the range of the data set: 1, 9, 15, 4, 28, 7, 32.

d. Which data set has the data points closest together?

e. Which data set has the data points spread apart the most?

Quick Tip

In statistics, the mean, median, and mode are all considered to be types of averages (also known as measures of center). In general use, people tend to only refer to the mean as the average. So, be sure what is meant when reading about an average.

Measures of Center

The **mean** of a data set is the sum of all the data divided by the number of data items.

The **median** of a data set is the middle data item, or average of the two middle data items, when all the data is ordered from least to greatest.

The **mode** of a data set is the data item which appears the greatest number of times.

2. In Problems 1 through 3 of Section 1.7, we explored what is meant when the mean is referred to as the middle, or center, of a set of numbers. What conclusion did you reach for the mean being a type of middle, or center, number?

Finding the Median of a Data Set

1. Order the data items from least to greatest.
2. If there is an odd number of data items, then the median is the middle data item.
3. If there is an even number of data items, the median is the mean of the two middle data items.

3. Let's explore what the median is.

a. Find the median of the data set: 6, 1, 3, 4, 7.

b. Find the median of the data set: 7, 1, 11, 3.

c. Is the following statement true or false? If it is false, rewrite the statement so that it is true.

The median of a data set is always a number in the data set.

d. How do the medians found in parts **a.** and **b.** compare to the middle number of the range of each data set. (**Note:** To find the middle number of the range of a set of data, find the mean of the smallest and largest data items.)

e. When you are told that the median is a measure of center, what do you think this means?

Name: Date:

Quick Tip

A data set can have no mode, one mode, or multiple modes.

Finding the Mode of a Data Set

1. Order the data items from least to greatest.
2. Determine which data item appears the greatest number of times. If all data items appear the same number of times, there is no mode.

4. Next, we'll explore what the mode is.

a. Find the mode of the data set: 2, 4, 6, 2, 5, 7.

b. Find the mode of the data set: 3, 2, 5, 3, 5, 7, 9, 5.

c. Is the following statement true or false? If it is false, rewrite the statement so that it is true.

If a data set has a mode, then the mode is always a number in the data set.

d. Compare the modes found in parts **a.** and **b.** to the middle number of the range of the data set. (**Note:** To find the middle number of the range of a set of data, find the mean of the smallest and largest data item.)

e. When you are told that the mode is a measure of center, what do you think this means?

Lesson Link

There is also another important measure of data called the **standard deviation**. This measures the spread of the set of data around the mean of the data set. Standard deviation is covered in Section A.10.

5. The range, mean, median, and mode work together to describe the data set. Answer the following questions by using the data set: 12, 7, 11, 13, 7, 9, 18, 8, 10, 5.

a. What is the range of the data set?

b. What is the mean of the data set?

c. What is the median of the data set?

d. What is the mode of the data set?

> ***Quick Tip***
>
> Consider how the mean, median, and mode compare. Think about what information they give about the data set.

e. Plot the data items from the set along with the mean, median, and mode on the number line. Be sure to label which values are the mean, median, and mode.

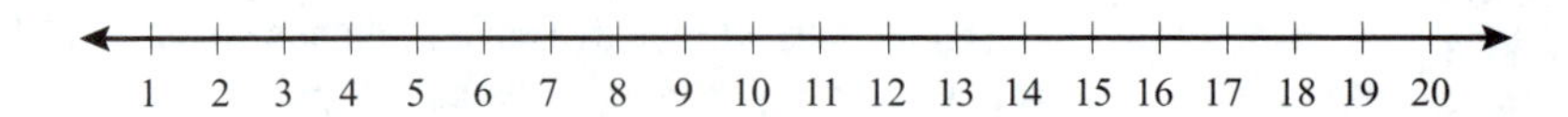

Read the following paragraph about weighted means, then work through the problems.

In calculating a typical mean, you find the sum of all the values in the data set and divide by the number of values or data items you have. In this case, all data items are equally important and carry the same weight. In some situations, certain data items have more weight than others. A common example of weighted means occurs when instructors assign grades. Typically quizzes and tests will carry more weight in determining your grade than homework or class work does.

> ***Quick Tip***
>
> The total sum of the weights will not always add up to one, which is why Step 4 is important.

Finding the Weighted Mean of a Data Set

1. Change the weights for each category from a percent to a decimal, if necessary.
2. Multiply the data items in each category by the weight assigned to that category.
3. Find the sum of the products from Step 2.
4. Divide the sum from Step 3 by the total sum of the weights in decimal number form.

> ***Quick Tip***
>
> In Problem 6, the categories are the grade items.

6. Suppose your instructor handed out a syllabus with the following grade distribution: exams: 40%, quizzes: 30%, homework: 15%, classwork: 15%. After Chapter 1, you have the following scores: 82 on an exam, 89 on a quiz, 96 on homework, and 97 on classwork.

a. Fill in the following table. Round answers to the nearest tenth, if necessary.

Grade Item	Weight	Chapter 1 Score	Weight · Score
Exams			
Quizzes			
Homework			
Classwork			
Totals:			

b. Use the data from the table to calculate the weighted mean of the Chapter 1 scores.

c. Find the typical mean of the Chapter 1 scores as if each grade item has the same weight.

d. Compare the typical mean of the scores from Chapter 1 with the weighted mean of the scores from Chapter 1. Which mean do you think is a better representation of a student's level of understanding? Explain why you chose this mean.

Name: ______________________ Date: ______________________

Skill Check

Go to Software Work through Practice in Lesson 6.1 of the software before attempting the following exercises.

Find the mean, median, mode, and range of each data set.

7. 75, 83, 93, 65, 85, 85, 88, 90, 55, 71

8. 14.3, 13.6, 10.5, 15.5, 20.1, 10.9, 12.4, 25.0, 30.2, 32.5

Apply Skills

Work through the exercises in this section to apply the skills you have learned related to statistics.

Quick Tip

To improve accuracy when calculating the mean of a data set, always carry one more decimal place in the calculations and final result than the original data. So, a data set with only integers will have a more accurate mean if it is rounded to the nearest tenth.

9. A math class performed an experiment with fun-size packs of M&Ms. One step of the experiment was to determine how many blue M&Ms each student had in their pack of M&Ms. One group of students came up with the data set: 5, 1, 4, 3, 1, 2, 1, 3, 4, 2.

a. Find the mean, median, mode, and range for the given data set. Round to the nearest tenth if necessary.

b. How do the three measures of center (mean, median, and mode) compare to each other?

c. Based on this analysis, how many M&Ms would you expect to get in each pack?

10. The final grades for students during a summer session of a math class are given in the data set 96, 97, 60, 75, 96, 95, 92, 98, 96, 100, 91.

a. Find the mean, median, mode, and range for the given data set. Round to the nearest tenth if necessary.

b. How do the three measures of center (mean, median, and mode) compare to each other?

c. Based on this analysis, what numerical grade would you expect a student in this class to earn?

11. Final grades for an economics class are based on homework, quizzes, and exam scores. The homework counts as 20% of the final grade, the quizzes count as 30% of the final grade, and the exams count for 50% of the final grade.

a. By midterm, a student has the following average grades: homework: 75, quizzes: 95, exams: 75. Determine the student's current average grade by calculating the weighted mean.

b. Another student has the following average grades: homework: 95, quizzes: 75, exams: 75. Determine the student's current average grade by calculating the weighted mean.

c. A third student has the following average grades: homework: 75, quizzes: 75, exams: 95. Determine the student's current average grade by calculating the weighted mean.

d. Compare the grades of the three students. Why do you think the students' grades differ in the way they do even though the students have an average of 75 in two categories and 95 in another?

12. Lyle works for a food distributor and needs to determine the current market price for various items. After calling around to several local grain elevators, he compiles the following data sets. Find the mean, median, and mode for each type of crop.

a. Prices per bushel for wheat: $6.28, $6.25, $6.40, $6.35, $6.29, $6.30

b. Prices per bushel for soybeans: $16.30, $16.25, $16.35, $16.50, $16.30, $16.30

c. Prices per bushel for corn: $6.41, $6.71, $6.76, $6.93, $6.90, $6.76, $6.20

d. Based on the calculations from parts **a.**, **b.**, and **c.**, how much should Lyle charge for each item? Explain why you chose to use the mean, median, or mode as the price.

Name: Date:

6.2 Reading Graphs

Objectives

Learn the purposes and properties of graphs.

Read bar graphs.

Read circle graphs.

Read line graphs.

Read histograms.

Read pictographs.

Success Strategy

If you read news articles in print or online, you will likely find that graphs are often used to summarize data. After working through this section, find a graph in an article and see if you can find anything incorrect with the graph or point out what you would have done differently to make the presentation of the data easier to understand.

Understand Concepts

Go to Software First, read through Learn in Lesson 6.2 of the software. Then, work through the exercises in this section to expand your understanding of the concepts related to reading graphs.

Graphs are a visual way to display data. Graphs are seen in various forms in our daily lives, from media to school to careers. Understanding how to read and interpret the data in a graph are important skills. Graphs often describe or summarize data very quickly and with fewer words than if you described the data using sentences. As the saying goes, "A picture is worth a thousand words."

Quick Tip

Circle graphs are commonly called **pie graphs** or **pie charts**.

Four Common Types of Graphs

Type of Graph		Definition
Circle Graph		A **circle graph** is used to compare parts of a whole or percents.
Bar Graph		A **bar graph** is used to emphasize comparative amounts.
Histogram		A **histogram** is used to indicate the frequency of data points in classes, or intervals.
Line Graph		A **line graph** is used to indicate trends over time.

Properties of Graphs

Well-made graphs should have the following three properties.

1. Graphs should be clearly labeled.
2. Graphs should be easy to read.
3. Graphs should have appropriate titles.

1. Bar graphs are used to compare amounts among different categories. The amount of money a movie makes at the box office is commonly used to determine how successful the movie was. The bar graph here shows the amount of money earned by the top-earning movies from May 2013. Source: Information courtesy of Box Office Mojo. Used with permission. http://www.boxofficemojo.com

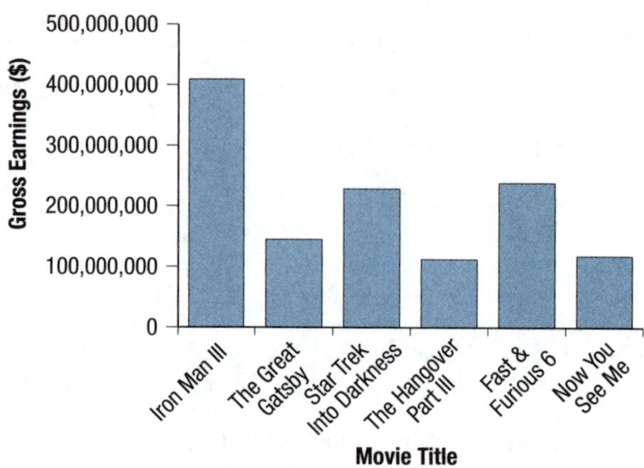

a. What does the vertical axis represent?

b. What does the horizontal axis represent?

c. How can you tell which movie from the graph had the highest gross earnings?

2. Circle graphs are used to compare parts of a whole. The total tuition you pay to attend college is made up of base tuition and various fees. A representation of the tuition cost to attend the College of Charleston for the 2013–2014 school year is shown here. The estimated cost for the full year for non-residential students is $38,131. Source: http://admissions.cofc.edu

College of Charleston Tuition Schedule, 2013–2014

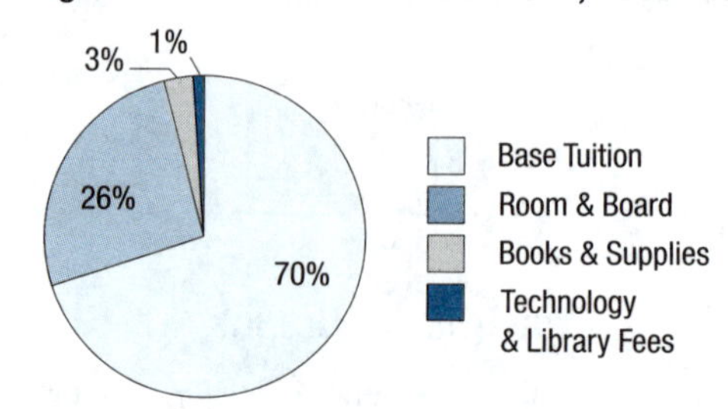

a. The sum of all of the percents given in the circle graph must add up to 100%. Verify this is true for the given circle graph.

b. Does the circle graph represent how the total tuition is divided or just part of the tuition?

c. How many categories is the data divided into?

3. A line graph is a graph that uses points that are connected by lines or line segments to show how an amount changes in value as time goes by. A line graph showing the change in yearly average tuition price at public 4-year colleges is shown here. The amounts are based on the value of the dollar during the 2009–2010 academic year. Source: US Department of Education. Institute of Education Sciences, National Center for Education Statistics

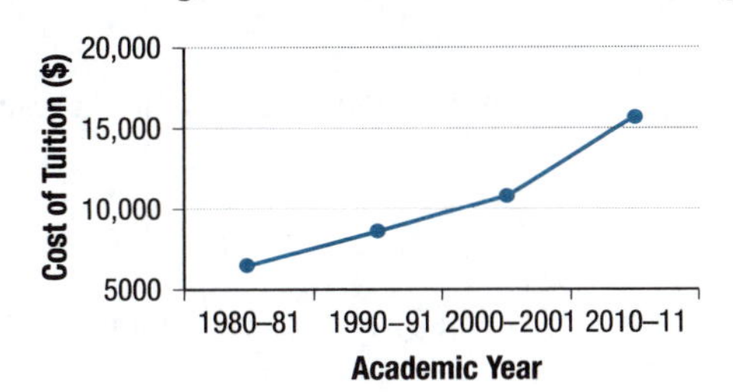

a. What does the vertical axis represent?

b. What does the horizontal axis represent?

c. If a trend is present in the data, it can be seen in the line graph. Is there a trend in this data? If yes, describe the trend.

Read the following paragraph about histograms and work through the problems.

The only one of the four common types of graphs that you haven't seen yet in this workbook is the **histogram**. The histogram is a common type of graph in statistics used to summarize large data sets where the data are typically from measurements. This type of graph is commonly called a **frequency histogram** since it illustrates the frequency at which data fall within certain ranges.

Bar graphs are often used with data that fall into categories and have repeated values, whereas histograms are often used for measurement data that don't repeat often and, therefore, you have to use intervals, or bins, to collect the data.

Histogram Key Terms

A **class** is an interval (or range) of values that contains data items.

A **class boundary** is the value that is halfway between the largest member of one class and the smallest member of the next class.

The **class width** is the difference between the class boundaries of a class.

The **frequency** of a class is the number of data items in that class.

Quick Tip

To help understand how the size of class intervals affect frequency and the appearance of a histogram, try experimenting with the histograms found here: http://www.shodor.org/interactivate/activities/Histogram/

4. A group of randomly selected people were asked to report the size of their weekly income. The results are reported in the histogram shown here.

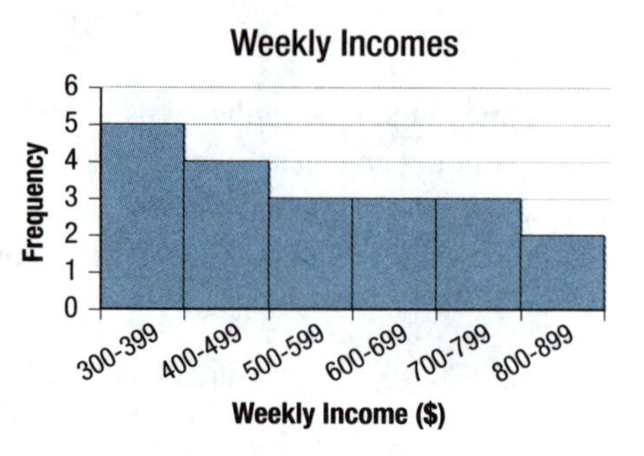

a. What does the vertical axis represent?

b. How many classes of income are there?

c. What is the class width?

5. State which graph type would be best suited to display the data for each situation.

a. Inventory amounts of six different items currently in stock.

b. The amount of income a company had at the end of each month for a year.

c. The weights of 100 gala apples grown in the same apple orchard.

d. The division of your paycheck between federal income tax, state income tax, health benefits, and take-home pay.

Read the following paragraph about pictographs and work through the problems.

A **pictograph** is a graph which uses images or icons to stand for a set amount of items. Pictographs have a similar style to bar graphs except instead of bars they use images to represent the amount in each category. As with bar graphs, the size of the icons used in a pictograph should be the same for each column. If the icons are different sizes, the representation of the data could be misleading.

Name: Date:

6. The pictograph shown here displays the approximate number of copies of Bethesda Softworks' The Elder Scrolls V: Skyrim video game sold across three different gaming platforms.
Source: http://www.statisticbrain.com

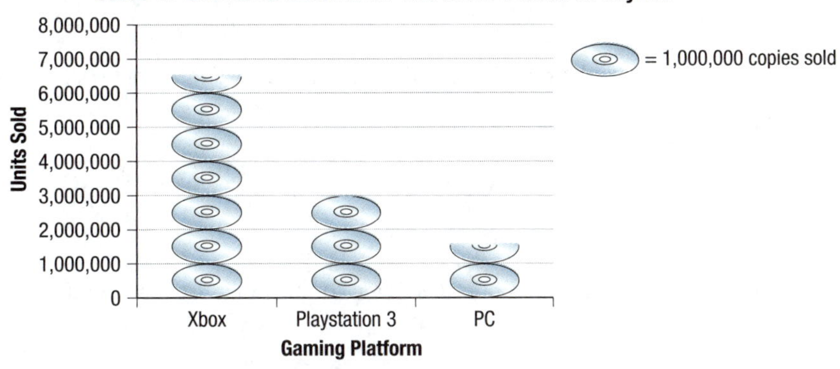

a. How many copies does each DVD symbol represent?

b. What does half of the DVD symbol represent?

c. Approximately how many copies of Skyrim were sold for Playstation 3?

d. What are some limitations of displaying data with a pictograph?

Skill Check

Go to Software Work through Practice in Lesson 6.2 of the software before attempting the following exercises.

A monthly income of \$2800 is budgeted as shown. Find the amount of money budgeted for the given category.

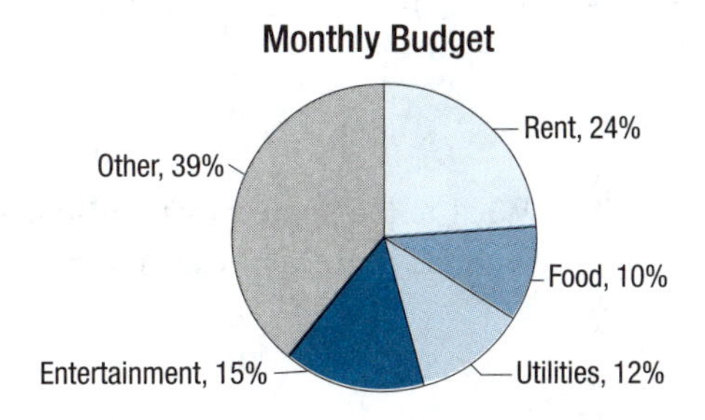

7. Amount budgeted for rent each month.

8. Amount budgeted for other expenses each month.

9. Amount budgeted for food and entertainment each month.

10. Total amount budgeted for all categories excluding other expenses.

Apply Skills

Work through the exercises in this section to apply the skills you have learned related to reading graphs.

11. Barbara's Bombtastic Bakery provides nutrition information for the different baked goods sold. A small poster on the wall displays the number of calories per serving of different cake flavors available for custom cakes (not including frosting).

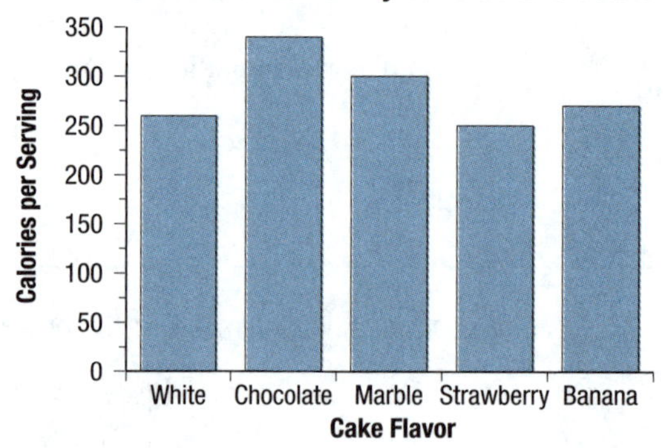

a. Which flavor of cake has the most calories per serving?

Quick Tip

To find the approximate value of a bar in a bar graph, draw a horizontal line from the bar to the vertical axis. Then, approximate the value of the intersection of the line with the vertical axis.

b. What is the approximate difference in number of calories per serving between the cake with the most calories per serving and the cake with the fewest calories per serving?

c. If the bar graph was adjusted to include the additional 80 calories per serving of frosting for each flavor of cake, would that change which flavor of cake has the most calories per serving?

Name: Date:

Quick Tip

Two or more data sets are often displayed on the same graph, as is done in Problem 12, to make it easier to compare them.

12. The line graph shows the highest and lowest recorded temperatures in Houston, TX, for the first six months of 2013. Use this data to answer the following questions. Source: weather.gov

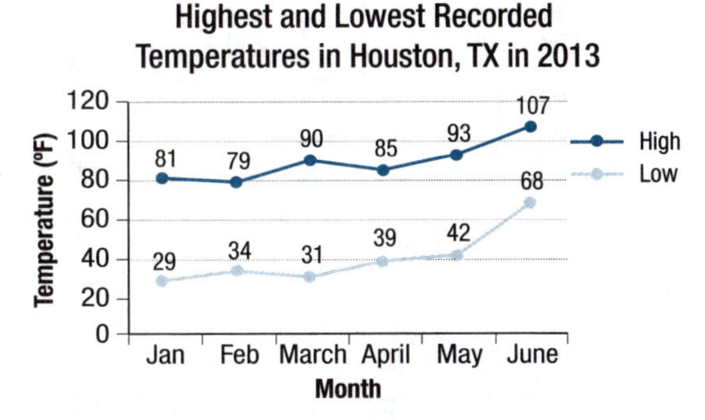

a. During which month was the difference between the highest and lowest temperatures the greatest?

b. What trend do you notice about each data line?

c. Between which two months did the high temperature decrease the most?

d. Between which two months did the low temperature increase the most?

13. The circle graph shows the sources for calories in one serving of a Greek style yogurt. Source: Nutritiondata.self.com

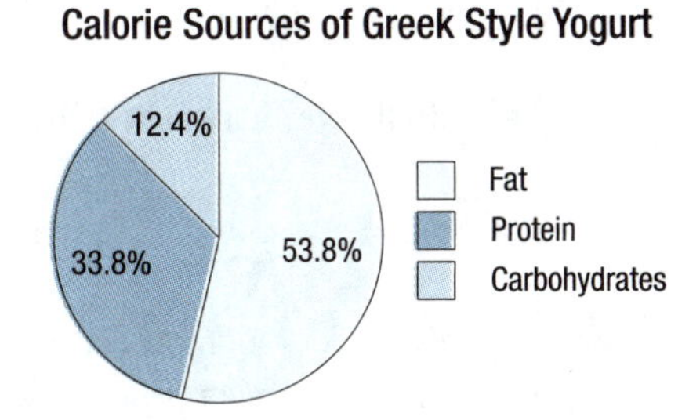

a. There are 130 calories in a serving of Greek style yogurt. How many calories come from carbohydrates?

b. What source provides the most calories per serving?

c. How many total calories come from protein and carbohydrates?

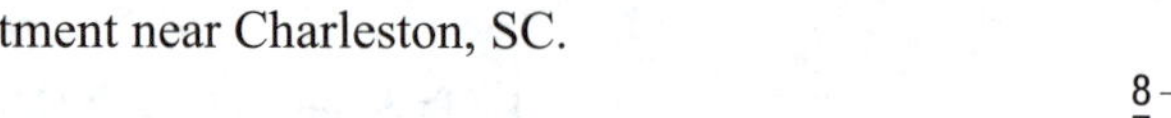

14. The histogram shows prices for a one-bedroom apartment near Charleston, SC.

a. How many classes are represented?

b. What is the width of each class?

c. How many apartments total are in the data set?

d. Which class has the highest frequency?

Quick Tip

Problem 15 will hopefully alert you to being cautious when reading and interpreting graphs, as the creator may sometimes be trying to make you reach a particular conclusion by presenting the data in a biased way.

15. Sometimes the scale of a graph can affect your interpretation of the data presented by the graph. The following data on the monthly unemployment rate in the United States taken from the Bureau of Labor Statistics website from September 2013 to February 2014 is as follows: 7.2, 7.2, 7.0, 6.7, 6.6, and 6.7.

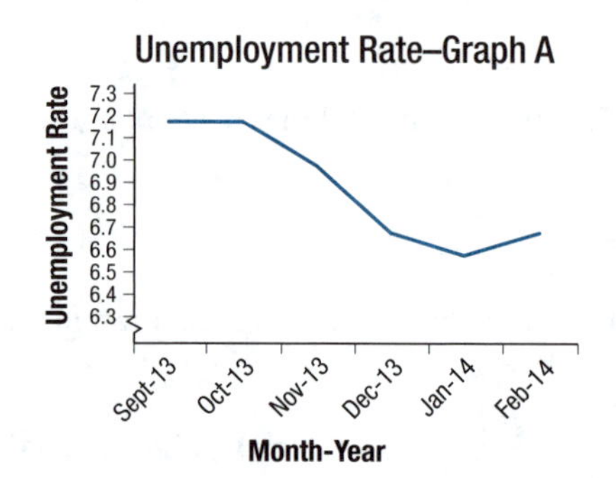

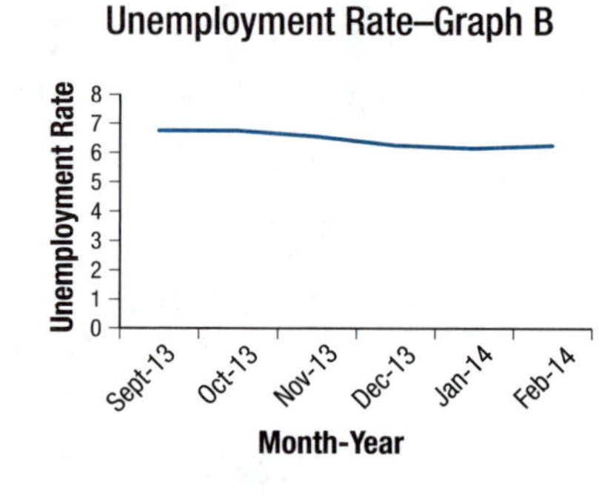

a. How do the vertical scales differ between the two graphs?

b. If you wanted to demonstrate how the unemployment rate has dropped significantly under the current administration, which graph would you choose to show?

c. If you wanted to demonstrate how the current administration has not significantly reduced the unemployment rate, which graph would you choose to show?

Quick Tip

When reading and analyzing graphs, be sure to look at all parts of the graph, particularly the scales on both axes. Make sure that the conclusions drawn from the graph make practical sense or do some additional research to find other data that confirm the findings.

Name: Date:

6.3 Constructing Graphs from Data Sets

Objectives

Organize and represent given data in the form of a bar graph.

Organize and represent given data in the form of a circle graph.

Success Strategy

Be sure to do some practice constructing graphs by hand and also using available technology such as a graphing calculator or Excel software.

Understand Concepts

Go to Software First, read through Learn in Lesson 6.3 of the software. Then, work through the exercises in this section to expand your understanding of the concepts related to constructing graphs from databases.

If you ever need to present data, whether it is for a class or on the job, creating graphs is an efficient way to display the data. Two common forms of displaying data are bar graphs and circle graphs. These graphs can be made by hand or created with software, such as Microsoft Excel. Both methods will be discussed in this section.

Steps to Create a Vertical Bar Graph

1. Draw the vertical and horizontal axes and label them.
2. Determine the scale on the vertical axis to represent the frequency of each category and mark it in equal intervals.
3. Mark the categories along the horizontal axis at equal distances.
4. Draw vertical bars (with equal widths) for each category where the height of the bar is equal to the frequency of the data in that category.
5. Verify that the bars have equal widths and do not touch each other.

Quick Tip

All graphs should have a title that describes the data being presented.

1. A bar graph was created using the following data set.

Calories per 12-oz Serving of Soft Drink	
Soft Drink	**Calories**
Coca-Cola	140
Pepsi	150
Sprite	148
Sierra Mist	140

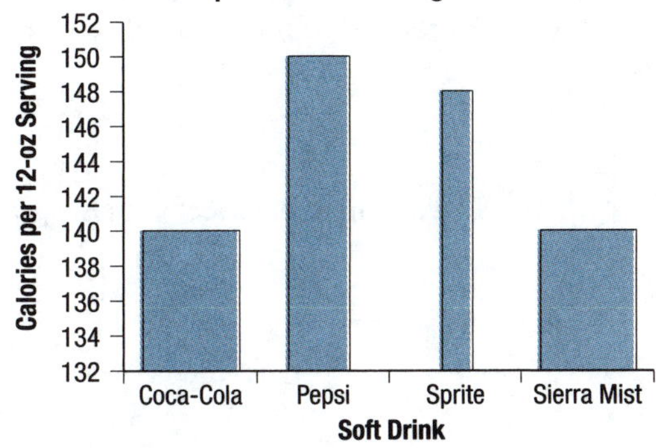

a. Is the title of the graph descriptive enough for anyone to understand what the graph represents? If not, create a new title for the graph.

b. List anything incorrect with the graph.

Quick Tip

The percent used in Step 1 should be in decimal form.

Steps to Create a Circle Graph

1. Multiply each data percent by 360° (the number of degrees in an entire circle) to determine the size of the angle that represents the category.
2. Draw a circle.
3. Use a protractor to divide the circle into wedges using the angle values from Step 1.
4. Label the wedges with the name of the category and the percent.

2. A circle graph was made from the following data set.

Calorie Sources of Dry Roasted Peanuts		
Source	**Calories**	**Percent of Total**
Carbohydrates	127	15%
Fat	607	71%
Protein	120	14%
Total	854	100%

Calorie Sources

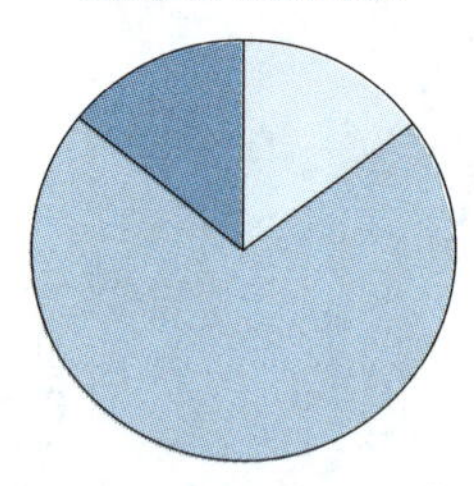

a. Is the title of the graph descriptive enough for anyone to understand what the graph represents? If not, create a new title for the graph.

b. List anything incorrect with the graph.

Name: Date:

The following Excel directions will be used to create a graph of the data set that represents the tuition schedule from Problem 2 in Section 6.2.

Tuition Schedule for the 2013-2014 School Year	
Tuition	$26,694
Room & Board	$9856
Books & Supplies	$1186
Technology & Library Fees	$395

Excel Directions for Creating Graphs

Quick Tip TECH

The Excel directions given here are based on Excel 2010. The steps may vary depending on the version of Excel being used.

1. Enter the categories for the data in cells A1 through A4.
2. Enter the amounts per category into cells B1 through B4

	A	B
1	Tuition	$26,694
2	Room & Board	$9,856
3	Books & Supplies	$1,186
4	Technology & Library Fees	$395

3. Highlight the data cells with the information entered in steps 1 and 2. Click on the Insert tab in the tool bar.

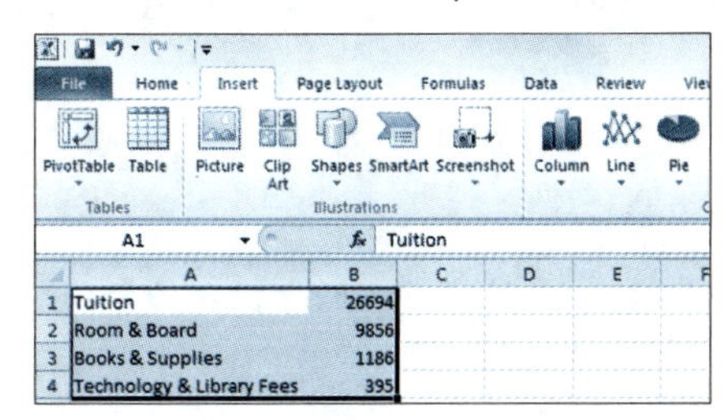

a. To create a bar graph, click the Column button. Then click the first 2-D Column option.

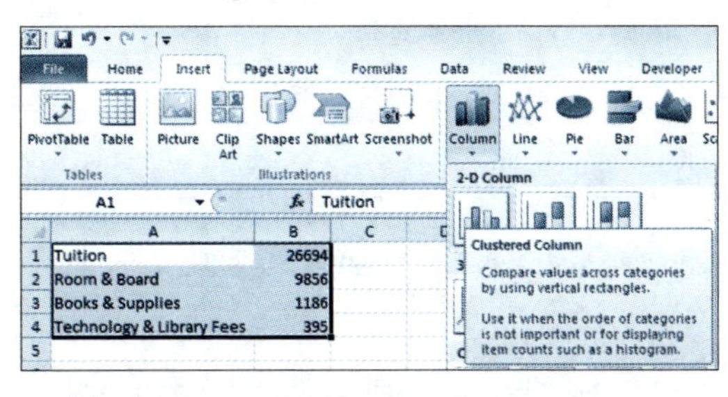

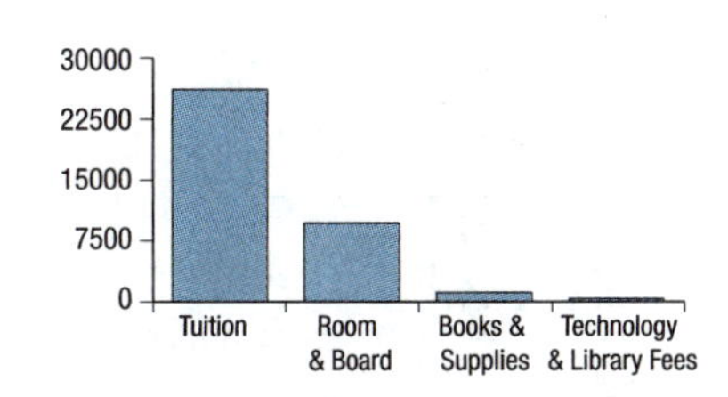

b. To create a circle graph, click the Pie button. Then click the first 2-D Pie option.

Quick Tip TECH

Excel will calculate the percents from the original data entered into the spreadsheet. If the final circle graph should display percents, the percent values should be entered and not the original data.

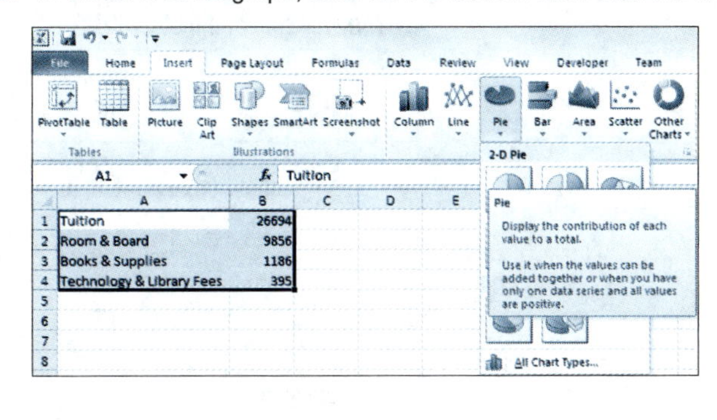

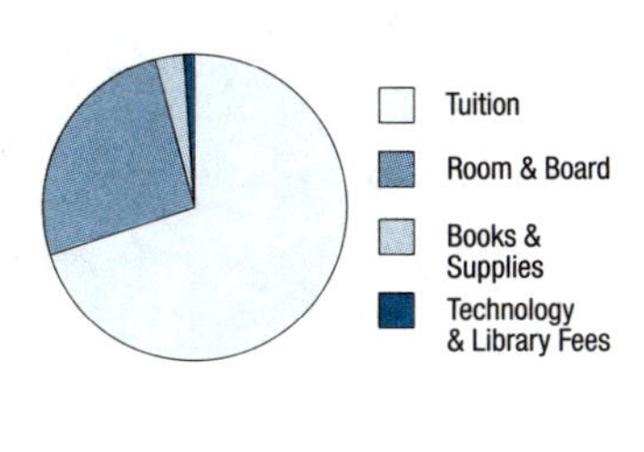

Quick Tip TECH

To find more tips and tricks for using Excel to make graphs, perform an Internet search with the key words "Microsoft Excel graphs". Be sure to pay attention to which version of Excel the tips and tricks are for.

4. To add a title and other labels to your chart, click on the Layout tab under Chart Tools.

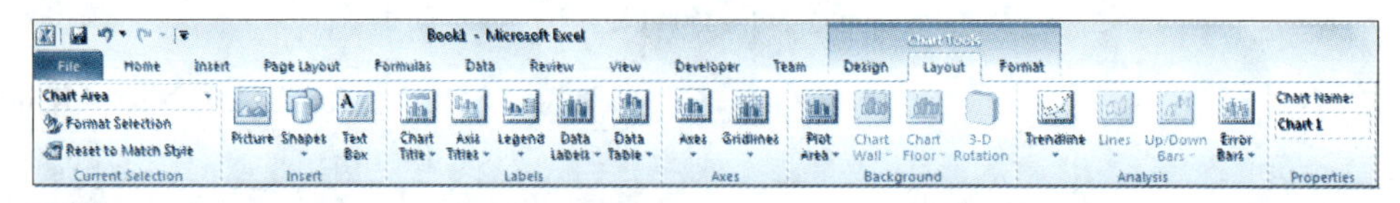

a. Chart Title allows you to add a title to the chart.

b. Axis Titles allows you to add titles to the axes of a bar graph.

c. Legend allows you to identify the pattern or color assigned to a data category.

d. Data Labels allows you to add more details to the labels.

Skill Check

Go to Software Work through Practice in Lesson 6.3 of the software before attempting the following exercises.

Find the given percent of 360°.

3. 15%

4. 25%

5. 55%

6. 74%

Apply Skills

Work through the exercises in this section to apply the skills you have learned related to constructing graphs from databases.

7. Around the world, the mean age when women give birth to their first child varies.

Mean Age for First Child Birth	
Country	**Mean Age**
United States	25.0
Canada	27.6
Germany	28.9
France	28.6
Japan	29.4
India	19.9
South Africa	22.2
Latvia	26.4
Source: Central Intelligence Agency	

a. Should you use a bar graph or a circle graph to represent this data? Explain why.

b. Create a graph based on the data.

8. In the United States, the following languages are spoken by the indicated percent of the population as their first language.

First Language of US Population	
Language	**Percent**
English	82.1%
Spanish	10.7%
Other Indo-European	3.8%
Asian and Pacific Island	2.7%
Other	0.7%
Source: Central Intelligence Agency	

a. Should you use a bar graph or a circle graph to represent this data? Explain why.

b. Create a graph based on the data.

9. In the United States, citizens can be broken down into the following age groups.

US Population by Age	
Age Range	**Percent of Population**
0–14 years	20%
15–24 years	13.7%
25–54 years	40.2%
55–64 years	12.3%
65 years and over	13.8%
Source: Central Intelligence Agency	

a. Should you use a bar graph or a circle graph to represent this data? Explain why.

b. Create a graph based on the data.

Name: Date:

6.4 Probability

Objectives

Understand the terminology related to probability.

Determine the sample space of an experiment.

Calculate the probability of an event.

Success Strategy

Many statistics courses and quantitative reasoning courses have a lot of topics in them that will require you to understand and be able to compute probabilities. Probabilities are typically expressed as fractions or percents, so you may want to briefly review these topics before starting this section.

Understand Concepts

Go to Software First, read through Learn in Lesson 6.4 of the software. Then, work through the exercises in this section to expand your understanding of the concepts related to probability.

The type of probability introduced in this section is called **theoretical probability** because you can count all of the possible outcomes without performing an experiment and determine the exact probability of each outcome. **Experimental probability** uses data obtained from performing experiments. The probabilities found from experiments are generally not exact. In order for the theoretical and experimental probabilities to be the same, you would need to perform the experiment an extremely large number of times.

Probability Key Terms

An **experiment** is an activity in which the result is random in nature.

An **outcome** is an individual result of an experiment.

A **sample space** is the set of all possible outcomes of an experiment.

An **event** is a set of some (or all) of the outcomes from a sample space.

Quick Tip

The **sample space** is the set of all possible outcomes of an experiment, not the number of possible outcomes.

1. Experiments can be repeated any number of times, but the results are usually a limited set of values. For example, the experiment of rolling a standard die can be repeated an unlimited number of times but there are only six outcomes in the sample space: 1, 2, 3, 4, 5, 6. Think of two more experiments and give the sample space for each.

Read the following paragraph about tree diagrams and work through the problems.

When determining the sample space of a probability experiment, it can be helpful to draw a tree diagram. Suppose you are performing an experiment that involves tossing a fair coin three times.

The sample space of tossing a single coin can be represented by the tree diagram shown here.

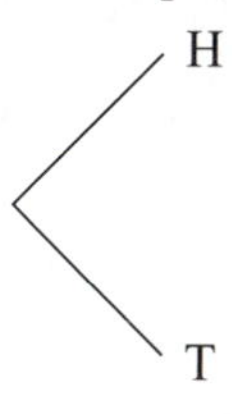

In this tree diagram, H stands for heads and T stands for tails. Since there are only two possible outcomes when tossing a coin, the tree diagram only has two branches.

2. To add another step to the experiment, another tree diagram is created at the end of **each** branch.

a. Complete the tree diagram to show the sample space for tossing a fair coin twice. One branch has been filled in for you.

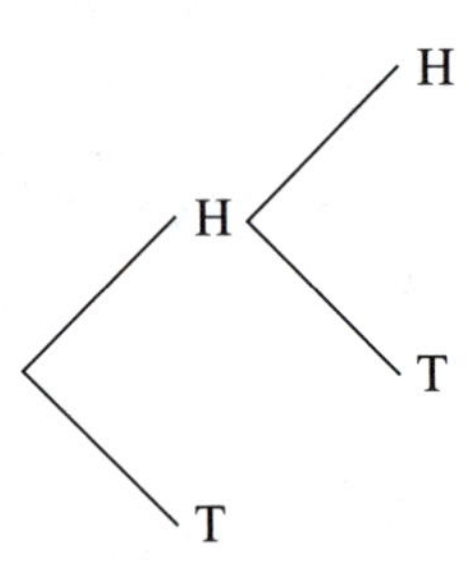

b. Draw a complete tree diagram for an experiment where a coin is tossed three times.

3. To list out the sample space described by a tree diagram, follow each path created by the branches of the tree diagram. For example, following the top path in part **b.** of Problem 2 may give you the outcome of HHH which means that the coin landed on heads on each of the three tosses. List out the complete sample space of the experiment where a fair coin is tossed three times.

Quick Tip

The probability of an event happening will always be between 0 and 1. A probability of 0 means the event is impossible and a probability of 1 means that the event is certain to happen.

Probability of an Event

The **probability of an event** P is equal to the number of outcomes in the event divided by the number of outcomes in the sample space.

$$P = \frac{\text{number of outcomes in event}}{\text{number of outcomes in sample space}}$$

The probability of an event is always a number between 0 and 1. The probability of an event is also commonly written as a percent between 0% and 100%.

Name: Date:

Quick Tip

A **fair coin** is a coin that is uniformly weighted so that neither side of the coin has an advantage. That is, the probability of the coin landing on heads is equal to the probability of the coin landing on tails.

4. Use the sample space created in Problem 3 to answer the following questions related to an experiment where a fair coin is tossed three times.

a. How many outcomes are in the sample space?

b. List the outcomes in the event where the coin lands on heads two times. How many outcomes are in this event space?

c. What is the probability of a coin landing on heads two times out of three tosses?

5. In a Pick 3 lottery, a three-digit number is chosen from the range 000 to 999. Suppose a person plays the numbers 314, 271, and 360.

a. How many outcomes are in the event of the numbers played by the person?

b. How many outcomes are in the sample space? (**Hint:** Each three-digit number between 000 and 999 counts as an outcome.)

c. If all three-digit numbers have the same probability of winning, what is the probability of the person winning with the numbers they chose to play?

Skill Check

Go to Software Work through Practice in Lesson 6.4 of the software before attempting the following exercises.

Find the probability for the specified events in the given sample space. Round to the nearest thousandth when necessary.

6. Sample space of size 12.

a. Event of size 4

b. Event of size 7

7. Sample space: 1, 2, 3, 4, 5, 6.

a. Event: 2, 4, 6

b. Event: 1, 3, 5, 6

Apply Skills

Work through the exercises in this section to apply the skills you have learned related to probability. It may be helpful to draw a tree diagram of each sample space.

8. Julie is going on a seven-day business trip and would like to wear a different outfit each day. She packs four tops: a red blouse, a yellow blouse, a green blouse, and a white blouse. She also packs three skirts: a black skirt, a grey skirt, and a black skirt. Finally, she packs two pairs of shoes: black high heels and black flats.

a. How many outfits can Julie make from the clothes she packed?

b. Does she have enough outfit options to wear a different outfit each day of her trip?

c. What is the probability that Julie wears an outfit that includes the yellow blouse?

Quick Tip

Interior refers to the inside of an object. **Exterior** refers to the outside of an object.

9. Hugo is buying a new car and has determined which make and model he wants to buy. He now has to decide on the colors that he wants for the exterior and interior of the car. The exterior colors he can choose from are blue, silver, red, burnt orange, and black. The interior colors he can choose from are beige, grey, and black.

a. How many color combinations does Hugo have to choose from?

b. What is the probability that Hugo chooses a burnt orange exterior with a beige interior? Round to the nearest thousandth.

c. What is the probability that Hugo chooses a green exterior?

Quick Tip

The birth order of the children is important. That means that {boy, girl} is a different event than {girl, boy}.

10. A family has three children.

a. List the possible gender combinations in the sample space. Use B to represent a boy and G to represent a girl.

b. What is the probability that the family has two girls and one boy (in any order)?

c. What is the probability that the family has three boys?

Name: Date:

Chapter 6 Projects

Project A: What's My *Average*?

An activity to investigate the use of different measures of central tendency in real life

If you are a college student, then grades are important to you. They determine whether or not you are eligible for scholarships or getting into a particular college or program of your choice. It is important to be able to calculate your grade point average in a class and to be able to determine the score you need on a test to reach your desired average. Professors have many different ways of calculating your *average* for a class. Measures of *average* are often referred to as *measures of central tendency*. For this project you will be working with two of these measures, the **mean** and the **median**.

Recall that the **mean** of a set of data is found by adding all the numbers in the set, and then dividing by the number of data items. The **median** is the middle number once you arrange the data in order from smallest to largest. If there is an even number of data items then the median is the mean of the two middle items. The median separates the data into two parts where approximately half of the data values are less than the median, and half are above the median.

Jonathan and Tristen are two students in Dr. Hawkes' Math 230 class. So far, Dr. Hawkes has given five tests and each student's scores are listed below.

Jonathan	Tristen
24	80
98	84
86	88
96	72
96	81

1. Calculate the mean and median of Jonathan's grades.

2. Calculate the mean and median of Tristen's grades.

3. Compare the two measures of *average* for each student.

a. Are the mean and median similar for Jonathan?

b. Are the mean and median similar for Tristen?

c. Based on the **mean**, who has the best *average* in the class?

d. Based on the **median**, who has the best *average* in the class?

4. In your opinion, which student has the most consistent test scores? Explain your reasoning.

5. If each student had scored 2 points higher on each test, how would this affect:

   a. the mean of their grades?

   b. the median?

6. Dr. Hawkes is planning to give one more test to the class. His grading scale is shown here.

Grade	Range
A	93–100
B	85–92
C	77–84
D	69–76
F	Below 69

   a. What is the lowest score Jonathan and Tristen can make on the test and still end up with a grade of C for the class (based on the **mean** of all test scores)?

   b. Who has to make the higher grade on the last test to get a C, Jonathan or Tristen?

   c. If the last test counts double (equivalent to two test grades) what is the lowest score each student can make on the test in order to make a B in the class (based on the **mean** of all test scores)? (Do not round the mean.)

   d. If the last test counts double, who has to make the higher grade on the last test to get a B, Jonathan or Tristen?

7. Based on the work you have done in Problems 1 through 6, which measure do you think is the best measure of a student's *average* grade—the mean or the median? (Explain your reasoning by looking at this question from both Jonathan and Tristen's point of view.)

Name: Date:

Project B: Going for Gold!

An activity to demonstrate the use of charts and graphs in real life

Use the website http://espn.go.com/olympics/summer/2012/medals to determine the number of gold medals won by each country in the 2012 Summer Olympics. Notice that the countries are listed in order by total medals won. To look at the top gold medal winners in order, click the letter G in the gold circle above the chart. List the top 15 countries in the chart, along with the number of gold medals won by each. Then, combine the remaining countries into one group called Other and place the total number of gold medals won by the remaining countries in the appropriate cell of the table.

Top 15 Gold Medal Winning Countries in the 2012 Summer Olympics			
Rank (largest to smallest)	**Country**	**Number of Gold Medals Won**	**Percent of Gold Medals Won**
1			
2			
3			
4			
5			
6			
7			
8			
9			
10			
11			
12			
13			
14			
15			
	Other		
Total			

1. Find the total number of gold medals awarded in 2012 by finding the sum of the items in column three of the table and place this value in the appropriate cell of the last row of the table.

2. Convert each value in column three to a percent by dividing it by the total from Problem 1 and multiplying by 100. (Round to the nearest tenth.) Place the results in column four of the table.

3. Construct a bar chart using the information from column four of the table and the axes below. Be sure to label both axes and title the chart.

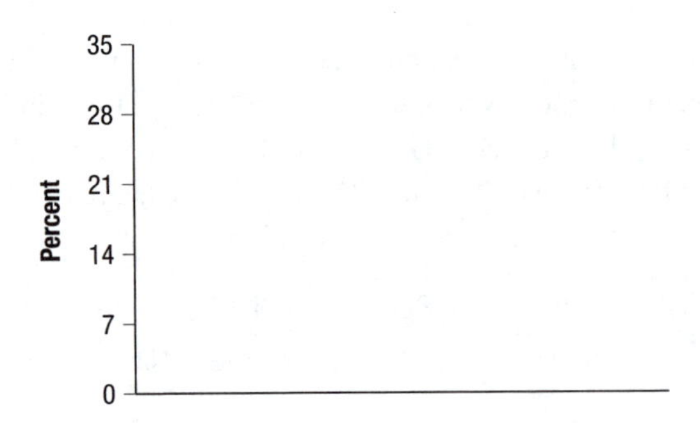

A **Pareto chart** is a bar chart in which the bars are arranged in descending order of height from left to right. It is often used in quality control situations to highlight the most common sources of defects, the most frequent reasons for customer complaints, etc. The Pareto chart is named after the Italian economist, Vilfredo Pareto.

4. Is the bar chart that you constructed in Problem 3 a Pareto chart? Why or why not? If not, describe what needs to be done to transform the bar chart into a Pareto chart.

5. Now construct a pie chart, or circle graph, of this data using the appropriate column from the table. (Remember that there are 360 degrees in a circle.) Be sure to title the graph and place labels and percentages on each wedge of the circle.

6. In your opinion, which chart—the bar chart or pie chart—displays the data best? Explain your reasoning.

Name: Date:

Math@Work

Statistician: Quality Control

Suppose you are a statistician working in the quality control department of a company that manufactures the hardware sold in kits to assemble book shelves, TV stands, and other ready-to-assemble furniture pieces. There are three machines that produce a particular screw and each machine is sampled every hour. A measurement of the screw length is determined with a micrometer, which is a device used to make highly precise measurements. The screw is supposed to be 3 inches in length and can vary from this measurement by no more than 0.1 inches or it will not fit properly into the furniture. The following table shows the screw length measurements (in inches) taken each hour from each machine throughout the day. The screw length data from each machine has also been plotted

Screw Length Measurements (in inches)

Sample Time	Machine A	Machine B	Machine C
8 a.m.	2.98	2.92	2.99
9 a.m.	3.00	2.94	3.00
10 a.m.	3.02	2.97	3.01
11 a.m.	2.99	2.96	3.03
12 p.m.	3.01	2.94	3.05
1 p.m.	3.00	2.95	3.04
2 p.m.	2.97	2.93	3.06
3 p.m.	2.99	2.92	3.08
Mean			
Range			

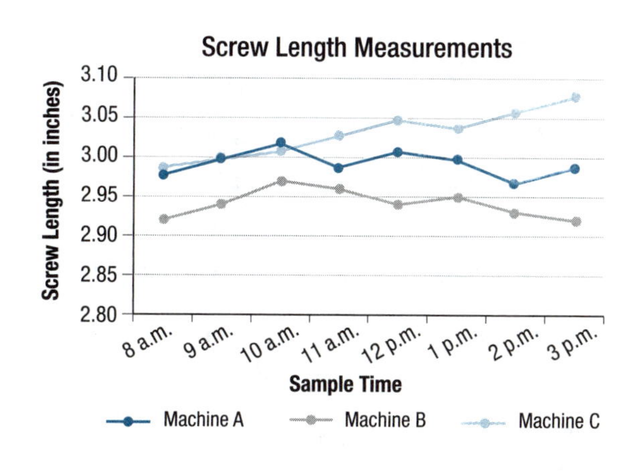

1. Calculate the mean and range of the data for each machine and place them in the bottom two rows of the table.

2. If the screw length can vary from 3 inches by no more than 0.1 inch (plus or minus), what are the lowest and highest values for length that will be acceptable? Place a horizontal line on the graph at each of these values on the vertical axis. These are the tolerance or specification limits for screw length.

3. Have any of the three machines produced an unacceptable part today? Are any of the machines close to making a bad part? If so, which one(s)?

4. Look at the graph and the means from the table that show the average screw length produced by each machine. Draw a bold horizontal line on the graph at 3 to emphasize the target length. Do all the machines appear to be making parts that vary randomly around the target of 3 inches?

5. Look at the range values from the table. Do any of the machines appear to have more variability in the length measurements than the others?

6. In your opinion, which machine is performing best? Would you recommend that any adjustments be made to any of the machines? If so, which one(s) and why?

Foundations Skill Check for Chapter 7

This page lists several skills covered previously in the book and software that are needed to learn new skills in Chapter 7. To make sure you are prepared to learn these new skills, take the self-test below and determine if any specific skills need to be reviewed.

Each skill includes an easy (**e.**), medium (**m.**), and hard (**h.**) version. You should be able to complete each problem type at each skill level. If you are unable to complete the problems at the easy or medium level, go back to the given lesson in the software and review until you feel confident in your ability. If you are unable to complete the hard problem for a skill, or are able to complete it but with minor errors, a review of the skill may not be necessary. You can wait until the skill is needed in the chapter to decide whether or not you should work through a quick review.

2.1 Reduce the fraction to lowest terms.

e. $\frac{70}{100}$ **m.** $\frac{24}{82}$ **h.** $\frac{154}{770}$

3.5 Simplify the expression using the order of operations.

e. $\frac{1}{4}+1.92$ **m.** $\left(1.5+4\frac{2}{5}+2\frac{1}{10}\right)\div 2^2$ **h.** $32\cdot\left(\frac{2}{8}\right)^2-1.35\div\frac{3}{2}$

3.5 Arrange the numbers in order from least to greatest.

e. $1, \frac{1}{3}, \frac{3}{4}, 0.5$ **m.** $\frac{5}{6}, 0.75, \frac{9}{10}, \frac{7}{8}$ **h.** $3.76, 3\frac{3}{5}, \frac{11}{3}, 3.66$

5.5 Calculate the volume of the shape.

e.

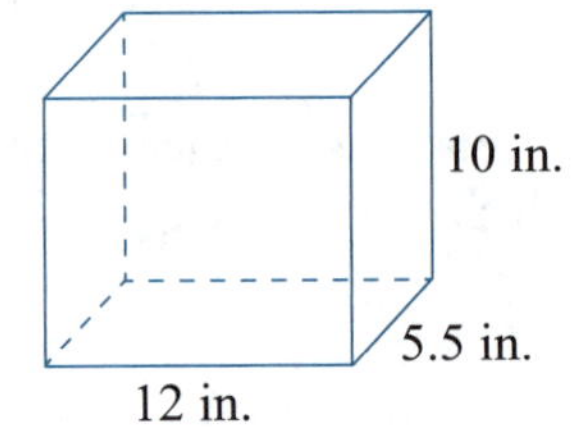

m.

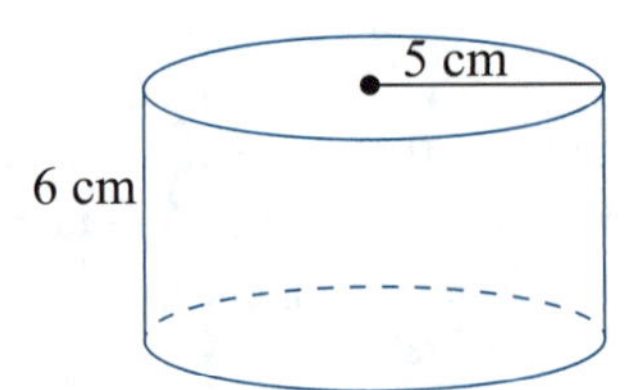

h.

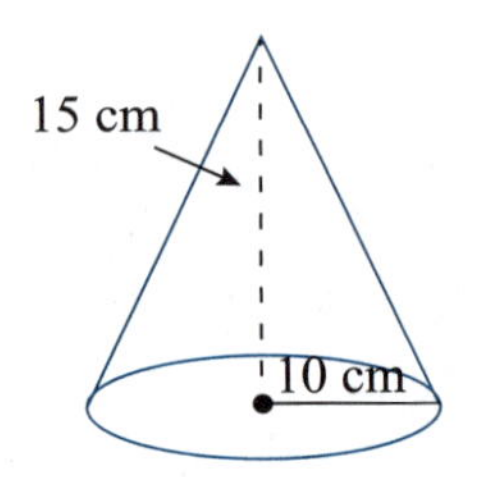

5.7 Calculate the value of the square root. Round to the nearest hundredth if necessary.

e. $\sqrt{25}$ **m.** $\sqrt{324}$ **h.** $\sqrt{7}$

Chapter 7: Introduction to Algebra

Study Skills

7.1 The Real Number Line and Absolute Value

7.2 Addition with Real Numbers

7.3 Subtraction with Real Numbers

7.4 Multiplication and Division with Real Numbers

7.5 Order of Operations with Real Numbers

7.6 Properties of Real Numbers

7.7 Simplifying and Evaluating Algebraic Expressions

7.8 Translating English Phrases and Algebraic Expressions

Chapter 7 Projects

Math@Work

Foundations Skill Check for Chapter 8

Math@Work

Introduction

If you plan on becoming a dental assistant, you can expect to spend your work day in a dental office, a public clinic, or a hospital. As a dental assistant, you will work closely with the dentist to ensure that the patients receive proper dental care. Part of your daily work may include reviewing a patient's file to determine if they are up to date on X-rays, cleanings, and other routine care. You may also be responsible for sterilizing and preparing equipment, taking X-rays, and assisting in dental procedures. Some dental assistants also help with the day-to-day running of the office in the form of managing inventory, answering phone calls, and processing payments. No matter what your role is as a dental assistant, you will be required to work with numbers on a frequent basis.

Suppose you are a dental assistant in a dental office. How will you determine when a patient is due for X-rays? If you need to order more dental supplies, how much will the cost be and did you stay within the budget? Also, how do you determine the amount owed at the end of a patient's visit? Finding the answers to these questions and many more requires several of the skills covered in this and previous chapters. At the end of this chapter, we'll come back to this topic and explore how math is used in dental assisting.

Study Skills

Practice, Patience, and Persistence!

Have you ever heard the phrase "Practice makes perfect"? This saying applies to many things in life. You won't become a concert pianist without many hours of practice. You won't become an NBA basketball star by sitting around and watching basketball on TV. The saying even applies to riding a bike. You can watch all of the videos and read all of the books on riding a bike, but you won't learn how to ride a bike without actually getting on the bike and trying to do it yourself. The same idea applies to math. Math is not a spectator sport.

Math is *not* learned by sleeping with your math book under your pillow at night and hoping for osmosis (a science term implying that math knowledge would move from a place of higher concentration—the math book—to a place of lower concentration—your brain). You also don't learn math by watching your professor do hundreds of math problems while you sit and watch. Math is learned by doing. Not just by doing one or two problems, but by doing many problems. Math is just like a sport in this sense. You become good at it by *doing* it, not by watching others do it. You can also think of learning math like learning to dance. A famous ballerina doesn't take a dance class or two and then end up dancing the lead in *The Nutcracker*. It takes years of practice, patience, and persistence to get that part.

Now, we don't recommend that you dedicate your life to doing math, but at this point in your education, you've already spent quite a few years studying the subject. You will continue to do math throughout college—and your life. To be able to financially support yourself and your family, you will have to find a job, earn a salary, and invest your money—all of which require some ability to do math. You may not think so right now, but math is one of the more useful subjects you will study.

It's not only important to practice math when taking a math course, but it's particularly important to be patient and don't expect immediate success. Just like a ballerina or NBA basketball star, who didn't become exceptional athletes overnight, it will take some time and patience to develop your math skills. Sure, you will make some mistakes along the way, but learn from those mistakes and move on.

Practice, patience, and persistence are especially important when working through applications or word problems. Most students don't like word problems and, therefore, avoid them. You won't become good at working word problems unless you practice them over and over again. You'll need to be patient when working through word problems in math since they will require more time to work than typical math skills exercises. The process of solving word problems is not a quick one and will take patience and persistence on your part to be successful.

Just as you work your body through physical exercise, you have to work your brain through mental exercise. Math is an excellent subject to provide the mental exercise needed to stimulate your brain. As we have stressed in other parts of this workbook, your brain and your capacity for learning are not static and can be increased. Your brain is flexible and it continues to grow throughout your life span—but only if provided the right stimuli. Studying mathematics and persistently working through tough math problems is one way to promote increased brain function. So, when doing mathematics, remember the 3 P's—Practice, Patience, and Persistence—and the positive effects they will have on your brain!

Name: Date:

7.1 The Real Number Line and Absolute Value

Objectives

Identify types of numbers.

Graph numbers on number lines.

Use inequality symbols such as < and >.

Know the meaning of absolute value.

Graph absolute value inequalities.

Success Strategy

In studying and working with negative numbers, it may help to put the problem in a real life context. For example, a value of –5 can mean that the temperature outside has dropped 5° degrees below zero or that you have overdrawn your bank account by $5.

Understand Concepts

Go to Software First, read through Learn in Lesson 7.1 of the software. Then, work through the problems in this section to expand your understanding of the concepts related to the real number line and absolute value.

Understanding the difference between finite and infinite is important when working with types of numbers.

Finite and Infinite

The term **infinite** means that a list or collection of items goes on forever. The number of items in an infinite list cannot be represented by a natural number. The size of this collection of numbers is represented by the symbol ∞.

For example, there are an infinite amount of counting numbers, $\{1, 2, 3, \ldots\}$. This means there is no largest counting number.

The term **finite** is the opposite of infinite. The number of items in a finite list or collection can be represented by a natural number.

For example, the list of numbers $\{2, 4, 8\}$ has a finite number of items. More specifically, the size of the list is 3.

Lesson Link

The notation { } represents a **set**, which means the elements inside are part of the same list. This notation will be further explained in Section 8.7.

Types of Numbers

Integers are the set of numbers consisting of the whole numbers and their opposites.

Rational numbers are the set of numbers which can be written in the form $\frac{a}{b}$, where $b \neq 0$. In decimal form, a rational number can be a **finite terminating** decimal number, such as $\frac{1}{4} = 0.25$, or an **infinite, repeating** decimal number, such as $\frac{1}{3} = 0.333\ldots$.

Irrational numbers are any number which cannot be written in the form $\frac{a}{b}$. They are infinite, nonrepeating decimal numbers. The real number π is a common example of an irrational number.

The set of **Real numbers** is the union of the set of rational numbers and the set of irrational numbers.

Quick Tip

The **ellipsis** "…" in "0.333…" means that the digit 3 repeats forever. Another way to write a repeating decimal is to place a bar over the repeating string of digits. For example, $0.333\ldots = 0.\overline{3}$ since the repeating digit is 3.

1. Problem 1 of Section 1.1 of the workbook asked you to write down the types of numbers that you remembered from previous math courses you have taken and write a short description of each type of number.

   a. How does your list compare to the different subsets of real numbers mentioned in this section?

   b. Did you forget any types of numbers? If so, which types of numbers?

   c. Did you include any type of number which isn't included in the real numbers?

Lesson Link

Venn diagrams were discussed in Section 2.3 in relation to finding the least common multiples. Venn diagrams can be made of circles or squares and used in a variety of contexts.

One way to represent the set of real numbers is with the following Venn diagram, which contains each number type along with two examples. One thing to keep in mind is that when a smaller circle or square in a Venn diagram is completely contained within another larger circle or square, it means that the smaller set is a **subset** of the larger set. When a smaller set is contained in a larger set, all elements of the smaller set are also elements of the larger set which contains it.

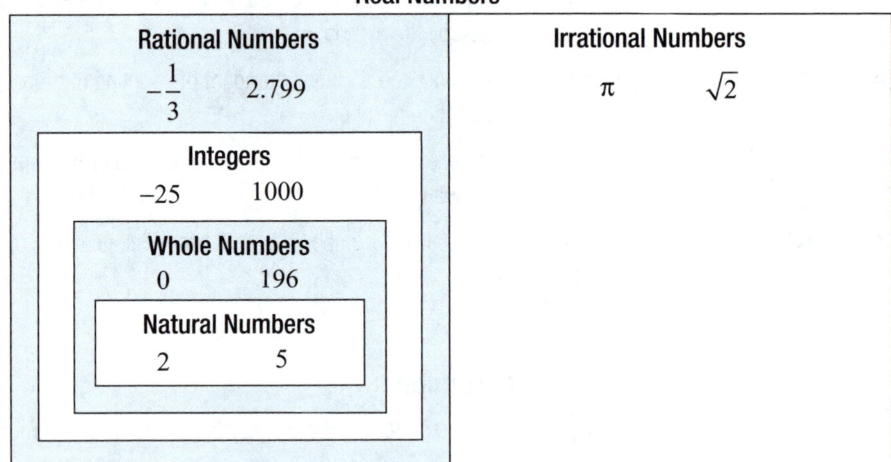

True or False: Determine whether each statement is true or false. Rewrite any false statement so that it is true. (There may be more than one correct new statement.)

2. All natural numbers are also rational numbers.

3. All irrational numbers can be written in fractional form.

Name: Date:

4. Every rational number is also a whole number.

5. Not every integer is a natural number.

6. The set of rational numbers and the set of irrational numbers have numbers in common.

Absolute Value

The **absolute value** of a number, written with the notation $|a|$, represents the distance between the number a and the point 0 on the number line. This value is **nonnegative**, meaning it is either positive or equal to zero.

$|a| = a$ when $a \geq 0$. For example, $|12| = 12$.

$|a| = -a$ when $a < 0$. For example, $|-7| = -(-7) = 7$.

Quick Tip

The **absolute value** of a number indicates the distance the number is from 0. It doesn't give any information about the direction of the number from 0 on the real number line.

True or False: Determine whether each statement is true or false. Rewrite any false statement so that it is true. (There may be more than one correct new statement.)

7. The absolute value of a number is always the opposite of that number.

8. The absolute value of 0 is undefined.

Inequalities

The **inequality symbols** are $<, >, \leq, \geq$, and $\neq$.

An inequality is **true** if it describes a true statement. Otherwise, it is **false**. For example, $3 < 7$ is true while $3 > 7$ is false.

Quick Tip

Inequalities can be read from left to right or from right to left. That is, the inequality $3 < 7$ can be read as "three is less than seven" or "seven is greater than three."

The solution set of an inequality is a set of values that make the inequality true. For example, $x > 5$ has the solution set of integers $S = \{6, 7, 8, 9, \ldots\}$. Notice that 5 is not included in the solution set. This is because $5 > 5$ is not a true statement.

Combining absolute value bars with inequalities results in two different situations. The next two problems explore the combination of absolute value bars with inequalities.

9. First, let's explore what the inequality $|x| > 5$ represents.

a. Looking at positive integers only, which integers have an absolute value greater than 5?

b. How would you write this as an inequality with no absolute value symbols?

c. Looking at the negative integers only, which integers have an absolute value greater than 5?

d. How would you write this as an inequality with no absolute value symbols?

e. Graph the solution set for $|x| > 5$ by graphing the results from parts **a.** and **c.** on the number line.

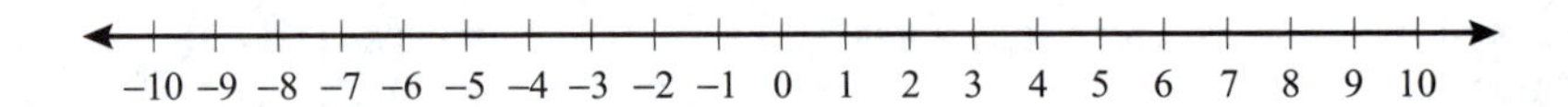

10. Now, let's explore what the inequality $|x| < 5$ represents.

a. Looking at positive integers only (and zero), which integers have an absolute value less than 5?

b. How would you write the as an inequality with no absolute value symbols?

c. Looking at the negative integers only, which integers have an absolute value less than 5?

d. How would you write this as an inequality with no absolute value symbols?

e. Graph the solution set for $|x| < 5$ by graphing the results from parts **a.** and **c.** on the number line.

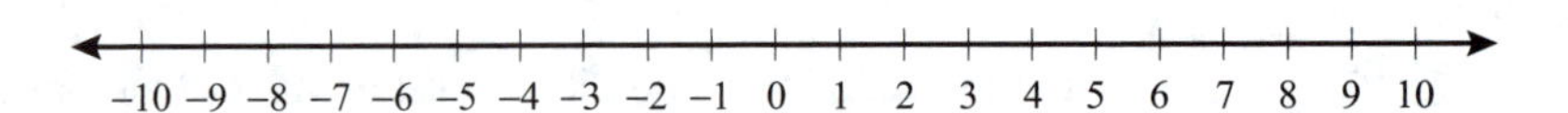

Name: Date:

Skill Check

Go to Software Work through Practice in Lesson 7.1 of the software before attempting the following exercises.

Graph each set of real numbers on a real number line.

11. 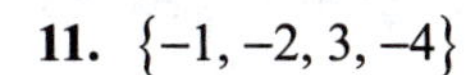

$\{-1, -2, 3, -4\}$

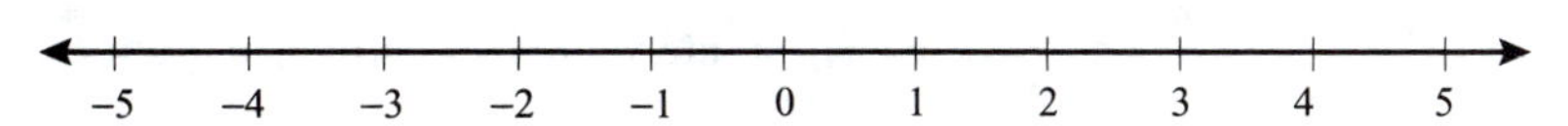

12. $\left\{\frac{1}{2}, -1\frac{1}{2}, 3, -2\right\}$

–4 –3.5 –3 –2.5 –2 –1.5 –1 –0.5 0 0.5 1 1.5 2 2.5 3 3.5 4

13. All whole numbers less than 5

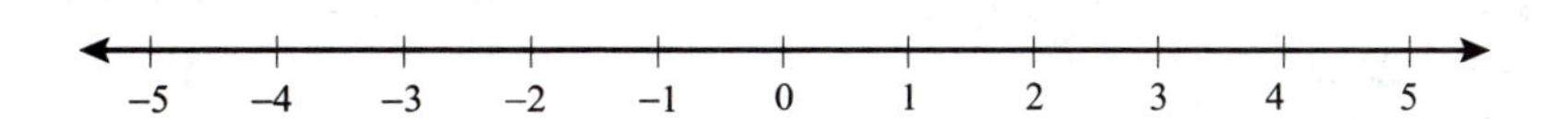

14. $\left\{0.2, -0.8, \frac{4}{5}, -\frac{2}{5}\right\}$

–1 –0.8 –0.6 –0.4 –0.2 0 0.2 0.4 0.6 0.8 1

Lesson Link

The integer answers for these exercises are not the complete solution set to these inequalities. We will look at this topic more in depth in Section 8.7.

List the integers in the solution set for each absolute value inequality. Use an ellipsis to indicate an infinite set.

15. $|x| \leq 7$

16. $2 < |x|$

Apply Skills

Work through the problems in this section to apply the skills you have learned related to the real number line and absolute value.

17. Terrence placed a drop of colored water on the center of a white strand of yarn and measured how much the color spread. Before placing the drop, he predicts that the color will spread no more than 3 inches away from the initial drop.

a. Write an absolute value inequality using the variable x to represent the predicted spread.

b. Graph the solution set of integers for the absolute value inequality from part **a.** on the given number line, placing the initial drop at the point 0.

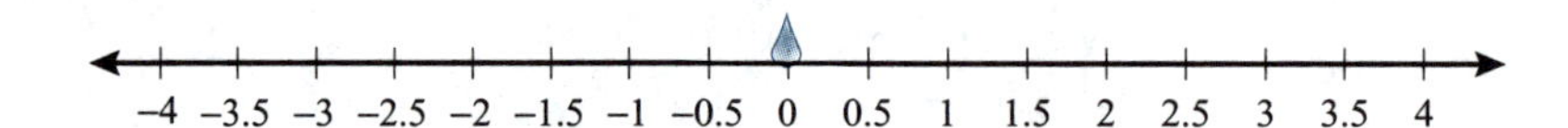

Quick Tip

In manufacturing and engineering, a **tolerance** is specified to allow the final product to vary slightly without affecting performance quality. The tolerance is a range of values that the final measurement can fall within. A bolt with a length of 25 cm with a tolerance of 1 cm can vary between 24 cm to 26 cm in length.

18. A ready-to-assemble bookcase contains wooden boards that have predrilled holes along with the screws and washers needed for assembly. The screws used to assemble the bookcase need to have a length of 38 mm with a tolerance of 2 mm. If the screw is too short, it won't be able to hold the pieces of wood together. If the screw is too long, it might stick out of the other end of the board.

a. What is the largest length the screws can have before they are too long?

b. What is the smallest length the screws can have before they are too short?

c. Graph the tolerance of the screw. (Graph only the integers in the tolerance range.)

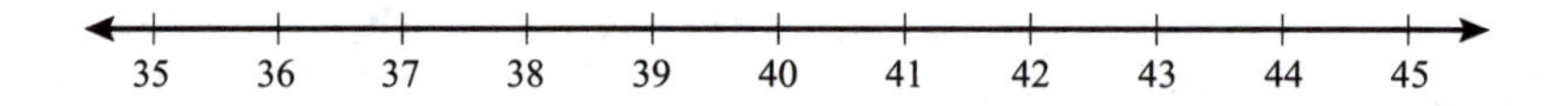

19. A freezer in a biology lab is supposed to be kept at 0 °C. A lab assistant places a thermometer in the freezer and marks down the temperature every half hour. She records the following temperatures: $\left\{2°, -\frac{1}{2}°, \frac{1}{2}°, 1\frac{1}{2}°, -\frac{5}{2}°\right\}$.

a. Graph the set of temperatures on a number line.

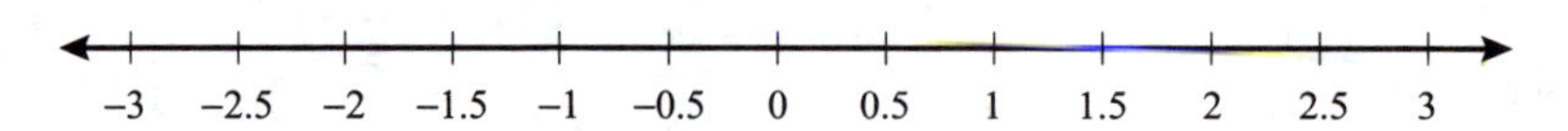

b. Which value is the furthest away from 0?

Name: Date:

7.2 Addition with Real Numbers

Objectives

Add real numbers.

Add real numbers in a vertical format.

Determine if a given real number is a solution to an equation.

Success Strategy

Using a real number line initially to add positive and negative real numbers together will help you visualize the process.

Understand Concepts

Go to Software First, read through Learn in Lesson 7.2 of the software. Then, work through the problems in this section to expand your understanding of the concepts related to addition with real numbers.

Addition with Real Numbers

With Like signs

1. Add the absolute values of the numbers.
2. Use the common sign.

With Unlike signs

1. Subtract the absolute values of the numbers, the smaller from the larger.
2. Use the sign of the number with the larger absolute value.

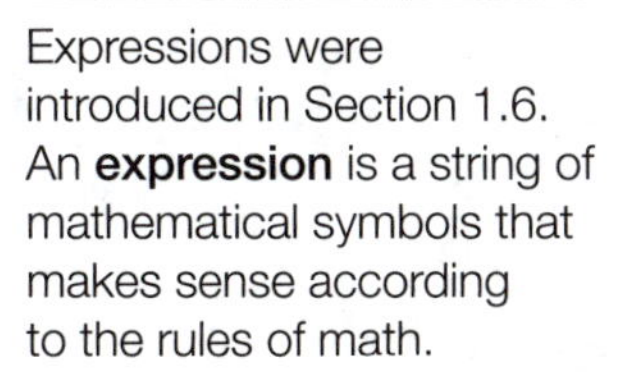

Lesson Link

Expressions were introduced in Section 1.6. An **expression** is a string of mathematical symbols that makes sense according to the rules of math.

1. The first step when adding two real numbers with unlike signs is to subtract their absolute values. When doing this, it is important to use the expression $|a|-|b|$ and not the expression $|a-b|$. *These two expressions do not always give the same result.* Let's try a few values for a and b to show that $|a|-|b| \neq |a-b|$ when the numbers have unlike signs.

a. Substitute $a=-5$ and $b=4$ into each expression and simplify. Are the expressions equal?

b. Pick different values than those used in part **a.** for a and b so that a and b have unlike signs. Test these two values in the expressions. Are the expressions equal?

Equations and Solutions

An **equation** is made by joining two expressions by an equal sign. The two expressions are considered to be equal to each other.

A **variable** is a symbol used in mathematics to represent an unknown number. Any symbol can be used, but it is commonly a letter of the alphabet.

A **solution** is a number that can be substituted for a variable in an equation to make the equation true. If the number does not make the equation true when substituted for a variable, it is not a solution. If an equation is never true, it will have no solutions. Inequalities typically have more than one solution.

An equation is considered to be true if both sides represent the same numerical value. If the two sides represent different values, the equation is considered to be false. Determine if each equation is true or false by simplifying both sides of the equation and comparing the results.

2. $7 = 5 - 2$

Lesson Link

Evaluating expressions by substituting values for variables was covered in Sections 4.5 and 5.2.

3. $x - 12 = -9$ given that $x = 3$

4. $x + 3 = x - 2$ given that $x = 1$

Quick Tip

If a value substituted for a variable results in a false equation, the value is not a solution of the equation.

Determine if any mistakes were made while determining if the equations are true for the indicated value. If any mistakes were made, describe the mistake and then correctly simplify to find the actual result.

5. $15 - x = 7$; given that $x = -8$

$15 - 8 = 7$

$7 = 7$; true

6. $6 = 24 - 2x$; given that $x = -9$

$6 = 24 - 2 - 9$

$6 = 13$; false

Skill Check

Go to Software Work through Practice in Lesson 7.2 of the software before attempting the following exercises.

Find the sum.

7. $\begin{array}{r} -32 \\ +\ 8 \\ \hline \end{array}$

8. $\begin{array}{r} -108 \\ -105 \\ + -330 \\ \hline \end{array}$

9. $-19.6 + 4.1$

10. $\frac{3}{4} + \left(-\frac{1}{8}\right)$

Name: Date:

11. $-\frac{5}{2}+\frac{3}{4}$

12. $9.7+\left(-12\frac{1}{5}\right)$

Apply Skills

Work through the problems in this section to apply the skills you have learned related to addition with real numbers.

13. For 2014, a business reports a profit of $45,000 during the first quarter, a loss of $8000 during the second quarter, a loss of $2000 during the third quarter, and a profit of $15,000 during the fourth quarter.

Quick Tip

In business, a **profit** is a positive amount of money earned during a period of time and a **loss** is a negative amount of money earned during a period of time. (In other words, a loss means that more money was spent than was made.)

a. Write an addition expression to represent the total profit made by the company in 2014. Do not simplify.

b. Simplify the expression from part **a.**

14. A climatologist takes weekly measurements of the height of a glacier near the North Pole. She keeps track of how much the glacier's height either increased or decreased during the week. Her results are presented in the table.

Week	Increase	Decrease
1	0.25 cm	
2		0.3 cm
3		0.1 cm
4	0.17 cm	

a. Which measurements in the table would have a negative value?

b. Calculate the total change in height of the glacier over the four weeks. (**Hint:** Find the sum.)

c. Did the total height of the glacier increase or decrease by the end of the four weeks?

15. Charlotte is a zoologist and part of her job is to keep track of the growth rate of a recently born koala. She writes in her report that the koala weighs 5.6 ounces more than it did when it was born a month ago and the current weight is 28.4 ounces. This can be translated into a mathematical equation as $w + 5.6 = 28.4$, where w is the weight in ounces of the koala at birth. Determine the birth weight of the koala by substituting each of the following values into the equation to find the solution: 22.7 ounces, 23.2 ounces, 22.8 ounces, 23.8 ounces.

16. Part of Noam's job as an accountant is to keep track of the amount of money in the reserve fund. Last week the fund started with \$1253.75 and only one transaction was made. This week the fund started with \$1155.89. This change in value can be written in equation form as $\$1253.75 + t = \$1155.89$, where t is the amount of the transaction. Determine the amount of the transaction by substituting each of the following values into the equation to find the solution: –\$150.14, –\$97.86, –\$89.86, –\$97.14

17. Trevor is installing a hardwood floor in a customer's living room. The length of the room is $17\frac{3}{8}$ feet. Since the flooring comes in 12-foot pieces, Trevor needs to determine the length he must cut off of one of the boards to make it fit. The amount he needs to trim off of one of the boards can be represented by the equation $12 + (12 - r) = 17\frac{3}{8}$, where r is the amount to be removed from one of the boards. Determine the amount that needs to be trimmed off of one of the boards by substituting each of the following values into the equation to find the solution: $5\frac{5}{8}$ feet, $5\frac{3}{8}$ feet, $6\frac{5}{8}$ feet, $6\frac{3}{8}$ feet.

Name: Date:

7.3 Subtraction with Real Numbers

Objectives

Subtract real numbers.

Subtract real numbers in a vertical format.

Find the change in value between two real numbers.

Determine if a given real number is a solution to an equation.

Success Strategy

When calculating the difference between two numbers, you have been subtracting the smaller number from the larger number because we didn't have negative numbers until this chapter. Keep in mind that the order in which numbers are subtracted is important.

Understand Concepts

Go to Software First, read through Learn in Lesson 7.3 of the software. Then, work through the problems in this section to expand your understanding of the concepts related to subtraction with real numbers.

Additive Inverses and Opposites

The **additive inverse** of a number a is the number that when added to a results in a sum of 0. In mathematical notation, $a+(-a)=0$, where a is a real number.

For example, -3 is the additive inverse of 3 because $3+(-3)=0$. Likewise, 3 is the additive inverse of -3.

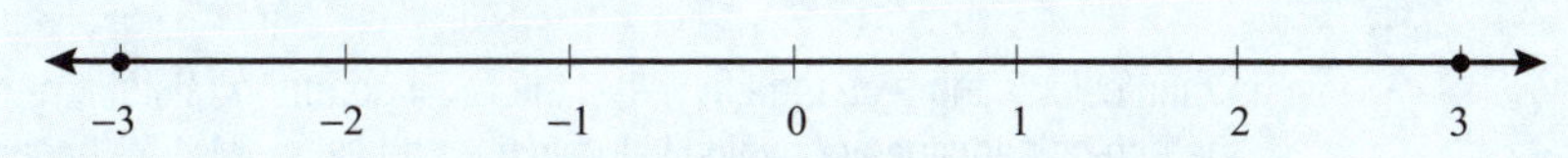

The additive inverse is commonly referred to as the number's **opposite**.

Quick Tip

The value of a in the definition of additive inverse does not have to be positive. It may be negative or equal to 0.

Quick Tip

Try drawing a number line and plotting numbers along with their absolute values. Do this with several positive values and negative values.

1. For which values of a is the absolute value of a equal to its additive inverse?

Subtraction with Real Numbers

For real numbers a and b, to subtract b from a, add the additive inverse of b to a. In mathematical notation, $a-b=a+(-b)$.

2. Write four key words for subtraction that you have seen in the word problems so far in this workbook.

True or False: Determine whether each statement is true or false. Rewrite any false statement so that it is true. (There may be more than one correct new statement.)

3. Subtracting a number is the same as adding its additive inverse.

Quick Tip

0 is an **unsigned number**. This means that $0 = +0 = -0$.

4. The number 0 does not have an additive inverse.

5. A number is the same distance from zero as its additive inverse.

Quick Tip

Absolute value is a distance from zero which means that it is always nonnegative.

6. $|a| = |-a|$.

Change in Value

The **change in value** between two numbers is the difference between the end value and the beginning value. In mathematical notation, *change in value* = (*end value*) – (*beginning value*).

When calculating a change in value, the most common mistake happens when determining which value is the beginning value and which value is the end value. Mixing these two values up will give an answer with the incorrect sign. Work through the next problems to find the change in value for two situations.

Lesson Link

When working with a change in value, it is important to determine the reference value. Reference values were introduced in Section 4.7.

7. When Allison went to work, the temperature outside was 55 °F. During Allison's lunch break, the temperature outside was 68 °F.

   a. What is the beginning value?

   b. What is the end value?

   c. What is the change in value of the temperature outside?

8. At the beginning of the work day, the company's auxiliary fund had $1528.50. At the end of the work day, the company's auxiliary fund had $1349.28.

   a. What is the starting value?

   b. What is the end value?

Name: Date:

c. What is the change in value of the auxiliary fund?

Skill Check

Go to Software Work through Practice in Lesson 7.3 of the software before attempting the following exercises.

Find the difference. Reduce if possible.

9. $\begin{array}{r} 27 \\ \underline{-(+42)} \end{array}$

10. $\begin{array}{r} -1.9 \\ \underline{-(-2.6)} \end{array}$

11. $0-(-12)$

12. $-8.5-7.1$

13. $\frac{5}{16}-\frac{9}{16}$

14. $\frac{9}{20}-\left(-\frac{1}{4}\right)$

Apply Skills

Work through the problems in this section to apply the skills you have learned related to subtraction with real numbers.

15. The highest temperatures recorded per month from July 2012 through June 2013 in Marquette, MI, are shown. Use the graph to answer the questions. Source: weather.gov

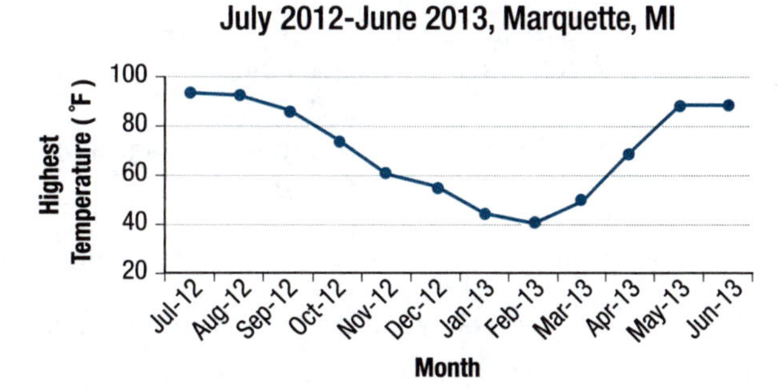

a. Was the change in temperature negative or positive from October 2012 to November 2012?

b. Was the change in temperature negative or positive from April 2013 to May 2013?

c. Between which two months did the smallest change in high temperature occur?

d. Between which two months did the largest change take place?

16. The manager of a music store is required to keep track of the inventory for each instrument that the store rents out. The store had twenty-five trumpets in stock at the beginning of the school year before twelve trumpets were rented out. Five trumpets were returned by the end of the first semester. Seven additional trumpets were rented at the beginning of the second semester. Eight trumpets were returned at the end of the second semester.

a. How many trumpets did the store have in stock at the end of the second semester?

b. What was the change in the trumpet inventory between the beginning of the school year and the end of the second semester?

17. A chemist is determining the water content of different foods in his lab by completely dehydrating food samples (that is, he is removing all of the water). He started an experiment with a fresh white potato that had a mass of 115 g. The mass of the potato after being dehydrated was 24.15 g. This situation can be described by the equation $115 - w = 24.15$, where w is the amount of water mass the potato lost in grams. Determine the water mass of the fresh potato by substituting each of the following values into the equation to find the solution: 90.15 g, −90.15 g, 90.85 g, −90.85 g

18. Grace keeps track of her grade average in Introduction to Anthropology throughout the semester. After the first exam, she had a 93.5%. A project decreased her average grade by 1.5%. The next exam decreased her grade by 4.7%. A term paper increased her grade by 2.3%. The final increased her grade by 3.1%.

a. What was Grace's grade average at the end of the semester?

b. What was the change in Grace's grade average during the semester?

Name: Date:

7.4 Multiplication and Division with Real Numbers

Objectives

Multiply real numbers.

Divide real numbers.

Find the average (mean) of a set of real numbers.

Success Strategy

Do not get multiplication and division of negative numbers confused with addition and subtraction of negative numbers when determining the sign of the result. Extra practice in the software is recommended.

Understand Concepts

Lesson Link

Multiplication with numbers has been covered in Sections 1.3, 2.2, and 3.3. Division with numbers has been covered in Sections 1.4, 2.2, and 3.4.

Lesson Link

The zero-factor law was introduced in Section 1.3 and the rules for division by 0 were introduced in Section 1.4.

Go to Software First, read through Learn in Lesson 7.4 of the software. Then, work through the problems in this section to expand your understanding of the concepts related to multiplication and division with real numbers.

The process for multiplying and dividing real numbers is almost the same as multiplying and dividing all of the previous number types we have covered, with the exception of the sign of the result.

1. The product or quotient of two numbers with the same sign is positive. The product or quotient of two numbers with opposite signs is negative. Fill in the table with the resulting sign to help keep track of the signs of products and quotients of real numbers.

	Negative	**Positive**
Negative		
Positive		

2. Knowing what happens when a number is multiplied or divided by zero is important when working with numbers.

a. What is the result when a number is multiplied by zero?

b. What is the result when a number is divided by zero?

3. The fact that *a negative number times a negative number is equal to a positive number* is a difficult thing for many people to visualize or understand. Use the key words "negative times negative" to perform an Internet search to find different explanations for why a negative number multiplied by a negative number is equal to a positive number. Write down a summary of the explanation which makes the most sense to you.

4. So far in this workbook, all of the values in our data sets have been positive. Data values can also be negative depending on the situation. An example of a situation where data values may be negative occurs with extremely cold temperatures. Another example occurs when a company loses money, in this case a loss can be considered a negative profit.

a. Think of some other situations where the data values can have a negative value.

b. Think of some situations where the data values cannot be negative?

Lesson Link

Mean was introduced in Section 1.7.

5. Let's look at how negative values in a set of data can affect the mean. Round your answer to the nearest hundredth when necessary.

a. How do you think negative values affect the mean of a data set?

b. Find the mean of the following data: 2, 3, 5, 7, 9, 10.

c. Find the mean of the following data: –2, –3, –5, –7, –9, –10.

d. Find the mean of the following data: –2, 3, 5, 7, –9, 10.

e. Explain how negative numbers affect the mean of a data set by comparing your answers from parts **b.**, **c.**, and **d.**

Name: Date:

True or False: Determine whether each statement is true or false. Rewrite any false statement so that it is true. (There may be more than one correct new statement.)

6. A data set that contains only negative values will always have a negative mean.

7. A data set that contains only positive values will have the same mean if all the values are turned into negative values.

8. Adding negative values to a data set with all positive values will increase the value of the mean.

Since negative values can be a part of a data set, they can also be used when creating graphs. The next problem explores how to read a graph with negative data.

9. When opening a business, it isn't uncommon to experience a loss within the first few years as the company "starts up." Suppose a small business showed the following profit during its first year.

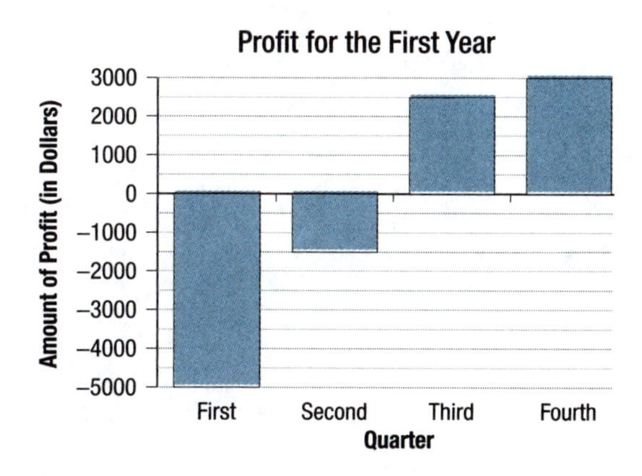

a. During which quarters did the company run at a loss?

b. What was the average profit per quarter during the first year?

c. What was the average profit for the first half of the year?

d. What was the average profit for the second half of the year?

e. If you take the average of the two half-year averages from parts **c.** and **d.**, what is the result?

f. How does the answer from part **e.** compare to the answer from part **b.**? Write a short explanation of why you think this is true.

Skill Check

Go to Software Work through Practice in Lesson 7.4 of the software before attempting the following exercises.

Find the products. Reduce any fractions to lowest terms.

10. $-11(-2)$

11. $(-6)(-3)(-9)$

12. $-\frac{3}{4}\cdot\left(-\frac{6}{7}\right)$

13. $-27\cdot 0$

Find the quotients. Reduce any fractions to lowest terms.

14. $(-26)\div(-13)$

15. $-15\div 0$

16. $-\frac{2}{15}\div\frac{8}{5}$

17. $0\div -15$

Name: Date:

Apply Skills

Work through the problems in this section to apply the skills you have learned related to multiplication and division with real numbers.

Quick Tip

A **credit** adds money to an account. A **debit** removes money from an account.

18. During one day, Sebastian made several transactions with his checking account. He deposited \$150 at the beginning of the day, bought groceries for \$45.50, filled his car with gas for \$39, bought tutoring supplies for \$15, and deposited \$120 at the end of the day.

a. Which amounts are credits to Sebastian's account? Be sure to include the sign.

b. Which amounts are debits to his account? Be sure to include the sign.

c. What was the average transaction amount that Sebastian made during the day? (**Hint:** Be sure to use the amounts from parts **a.** and **b.**)

19. An auto tire manufacturer recommends using 35 psi of air pressure in their standard tires. A tire has a leak that causes the air pressure of the tire to change at a rate of –2 psi per hour.

a. How much will the tire's air pressure change after 4 hours?

b. A tire with a standard air pressure of 35 psi is considered to be flat when it has only 24.5 psi of pressure. What change in air pressure will cause the tire to be considered flat?

c. How long will it take the tire to lose the amount of air pressure determined in part **b.**?

20. In King Salmon, Alaska, the lowest monthly temperature was recorded for several months as shown in the table. Use the data to answer the following questions. Round your answers to the nearest hundredth. Source: weather.gov

Lowest Monthly Temperature in King Salmon, Alaska	
Month	**Temperature**
October 2012	4 °F
November 2012	–4 °F
December 2012	–22 °F
January 2013	–11 °F
February 2013	–14 °F
March 2013	–18 °F

a. What was the average of the low temperatures over these months?

b. The lowest temperature recorded for April 2013 was 1 °F. Will the average low temperature increase or decrease if this data value is now used in calculating the average? Do not calculate the average.

c. Find the average lowest temperature from October 2012 through April 2013. (**Hint:** Part **b.** gives the lowest temperature for April 2013.)

d. How do the average temperatures from parts **a.** and **c.** compare?

e. What is the range of the lowest monthly temperatures for the months given in the table?

Name: Date:

7.5 Order of Operations with Real Numbers

Objectives

Use the rules for order of operations to evaluate expressions.

Success Strategy

Following the order of operations when working with negative numbers is slightly more complicated than when working with only positive values. Pay close attention to the signs on the numbers and make sure to do extra practice in the software.

Understand Concepts

Lesson Link

The order of operations has been introduced in Sections 1.6, 2.6, and 3.4.

Quick Tip

Once the expression is completely simplified, no grouping symbols should remain.

Go to Software First, read through Learn in Lesson 7.5 of the software. Then, work through the problems in this section to expand your understanding of the concepts related to order of operations with real numbers.

The simplification of expressions with real numbers follows the same order of operations that we have been using since Chapter 1. This section will serve mostly as a review of the order of operations and as practice for working with negative values.

Order of Operations

1. Simplify within grouping symbols. If there are multiple grouping symbols, start with the innermost grouping.
2. Evaluate any exponents.
3. Perform multiplication and division in the order they appear from left to right.
4. Perform addition and subtraction in the order they appear from left to right.

A common mistake with absolute value bars is to treat them exactly like standard grouping symbols. While they *are* evaluated first in the order of operations, there is a difference in how they behave. Similar to grouping symbols, the expression inside must be simplified completely before removing the absolute value bars. Unlike grouping symbols, you can't distribute a negative number from outside of the bars to what's inside. When evaluating expressions that contain parentheses, brackets, or braces, a negative sign outside of the pair of grouping symbols can be distributed to the inside, such as $-(-5)=+5$. With absolute value bars, you must evaluate the absolute value of the number inside before you can apply any negative signs on the outside. For example, $-|-5|=-(5)=-5$.

True or False: Determine whether each statement is true or false. Rewrite any false statement so that it is true. (There may be more than one correct new statement.)

1. Absolute value bars are evaluated in the same order as exponents.

2. The order of operations with real numbers is the reverse of the order of operations with whole numbers.

3. Multiplication should always be evaluated before division.

4. Addition and subtraction are evaluated left to right after multiplication and division are evaluated.

Determine if any mistakes were made while simplifying each expression. If any mistakes were made, describe the mistake and then correctly simplify the expression to find the actual result.

5. $$\begin{aligned}-2\cdot|3-5|+2\cdot 3^2 &= (-6+10)+2\cdot 9\\ &= 4+18\\ &= 22\end{aligned}$$

6. $$\begin{aligned}\frac{(-4)^2-|-5-3|}{2} &= \frac{-16-(-8)}{2}\\ &= \frac{-16+8}{2}\\ &= \frac{-8}{2}\\ &= -4\end{aligned}$$

Determine if any mistakes were made while evaluating each expression. If any mistakes were made, describe the mistake and then correctly evaluate the expression to find the actual result.

7. $14+x^2$ given that $x=-2$.

$$\begin{aligned}14-2^2 &= 14-4\\ &= 10\end{aligned}$$

8. $-15\div\left(x-\frac{7}{8}\right)$ given that $x=\frac{1}{4}$.

$$\begin{aligned}-15\div\left(\frac{1}{4}-\frac{7}{8}\right) &= -15\div\left(-\frac{5}{8}\right)\\ &= -15\cdot\left(-\frac{8}{5}\right)\\ &= 24\end{aligned}$$

Skill Check

Work through Practice in Lesson 7.5 of the software before attempting the following exercises.

Simplify each expression by using the order of operations.

9. $15\div(-3)\cdot 3-10$

10. $-\frac{5}{6}\cdot\frac{3}{4}+\frac{1}{3}\div\frac{1}{2}$

Name: Date:

11. $9-6\left[(-21)\div7\cdot2-(-8)\right]$

12. $\left(\frac{1}{2}-1\frac{3}{4}\right)\div\left(\frac{2}{3}+\frac{3}{4}\right)$

Apply Skills

Work through the problems in this section to apply the skills you have learned related to order of operations with real numbers.

13. Madeline sells homemade aprons online and needs to determine how to charge for each apron. To create each apron, she spends \$8.50 on supplies and it takes her $1\frac{1}{4}$ hours to cut and sew each one. Madeline wants to charge \$11 per hour of work plus the cost of supplies.

a. Write an expression to describe how much each apron will cost.

b. Evaluate the expression to determine the selling cost of each apron.

c. Madeline will sew a name or initials onto the apron for an additional charge of \$1.75 per letter. If Kathy orders an apron and wants her name sewn onto it, how much will the apron cost?

14. The Matthews family, a family of 4, is planning a trip to New York City. During their visit, they want to see the Broadway play *Matilda*. The tickets cost \$102 each. The Matthews purchase the tickets online and the website charges a service fee of \$7.50 per ticket. The website is running a sale where the Matthews can get 10% off of their entire purchase.

a. Write an expression to describe how much of a discount the Matthews will receive on their purchase.

b. What is the final purchase price of the tickets?

15. Dennis overdrew his checking account and ended up with a balance of –\$42. The bank charged a \$35 overdraft fee and an additional \$5 fee for every day the account was overdrawn. Dennis left his account overdrawn for 3 days.

a. Write an expression to show the balance of Dennis's checking account after 3 days.

b. Simplify the expression in part **a.** to find the balance of Dennis's checking account after 3 days.

16. Camila is a seamstress and is creating bridesmaid dresses. She has 115 yards of satin fabric. For each dress, the skirt requires 3 yards of satin and the bodice requires 1.5 yards of satin. She plans to make 20 dresses.

a. Write an expression to show how much fabric Camila will have left over after making the dresses.

b. Simplify the expression in part **a.** to determine how much fabric Camila will have left over.

c. Camila wants to make shawls from the leftover fabric. Each shawl requires 1.25 yards of satin. Can she make 15 shawls?

17. During harvest season, farmers donate fresh food to a local food kitchen. To make sure the food doesn't spoil, the food kitchen distributes the food between themselves and 5 other food kitchens in the area. One farmer donates $12\frac{1}{2}$ pounds of potatoes, another farmer donates $15\frac{3}{4}$ pounds of potatoes, and a third farmer donates $11\frac{3}{4}$ pounds of potatoes. The food kitchen finds that $1\frac{1}{4}$ pounds of the donated potatoes are rotten.

a. Write an expression to show how many pounds of potatoes each food kitchen will receive.

b. Simplify the expression from part **a.** to determine how many pounds of potatoes each food kitchen will receive.

Name: Date:

7.6 Properties of Real Numbers

Objectives

Apply the properties of real numbers.

Identify properties that justify given statements.

Success Strategy

Understanding the properties of real numbers will be very helpful when you start simplifying variable expressions in Section 7.7 and solving equations in Chapter 8. Make sure to do extra practice in the software to become familiar with these properties.

Understand Concepts

Go to Software First, read through Learn in Lesson 7.6 of the software. Then, work through the problems in this section to expand your understanding of the concepts related to properties of real numbers.

Properties of numbers were first introduced in Sections 1.2 and 1.3. They are summarized, for addition and multiplication, in this box.

Properties of Real Numbers for Addition and Multiplication

Commutative Property: The order of the numbers in the operation can be reversed.

Associative Property: The grouping of numbers can be changed.

Identity Property: The operation by this number leaves the original number unchanged.

Zero-Factor Law: The product of 0 and a number is always equal to 0.

Distributive Property: Multiplication can be distributed over addition.

Lesson Link

Additive inverse was introduced in Section 7.3.

1. The multiplicative inverse is the only new property introduced in this section. Fill in the table by describing each property in words and then writing two examples of the property.

Property	Algebraic Description	In Words	Examples
Additive Inverse	$a+(-a)=0$		
Multiplicative Inverse	$a\cdot\frac{1}{a}=1$		

Quick Tip

The multiplicative inverse is true for all values of a except $a=0$ since division by 0 is not valid.

2. Subtraction does not follow the commutative property. This means that $a-b\neq b-a$ when $a\neq b$. A numerical example would be $5-7\neq 7-5$. If you need to rearrange a subtraction problem, one way to do this is to rewrite the subtraction of the number as the addition of the additive inverse of the number.

   a. First, verify that $5-7\neq 7-5$. What is the simplified value for each side of the equation?

   b. Rewrite $5-7$ as an expression using addition.

   c. Evaluate $5-7$ and the rewritten expression from part **b.** How do the results compare?

d. Rewrite $5 - 7$ using part **b.** and the commutative property of addition.

3. Similarly, division does not follow the commutative property. This means that $a \div b \neq b \div a$. A numerical example would be $3 \div 12 \neq 12 \div 3$. If you need to rearrange a problem with division, one way to do this is to change the division by the number into the multiplication by the multiplicative inverse of the number.

a. First, verify that $3 \div 12 \neq 12 \div 3$. What is the simplified value for each side of the equation?

b. Rewrite $3 \div 12$ as an expression using multiplication and the multiplicative inverse of 12.

c. Evaluate $3 \div 12$ and the rewritten expression from part **b.** How do the results compare?

d. Rewrite $3 \div 12$ using part **b.** and the commutative property of multiplication.

For each problem, determine which property is demonstrated.

4. $5+(1+3)=(1+3)+5$

5. $(5+1)+3=5+(1+3)$

6. $0=4\cdot 0$

7. $15\cdot 7\cdot 2=7\cdot 15\cdot 2$

8. $12\cdot(3+10)=12\cdot 3+12\cdot 10$

9. $15+0=15$

Determine if any mistakes were made while simplifying each expression. If any mistakes were made, describe the mistake and then correctly simplify the expression to find the actual result.

10.
$$\begin{aligned} 0\cdot(5+2)-2\cdot 3\cdot 4 &= 0\cdot 5+2-2\cdot 12 \\ &= 0+2-24 \\ &= -22 \end{aligned}$$

Name: ______________________ Date: ______________________

11. $(24 \div 6) \div 2 = 24 \div (6 \div 2)$
$= 24 \div 3$
$= 8$

Skill Check

Go to Software Work through Practice in Lesson 7.6 of the software before attempting the following exercises.

Complete the expression using the given property. Do not simplify.

12. $7 + 3 =$ ________	commutative property of addition
13. $(6 \cdot 9) \cdot 3 =$ ________	associative property of multiplication
14. $6(5+8) =$ ________	distributive property
15. $5.2 +$ _____ $= 0$	additive inverse property
16. $\frac{1}{5} \cdot$ _____ $= 1$	multiplicative inverse property
17. $8.25 \cdot$ _____ $= 0$	zero-factor law

Apply Skills

Work through the problems in this section to apply the skills you have learned related to properties of real numbers.

18. Jessica works part-time at a retail store and makes $11 an hour. During one week, she worked $6\frac{1}{2}$ hours on Monday and $4\frac{1}{4}$ hours on Thursday.

a. Determine the amount of money she earned during the week by evaluating the expression $\$11 \cdot \left(6\frac{1}{2} + 4\frac{1}{4}\right)$.

b. Rewrite this expression to remove the parentheses using one of the properties talked about in this section.

c. What property did you use in part **b.** to rewrite the expression?

19. Robin went to the grocery store to buy a few items she needed in order to cook dinner. She bought milk for \$3.99, rolls for \$2.25, a package of steaks for \$12.01, and some marinade for \$1.75. Before getting to the checkout line, Robin remembered that she only had \$20 in her purse. Did she have enough money to buy the food items if the store does not charge sales tax on food?

a. Write an expression to find the total of Robin's food purchases. Do not simplify.

b. Robin doesn't have a calculator to determine the total cost of her items. She wants to make sure that she has enough money to buy them. Rearrange the expression from part **a.** so that she could quickly find the total using mental math.

c. What properties did you use in part **b.** to rewrite the expression?

d. Did Robin have enough money to purchase all of the items?

20. Jordan didn't balance his checking account during the week and ended up overdrawing his account. He had a starting balance of \$85.04 and wrote checks for two bills for the amounts of \$28.79 and \$50.00. He also used his debit card to purchase lunch for \$12.16. In order to avoid an overdraft fee, Jordan must deposit enough money today to bring his balance back to a minimum of zero.

a. Write an expression to find the current balance of Jordan's checking account. Do not simplify.

b. Evaluate the expression from part **a.** to determine the current balance of Jordan's checking account.

c. Write an equation to show Jordan's current checking account balance plus the amount he must deposit today to bring the balance to zero.

d. What property is illustrated in part **c.**?

Name: Date:

7.7 Simplifying and Evaluating Algebraic Expressions

Objectives

Identify like terms.

Simplify algebraic expressions by combining like terms.

Evaluate expressions for given values of the variables.

Success Strategy

There is a big difference between the words simplify and evaluate—they do not mean the same thing. Be sure to pay close attention to the directions for problems in this section.

Understand Concepts

Lesson Link

Variables were defined in Section 1.3. A **variable** is a symbol which is used to represent an unknown number or any one of several numbers.

Quick Tip

Remember that the operations of real numbers are addition, subtraction, multiplication, and division.

Go to Software First, read through Learn in Lesson 7.7 of the software. Then, work through the problems in this section to expand your understanding of the concepts related to simplifying and evaluating algebraic expressions.

Definitions for Algebraic Expressions

A **term** can be a constant, a variable, or the product or quotient of constants and variables.

A **coefficient** is the number multiplied by the variable. It's usually placed in front of the variable.

Like terms are terms that are constants or terms that contain the same variable, or variables, raised to the same powers.

The **degree** of a term is the sum of the exponents on all of the variables of the term.

An **algebraic expression** is a combination of variables and numbers using any of the operations of real numbers.

1. When determining if terms are like terms or not, it is important to look at each part of the term. The variables and degree of each variable need to be the same, but the coefficients may be different. Fill in the following table. In the bottom row, write two terms which would be considered like terms to the term described in the column.

	$14x$	$-15xy$	$2x^2z^3$
Coefficient			
Variables			
Degree of Each Variable			
Degree of the Term			
Like Terms			

The Invisible 1

There are several places where the number 1 is understood to be present, but is not visible.

a. In front of variables: $x = 1 \cdot x$, $x^2 = 1 \cdot x^2$

b. With a negative sign: $-xy = (-1) \cdot xy$, $-7 = (-1) \cdot 7$

c. As an exponent of variables and constants: $x = x^1$, $15 = 15^1$

d. As a divisor of whole numbers and variables: $y = \frac{y}{1}$, $12 = \frac{12}{1}$

2. Consider the algebraic expression $17xy^2 + 8x - 19y^2$.

a. How many terms are in the expression?

b. Which term has the greatest coefficient?

c. What is the degree of each term?

Combining Like Terms

To combine like terms,

1. Determine which terms are alike and group them together.
2. Find the sum of the coefficients of the like terms (be sure to keep track of negative coefficients).
3. Attach the common variable expression to the sum of the coefficients.

True or False: Determine whether each statement is true or false. Rewrite any false statement so that it is true. (There may be more than one correct new statement.)

3. When combining like terms, all terms with no coefficient are ignored.

4. The degree of a term is equal to the largest coefficient in the term.

5. When combining like terms, the variables stay the same and the coefficients are added together.

Name: Date:

Determine if any mistakes were made while combining like terms. If any mistakes were made, describe the mistake and then correctly combine like terms to find the actual result.

6. $12x + x - 4x$
$= (12 + 0 - 4)x$
$= 8x$

7. $25xy - 20xy - 10y + 6y$
$= (25 - 20)xy - (10 + 6)y$
$= 5xy - 16y$

Evaluating Algebraic Expressions

To evaluate algebraic expressions,

1. Combine like terms, if possible.
2. Substitute the given values for any variables.
3. Simplify.

Determine if any mistakes were made while evaluating the algebraic expression. If any mistakes were made, describe the mistake and then correctly evaluate the algebraic expression to find the actual result.

8. $4xy^2 - 2x - y$; given $x = 2, y = -1$.

$$\begin{aligned} 4(2)(-1)^2 - 2(2) - 1 &= 4(2)(1) - 4 - 1 \\ &= 8 - 4 - 1 \\ &= 3 \end{aligned}$$

9. $2xy + 7y - 12x$; given $x = -\frac{1}{2}, y = 3$.

$$\begin{aligned} 2\left(-\frac{1}{2}\right)(3) + 7(3) - 12\left(-\frac{1}{2}\right) &= -1(3) + 21 + 12\left(\frac{1}{2}\right) \\ &= -3 + 21 + 6 \\ &= 24 \end{aligned}$$

Skill Check

Go to Software Work through Practice in Lesson 7.7 of the software before attempting the following exercises.

Simplify each expression by combining like terms.

Quick Tip

When simplifying an expression, combine like terms together. The result is typically an algebraic expression, but can be a real number.

10. $8x + 10x$

11. $5x + 4 - 8x + 7$

12. $4y + 9x + 12y - 3x$

13. $2\frac{1}{2}y - 12 + 3\frac{3}{4} + 2\frac{3}{4}y$

Simplify each expression and then evaluate the expression for $a = -4$ and $b = 2.5$.

Quick Tip

It usually takes fewer calculations to evaluate an expression if it is simplified first.

14. $5a + 20 + 4b + 6b - 12$

15. $-5(a+b)+2(a-b)$

Apply Skills

Work through the problems in this section to apply the skills you have learned related to simplifying and evaluating algebraic expressions.

16. Bryan works at a moving company and is in charge of keeping inventory on the number of boxes they have. The company starts the week with 72 bundles of small boxes and 50 bundles of medium boxes. During the week, they use 25 bundles of small boxes and 32 bundles of medium boxes. At the end of the week, they purchase 125 bundles of medium boxes. There are more boxes in a bundle of small boxes than a bundle of medium boxes.

a. Write an algebraic expression to show how many boxes the moving company has at the end of the week. Use the variable s to represent the number of boxes in a bundle of small boxes and the variable m to represent the number of boxes in a bundle of medium boxes.

b. Simplify the expression from part **b.** to determine how many boxes were left at the end of the week.

c. If a bundle of small boxes contains 25 boxes and a bundle of medium boxes contains 20 boxes, how many boxes will the moving company have at the end of the week?

17. Vince is playing a game of Magic the Gathering with a friend. They decide to make new decks out of booster packs of cards. Vince brings 7 packs and his friend Tom brings 9 packs. Each pack has c cards.

a. Write an algebraic expression using the variable c to represent the number of cards the two friends have if they combine their packs.

b. Simplify the algebraic expression in part **a.**

c. If each pack has 16 cards, how many cards do the two friends have?

18. An apartment management company owns a property with 100 units. The company has determined that the profit made each month from the property can be calculated using the equation $P = -10x^2 + 1500x - 6000$, where x is the number of units rented.

a. Determine the amount of profit for a month when 60 units are rented.

b. Determine the amount of profit for a month when 80 units are rented.

c. Determine the amount of profit for a month when 90 units are rented.

d. Do you notice a trend in the profit per month when the number of units rented increases? If so, give an explanation for why you think this trend exists.

Quick Tip

A **cookie bouquet** is an arrangement of decorated cookies on sticks. The arrangement typically resembles a bouquet of flowers.

19. To determine how much to charge for a cookie bouquet, Barbara's Bombtastic Bakery uses the expression $\$6.50h + \$1.25c$, where h is the number of hours it takes to bake, decorate, and assemble the cookie bouquet and c is the number of cookies in the bouquet. All prices are rounded to the nearest cent.

a. A customer wants a cookie bouquet with a dozen cookies decorated to look like roses. The decorator at the bakery estimates that this bouquet will take 1.5 hours to create. How much should the bakery charge for this cookie bouquet?

b. Another customer orders a cookie bouquet with 18 cookies decorated with the local football team logo. The decorator at the bakery estimates that this bouquet will take $2\frac{3}{4}$ hours to create. How much should the bakery charge for this cookie bouquet?

Quick Tip

A **board-foot** is a unit of measurement for the volume of lumber in the United States and Canada. A board-foot is equal to the volume of a board that is 1 foot long, 1 foot wide, and 1 inch thick.

20. One of the more accurate rules for estimating the amount of board-feet in a log is the International ¼-Inch Log Rule. For a log that is 16 feet in length, the formula is $BF = 0.796D^2 - 1.375D - 1.23$, where D is the diameter of the log in inches after the bark is removed and BF is the amount of board-feet in the log.

a. Determine the amount of board-feet in a red oak log that is 16 feet long and 12 inches in diameter after bark removal. Round to the nearest tenth.

Lesson Link

Solving proportions was covered in Section 4.1.

b. If the price paid for red oak timber is \$400 per thousand board-feet, what is the value of the log?

c. A tree farmer estimates that he has approximately 2 dozen red oak trees per acre that have matured enough to each yield a log that is approximately 16 feet in length and 12 inches in diameter after bark removal. If he has 85 acres of land, what is the approximate value of the red oak timber that can currently be harvested on his property?

Name: Date:

7.8 Translating English Phrases and Algebraic Expressions

Objectives

Translate English phrases into algebraic expressions.

Translate algebraic expressions into English phrases.

Success Strategy

When changing English words or phrases to expressions, underline the key words that represent the type of operation being performed: addition, subtraction, multiplication, or division.

Understand Concepts

Go to Software First, read through Learn in Lesson 7.8 of the software. Then, work through the problems in this section to expand your understanding of the concepts related to translating English phrases and algebraic expressions.

Key Words for Translating English Words into Algebraic Expressions

Addition	Subtraction	Multiplication	Division	Exponents	Parentheses	Equals
add	subtract (from)	multiply	divide	square of	the quantity	equals
sum	difference	product	quotient	cube of	the amount	is
plus	minus	times	per	to the power	sum of	gives
more than	less than	twice				yields
increased by	decreased by	of				

1. In Chapter 1, you were introduced to several key words for each operation. This section presents the list again along with some key words for exponents. Can you think of any other key words that are not listed in the table? If so, list them here or add them to the table.

Key words for operations are very useful when talking about math (in the classroom, for homework, giving directions, etc.). If you don't know how to properly translate English phrases into mathematical notation (or the other way around), you don't have a full understanding of math. When describing math expressions, it is important to use words that are not ambiguous. Their meaning must be clear so that the correct expression is written in math symbols.

Quick Tip

Ambiguous means that the phrase or sentence can be understood in more than one way.

For each algebraic expression, an incorrect English phrase is given to describe it. Rewrite the phrase so that it is correct and not ambiguous. (There may be more than one correct new phrase.)

2. $5(n+1)$ Five times a number plus one.

3. $3-\frac{n}{2}$ Three less than the quotient of a number and two.

4. $5 \cdot (3x - 4)$ Five times three times a number minus four.

5. $(4x)^2$ Four multiplied by a number squared.

Skill Check

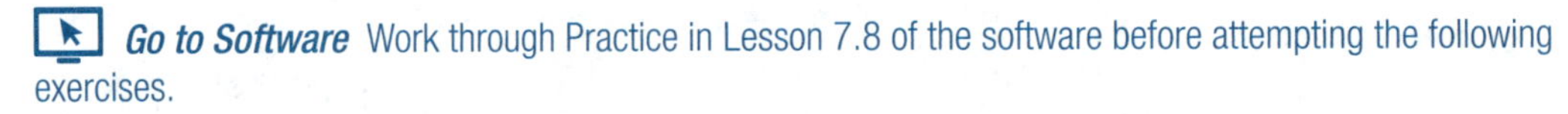

Go to Software Work through Practice in Lesson 7.8 of the software before attempting the following exercises.

Translate each English phrase into an algebraic expression.

6. four less than a number

7. five more than three times a number

8. the quotient of seven and the difference between three and a number

9. the sum of a number and six times the number

10. a. six less than a number **b.** six less a number

11. a. four less than twice a number **b.** four less twice a number

Name: Date:

Apply Skills

Work through the exercises in this section to apply the skills you have learned related to translating English phrases and algebraic expressions.

12. A group of seven friends are planning a party.

a. They want to buy eight sandwich party trays, which cost x dollars each, and evenly split the cost. Write an algebraic expression to describe how much will each friend contribute.

b. Due to a miscommunication, two of the friends bought cups. One friend bought three packs of cups and the other friend bought five packs of cups. Each pack contains c cups. Write an algebraic expression to describe how many cups the two friends brought.

13. Alisha runs her own computer repair business. She charges \$75 per house call and \$45 per hour to repair the computer.

a. Write an algebraic expression to determine how much she will charge if it takes her h hours to repair a computer on a house call.

b. Use the expression from part **a.** to determine how much Alisha would charge a client if it takes her 4 hours to repair the computer during a house call.

14. Andy orders pizza for himself and his friends. The cost for a large cheese pizza is \$9.99 plus t dollars for each topping on the pizza. He orders 4 large pizzas, two with three toppings, one with two toppings, and one with five toppings.

a. Write an unsimplified expression for the total cost of the pizza.

b. Simplify the expression from part **a.**

c. If each topping on a large pizza costs \$0.75, how much will Andy pay for his pizza order?

15. Gus is a pharmaceutical sales representative and he makes a salary of $500 a week plus a 5% commission of his total sales.

a. If Gus sold m dollars of medical supplies this week, then write an expression showing his total earnings for the week.

b. If Gus sold $15,000 of pharmaceuticals during the week, how much will his paycheck be worth?

Name: Date:

Chapter 7 Projects

Project A: Going to Extremes!

An activity to demonstrate the use of signed numbers in real life

When asked what the highest mountain peak in the world is, most people would say Mount Everest. This answer may be correct, depending on what you mean by highest. According to geology.com, there may be other contenders for this important distinction.

The peak of Mount Everest is 8850 meters (or 29,035 feet) above sea level, giving it the distinction of being the mountain with the highest altitude in the world. However, Mauna Kea is a volcano on the big island of Hawaii whose peak is over 10,000 meters above the nearby ocean floor, which makes it taller than Mount Everest. A third contender for the highest mountain peak is Chimborazo, an inactive volcano in Ecuador. Although Chimborazo only has an altitude of 6310 meters (or 20,703 feet) above sea level, it is the highest mountain above Earth's center. Most people think that the Earth is a sphere, so how could a mountain that is only 6310 meters tall be higher than a mountain that is 8850 meters tall? Since the Earth is not actually a sphere but an "oblate spheroid," it is widest at the equator. Chimborazo's location is 1° south of the equator, which makes it about 2 kilometers farther from the Earth's center than Mount Everest.

What about the other extreme? What is the lowest point on Earth? As you might have guessed, there is more than one candidate for that distinction as well. The lowest exposed area of land on Earth's surface is on the Dead Sea shore at 413 meters below sea level. The Bentley Subglacial Trench in Antarctica is the lowest point on Earth that is not covered by ocean, but it is covered by ice. This trench reaches 2555 meters below sea level. The deepest point on the ocean floor occurs 10,916 meters below sea level in the Mariana Trench in the Pacific Ocean.

1. Calculate the **difference** in elevation between Mount Everest and Chimborazo in both meters and feet. What operation does the word **difference** imply?

2. Write an expression to calculate the **difference** in elevation between the peak of Mount Everest and the lowest point on the Dead Sea shore in meters and simplify.

3. If you were to travel a distance equivalent to the distance from the bottom of the Mariana Trench to the top of Mount Everest, how many meters would you travel?

4. If Mount Everest were magically moved and placed at the bottom of the Mariana Trench, how many meters of water would lie above Mount Everest's peak?

5. How much farther below sea level (in meters) is the Mariana Trench as compared to the Dead Sea shore?

6. Add the elevations (in meters) together for Mount Everest, Chimborazo, the Dead Sea Shore, the Bentley Subglacial Trench, and the Mariana Trench and show your result. Is this number positive or negative? Would this value represent an elevation above or below sea level?

7. Convert the results in Problems 2 through 4 to feet using the conversion factor 1 meter = 3.28 feet. Do not round your answers.

8. Convert the results in Problem 7 from feet to miles using the conversion factor 1 mile = 5280 feet. Round your answers to the nearest thousandth.

9. Using the height of Mount Everest in meters as an example, the conversions in Problems 7 and 8 could have been combined to do the conversion from meters directly to miles by using the following sequence of conversion factors:

$$8850 \cancel{\text{m}} \cdot \frac{3.28 \cancel{\text{ft}}}{1 \cancel{\text{m}}} \cdot \frac{1 \text{ mile}}{5280 \cancel{\text{ft}}} = 5.498 \text{ miles}$$ **(A mountain peak over 5 miles high!)**

Notice that in doing the conversions, the units for meters and feet cancel out since they appear in both the numerator and denominator, leaving only the unit of miles in the numerator of your result. This is called **unit analysis** and is extremely helpful in converting measurements to make sure you end up with the correct answer and the correct units on your result.

Verify that this sequence of conversions works by taking the results from Problems 2 through 4 and applying both conversion factors above. How do your results compare to the results from Problem 8? Round your answers to the nearest thousandth.

10. There is more than one way that this conversion could have been performed. Using the conversion factors 1 kilometer = 1000 meters and 1 mile = 1.61 kilometers convert the results in Problems 2 through 4 from meters to miles by using these factors in sequence similar to Problem 9 and performing a **unit analysis**. Round your answer to the nearest thousandth. Do you get **exactly** the same results? Why do you think this is so?

Name: Date:

Project B: Rocky Mountain Highs and Lows

An activity to demonstrate the use of signed numbers in real life

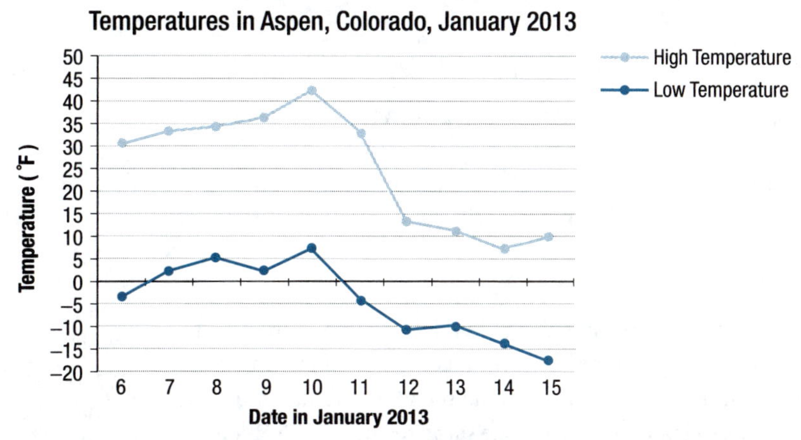

The graph above is a line graph of the high and low temperatures recorded during a 10-day period in Aspen, Colorado, in January of 2013. The daily high and low temperatures from the graph are recorded in the table.

Date	Low Temperature	High Temperature	Difference (High–Low)
6-Jan-13	–3	31	
7-Jan-13	3	34	
8-Jan-13	6	35	
9-Jan-13	3	37	
10-Jan-13	8	43	
11-Jan-13	–3	34	
12-Jan-13	–10	14	
13-Jan-13	–9	12	
14-Jan-13	–13	8	
15-Jan-13	–17	10	
Means			

1. Calculate the average low temperature during the 10-day period by calculating the mean of the temperatures in column two of the table.

2. Calculate the average high temperature during the 10-day period by calculating the mean of the temperatures in column three of the table.

3. Which was more difficult to calculate, the mean low temperature or the mean high temperature? Explain why.

4. Calculate the difference between the mean high and mean low temperatures.

5. Mark the mean high temperature and the mean low temperature on the thermometer figure. How many degrees is it from the mean high temperature to zero? How many degrees is it from zero to the mean low temperature? If you add these two results together do you get the same result as in Problem 4?

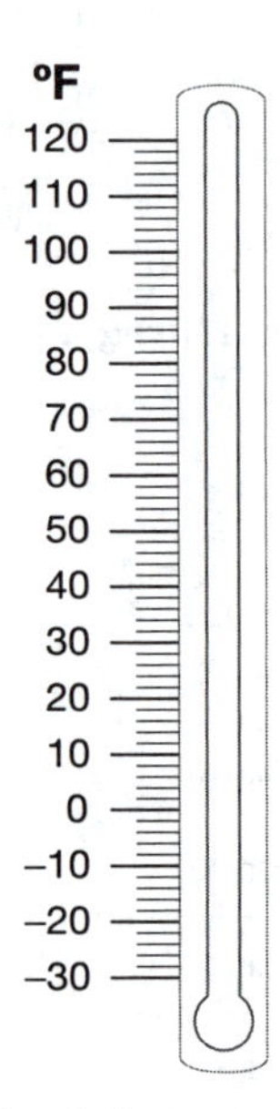

6. Calculate the difference between the **daily** high and low temperatures and place in column four of the table.

7. Calculate the mean of the differences in the last column. How does the mean of the daily temperature differences compare to the result found in Problem 4? Why do you think this happens?

8. The Channel 12 weatherman in Aspen reports on the 6 p.m. news that the current temperature is –7° F and will be steadily falling at a rate of 2 °F per hour for the next 8 hours. Write an algebraic expression to express the temperature for the next 8 hours. Use the variable h to represent the number of hours that has passed since the weather report.

9. Predict what the temperature will be at midnight using the expression in Problem 8.

Name: Date:

Math@Work

Dental Assistant

As a dental assistant, your job duties will vary depending on where you work. Suppose you work in a dental office where you assist with dental procedures and managing patients' accounts. When a patient arrives for their appointment, you will need to review their chart and make sure they are up to date on preventive care, such as X-rays and cleanings. When the patient leaves, you will need to fill out an invoice to determine how much to charge the patient for their visit.

Dental patients generally have a new X-ray taken yearly. Cleanings are performed every 6 months, although some patients have their teeth cleaned more often. The following table shows the date of the last X-ray and cleaning for three patients that are visiting the office today. (**Note:** All dates are within the past year.)

Patient Histories		
Patient	**Last X-ray**	**Last Cleaning**
A	April 15	October 20
B	June 6	January 12
C	October 27	October 27

During Patient A's visit, she received a fluoride treatment and a cleaning. Patient A has no dental insurance. During Patient B's visit, he received a filling on one surface of a tooth. Patient B has dental insurance which pays for 60% of the cost of fillings. During Patient C's visit, he had a cleaning, a filling on one surface of a tooth, and a filling on two surfaces of another tooth. Patient C has dental insurance which covers the full cost of cleanings and 50% of the cost of fillings.

Fee Schedule	
Procedure	**Cost**
Cleaning	$95
Fluoride treatment	$35
Filling, One surface	$175
Filling, Two surfaces	$235
X-ray, Panoramic	$110

1. Using today's date, determine which of the three patients are due for a dental cleaning in the next two months?

2. Using today's date, determine which of the patients will require a new set of X-rays during this visit.

3. Determine the amount each patient will be charged for their visit (without insurance). Don't forget to include the cost of any X-rays that are due during the visit.

4. Use the insurance information to determine the amount that each patient will pay out-of-pocket at the end of their visit.

Foundations Skill Check for Chapter 8

This page lists several skills covered previously in the book and software that are needed to learn new skills in Chapter 8. To make sure you are prepared to learn these new skills, take the self-test below and determine if any specific skills need to be reviewed.

Each skill includes an easy (**e.**), medium (**m.**), and hard (**h.**) version. You should be able to complete each problem type at each skill level. If you are unable to complete the problems at the easy or medium level, go back to the given lesson in the software and review until you feel confident in your ability. If you are unable to complete the hard problem for a skill, or are able to complete it but with minor errors, a review of the skill may not be necessary. You can wait until the skill is needed in the chapter to decide whether or not you should work through a quick review.

6.1 Find the mean of each set of numbers.

e. 2, 4, 6

m. 10.8, 8.6, 9.6, 11.4

h. $1\frac{1}{2}, 2, 5\frac{3}{4}, 6$

5.3 Find the area of each shape.

e.

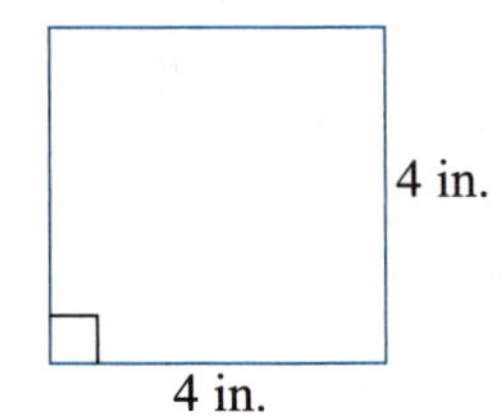

m.

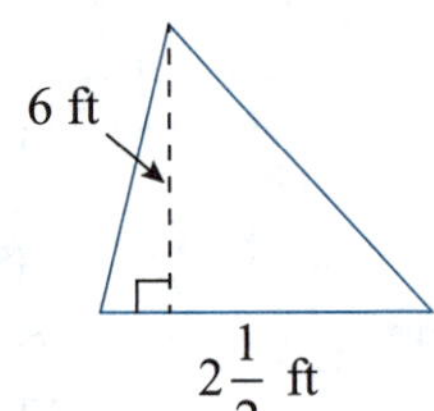

h.

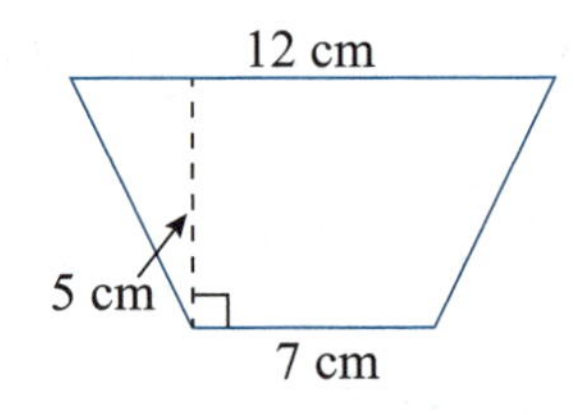

7.5 Simplify the expression using the order of operations.

e. $5 \cdot 20 \div 2$

m. $\frac{2}{5} \div \frac{6}{8} \cdot \frac{4}{3}$

h. $\frac{7}{6} \cdot 2^2 - \frac{2}{3} \div 3\frac{1}{5}$

7.7 Simplify the expression by combining like terms.

e. $4x + 6x$

m. $5x - 7 - 8x + 12$

h. $4(x - y) + 5(y - x)$

7.8 Evaluate the expression for the given value.

e. $12x - 9; x = 2$

m. $2.5 + 3y; y = 1.5$

h. $\frac{1}{2}x - 3\frac{3}{4}; x = \frac{1}{8}$

Chapter 8: Solving Linear Equations and Inequalities

Study Skills

8.1 Solving Linear Equations: $x + b = c$ and $ax = c$

8.2 Solving Linear Equations: $ax + b = c$

8.3 Solving Linear Equations: $ax + b = cx + d$

8.4 Applications: Number Problems and Consecutive Integers

8.5 Working with Formulas

8.6 Applications: Distance-Rate-Time, Interest, Average

8.7 Linear Inequalities

Chapter 8 Projects

Math@Work

Foundations Skill Check for Chapter 9

Math@Work

Introduction

If you are thinking about a career as a financial advisor, then you should consider getting a degree in accounting, finance, economics, business, mathematics, or law. The goal of a financial advisor or financial planner is to help people better manage their money. Financial advisors help people make budget plans, investment decisions, and choose the best insurance to meet their needs. A financial advisor can help people reach their financial goals of buying a house, putting children through college, or retiring comfortably. In your career as a financial advisor, you will need to keep up with changing financial markets, constantly monitor your clients' investment portfolios, and stay current on new investment strategies and funds. When it comes to helping people with investments, financial advisors must be confident about making decisions involving uncertainty, must be able to work under pressure, and have excellent people and communication skills.

Suppose you choose a career as a financial advisor and start by initially working for a large financial services firm, but hope to eventually own your own business. How do you determine a client's financial goals? How do you determine your client's tolerance for risk in order to choose appropriate investments for them? What investment products are appropriate for a particular client? How much money will your client earn on his investment? Finding the answers to these questions requires several of the skills you will learn in this chapter as well as skills from previous chapters. At the end of this chapter, we'll come back to the topic of investments and explore how math is used in financial advising.

Study Skills

Using Class Time Effectively

1. To effectively use your time in class, it is always a good idea to read over the material the instructor plans to cover beforehand. Many instructors have a schedule in their syllabus that shows the material to be covered during each class of the semester. Look over the headings in the scheduled section of the textbook and read through the examples to get an idea of what will be covered in class.

2. Make sure to get to class on time, or a few minutes early, so you can get a seat close to the front or center of the classroom. Always try to sit near the front of the classroom or close to the instructor so that you can hear better and see the whiteboard or projector screen more clearly. Getting to class early will also give you a few minutes to look over your notes from the last class to refresh your memory and get you in the right mind-set for the day's class.

3. Come to class prepared to actively participate. Bring your textbook, paper for note-taking, pencil, highlighter, and calculator, if your instructor allows. Always use a pencil for math so you can erase if you need to.

4. Turn off your cell phone so it will not be a distraction to you, your classmates, or your instructor during class. Most instructors have a strict policy regarding the use of cell phones in class. If you are paying attention to your phone, you are not paying attention to your instructor and the information being presented.

5. If you have any questions from the previous class session, come prepared to ask questions. Since math tends to build on previously learned concepts, you often must have a good understanding of past material to be able to do well going forward.

6. Take special note of any problems, procedures, or terms that the instructor emphasizes in class. These will most likely appear on quizzes or tests in the future.

7. As an instructor works through a problem, anticipate what step should come next. If the instructor pauses during class for you to practice working through problems, be sure to do so. If you have any difficulties with the problems, then you can ask a question immediately. Often students think they understand the material in class while the instructor is working the problems, but then have difficulty doing the work once they are on their own.

8. If you miss an important point in class, write down what you remember and make a note in the margin to ask a classmate or the instructor after class so that you can fill in what you missed.

9. Mark up your textbook as needed to draw attention to important terms or concepts noted by the instructor in class. Use annotations (notes in your text that explain or provide comments about the material) to add important ideas or points that the instructor mentions. If you plan to resell your textbook, use a pencil to do so. You can erase the marks once the class is over.

10. Do not be afraid to ask questions during class. It is very likely that if you don't understand a problem or concept, there are other students in the class who also don't understand it.

11. If you have time after class, reread your notes and try to fill in any gaps you missed with your textbook. If your notes are hard to read, you might want to consider rewriting them. If you wait till the next day, you may forget important points.

12. If you are slow at note-taking or not a good note-taker, ask the instructor if you can record the class session. This can be very helpful if you are an auditory learner.

13. Make sure you write down any assignments due for the next class period. Also be sure to record the dates for any upcoming quizzes or tests that the instructor mentions in class.

14. Be sure to do the assigned problems before the next class period so you will be prepared and can ask questions about problems or concepts that you do not understand.

Name: Date:

8.1 Solving Linear Equations: $x + b = c$ and $ax = c$

Objectives

Define the term linear equation.

Solve equations of the form $x + b = c$.

Solve equations of the form $ax = c$.

Success Strategy

Addition and subtraction are *opposite* operations because they undo each other. Likewise, multiplication and division are *opposite* operations. Keep this in mind as you progress through this section.

Understand Concepts

Go to Software First, read through Learn in Lesson 8.1 of the software. Then, work through the problems in this section to expand your understanding of the concepts related to solving linear equations of the form $x + b = c$ and $ax = c$.

Equations and Solutions

An **equation** is a statement that two algebraic expressions are equal.

A **solution** to the equation is any number which gives a true statement when substituted for the variable in the equation.

A **linear equation** is an equation with degree 1.

Lesson Link

The equations covered in this chapter have one variable. Linear equations with more than one variable will be introduced in Chapter 9.

In this section, we will be working with "one-step equations." These equations have this name because they require only one algebraic step to solve them. Such equations simplify to the forms

$$x + b = c \quad \text{or} \quad ax = c,$$

where x is the variable and a, b, and c are constants. To solve this type of equation you first need to learn the two basic steps for solving linear equations.

Quick Tip

The addition principle of equality is also known as the **subtraction principle of equality** when negative values are added. The multiplication principle of equality is also known as the **division principle of equality** when the equation is multiplied by reciprocals.

Principles of Equality

Addition Principle of Equality

If the same algebraic expression is added to both sides of an equation, then the new equation has the same solution as the original equation.

For example, $x = a$ has the same solution as $x + b = a + b$.

Multiplication Principle of Equality

If both sides of an equation are multiplied by the same nonzero constant, then the new equation has the same solution as the original equation.

For example, $x = a$ has the same solution as $cx = ca$, where $c \neq 0$.

1. Identify which principle of equality is being demonstrated in each pair of equations.

   a. $$\begin{aligned} x &= 2 \\ x+5 &= 2+5 \end{aligned}$$

   b. $$\begin{aligned} x &= 4 \\ \frac{x}{2} &= \frac{4}{2} \end{aligned}$$

   c. $$\begin{aligned} m &= p \\ c \cdot m &= c \cdot p \end{aligned}$$

   d. $$\begin{aligned} m &= p \\ m-c &= p-c \end{aligned}$$

   e. $$\begin{aligned} x &= a \\ 3 \cdot x &= 3 \cdot a \end{aligned}$$

   f. $$\begin{aligned} x &= a \\ x+8 &= a+8 \end{aligned}$$

Read the following paragraph about creating equations and work through the problems.

Equivalent equations can be created by using either the addition or multiplication principles of equality. Let's start with the equation $x = 7$. Notice that this equation also identifies the solution.

Applying the addition principle by adding 3 to both sides of the equation gives us the following.

$$\begin{aligned} x &= 7 \\ x+3 &= 7+3 \\ x+3 &= 10 \end{aligned}$$

Applying the multiplication principle with 2 as a factor gives us the following.

$$\begin{aligned} x &= 7 \\ 2 \cdot x &= 2 \cdot 7 \\ 2x &= 14 \end{aligned}$$

2. Create two more equations that are equivalent to $x = 7$.

   a. Create an equation by using the addition (or subtraction) principle of equality.

   b. Create an equation by using the multiplication (or division) principle of equality.

Name: Date:

Solving Linear Equations

Equations that Simplify to the Form $x + b = c$

1. Combine any like terms on each side of the equation.
2. Use the addition principle of equality to add the opposite of the constant b to both sides and then simplify. This will isolate the variable on one side of the equation.
3. Check your answer by substituting it for the variable in the *original* equation.

Equations that Simplify to the Form $ax = b$

1. Combine any like terms on each side of the equation.
2. Use the multiplication principle of equality to multiply each side of the equation by the reciprocal of the coefficient of the variable and then simplify. This will isolate the variable on one side of the equation.
3. Check your answer by substituting it for the variable in the *original* equation.

Lesson Link

Like terms were discussed in Section 7.7.

Quick Tip

Adding the opposite of a number is the same as adding the additive inverse of the number. Additive inverses were introduced in Section 7.3.

3. To solve the equation $x + 3 = 10$, we need to do the "opposite" of what we did when building the equation. That is, we need to do the opposite of adding 3. This can be thought of as adding -3, or subtracting 3. Solve this equation and verify that it has the same solution as the equation we started with when building the equation.

Quick Tip

Remember that multiplying by $\frac{1}{a}$ is the same as dividing by a, and dividing by $\frac{1}{a}$ is the same as multiplying by a. Multiplication and division are inverse operations.

4. Consider the equation $2x = 14$.

a. What step do we need to take to solve this equation?

b. Solve this equation and verify that it has the same solution we started with when building the equation.

5. Solve the equations that you created in Problem 2. Be sure to show your work.

Read the following paragraph about solving word problems using equations and work through the problem.

Quick Tip

Practice will make it easier to determine when arithmetic or algebra is more useful to solve a problem.

Knowing how to build an equation from the information provided in a word problem is an important skill. For simple problems, it may seem easier to set up the problems as an expression without variables. However, writing single variable equations for these types of problems is good practice to learn the steps and notation needed to create equations for more complicated problems. The following problem will guide you through the process of writing single-variable equations based on word problems.

6. Wynona is making party invitations. She needs a total of 50 invitations. She has already made 22 invitations. She wants to know how many more invitations she needs to make. Create an equation of the form $x + a = b$ to describe the situation.

a. The variable x represents the unknown value. What is unknown value in the problem?

b. The constant a represents a value that is added to the unknown value. What does this constant represent in this problem and what is its value?

c. The constant b represents the sum of two values. What does this constant represent in this problem and what is its value?

d. Create a mathematical equation of the form $x + a = b$ using the values from parts **b.** and **c.**

Quick Tip

When solving linear equations, the goal is to get the variable by itself on one side of the equation and a constant on the other side.

e. Solve the equation from part **d.** for the variable x.

f. What does the answer from part **e.** mean? Write a complete sentence.

Skill Check

Go to Software Work through Practice in Lesson 8.1 of the software before attempting the following exercises.

Solve for the variable.

Quick Tip

When solving a linear equation, be sure to rewrite each step on the line below the previous step. This will make the work organized and easy for anyone to follow.

7. $x - 6 = 1$

8. $32 = 4y$

9. $y + 1.6 = -3.7$

10. $-7.5x = -37.5$

Name: Date:

11. $6x - 5x + \frac{3}{4} = -\frac{1}{12}$

12. $1.7x = -5.1 - 1.7$

Apply Skills

Work through the problems in this section to apply the skills you have learned related to solving linear equations of the form $x + b = c$ and $ax = c$.

Quick Tip

Show all work when solving equations, even if the equation can be solved mentally. Remember, instructors aren't mind readers!

13. A nurse must give a patient 800 milliliters of intravenous solution over 4 hours. This can be represented by the equation $4x = 800$, where x represents the amount of solution the patient receives per hour in milliliters.

a. Why was multiplication chosen in the equation?

b. Solve the equation to determine the value of x.

c. What does the answer to part **b.** mean? Write a complete sentence.

14. The inventory manager's computer crashed and he did not have a backup of his data. The company manager is requesting an inventory report for the week for a specific item. The inventory manager knows that there are currently 1472 of that item in stock. During the week, a shipment arrived with 1500 of the item. The company also shipped out 975 of the item during the week. This situation can be represented by $x + 1500 - 975 = 1472$, where x is the number of items in the inventory at the beginning of the week.

a. Why were the operations of addition and subtraction chosen in this equation?

b. Solve the equation to determine the value of x.

c. What does the answer to part **b.** mean? Write a complete sentence.

15. John is making a garden in his backyard. He buys enough topsoil to cover 300 square feet. John wants the garden to go along the side of his garage, which is 24 feet in length. To determine how wide the garden needs to be, John uses the equation $24x = 300$, where x is the width of the garden in feet.

a. Why was multiplication chosen in this equation?

b. Solve the equation to determine the value of x.

c. What does the answer to part **b.** mean? Write a complete sentence.

16. Clara has \$4200 saved to use as a down payment on the new car she is buying that costs \$15,750. She will have to get a loan to pay for the rest of the cost. This situation can be modeled by $4200 + x = 15{,}750$, where x is the amount of the loan in dollars.

a. Why was the operation of addition chosen in this equation?

b. Solve the equation to determine the value of x.

c. What does the answer to part **b.** mean? Write a complete sentence.

Name: Date:

8.2 Solving Linear Equations: *ax* + *b* = *c*

Objectives

Solve equations of the form $ax + b = c$.

Success Strategy

The order in which math steps are taken in this section are important and can make solving more complex linear equations much easier if followed.

Understand Concepts

Go to Software First, read through Learn in Lesson 8.2 of the software. Then, work through the problems in this section to expand your understanding of the concepts related to solving linear equations of the form $ax + b = c$.

In this section we will be working with "two-step equations." These equations have this name because they require two algebraic steps to solve them. Such equations simplify to the form

$$ax + b = c,$$

where x is the variable and a, b, and c are constants.

To solve these equations, both the addition principle of equality and the multiplication principle of equality need to be used.

Quick Tip

Since the equations $x + b = c$ and $ax = c$ can be created by plugging in values for the constants in the equation $ax + b = c$, these equations are **special forms** of $ax + b = c$.

True or False: Determine whether each statement is true or false. Rewrite any false statement so that it is true. (There may be more than one correct new statement.)

1. For the equation $ax + b = c$ to match $x + b = c$, the constant a should be equal to 0.

2. For the equation $ax = c$ to match $ax + b = c$, the constant b should be equal to 0.

Read the following paragraph about creating equations and work through the problems.

Equations of the form $ax + b = c$ can be created by using both the addition and multiplication principles of equality. Let's start with the equation $x = 2$.

First, apply the addition principle by adding (-7) to both sides of the equation.

$$x = 2$$
$$x + (-7) = 2 + (-7)$$
$$x - 7 = -5$$

Then, apply the multiplication principle to $x - 7 = -5$ with 3 as a factor.

$$3(x - 7) = 3(-5)$$
$$3x - 21 = -15$$

This equation has the same solution as $x = 2$. This can be verified by substituting 2 for x in the equation just created.

Quick Tip

Never leave two operators such as + – next to one another in an expression. Always simplify them to one operator. In this case $x + (-7) = x - 7$.

3. The order in which you use the principles of equality makes a difference. Let's create two more equivalent equations that have the same solution $x = 2$. Use the number 5 with the addition principle and the number 10 with the multiplication principle. Be sure to simplify after using each principle.

Quick Tip

Remember that the **multiplication principle of equality** states that both sides must be multiplied by the same value, which means *every* term on both sides need to be multiplied by the same value.

a. First, use the addition principle of equality and then use the multiplication principle of equality.

b. Now, use the multiplication principle of equality first and then use the addition principle of equality.

c. How do your answers from parts **a.** and **b.** compare?

Quick Tip

Solving equations is similar to performing the order of operations in reverse. When solving two-step equations, perform addition or subtraction before multiplication or division.

Solving Linear Equations that Simplify to the Form $ax + b = c$

1. Combine like terms on both sides of the equation.
2. Use the addition principle of equality to add the opposite of the constant b to both sides and simplify.
3. Use the multiplication principle of equality to multiply both sides of the equation by the reciprocal of the coefficient of the variable and simplify.
4. Check your answer by substituting it for the variable in the original equation.

4. To solve the equation created in Problem 3, we need to "undo" the steps that were taken to build it. However, we may not undo the steps in the order that they were used to create the problem.

a. Solve the equation from part **a.** of Problem 3.

b. Solve the equation from part **b.** of Problem 3.

c. The two equations were built using different steps but solved using the same steps. Why do you think that is?

Name: Date:

5. When solving equations with decimal numbers, such as $2.5x + 4.2 = 12.7$, you may find it easier to work with integers than to work with decimals.

a. Multiples of what number can be used to change the decimal numbers in the equation $2.5x + 4.2 = 12.7$ into integers? (**Hint:** This is from Section 3.3.)

b. Use the concept from part **a.** to rewrite the equation $2.5x + 4.0 = 12.75$ with integers.

Lesson Link

Finding the LCD of a list of fractions was introduced in Section 2.3.

6. When solving equations with fractions, such as $\frac{1}{2}x + \frac{7}{8} = 2\frac{3}{8}$, you may find it easier to work with integers than to work with fractions.

a. Multiplying the entire equation by the LCD will change the fractions in the equation $\frac{1}{2}x + \frac{7}{8} = 2\frac{3}{8}$ into integers. Find the LCD of the equation.

b. Use the LCD from part **a.** to rewrite the equation $\frac{1}{2}x + \frac{7}{8} = 2\frac{3}{8}$ without fractions.

Skill Check

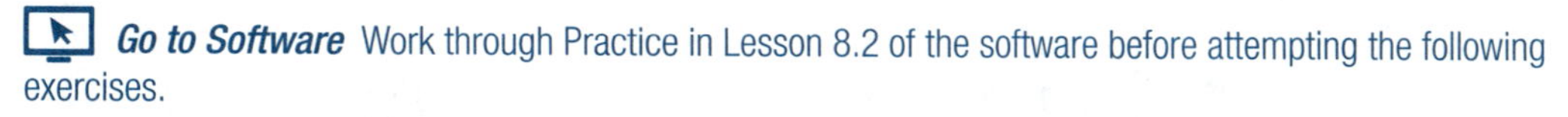

Go to Software Work through Practice in Lesson 8.2 of the software before attempting the following exercises.

Solve each linear equation.

7. $3x + 11 = 2$

8. $14 + 9t = 5$

9. $0.2n - 1.2 + 0.1n = 0$

10. $\frac{1}{2} - \frac{3}{8}x = \frac{5}{6}$

11. $4.7 - 0.5x - 0.3x = -0.1$

12. $\frac{5}{8}x - \frac{1}{4}x + \frac{1}{2} = \frac{3}{10}$

Apply Skills

Work through the problems in this section to apply the skills you have learned related to solving linear equations of the form $ax + b = c$.

13. Starbucks sells cake pops individually and in packages of 4. At the beginning of the day, Starbucks had 114 cake pops in stock. They sold 34 individual cake pops and several packages of cake pops. At the end of the day, there were 8 cake pops left. This situation can be modeled by the equation $114 - 34 - 4x = 8$, where x is the number of packages of cake pops sold.

a. Explain what each term in the equation $114 - 34 - 4x = 8$ represents in the situation.

b. Solve the equation to determine the value of x.

c. What does the answer to part **b.** mean? Write a complete sentence.

14. The lowest temperature of the night was reported to be 24 °F. The weather report mentioned that the temperature has steadily risen 1.5 degrees per hour since the lowest temperature of the day and it is currently 30 °F. This situation can be modeled by the equation $24 + 1.5x = 30$, where x is the time in hours since the lowest temperature was recorded.

a. Explain what each term in the equation $24 + 1.5x = 30$ represents in the situation.

b. Solve the equation to determine the value of x.

c. What does the answer to part **b.** mean? Write a complete sentence.

15. While taking inventory, a nurse records that there are $\frac{3}{5}$ of a box of syringes in one closet, two boxes that are $\frac{1}{8}$ full in another closet, and 24 syringes in the supply cart. He calculates the total to be 194 syringes. The staff member who reorders supplies is new and doesn't know how many syringes are in the boxes that the clinic uses, so she sets up the equation $\frac{3}{5}x + 2 \cdot \left(\frac{1}{8}x\right) + 24 = 194$, where x is the number of syringes in a box.

a. Solve the equation to determine the value of x.

b. What does the answer to part **a.** mean? Write a complete sentence.

Name: Date:

8.3 Solving Linear Equations: *ax* + *b* = *cx* + *d*

Objectives

Solve equations of the form $ax + b = cx + d$.

Identify conditional equations, identities, and contradictions.

Success Strategy

You will be learning new terminology related to linear equations in this section. Make sure you understand the difference between the three types of linear equations by doing extra practice in the software.

Understand Concepts

Go to Software First, read through Learn in Lesson 8.3 of the software. Then, work through the problems in this section to expand your understanding of the concepts related to solving linear equations of the form $ax + b = cx + d$.

The most general form for linear equations with one variable is

$$ax + b = cx + d,$$

where x is the variable and a, b, c, and d are constants. As in the previous two sections, most equations will not appear exactly in this form. Equations that fall into this category may have many like terms to be combined and may have grouping symbols that need to be simplified with the distributive property. To solve these equations, both the addition principle of equality and the multiplication principle of equality will be used.

Lesson Link

The distributive property was introduced in Section 7.6.

1. Combine all like terms on each side of the equation and determine if this equation simplifies to the form $ax + b = cx + d$, $ax + b = c$, $x + b = c$, or $ax = c$.

a. $2x + 7 - 3x + 5 = 18 - 18$

b. $4 + 12x - 4 = 3x + 5 - 3x$

c. $3x + 24 + 7x - 15 = 10x - 2 + 3$

d. $-2x - 1 + 3x = 52 - 42$

Read the following paragraph about creating equations and work through the problems.

Equations of the form $ax + b = cx + d$ can be created by using both the addition and multiplication principles of equality. Let's start with the equation $x = 1$.

Apply the addition principle of equality by adding 9 to both sides.

$$x = 1$$
$$x + 9 = 1 + 9$$
$$x + 9 = 10$$

Apply the multiplication principle of equality to $x + 9 = 10$ by multiplying both sides by 2.

$$2(x + 9) = 2 \cdot 10$$
$$2x + 18 = 20$$

Apply the addition principle of equality again by adding $(-4x)$ to both sides.

$$2x+18=20$$
$$2x+18+(-4x)=20+(-4x)$$
$$-2x+18=20-4x$$

2. Follow the procedure demonstrated above to create another equation of the form $ax+b=cx+d$ with $x=1$ as the solution. Use different values when applying the addition and multiplication principles than were used in the example.

Quick Tip

For Step 2, it is usually easier to use the addition principle of equality on the variable term with the smallest coefficient.

Solving Linear Equations that Simplify to the Form $ax+b=cx+d$

1. Simplify both sides of the equation and combine like terms.
2. Use the addition principle of equality to get all variable terms on one side of the equation. Simplify both sides of the equation.
3. Use the addition principle of equality again to get all of the constant terms on the other side of the equation from the variable terms. Simplify both sides of the equation.
4. Use the multiplication principle of equality to multiply both sides of the equation by the reciprocal of the coefficient of the variable and simplify.
5. Check your answer by substituting it for the variable in the original equation.

Quick Tip

Remember, the goal when solving linear equations is to get a single variable with a coefficient of positive 1 on one side of the equation and a constant on the other side.

3. To solve the equation created in Problem 1, we need to "undo" the steps that were taken to build it. However, we may not undo the steps in the order that they were used to create the problem.

a. Solve the equation $18-2x=20-4x$ for x to verify that the solution is $x=1$. Be sure to show all of your work.

b. Solve the equation you created in Problem 2 to verify that the solution is $x=1$.

Lesson Link

The term **infinite** was introduced in Section 7.1. Infinite means that a list or collection of items goes on forever. The number of items in the list cannot be represented by a natural number.

Types of Linear Equations

Conditional equations have exactly one solution. These equations simplify to the form $x=a$, where a is a real number and x is the variable.

Identity equations have an infinite number of solutions. Identities are equations which are true for all real numbers. These equations simplify to the form $0=0$.

Contradiction equations have no solutions. Contradictions are equations which are always false. These equations simplify to the form $a=b$, where a and b are different real numbers.

Name: Date:

Quick Tip

The m-w.com definition of **contradict** is to disagree with (something) in a way that shows or suggests that it is false, wrong, etc.

4. Simplify or solve each equation and then identify it as conditional, an identity, or a contradiction.

a. $5 = 7 - x$

b. $11x + 4 = 11x + 3$

c. $x + 7 = 2x - 5$

d. $x + 9 + 3(x + 1) = 6x + 12 - 2x$

Skill Check

Go to Software Work through Practice in Lesson 8.3 of the software before attempting the following exercises.

Solve each linear equation.

5. $4n - 3 = n + 6$

6. $2(z + 1) = 3z + 3$

7. $\frac{y}{5} + \frac{3}{4} = \frac{y}{2} + \frac{3}{4}$

8. $0.4(x + 3) = 0.3(x - 6)$

Apply Skills

Work through the problems in this section to apply the skills you have learned related to solving linear equations $ax + b = cx + d$.

9. A company has two packaging options for shipping quantities of a certain inventory item. Option A uses 20 boxes and there are 5 items unpacked. Option B requires more filler and uses 23 boxes where each box holds 2 less items than Option A and there are only 3 items unpacked. This situation can be represented by $20x + 5 = 23(x - 2) + 3$, where x is the number of items that can fit in the box used for Option A.

a. What does $20x + 5$ represent in the equation?

b. What does $x - 2$ represent?

c. Solve the equation for x.

d. Check the solution.

e. What does the answer from part **c.** mean? Write a complete sentence.

Lesson Link

The **area of a rectangle** is $A = l \cdot w$. Area formulas were introduced in Section 5.3.

10. Two advertisement flyers have the same area. The first flyer has a length of 12 inches and a width of x inches. The second flyer has a length of 4 inches and a width that is 10 inches more than x. This situation can be represented by $12x = 4(10 + x)$, where x is the width of the first flyer.

a. What does $12x$ represent in the equation?

b. What does $10 + x$ represent?

c. Solve the equation for x.

d. Check the solution.

e. What does the answer from part **c.** mean? Write a complete sentence.

Quick Tip

100% of cost + 9% of cost for taxes is equal to 109% of cost. This is written as 1.09 in decimal number form.

11. The manager of a café wants to list a price for the weekly featured combo that includes tax. He wants to sell a medium house-blend coffee with a pastry for a total of \$5.45. He doesn't know which pastry to sell with the coffee to avoid losing money on the combo. The medium coffee costs \$2.75 and the tax is 9%. He uses the equation $1.09(2.75 + x) = 5.45$ to determine the price of the pastry, which is represented by the variable x.

a. What does the sum $2.75 + x$ represent?

b. Solve the equation for x.

c. Which of the following pastries would you choose to be a part of the combo? Explain why you made your choice.

cherry pie for \$2.50, coffee cake for \$2.25, bagel for \$2.00

Name: Date:

8.4 Applications: Number Problems and Consecutive Integers

Objectives

Solve word problems involving translating number problems.

Solve word problems involving translating consecutive integers.

Solve word problems involving a variety of other applications.

Success Strategy

The key to solving number problems and integer problems is in setting up the equation correctly. Be sure to work through the examples in the software and the workbook carefully. Always check your solutions against the original wording in the problem to make sure they make sense.

Understand Concepts

Lesson Link

Solving number problems was first introduced in Section 1.7. Setting up the equations for number problems follows the same basic strategy that was previously introduced for all word problems.

Go to Software First, read through Learn in Lesson 8.4 of the software. Then, work through the problems in this section to expand your understanding of the concepts related to applications of solving linear equations.

There are several different types of word problems. The majority of the word problems presented in this workbook are real-life application problems. Number problems are typically not based on real-life scenarios, but still require you to translate English words into algebraic equations and then solve them. Number problems are like riddles that use numbers and variables to describe relationships among expressions or a set of numbers. A specific type of number problem is the consecutive integer problem which requires you to solve for a variable to determine several consecutive numbers that meet certain criteria.

Consecutive Integers

Consecutive integers have the form $n, n+1, n+2, \ldots$, where n is an integer.

Consecutive even or odd integers have the form $n, n+2, n+4, \ldots$, where n is an integer.

Notice that n is the smallest integer in the sequence of consecutive integers.

True or False: Determine whether each statement is true or false. Rewrite any false statement so that it is true. (There may be more than one correct new statement.)

1. The odd integers 3, 5, and 7 can be represented by x, $x+1$, and $x+3$, where $x = 3$.

2. The integers -5, -4, and -3 can be represented by n, $n+1$, and $n+2$, where $n = -3$.

3. The even integers -2, 0, and 2 can be represented by y, $y+2$, and $y+4$, where $y = -2$.

When solving consecutive integer problems, you need to set up an equation using the appropriate consecutive integer pattern. Then you solve for the variable. The next problem will guide you through the process.

4. Three consecutive integers have a sum of 24. What are the integers?

a. Which pattern would you use to represent the three consecutive integers?

b. Find the sum of the consecutive integers from part **a.** and combine like terms.

c. Write an equation using part **b.** to represent the sum of the integers being equal to 24.

d. Solve the equation from part **c.**

e. What does the answer to part **d.** mean? Write a complete sentence.

f. What are the three consecutive integers?

Skill Check

Go to Software Work through Practice in Lesson 8.4 of the software before attempting the following exercises.

Solve each number problem.

5. Five less than a number is equal to thirteen decreased by the number. Find the number.

6. Two added to the quotient of a number and seven is equal to negative three. What is the number?

Quick Tip

Remember the basic steps for problem-solving are as follows.

1. Understand the problem.
2. Devise a plan.
3. Carry out the plan.
4. Look back over your results.

Name: Date:

7. The sum of three consecutive even integers is 78. What are the integers?

8. Find three consecutive integers whose sum is one hundred sixty-eight more than the second number.

Apply Skills

Work through the problems in this section to apply the skills you have learned related to applications of solving linear equations.

Quick Tip

A **collect call**, also known as a **reverse-charge call**, is a telephone call where the person receiving the call is charged for the call. The person receiving the call has the option to accept or refuse the call.

9. A collect call from a landline in Ohio to another landline in Ohio has a connection fee of \$2.75 and a charge of \$0.36 per minute. Mr. Anderson made a collect call which cost the receiver of the call \$9.95. This situation can be modeled by $\$9.95 = \$2.75 + \$0.36m$.

a. The unknown value is represented by the variable m in the equation. What is the unknown value in this situation?

b. Solve the equation for the variable.

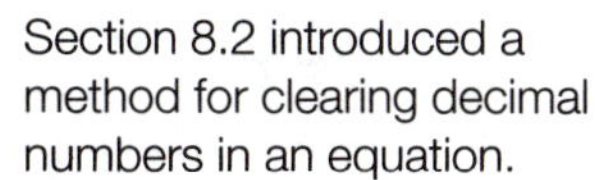

Section 8.2 introduced a method for clearing decimal numbers in an equation.

c. What does the answer to part **b.** mean? Write a complete sentence.

10. Robin is in charge of purchasing desserts for a dinner party that her nonprofit organization is throwing. She decides to buy a cake and several specialty cupcakes from Barbara's Bombtastic Bakery. She needs to buy one 8-inch round cake which costs \$19.50. She has \$45 to spend and will spend the leftover amount on cupcakes, which are \$8.50 for a box of 4. How many boxes of cupcakes can Robin purchase?

a. What is the unknown value in this problem? Let the variable c represent this unknown value.

b. Write an equation to represent this situation.

c. Solve the equation for the variable.

d. What does the answer to part **c.** mean? Write a complete sentence.

11. For his Superbowl party, John bought 3 large pizzas: a pepperoni, a sausage and mushroom, and a Hawaiian pizza with ham and pineapple. Each pizza was cut into the same number of slices. After the party was over, there was $\frac{1}{4}$ of the pepperoni pizza left, $\frac{1}{2}$ of the sausage and mushroom pizza left, and $\frac{3}{8}$ of the Hawaiian pizza left. There were a total of 9 pieces of pizza leftover. How many slices was each pizza cut into?

a. What is the unknown value in this problem? Let the variable p represent this unknown value.

b. Write an equation to represent this situation.

c. Solve the equation for the variable.

d. What does the answer to part **c.** mean? Write a complete sentence.

Name: Date:

8.5 Working with Formulas

Objectives

Use common formulas.

Evaluate formulas for given values of the variables.

Solve formulas for specified variables in terms of other variables.

Success Strategy

When solving a formula for a variable, circling the variable you are solving for can help you keep track of it as you try to isolate it on one side of the equation.

Understand Concepts

Go to Software First, read through Learn in Lesson 8.5 of the software. Then, work through the problems in this section to expand your understanding of the concepts related to working with formulas.

In commonly used formulas, the letter chosen for the variable usually relates to the value it represents. For instance, the variable t is often used to represent an amount of time and the variable r is often used to represent the length of a radius in geometric formulas. Formulas in which the variables have no operators between them imply multiplication, for example $I = Prt$ means $I = P \cdot r \cdot t$.

1. Fill in the table with the meanings of each variable in the given equations.

Formula Name	Formula	Variable Meanings
Simple Interest	$I = Prt$	I = interest, P = principle, r = rate, t = time
Temperature	$C = \frac{5}{9}(F - 32)$	
Distance Traveled	$d = rt$	
Perimeter of Rectangle	$P = 2l + 2w$	
Lateral Surface Area of Cylinder	$L = 2\pi rh$	
Force	$F = ma$	
Perimeter of a Triangle	$P = a + b + c$	

Quick Tip

The **lateral surface area** of a 3-D object is the sum of the surface area of all its sides except the top and bottom bases.

It's important to understand formulas and not just memorize them. If you understand what the formula means and what the different variables in the formula represent, you might be able to build the formula through reasoning rather than memorizing it. Having a better understanding of a formula will also make it easier to determine which values should be substituted for each of the variables.

Perimeter was discussed in depth in Section 5.2.

2. Let's look at the formula for the perimeter of a rectangle, $P = 2l + 2w$.

a. What is the definition of perimeter?

b. What do the variables l and w represent?

c. Why does the formula have the number 2 in it twice?

Read through the following information about solving formulas for different variables and work through the problem.

Solving a formula for a different variable can be useful. For example, if you constantly need to change temperature from degrees Celsius to degrees Fahrenheit, repeatedly plugging in values for C into the formula $C = \frac{5}{9}(F - 32)$ and then solving for F is not the most efficient approach. Solving the formula for F first and then substituting in values for C will save you time.

Quick Tip

Before solving the formula for a specific variable, determine which variable the formula is currently solved for.

Solving Formulas for a Variable

1. Determine which variable the equation needs to be solved for.
2. Treat all other variables as constants.
3. Solve the equation for the variable from Step 1 by using the properties of real numbers and the principles of equality to rearrange the formula so that the variable to be solved for is isolated on one side of the equation.

3. This problem will guide you through solving an equation for another variable.

a. What variable is the equation $C = \frac{5}{9}(F - 32)$ currently solved for?

b. Which variable do you need to solve for in the formula $C = \frac{5}{9}(F - 32)$ to make it easier to convert *from* degrees Celsius *to* degrees Fahrenheit?

c. When solving a formula for one variable, all other variables in the equation will be treated as constants. Which variable will be treated as a constant?

Lesson Link

Reciprocal was explained in Section 2.2.

d. Solve the equation for the variable from part **b.** (**Hint:** First change the fraction into an integer by multiplying both sides of the equation by its reciprocal.)

e. Substitute 20 degrees Celsius for C into the formula $C = \frac{5}{9}(F - 32)$ and solve for F. Write the solution you get for F here. (Do not convert the fraction to a decimal.)

f. Substitute 20 degrees Celsius for C into the equation from part **d.** Does the value you get for F match the value you found in part **e.**?

Name: Date:

g. Which equation, the one in part **b.** or the one in part **d.**, was easier to use to solve for the value of F given 20 degrees Celsius for the value of C?

Skill Check

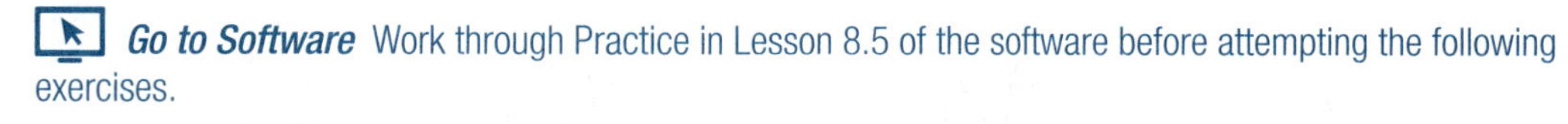

Go to Software Work through Practice in Lesson 8.5 of the software before attempting the following exercises.

Evaluate each formula for the given values. Use $\pi = 3.14$ when needed.

4. $F = ma$; $m = 250$ kg; $a = 9.8\text{ m/s}^2$

5. $L = 2\pi rh$; $r = 15$ cm; $h = 20$ cm

Solve each equation for the indicated variable.

6. $I = Prt$; Solve for t

7. $V = \frac{1}{3}bh$; Solve for h

Apply Skills

Work through the problems in this section to apply the skills you have learned related to working with formulas.

8. Samantha uses a credit promotion at a home improvement store where she doesn't have to pay any interest on her purchase as long as she pays off the entire balance within 6 months. She purchases $8000 in merchandise. If she fails to pay off the balance within 6 months, then she will be charged $600 in interest. Samantha lost the paper work and wants to determine the interest rate on her purchase.

a. Which formula from Problem 1 fits this situation?

b. Match the variables in the formula from part **a.** to the information provided.

c. The formula from part **a.** needs to be solved for which variable?

d. What is the interest rate on her purchase? (Remember to convert to a percent.)

Lesson Link

Changing decimal numbers to percents was covered in Section 4.3.

9. In a physics lab, a ball is rolled down an incline that has a machine at the bottom which calculates the force of impact. The ball has a mass of 1.5 kilograms. After several trials, the average force of impact is calculated to be 12.75 kg · m/s^2. The researchers need to determine the average acceleration of the ball at the moment it struck the machine.

a. Which formula from Problem 1 fits this situation?

b. Match the variables in the formula from part **a.** to the information provided.

c. The formula from part **a.** needs to be solved for which variable?

d. What was the average acceleration of the ball in m/s^2?

10. Charles is experimenting with a new sail design for his sailboat and needs to keep the total area of the triangular sail to 150 square feet. The base of the sail must be exactly 3 times the height of the sail.

Lesson Link

Area formulas for geometric figures were introduced in Section 5.3.

a. What geometric formula for area should be used?

b. Write an expression for the base of the formula using the variable h for height.

c. Substitute the expression from part **b.** into the area formula for the base.

d. Solve this formula for the height squared.

e. What would you have to do to both sides of the equation in part **d.** to solve the formula for the height?

f. Substitute 150 for the area of the sail in the formula from part **d.** and solve for the height of the sail.

g. What is the length of the base of the sail?

Name: Date:

8.6 Applications: Distance-Rate-Time, Interest, Average

Objectives

Solve distance-rate-time problems by using linear equations.

Solve simple interest problems by using linear equations.

Solve average problems by using linear equations.

Solve cost problems by using linear equations.

Success Strategy

Before starting this section, you may find it helpful to review how to calculate discounts from Section 4.7 and how to calculate simple interest for a single investment from Section 4.8.

Understand Concepts

Go to Software First, read through Learn in Lesson 8.6 of the software. Then, work through the problems in this section to expand your understanding of the concepts related to applications of formulas.

When solving problems, tables can be useful to help organize and set up an equation. In this section, tables are introduced for two particular problem types: distance-rate-time problems and interest problems. Both of these problem types involve using the information from two equations to solve for unknown variables.

Distance-rate-time problems use the equation $d = rt$. The word problem will describe two situations or two parts of the same situation. This information will be used to set up an equation. The next problem will guide you through the steps to solve a distance-rate-time problem.

1. A motorist averaged 45 mph for the first part of a trip and 54 mph for the last part of the trip. If the total trip was 303 miles and took 6 hours, how long did each part of the trip take the motorist?

a. Does the problem statement give you values for the rates, times, or distances for **both parts** of the trip? If so, what is given? Place these values in the appropriate cells of the table below.

b. The problem is asking for the time for each part of the trip. Fill in the time column of the table by using the variable t to represent the time for the first part of the trip. The time for the last part of the trip will be equal to the difference between the total time for the trip and the time for the first part of the trip.

c. We now have the rates and times for both parts of the trip. Multiply the rate by the time in each row to get an expression for distance and place in the distance column.

	Rate (mph)	·	Time (min)	=	Distance (miles)
First Part					
Last Part					

d. We now have terms to represent the distance of each part of the trip. What was the total distance of the trip?

Quick Tip

Remember to use the distributive property when you multiply a constant by an expression with more than one term. For example, $5(4 - t) = 20 - 5t$.

e. The sum of the terms in the distance column is equal to the total distance of the trip. Write an algebraic equation to represent this situation.

f. Solve the equation from part **e.** for t, which represents the time for the *first* part of the trip. Write your answer in fractional form.

g. What was the time for the *last* part of the trip? Write your answer in fractional form.

h. Determine the length in miles of each part of the trip.

i. Do the miles from part **h.** add up to the total miles traveled on the trip?

Interest problems use the equation $I = Prt$. For the following problems, time will be equal to 1 year. This means we will work with the equation $I = Pr$. The next problem will guide you through the steps to solve an interest problem.

Quick Tip

Remember that rates need to be written in decimal number form when solving interest problems.

2. Amanda invests $25,000 for a year in a bond fund that pays 5% interest and a stock fund that pays 6% interest. The annual return (or interest) on the 5% investment exceeds the annual return on the 6% investment by $40. How much did she invest at each rate?

a. Does the problem statement give you values for the principal, rates, or interest for both parts of the investment? If so, what is given? Place these values in the appropriate cell of the table.

b. The problem is asking for the amount invested in each type of fund. The total of the two investments has to add up to $25,000. Fill in the principle column of the table by using the variable P to represent the principal for the bond fund. The stock fund will be equal to the difference between the total investment amount and the principle for the bond fund.

c. We now have the principals and rates for both parts of the investment. Multiply the principal by the rate in each row to get an expression for interest and place in the interest column of the table.

	Principal ($)	·	Rate	=	Interest ($)
Stock Fund					
Bond Fund					

Name: Date:

Quick Tip

Notice that Problem 1 involves a sum of terms and Problem 2 involves a difference of terms. Be sure to carefully read the problem statement to determine if a sum or difference needs to be used to find the solution.

d. We know that the difference between the interest earned on the bond fund and the interest earned on the stock fund is \$40. Write an algebraic equation to represent this situation.

e. Solve the equation from part **d.** for P, which represents the amount of money invested at 5%.

f. Determine the amount of money invested at 6%.

Read the following paragraph about average problems and work through the problems.

Average problems in this section are similar to the ones you have seen before in the workbook. The only difference is that one piece of data is missing. The missing piece of data will be represented by a variable.

3. Kayleigh needs an exam average of 82 to pass her history course. All five exams are weighted equally and are worth 100 points each. She has earned the following scores on exams so far in the course: 62, 90, 89, 80. What does she need to earn on the final exam to have an exam average of 82 for the course?

a. What value is unknown? Let the variable E represent this unknown value.

b. Write an expression to find the average of the exam scores. Be sure to include the variable from part **a.**

Lesson Link

A method to clear an equation of fractions was introduced in Section 8.2.

c. Create an equation by setting the expression from part **b.** equal to the average score that Kayleigh needs to pass.

d. Solve the equation from part **c.** for the variable.

e. What does the answer to part **d.** mean?

Lesson Link

Weighted means were introduced in Section 6.1.

f. Suppose the final exam counted double in the average. What would Kayleigh need to earn on the final exam to have an exam average of 82?

g. Compare the values from parts **d.** and **f.** Which scenario requires that Kayleigh make a higher grade on the final exam? Why do you think this is so?

Skill Check

Go to Software Work through Practice in Lesson 8.6 of the software before attempting the following exercises.

Solve each linear equation.

4. $12(5.5-t)=10t$

5. $\frac{1}{2}x=3(8-x)$

6. $0.055x-251=0.06(10{,}000-x)$

7. $\frac{75+82+90+85+77+x}{6}=80$

Apply Skills

Quick Tip

Problem 8 is based on one of Zeno's paradoxes of motion. Lewis Carroll wrote a short story "What the Tortoise Said to Achilles" based on this paradox.

Work through the problems in this section to apply the skills you have learned related to applications of formulas.

8. Achilles is racing a tortoise and gives him a 2-hour head start. The tortoise runs at a pace of 10 miles per hour and Achilles runs at a pace of 25 miles per hour. How long will it take Achilles to catch up to the tortoise?

a. Fill out the $d=r\cdot t$ table. Let the variable t represent the amount of time that the tortoise has traveled.

	Rate (mph)	·	Time (min)	=	Distance (miles)
Tortoise					
Achilles					

b. When Achilles catches up to the tortoise, they will have traveled the same distance. Set up a linear equation using the information in the table.

c. Solve the equation from part **b.** for the variable.

d. How long will it take Achilles to catch up to the tortoise?

e. If the race is 35 miles long, will Achilles pass the tortoise before crossing the finish line? Show work to support your answer.

9. Savannah invests \$3600 per year into her retirement account, a portion of which is a contribution match from her employer. Savannah invests the employer match in a high-risk fund that averages a return of 8% and invests the rest in a low-risk account that averages a return of 4%. She wants to earn a total of \$198 in interest for the year. How much should be invested in each fund?

a. Fill out the $I = P \cdot r$ table. Let the variable P represent the amount of money invested in the high-risk fund.

	Principal (\$)	·	Rate	=	Interest (\$)
High-Risk Fund					
Low-Risk Fund					

b. Write an equation to represent the total interest earned for the year by using the information in the table.

c. Solve the equation from part **b.** for the variable.

d. How much should be invested in each fund?

e. Verify that investing the amounts from part **d.** at the given rates yields \$198 total interest in Savannah's retirement account.

10. Kevin consulted a dietician who told him to consume an average of 2100 calories per day based on his age, current weight, activity level, and weight goals. Kevin kept track of his calorie intake for several days. He consumed 2050 calories on Monday, 2200 calories on Tuesday, 2300 calories on Wednesday, and 2400 calories on Thursday. How many calories would he need to consume on Friday to have an average calorie intake of 2100 for the five days?

Quick Tip

The amount of calories a person needs per day varies depending on several factors, such as their sex, age, current weight, activity level, and weight goals. A doctor should be consulted before starting any diet or exercise regimen.

a. Set up an equation to solve for the amount of calories Kevin would need to consume on Friday. Use the variable x to represent the number of calories needed.

b. Solve the equation from part **a.** for the variable.

c. It is recommended that active men consume more than 1500 calories per day to avoid triggering "starvation mode" in the body. Can Kevin stay above this calorie amount and meet his recommended average for the 5 days?

d. Do you think this is a smart way for Kevin to adjust his average calorie intake? If not, what are some alternatives?

11. Robin's Refurbished Wrecks purchased a used car for $2850. For the upcoming Labor Day sale, the car dealership would like to offer a 5% discount off the posted selling price of the car, but would still like to make a 40% profit. What price should the car dealership advertise for the car?

a. Use the purchase price of the car to determine how much a 40% profit will be.

b. Use the variable x to represent the actual selling price of the car. Write an expression to represent the selling price of the car after the 5% discount.

c. Write an equation the represents the situation by using the answers from parts **a.** and **b.** along with the equation *selling price* – *cost* = *profit.*

Lesson Link

Solving distance-rate-time problems will be covered in more depth in Chapter 10.

d. Solve the equation from part **c.**

e. What does the answer from part **d.** mean? Write a complete sentence.

Name: Date:

8.7 Linear Inequalities

Objectives

Use set-builder notation.

Use interval notation.

Solve linear inequalities.

Solve compound inequalities.

Learn how to apply inequalities to solve word problems.

Success Strategy

When writing interval notation to describe the solution sets of inequalities in this section, it is helpful to first graph the solution set on the real number line. This will help you determine the lowest value and the highest value in the solution set.

Understand Concepts

Go to Software First, read through Learn in Lesson 8.7 of the software. Then, work through the problems in this section to expand your understanding of the concepts related to linear inequalities.

Sets

A **set** is a collection of objects or numbers.

An **element** is an item in a set.

Quick Tip

In roster form, the order of the elements is not important and all elements must be distinct. For example, the set of letters forming MISSISSIPPI could be written in roster notation as {M, I, S, P} or {I, M, P, S}.

The set $A = \{1, 2, 3, 4\}$ is said to be written in **roster form** because each element of the set is listed. The symbol $\in$ is used to indicate that a particular item is an element of a set. That is, $2 \in A$ means that "2 is an element of the set A." The symbol $\notin$ means that a particular item is not an element of the set. That is, $5 \notin A$ means that "5 is not an element of the set A."

True or False: Determine whether each statement is true or false. Rewrite any false statement so that it is true. (There may be more than one correct new statement.)

Let $B = \{4, 8, 12, 16\}$ and $C = \{1, 4, 10, 17\}$.

1. 12 is not an element of the set B.

2. $10 \in C$

3. 4 is an element of both set B and set C.

4. 5 is not an element of set B or set C.

5. 8 is not an element of set B or set C.

Read the following paragraph about roster notation and work through the problems.

Lesson Link

The definitions for finite and infinite were first introduced in Section 7.1.

Each set listed in roster notation so far is said to be **finite**. If you can indicate exactly how many elements are in a set with a whole number, then the set is finite. Any set which is not finite is called **infinite**. A set which contains no elements is called an **empty set**, or **null set**. The empty set is indicated by either a set of empty brackets $\{\ \}$ or the symbol $\varnothing$. A series of three dots, called an ellipsis, is often used when working with sets to denote that a list of numbers continues in the established pattern such as $\{2, 4, 6, \ldots\}$ or $\{2, 4, 6, \ldots, 20\}$.

6. Determine if each set below is finite or infinite.

a. The set of natural numbers, $\mathbb{N} = \{1, 2, 3, \ldots\}$

b. The empty set $\{\ \}$

c. The set $P = \{1, 2, 4, 8, 16, 32, 64, 128\}$

d. The set of even integers, $E = \{\ldots, -4, -2, 0, 2, 4, \ldots\}$

e. The set of even numbers from 2 to 20, $B = \{2, 4, 6, \ldots, 20\}$

Read the following paragraph about set builder notation and work through the problems.

Another way to describe a set of numbers is with **set-builder notation**. Set-builder notation uses a combination of words and mathematical symbols to describe a set.

An example of set-builder notation is $S = \{x \mid x < 5 \text{ and } x \text{ is an integer}\}$. The name of this set is S, the variable is x, and the rules that describe the elements of the set are "$x < 5$" and "x is an integer." The vertical bar is read as "such that" and separates the variable from the rules. To describe the set S in words, we would read the set-builder notation as "S is the set of all values of x such that x is less than 5 and x is an integer." This set can be written in roster notation as $S = \{\ldots, -2, -1, 0, 1, 2, 3, 4\}$.

Quick Tip

If the set is an infinite set, remember to use an ellipsis. It is impossible to write out every element in an infinite set in roster notation.

7. The following sets are described using set-builder notation. Write the sets in roster notation.

a. $A = \{a \mid a \text{ is a whole number and } a \geq 2,\ a < 10\}$

b. $B = \{b \mid b \text{ is a multiple of 3 and } b > 17\}$

Name: Date:

8. Write the following sets described in words using set-builder notation and answer the questions.

a. The set E contains all even integers that are greater than or equal to ten.

b. The set Q contains all *real numbers* that are less than five.

c. How does the set Q in part **b.** compare to the set S described in the example before Problem 7?

Read the following paragraph about linear inequalities and work through the problems.

Quick Tip

The notation for open intervals and the notation for ordered pairs look similar. For example, writing (1, 2) can be confusing if the context isn't clear.

Interval notation is used when a set contains all real numbers between two values. Interval notation uses parentheses ( ) or square brackets [ ] or a combination of both. A parenthesis means that the end point is not a part of the set of numbers and a square bracket means that the end point is included.

Linear Inequalities

A **linear inequality** is an inequality of the form

$$ax + b < c,$$

where a, b, and c are real numbers and x is a variable. The inequality symbol can be any of the following: $<$, $>$, $\leq$, or $\geq$.

A **compound linear inequality** has the form

$$c < ax + b < d,$$

where a, b, c, and d are real numbers and x is a variable. As long as the pair of inequalities have the same direction, the inequality symbols can be any of the following: $<$, $>$, $\leq$, or $\geq$.

Quick Tip

Open intervals include neither endpoint, **half-open intervals** include only one end point, and **closed intervals** include both end points.

9. Linear inequalities can have solutions which are open, half-open, or closed intervals. Write the solution set of each graph in algebraic notation and interval notation. The first one has been filled in for you.

Graph	Type of Interval	Algebraic Notation	Interval Notation
	Open Interval	$a < x < b$	(a, b)
	Closed Interval		
	Half-Open Interval		
	Open Interval		
	Half-Open Interval		

Read the following paragraph about solving linear inequalities and work through the problems.

Quick Tip

The solution to a linear inequality is typically an interval of real numbers and not a single value, which occurs with linear equations.

Solving linear inequalities follows the same rules as solving linear equations with the exception of what happens when you multiply or divide by a negative number. Multiplying or dividing both sides of an inequality by a negative number causes the direction of the inequality symbol to reverse. That is, if the inequality was <, then multiplying or dividing both sides of the inequality by a negative would cause the inequality to change to >.

10. Let's explore why multiplying an inequality by a negative number switches the inequality symbol.

a. Start with the inequality $2 < 5$. Is this inequality true?

b. Multiply both sides of the inequality by negative one, but leave the inequality symbol alone.

c. Graph the two values from part **b.** on the number line.

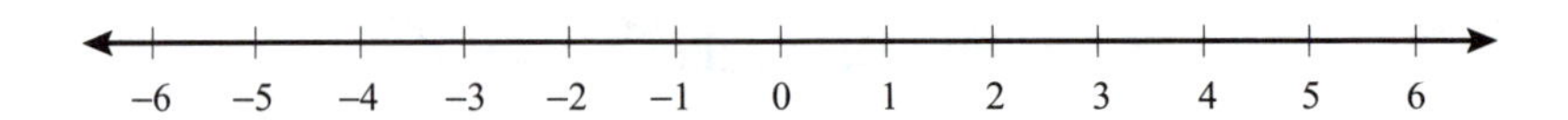

d. Is the result from part **b.** a true statement? Explain why or why not.

Rules for Solving Linear Inequalities

1. Simplify each side of the inequality.
2. Use the addition principle of equality to add the opposite of the constant term and/or the variable term to each side. At the end of this step, all terms with variables should be on only one side of the inequality.
3. Use the multiplication principle of equality to multiply each side of the inequality by the reciprocal of the coefficient of the variable. Remember that multiplying or dividing by a negative value reverses the inequality.
4. Check the answer by substituting a value from the solution set for the variable in the original inequality.

Quick Tip

All of the solutions to an inequality cannot be checked since the solution set is usually infinite in size.

Determine if any mistakes were made while solving each inequality. If any mistakes were made, describe the mistake and then correctly solve the inequality to find the actual result.

11.
$$7x-3\le -6x+36$$
$$13x-3\le 36$$
$$13x\le 33$$
$$x\le \frac{33}{13}$$

12.
$$-2x+15-7>19-27$$
$$-2x+8>-8$$
$$-2x>-16$$
$$x>8$$

Name: Date:

The goal when solving compound inequalities is to get the variable isolated in the middle expression. When solving compound inequalities, the addition and multiplication principles of equality are still used. The main difference is that there are now three parts of the inequality to work with, so all three parts must be added to or multiplied by the same values. If a compound inequality is multiplied or divided by a negative value, both inequality symbols are reversed.

Determine if any mistakes were made while solving each inequality. If any mistakes were made, describe the mistake and then correctly solve the inequality to find the actual result.

13. $-4 < 2x - 4 < 6$
$0 < 2x < 10$
$x < 5$

14. $1 \leq \frac{2}{3}x - 1 < 9$
$2 \leq \frac{2}{3}x < 10$
$3 \leq x < 15$

Skill Check

Quick Tip

To do a quick check of the solution set of a linear inequality, pick a point in the interval which is not an end point and substitute it into the original inequality.

Go to Software Work through Practice in Lesson 8.7 of the software before attempting the following exercises.

Solve each linear inequality.

15. $2x + 3 < 5$

16. $\frac{x}{3} - x > 1 - \frac{x}{3}$

17. $0.9x - 11.3 \leq 3.1 - 0.7x$

18. $\frac{2(x-1)}{3} < \frac{3(x+1)}{4}$

19. $2 \leq -x + 2 \leq 6$

20. $0.9 < 3x + 2.4 < 6.9$

Apply Skills

Work through the problems in this section to apply the skills you have learned related to linear inequalities.

21. Jeph is in charge of buying office supplies for the non-profit organization he works for. He has \$400 to spend. He needs to buy a printer that costs \$150, a box of printer paper for \$60, and some ink cartridges for \$12.50 each. What is the maximum number of ink cartridges that Jeph can buy? (**Note:** Tax is not included in the sales price.)

a. Set up the linear inequality. Use the variable c to represent the number of ink cartridges.

b. Solve the equation from part **a.** for the variable.

c. What does the answer from part **b.** mean? Write a complete sentence.

Quick Tip

The months with 30 days are April, June, September, and November.

22. Sarah is participating in National Novel Writers Month where she has to write a rough draft of a novel with at least 50,000 words during the month of November. At the end of the day on November 20th, she has a total of 32,500 words. What is the minimum number of words that Sarah needs to write each day for the rest of the month to make the goal of 50,000 words?

a. Set up the linear inequality. Use the variable w to represent the number of words per day.

b. Solve the equation from part **a.** for the variable.

c. What does the answer from part **b.** mean? Write a complete sentence.

23. Andrew needs to earn at least a B in each class to keep his scholarship. The grade in his economics class is based on five exams that are equally weighted. On the first four exams, Andrew received the following scores: 92, 74, 80, 72. Andrew needs an average of at least 80 to earn a B for the class. What range of scores does he need on the fifth exam to keep his scholarship?

a. Set up the linear inequality. Use the variable E to represent the fifth exam score.

b. Solve the equation from part **a.** for the variable.

c. What does the answer from part **b.** mean? Write a complete sentence.

Name: Date:

Chapter 8 Projects

Project A: What's a Safe Dosage?

An activity to demonstrate the use of equation-solving in real life

If you have children, you may have found yourself in this predicament. Your child wakes up with a high fever in the middle of the night, but you do not have any children's fever-reducing medication on hand. The stores are closed and you want to make your child comfortable. Is it alright to give a child a fever-reducing medication that is made for adults?

It is never alright to give children medication that is prescribed to an adult. However, for over-the-counter medications (or OTCs), there are mathematical formulas that can be used to determine how much of an adult medication is safe for a child. To be safe, always check with a pediatric nurse advice line before giving a child any medication not specifically designed for children.

One method for determining a safe child's dosage is given by the following equation called **Young's Rule**:

$$d = \frac{a \cdot D}{a + 12},$$

where d is the child's dosage, D is the adult dosage, and a is the age of the child in years.

Another formula used to determine the safe dosage of a medication to give to a child is called **Cowling's Rule**:

$$d = \frac{D(a+1)}{24},$$

where d is the child's dosage, D is the adult dosage, and a is the age of the child in years.

Let's compare the results from using these two formulas. Round your answers to the nearest tenth if necessary.

1. If the adult dosage of a drug is 40 mL, how much should a 3-year-old child receive?

a. Use both formulas to calculate the dosage.

b. How do the results from using the two formulas compare?

c. What are the units for the child's dosage?

2. If the adult dosage of a drug is 170 milligrams, how much should a 5-year-old child receive?

a. Use both formulas to calculate the dosage.

b. How do the results from using the two formulas compare?

c. What are the units for the child's dosage?

3. Does doubling the age of the child double the dosage?

   a. Double the age of the child in Problem 1 and recalculate the dosage using both formulas. Show all work.

   b. Compare these results to those of Problem 1. Did either dosage double exactly?

   c. Why do you think this happens?

4. If the adult dosage of a drug is 170 milligrams, how much should a 12-year-old child receive?

   a. Use both formulas to calculate the dosage.

   b. How do the results from using the two formulas compare?

   c. How do these results compare to the answers found in Problem 2?

5. If you have both dosage formulas available to you and you want to take a conservative approach in determining the amount of medication to give a child, what could you do?

6. Let's take the formula for Cowling's Rule and solve this equation for the age of the child by rearranging the formula such that the variable a is on the left side of the equal sign by itself and all other variables and constants are on the right side.

7. Use the formula from Problem 6 to determine the age of a child if the adult dosage is 180 mg and the child's dosage is 60 mg.

Name: Date:

Project B: A Linear Vacation

An activity to demonstrate the use of solving linear equations in real life

The process of finding ways to use math to solve real-life problems is called **mathematical modeling**. In the following activity you will be using linear equations to model some real-life scenarios that arise during a family vacation. (See Section A.11 for more on mathematical modeling.)

For each question be sure to write a linear equation in one variable and then solve.

1. Penny and her family went on vacation to Florida and decided to rent a car to do some sightseeing. The cost of the rental car was a fixed price per day plus \$0.29 per mile. When she returned the car, the bill was \$209.80 for three days and they had driven 320 miles. What was the fixed price per day to rent the car?

2. Penny's son Chase wanted to go to the driving range to hit some golf balls. Penny gave the pro-shop clerk \$60 for three buckets of golf balls and received \$7.50 in change. What was the cost of each bucket?

3. Penny's family decided to go to the Splash Park. They purchased two adult tickets and two child tickets. The adult tickets were $1\frac{1}{2}$ times the price of the child tickets and the total cost for all four tickets was \$85. What was the cost of each type of ticket?

4. Penny's family went shopping at a nearby souvenir shop where they decided to buy matching T-shirts. If they bought four T-shirts and a $2.99 bottle of sunscreen for a total cost of $54.95, before tax, how much did each T-shirt cost?

5. Penny and her family went out to eat at a local restaurant. Three of them ordered a fried shrimp basket but her daughter, Meghan, ordered a basket of chicken tenders, which was $4.95 less than the shrimp basket. If the total order before tax was $46.85, what was the price of a shrimp basket?

6. While on the beach, Penny and her family decided to play a game of volleyball. Penny and her son beat her husband and daughter by two points. If the combined score of both teams was 40, what was the score of the winning team?

Name: Date:

Math@Work

Financial Advisor

As a financial advisor working with a new client, you must first determine how much money your client has to invest. The client may have a lump sum that they have saved or inherited, or they may wish to contribute an amount monthly from their current salary. In the latter case, you must then have the client do a detailed budget, so that you can determine a reasonable amount that the client can afford to set aside on a monthly basis for investment.

The second piece of information necessary when dealing with a new client is determining how much risk-tolerance they have. If the client is young or has a lot of money to invest, they may be willing to take more risk and invest in more aggressive, higher interest-earning funds. If the client is older and close to retirement, or has little money to invest, they may prefer less-aggressive investments where they are essentially guaranteed a certain rate of return. The range of possible investments that would suit each client's needs and goals are determined using a survey of risk-tolerance.

Suppose you have a client who has a total of $25,000 to invest. You determine that there are two investment funds that meet the client's investment preferences. One option is an aggressive fund that earns an average of 12% interest and the other is a more moderate fund that earns an average of 5% interest. The client desires to earn $2300 this year from these investments.

Investment Type	Principal Invested	·	Interest Rate	=	Interest Earned
Aggressive Fund	x				
Moderate Fund					

To determine the amount of interest earned you know to use the table above and the formula $I = Prt$, where I is the interest earned, P is the principal or amount invested, r is the average rate of return, and t is the length of time invested. Since the initial investment will last one year, $t = 1$.

1. Fill in the table with the known information. If x is the amount invested in the aggressive fund and the total amount to be invested is $25,000, create an expression involving x for the amount that will be left to invest in the moderate fund. Place this expression in the appropriate cell of the table.

2. Determine an expression in x for the interest earned on each investment type by multiplying the principal by the interest rate.

3. Determine the amount invested in each fund by setting up an equation using the expressions in column four and the fact that the client desires to earn $2300 from the interest earned on both investments.

4. Verify that the investment amounts calculated for each fund in the previous step are correct by calculating the actual interest earned in a year for each and making sure they sum to $2300.

5. Why would you not advise your client to invest all their money in the fund earning 12% interest, after all, it has the highest average interest rate?

Foundations Skill Check for Chapter 9

This page lists several skills covered previously in the book and software that are needed to learn new skills in Chapter 9. To make sure you are prepared to learn these new skills, take the self-test below and determine if any specific skills need to be reviewed.

Each skill includes an easy (**e.**), medium (**m.**), and hard (**h.**) version. You should be able to complete each problem type at each skill level. If you are unable to complete the problems at the easy or medium level, go back to the given lesson in the software and review until you feel confident in your ability. If you are unable to complete the hard problem for a skill, or are able to complete it but with minor errors, a review of the skill may not be necessary. You can wait until the skill is needed in the chapter to decide whether or not you should work through a quick review.

6.1 Find the mean of each set of numbers.

e. 5, 6, 10

m. 1.8, 1.3, 2.6, 1.5

h. $4\frac{1}{3}, 5\frac{1}{6}, 1\frac{1}{2}$

7.7 Simplify the expression by combining like terms.

e. $12n - 4n$

m. $10z + 8 - 15z + 22$

h. $2(x-y)-8(y-x)$

7.7 Evaluate the expression for the given value.

e. $5x - 3;\ x = 0$

m. $9.3 - 4y;\ y = 2.5$

h. $\frac{1}{3}x - 2\frac{1}{6};\ x = 10$

8.5 Solve the formula for the indicated variable.

e. $P = a + b + c$; solve for c

m. $A = \frac{1}{2}bh$; solve for b

h. $P = 2l + 2w$; solve for w

8.7 Solve the linear inequality for the variable.

e. $x - 17 < 22$

m. $2x + 3 > 5$

h. $14 - 3x \leq -25$

Chapter 9: Linear Equations and Inequalities in Two Variables

Study Skills

9.1 The Cartesian Coordinate System

9.2 Graphing Linear Equations in Two Variables: $Ax + By = C$

9.3 The Slope-Intercept Form: $y = mx + b$

9.4 The Point-Slope Form: $y - y_1 = m(x - x_1)$

9.5 Introduction to Functions and Function Notation

9.6 Graphing Linear Inequalities in Two Variables

Chapter 9 Projects

Math@Work

Foundations Skill Check for Chapter 10

Math@Work

Introduction

If you are thinking about a career as a market research analyst, you could potentially work in a wide variety of industries. Many industries hire market research analysts directly to help them market their products and services or you could work for a large consulting firm that specializes in doing marketing research and analysis for other businesses. A market research analyst studies market conditions to determine the potential sales of a new product or service. Through their research they help companies determine what products or services people want, the demographics of the people who would buy them, and the price of the product or service. A market research analyst may also gather data on a company's competitors and their products to help them to better position themselves in the marketplace and price their products and services.

Suppose you start your career with a manufacturing firm that plans to create a new product line similar to one that a competitor recently introduced into the market. There are several things the company needs to know before investing the money to purchase equipment and modify its facilities in order to make the new product. What is the potential demand for this product? Who are the people who would buy this product and what are their demographics: age, gender, income, education, ethnicity, employment status, etc.? What price should the product be sold at to be competitive yet generate a reasonable profit? Finding the answers to these questions requires several of the skills you will learn in this chapter as well as skills from previous chapters. At the end of this chapter, we'll come back to the topic of market research and explore how math is used by a marketing research analyst.

Paying Attention and Being Mindful

Has anyone ever told you to pay attention? Perhaps it was a parent at home, an instructor while in class, or a supervisor at work. Chances are that we have all been told to pay attention at one time or another. What does it mean to pay attention? How do you become more attentive?

In the late 1990s, advancements in brain structure imaging provided a way to measure brain activity that could be linked to intelligence and attention. Researchers, such as Dr. Richard Davidson, discovered the "neuroplasticity" of the brain, which refers to the brain's ability to continue developing throughout our lifetime. Their research changed the theory that the brain was fairly static, or showed little growth, after a certain age. From his research, Dr. Davidson noted the high level of brain activity found among people who meditated regularly. He discovered that even thirty minutes of meditation a day over a period of two weeks resulted in a measurable increase in brain activity.

Meditation refers to a broad range of practices that were developed for various reasons: to promote relaxation, reduce stress, create energy, or develop certain emotions such as patience, compassion, or forgiveness. Regular meditation leads to a state of "mindfulness" that can also be achieved by simply focusing on and living in the present moment. Instead of dwelling on the past or constantly looking to the future, you focus on the here and now. By paying more attention to what is going on around you in the present, and focusing on tasks at hand, you can improve your brain functioning and attentiveness.

When you are doing everyday activities such as sitting down at the table to eat, does your mind wander to things you did yesterday or what you need to do tomorrow? How much attention do you pay to the food you are eating and the others sitting around the table eating with you? Are you already thinking about your hectic schedule for tomorrow or upcoming deadlines for assignments at school? Are you constantly checking your e-mail or phone for the latest updates on future plans? Sometimes you may need to disconnect yourself from technology and the busyness of the world. Take a deep breath, slow down your mind, and just focus on the present moment and those around you. You will never get that moment in time back again.

This focusing technique works especially well for those who play sports. When you are playing basketball, for example, how often is your mind racing ahead, thinking, "If I just make this next basket, I will look good and put my team ahead. My coach will be impressed and possibly start me in the next game. My teammates will throw me the ball more often," and so on. Instead of concentrating on the task at hand—making the basket—and focusing on the moment, we put ourselves in jeopardy of not making the basket and being embarrassed and angry at ourselves. It may surprise you to know that Phil Jackson, one of the most successful NBA coaches in the history of basketball, has practiced meditation for years. In his book, *Sacred Hoops: Spiritual Lessons of a Hardwood Warrior*, he states the benefits of being mindful:

> "When players practice what is known as mindfulness—simply paying attention to what's actually happening—not only do they play better and win more, they also become more attuned with each other."

So, how does all of this apply to you and the math class you are currently taking? The next time you go to math class, instead of focusing on what you plan to do when class is over, focus on what is taking place in class right now. Concentrate your thoughts on what you can do right now to be successful: paying attention to the instructor, practicing math problems, taking notes, etc. It's okay to think ahead about the rewards of getting a good grade on the next test, passing the course, and getting your degree, but what you do in the present moment will affect all of those things. Focus your attention on the tasks at hand, become mindful, and you may find that you reach those future goals a little faster and with more confidence.

Sources: Sacred Hoops: Spiritual Lessons of a Hardwood Warrior by Phil Jackson, Hugh Delehanty (2006) Hyperion.

The Mindful Way to Study: Dancing With Your Books by Jake J. and Roddy O. Gibbs (2013) O'Connor Press.

The Emotional Life of Your Brain by Richard J. Davidson and Sharon Begley (2012) Penguin Group

Name: Date:

9.1 The Cartesian Coordinate System

Objectives

Learn about equations in two variables.

Graph and label ordered pairs of real numbers as points on a plane.

Find ordered pairs of real numbers that satisfy a given equation.

Locate points on a given graph of a line.

Success Strategy

It is highly recommended that you buy graph paper for this chapter and the next. Graph paper will allow you to graph points and linear equations with precision. Creating graphs which are more precise will enhance your understanding of the concepts presented.

Understand Concepts

Go to Software First, read through Learn in Lesson 9.1 of the software. Then, work through the problems in this section to expand your understanding of the concepts related to the Cartesian coordinate system.

1. The Cartesian coordinate system was created by the mathematician and philosopher René Descartes. Using his first and last name as key words, perform an Internet search to learn more about this historical figure in mathematics. Write a short paragraph describing an interesting aspect of his life.

Read the following paragraph about the Cartesian coordinate system and work through the problems.

Quick Tip

The **Cartesian coordinate system** is also referred to as the **Cartesian plane** or the **coordinate plane**.

The Cartesian plane is shown here. The horizontal axis is called the x-axis. The vertical axis is called the y-axis. The axes are made from two perpendicular real number lines that intersect each other at 0. On the x-axis, the positive values are to the right of zero (similar to a standard number line). On the y-axis, positive values are above zero. In this workbook, if no scale is indicated on the Cartesian plane, you can assume that it uses increments of 1 unit.

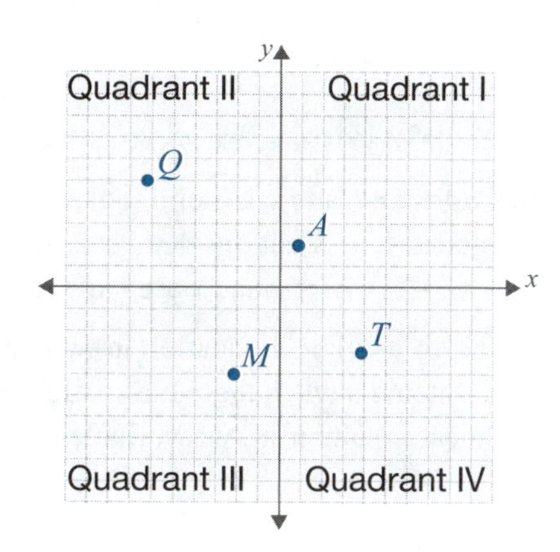

2. Each point on the Cartesian plane represents an ordered pair. An ordered pair of numbers has the form (x, y), where x and y are both real numbers. Within each quadrant, the signs of the values in every (x, y) ordered pair follow a pattern. For instance, in the first quadrant, all x-values and all y-values are positive. This means that the signs of ordered pairs in the first quadrant follow the pattern $(+, +)$. Fill in the table with the quadrant number that follows the given sign pattern.

Quadrant	(x, y)
	(+, +)
	(–, +)
	(–, –)
	(+, –)

The order of the numbers in an ordered pair is important. For example, point A on the coordinate plane on the previous page represents the ordered pair $(1, 2)$. In this ordered pair, 1 is the x-value and 2 is the y-value. The point $(1, 2)$ is a different point than $(2, 1)$.

3. In the coordinate plane on the previous page, determine the ordered pairs that are represented by each point.

   a. M

   b. Q

   c. T

4. Plot each point on the Cartesian plane.

   a. $(-1, 3)$

   b. $(0, 0)$

   c. $(-4, 2)$

   d. $(1, 3)$

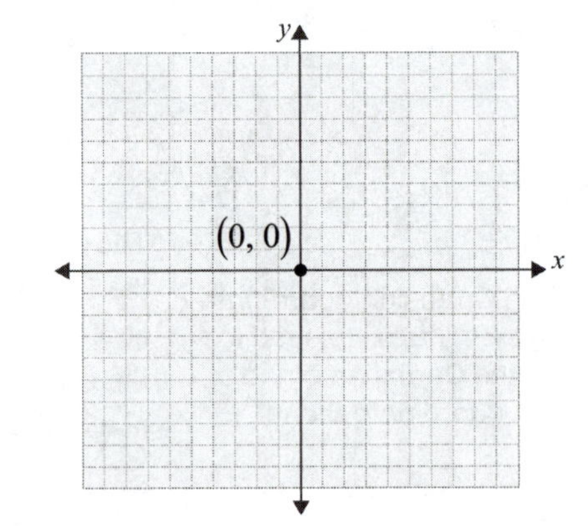

Quick Tip

In the Cartesian coordinate system, the point $(0, 0)$ is called the **origin**.

Read the following paragraph about linear equations and work through the problems.

The Cartesian plane is a useful tool in mathematics. It is commonly used as a way to visually display the solutions for equations that have two variables. An example of a linear equation in two variables is $y = 4x + 7$, where x and y are variables.

Lesson Link

Finding solutions for single-variable linear equations was covered in Sections 8.1, 8.2, and 8.3.

A **solution** to a linear equation in two variables is an ordered pair, such as $(1, 11)$, which results in a true statement when the x and y values are substituted into the equation. One way to find solutions to a linear equation in two variables is to create a table. The ordered pair $(1, 11)$ is a solution to $y = 4x + 7$ because

$$11 = 4(1) + 7$$
$$11 = 4 + 7$$
$$11 = 11$$

Name: Date:

Quick Tip

In this equation, y is called the **dependent variable** because the value of y depends on the value of x, which is the **independent variable**. Some people refer to x as the **input** and y as the **output**.

5. Work through the following problems to find solutions to the equation $y = 2x - 5$.

a. Substitute the value $x = 7$ into the equation to determine the corresponding value of y. Place this value in the appropriate cell of row 1 of the table.

	x	$y = 2x - 5$
row 1	7	
row 2		5
row 3		
row 4		

b. Substitute the value $y = 5$ into the equation and solve for x. Place this value in the appropriate cell of row 2 of the table.

c. Choose a value for x other than 7 and substitute it into the equation to find the corresponding value of y. Place these values in the appropriate cells of row 3 of the table.

d. Choose a value for y other than 5 and substitute it into the equation to find the corresponding value of x. Place these values in the appropriate cells of row 4 of the table.

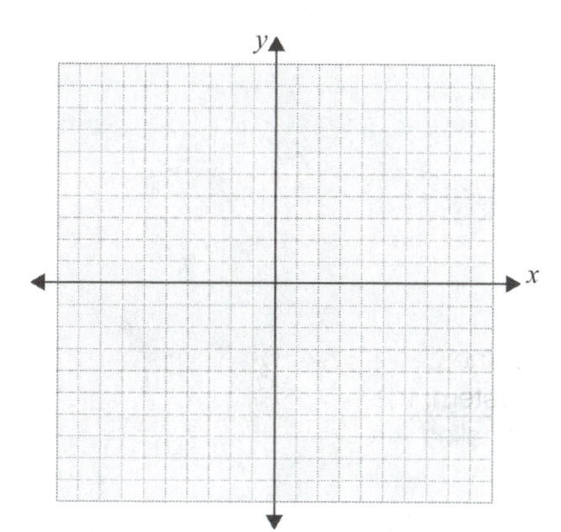

e. Plot the points from the table in part **a.** on the given Cartesian plane.

Read the following paragraph about solutions to linear equations in two variables and work through the problem.

Every solution to a linear equation in two variables represents a point on the line of the graph of the equation. A line contains an infinite number of points, so the linear equation in two variables that represents the line has an infinite number of solutions. Another way to determine the solutions to an equation is to look at the graph of the equation, if it is given.

6. Some of the points plotted on the Cartesian plane to the right are solutions to the equation $2y = 3x + 4$. Determine which points on the graph, and their corresponding ordered pairs, are solutions.

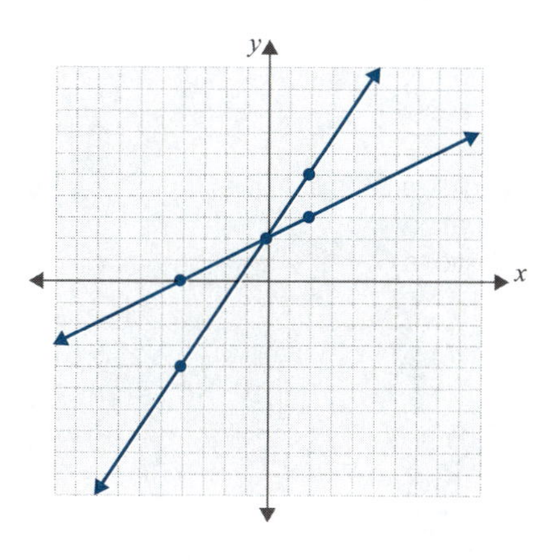

Skill Check

 ***Go to Software*** Work through Practice in Lesson 9.1 of the software before attempting the following exercises.

Complete the tables so that each ordered pair will satisfy the given equation.

7. $y = 3x$

x	y
0	
	–3
–2	
	6

8. $\frac{1}{2}x - \frac{1}{4}y = 12$

x	y
0	
	0
10	
	8

Apply Skills

Work through the problems in this section to apply the skills you have learned related to the Cartesian coordinate system.

9. One way to describe locations on a map is to lay a grid on top of the map and define locations with an ordered pair. A map with an overlaid grid is called a **grid reference**. Typically the origin of the coordinate system is placed at the bottom left corner of the map and vertical and horizontal lines are extended from each reference point on the axes.

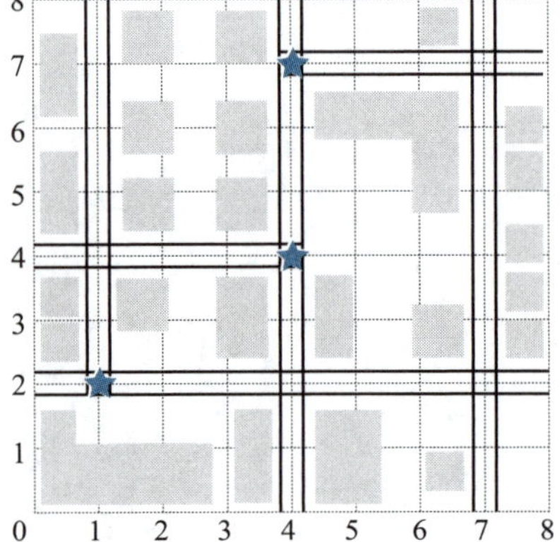

A city planning committee is using a map with a grid system to plot the locations in a portion of the city where the highest frequency of traffic accidents occurs. The map is shown here with the highest accident points indicated with a star. A proposal is made to replace stop signs with electric traffic signals at every intersection that has a high frequency of accidents.

a. Will a traffic signal be placed at any of the following points? If so, which points?

$(1, 7), (4, 4), (0, 5)$

b. Which points not listed in part **a.** will have traffic signals?

Name: Date:

Lesson Link

The equations in Problems 10 and 11 can also be considered **variation equations**. Variation problems will be covered in Section 13.6.

Quick Tip

When variables other than *x* and *y* are used, the *x*-axis and *y*-axis are relabeled to match the variables. For instance, the coordinate plane in this problem uses an *r*-axis and a *t*-axis.

10. The equation $t = \frac{d}{r}$ can be used to determine how long it will take you to travel a certain distance while traveling at a specified rate, or speed. Suppose you have to travel a distance of 150 miles. (**Hint:** Rate is in miles per hour, time is in hours, and distance is in miles.)

a. Complete the table of values.

r (mph)	*t* (hours)
25	
50	
60	

b. Graph the ordered pairs from the table. Try using a scale of 10 for each increment on the *r*-axis.

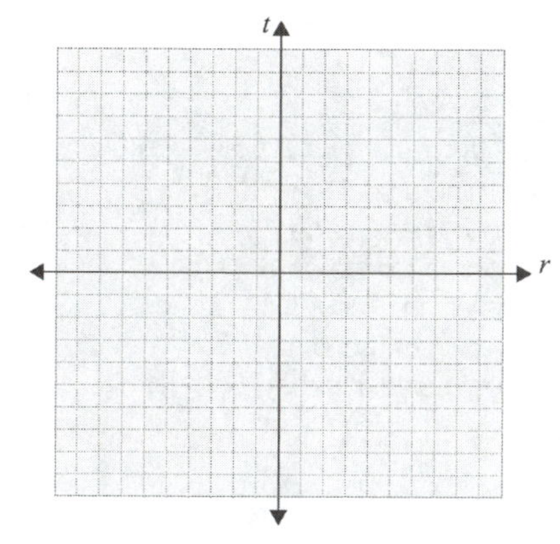

c. What happens to the value of time as the rate increases?

d. Complete the table of values. You may want to solve the given equation for *r* first. Remember that $d = 150$ miles.

r (mph)	*t* (hours)
	5
	2.5
	1
	0.5
	0.1

e. As the value of *t* gets smaller (and closer to zero), how does the rate change?

Quick Tip

Remember that the rate used in interest problems should be in decimal number form.

Quick Tip

A time of ¼ of a year is equivalent to 3 months and a time of ½ of a year is equivalent to 6 months.

Quick Tip

Determine a scale for the coordinate plane that works for the data values given. The x-axis and y-axis can use a different scale, if needed.

11. Suppose you deposit $1000 into an account that earns simple interest at a rate of 4%. Use the simple interest formula $I = Prt$ to solve the following problems.

a. Fill in the given table to determine the amount of interest earned if you keep the deposit in the account for 3 months, 6 months, 1 year, 2 years, and 3 years.

t (years)	I ($)
$\frac{1}{4}$	
$\frac{1}{2}$	
1	
2	
3	

b. Plot the ordered pairs from the table in part **a.** on the coordinate plane and draw a line through the points.

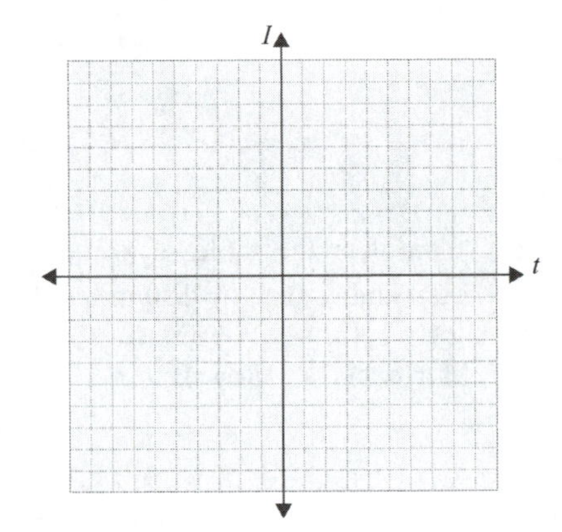

c. Since graphs are models of situations, they can be used to predict values. Use the graph to predict the amount of interest that will be earned after 9 months. (Remember to convert the time to years.)

d. Use the graph to predict the amount of interest that will be earned after one and a half years.

Name: Date:

9.2 Graphing Linear Equations in Two Variables: $Ax + By = C$

Objectives

Recognize the standard form of a linear equation in two variables: $Ax + By = C$.

Plot points that satisfy a linear equation and graph the corresponding line.

Find the y-intercept and x-intercept of a line and graph the corresponding line.

Success Strategy

You will be learning three different forms, or representations, of lines in Chapter 9. Make sure you understand each of the forms well, starting with the standard form in this section. You must be able to convert from one form to another starting in the next section.

Understand Concepts

Go to Software First, read through Learn in Lesson 9.2 of the software. Then, work through the problems in this section to expand your understanding of the concepts related to graphing linear equations in two variables.

In Section 9.1, you learned how to find points on a line from an equation and how to plot points. In this section, you will advance to plotting linear equations on a Cartesian plane. First, we need to discuss the purpose of graphs.

Graph of an Equation

An equation in two variables is a **model**, or mathematical description, of a situation. A visual representation of the model is the **graph of the equation**. The terms *equation* and *graph of the equation* are often used interchangeably.

1. Every linear equation in two variables can be represented by a line on the Cartesian plane.

a. In Section 5.1, you learned what a line is. Write the definition of a line and provide an example.

b. In Section 9.1, you learned how to determine if a point on a line was a solution to an equation. Write a short explanation of this process.

Read the following information about the standard form of an equation and work through the problems.

Lesson Link

Different forms of a linear equation in two variables will be introduced in Sections 9.3 and 9.4.

A line is made up of an infinite number of points. Each of these points can be represented by an ordered pair which satisfies an equation that describes the line. The equation that describes the line can take on many forms.

The **standard form** of a linear equation in two variables is

$$Ax + By = C,$$

where x and y are variables and A, B, and C are constants. Saying that an equation is in "standard form" means that the equation has the variables in alphabetical order on the left side of the equation with coefficients of A and B, respectively, and the constant C by itself on the right side.

True or False: Determine whether each statement is true or false. Rewrite any equation in each false statement so that the statement is true. (There may be more than one correct new statement.)

2. The equation $5x - 7y = 42$ is in standard form.

3. The equation $y = 5x - 7$ is in standard form.

4. The equation $14y + 6x - 3 = 12$ is in standard form.

Lesson Link

The **definition of a line** from Section 5.1 states that a line is labeled by two points. This is because only two points are needed to draw a line. A third point verifies that the other two points are correct.

Graphing an Equation in Two Variables

1. Find two points that satisfy the equation.
2. Plot those points on a coordinate plane.
3. Draw a straight line through the points.
4. (Optional check) Find a third point using the equation and verify that it falls on the line.

Quick Tip

In general, any values can be chosen for one of the variables to find the corresponding value of the other variable. Some values will be easier to work with than others. Smaller values will be easier to graph than larger values.

5. Follow the steps to graph the equation $2x + y = 8$ on the Cartesian plane.

a. Complete the table of solutions.

x	y
1	
	2

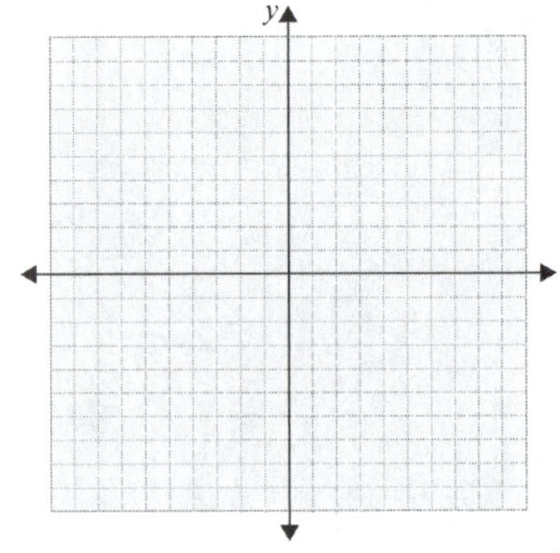

b. Write the solutions from part **a.** as ordered pairs.

c. Plot the solutions on the Cartesian plane.

d. Draw a line through the two plotted solutions.

e. Find a third point using the equation. Does this point fall on the line created in part **d.**?

Name: Date:

Intercepts

The ***x*-intercept** is the point where the graph of the equation crosses the x-axis. This point always has the form $(a, 0)$, where a is a real number.

To find the x-intercept of an equation, substitute $y = 0$ into the equation and solve for x.

The ***y*-intercept** is the point where the graph of the equation crosses the y-axis. This point always has the form $(0, b)$, where b is a real number.

To find the y-intercept of an equation, substitute $x = 0$ into the equation and solve for y.

6. Consider the equation $4x + 3y = 12$.

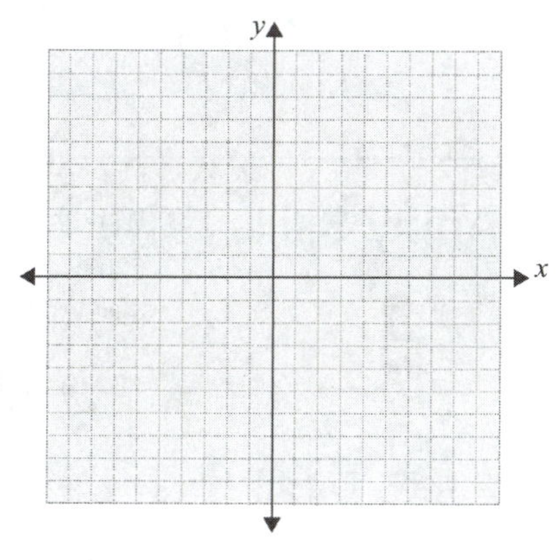

a. Find the x-intercept of the equation.

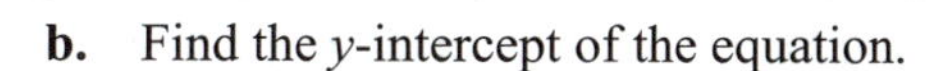

b. Find the y-intercept of the equation.

c. Graph the equation using the intercepts.

d. Find a third point using the equation. Does this point fall on the line created in part **c.**?

Quick Tip

It is a good idea to always check a third point on the graph, even when graphing by using the intercepts.

Skill Check

Go to Software Work through Practice in Lesson 9.2 of the software before attempting the following exercises.

Graph each equation according to either graphing method introduced in this section.

7. $x + y = 6$

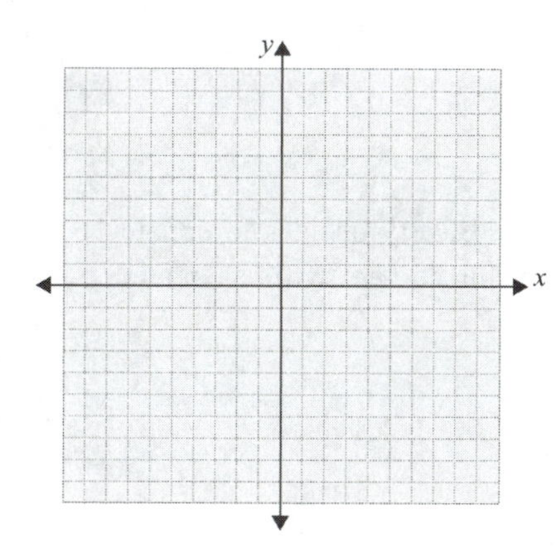

8. $2x - y = 2$

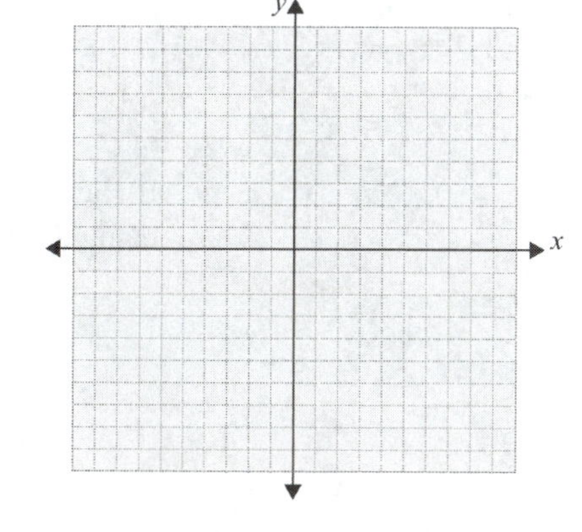

9. $2y = 2x - 5$

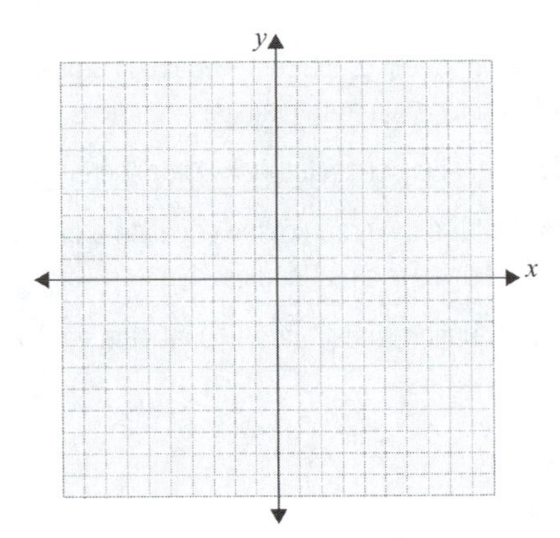

Apply Skills

Work through the problems in this section to apply the skills you have learned related to graphing linear equations in two variables. You may find the graphs provided to be too small for your needs. Try using a full sheet of graphing paper to graph the data for these problems.

10. Barbara's Bombtastic Bakery is donating cookies to a charity bake sale. The bakery decides to donate chocolate chip cookies and peanut butter cookies by the dozen, and they want to donate a total of 30 dozen cookies. To determine the possible combinations, the bakery uses the equation $C + P = 30$, where C is the number of dozens of chocolate chip cookies and P is the number of dozens of peanut butter cookies.

a. Find the intercepts of the given equation.

Quick Tip

The **graph of an equation** is a line which extends infinitely in both directions. In real-life situations, some points on the line may not make sense as a solution. As a result, some of the solutions to the equation may need to be discarded to fit the context of the problem.

b. Graph the equation using the intercepts.

c. Are there any solutions to the equation that do not make sense in the context of the problem? Explain why.

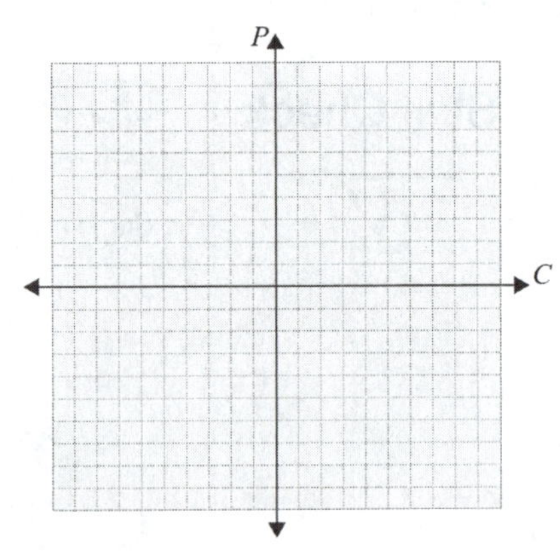

d. If Barbara's Bombtastic Bakery decides to donate 16 dozen peanut butter cookies, how many dozen chocolate chip cookies will be donated?

Name: Date:

Quick Tip

Always consult a doctor and read the package directions before giving medication to a child.

11. Ibuprofen can be given to a child to treat a fever. For a fever lower than 102.5 °F, the recommended dosage is 2.2 milligrams per pound of body weight. This can be modeled by the equation $D = 2.2w$, where D is the dosage of ibuprofen in milligrams and w is the weight of the child in pounds.

a. Create a table to determine several coordinate pairs that satisfy the equation.

w	*D*

b. Graph the equation using the coordinate pairs from part **a.**

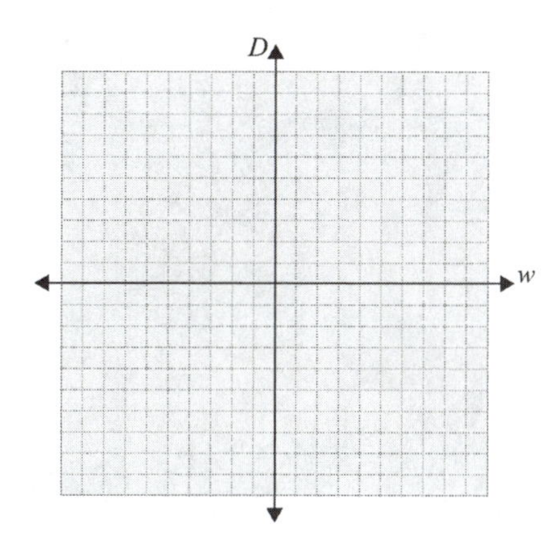

c. Are there any solutions from the graph that do not make sense in the context of the problem? Explain why.

d. How much ibuprofen should be given to a child that weighs 45 pounds?

12. Mason bought an MP3 player for \$20. The music that he downloads costs an average of \$1 per song. Mason wants to keep track of how much money he spends on the MP3 player and the music, so he creates the equation $C = 20 + 1n$, where C is the total cost in dollars and n is the number of songs purchased.

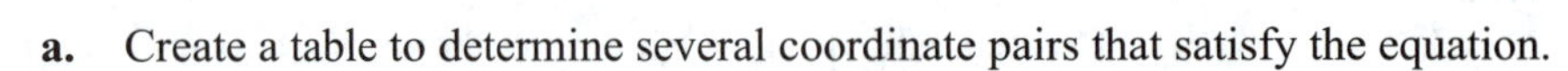

a. Create a table to determine several coordinate pairs that satisfy the equation.

n	C

b. Graph the equation using the coordinate pairs from part **a.**

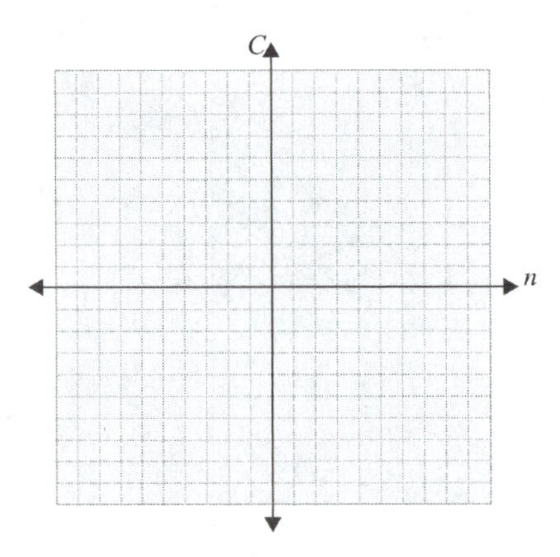

c. Are there any solutions from the graph that do not make sense in the context of the problem? Explain why.

d. If Mason buys 37 songs, what will the total cost be?

Name: Date:

9.3 The Slope-Intercept Form: $y = mx + b$

Objectives

Interpret the slope of a line as a rate of change.

Calculate the slope of a line given two points that lie on the line.

Find the slopes of and graph horizontal and vertical lines.

Recognize the slope-intercept form for a linear equation in two variables: $y = mx + b$.

Success Strategy

Slope is an important concept for linear equations. Make sure you understand how to calculate slope when given two points or a graph of a line. Do extra practice converting linear equations to slope-intercept form by solving for the variable y.

Understand Concepts

Go to Software First, read through Learn in Lesson 9.3 of the software. Then, work through the problems in this section to expand your understanding of the concepts related to the slope-intercept form of linear equations in two variables.

The **slope** of a line is a measure of how much the line increases or decreases as it moves from left to right. Some other words for slope are grade and pitch. **Grade** is often used in describing the steepness of a road or hill. Road signs are placed along a road to inform cars and trucks that there is a steep grade ahead. **Pitch** is often used in architecture when referring to the slope of a roof.

On a graph, slope can be thought of in terms of right triangles, where the hypotenuse represents a segment of a line. The height of the triangle represents the **rise** and the length of the base of the triangle represents the **run**. Graphs should be read from left to right. This means that if a line is sloping "downhill," the rise will be a negative value. If the line is sloping "uphill," the rise will be a positive value.

1. Use the triangles shown on the graph to answer the following questions about the basic components of the slope of a line.

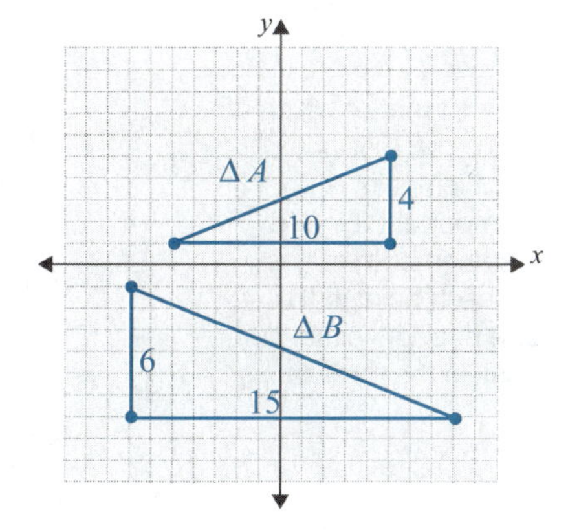

a. Is the slope of the hypotenuse of ΔA increasing or decreasing as it goes from left to right?

b. What is the value of the rise for ΔA?

c. What is the value of the run for ΔA?

d. Is the slope of the hypotenuse of ΔB increasing or decreasing as it goes from left to right?

e. What is the value of the rise for ΔB?

f. What is the value of the run for ΔB?

Quick Tip

When calculating slope, write the slope in reduced fraction form unless instructed to write it in decimal number form.

2. One way to calculate slope is to divide the rise by the run. Use the answers from Problem 1 to answer the following problems. Write each slope in reduced fraction form.

   a. What is the slope of the hypotenuse of ΔA?

   b. What is the slope of the hypotenuse of ΔB?

   c. What do you notice about the slope of the triangle that had the hypotenuse which decreased from left to right?

Quick Tip

Since the slope measures the amount that y changes as x changes from x_1 to x_2, it is also referred to as the average **rate of change** of y with respect to x.

Slope

The following equations are used to calculate the **slope** of a line going through the points (x_1, y_1) and (x_2, y_2).

$$\text{slope} = m = \frac{\text{rise}}{\text{run}}, \quad m = \frac{y_2 - y_1}{x_2 - x_1}, \quad m = \frac{y_1 - y_2}{x_1 - x_2}$$

The 3 equations are equivalent. The second and third equations are called the **slope formulas**.

Quick Tip

It doesn't matter which of the two points is (x_1, y_1) and which point is (x_2, y_2). The important thing to do is keep both values from each point in the same position (either first or last) in the formula.

Finding the Slope of a Line from a Graph

1. Find two points on the line.
2. Choose one of the points to represent (x_1, y_1). The other point will represent (x_2, y_2).
3. Substitute the coordinates into either slope formula and simplify.

3. Use the graph to determine the slope of the line.

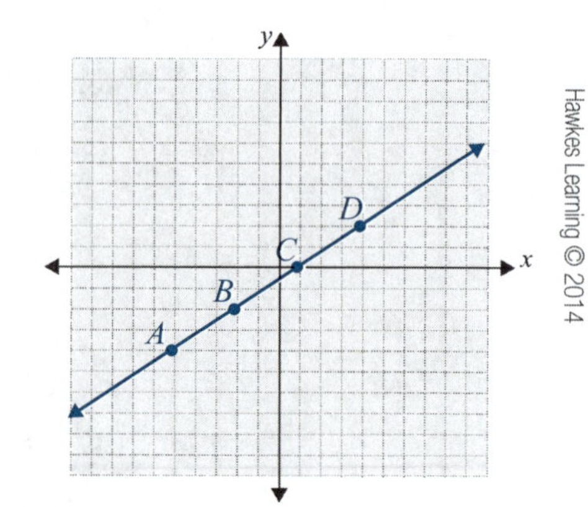

   a. Find the ordered pairs that correspond to each point indicated on the line.

   b. Use points A and B to find the slope of the line.

   c. Use points C and D to find the slope of the line.

   d. How do the slopes found in parts **b.** and **c.** compare?

Lesson Link

The slope formula uses the change in value formula from Section 7.3 twice. Once to find the change in y, or the rise, and once to find the change in x, or the run.

Name: Date:

e. Do you think it matters which points on a line are used to find the slope of the line? Explain why or why not.

Determine if any mistakes were made while calculating the slope of a line from two points. If any mistakes were made, describe the mistake and then correctly calculate the slopes to find the actual result.

4. Given $(2, 3)$ and $(4, 9)$,

$$m = \frac{2-4}{3-9} = \frac{2}{6} = \frac{1}{3}.$$

5. Given $(1, 4)$ and $(3, 5)$,

$$m = \frac{5-4}{1-3} = \frac{1}{-2} = -\frac{1}{2}.$$

Vertical and Horizontal Lines

Vertical lines have equations of the form $x = a$, where a is the x-coordinate of the x-intercept.

Horizontal lines have equations of the form $y = b$, where b is the y-coordinate of the y-intercept.

Quick Tip

Remember, dividing by zero results in an *undefined* value.

6. Horizontal and vertical lines have unique slopes. Use the graphs to determine the slopes of each line.

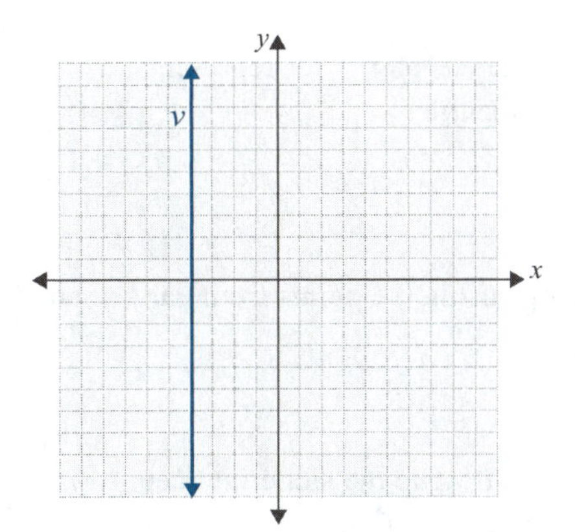

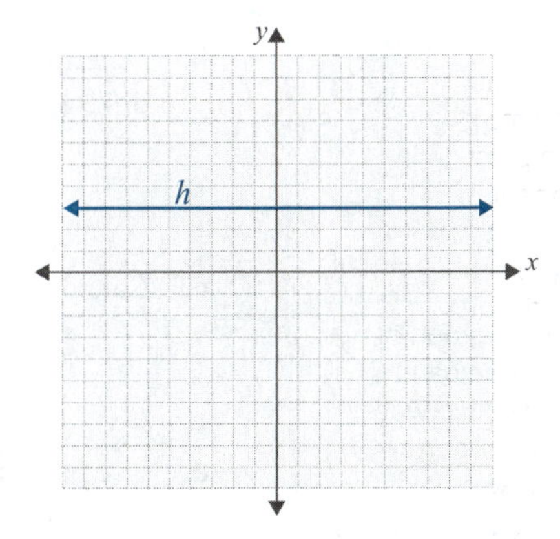

a. Find two points on the line v and determine the slope of the vertical line.

b. Find two points on the line h and determine the slope of the horizontal line.

c. Line v is a vertical line. Does it have a y-intercept?

Slope-Intercept Form

The **slope-intercept form** for an equation is $y = mx + b$, where m is the slope of the line and $(0, b)$ is the y-intercept.

Lesson Link

Three common forms for equations are the **standard form** (introduced in Section 9.2), **slope-intercept form** (introduced in this section), and **point-slope form** (introduced in Section 9.4)

Lesson Link

Solving an equation for a variable is similar to solving a formula for a variable. Solving formulas was covered in Section 8.5.

Each line on a coordinate plane is unique, but it can be represented by many "different" equations. Each equation that represents a specific line has the same solution set (which consists of all points on the line). Because these equations have the same solution set, they are technically the same equation. The "different" equations, which are called **forms**, are the same equation rearranged to look different. Certain forms have specific uses. For instance, when an equation is written in slope-intercept form, it is easy to determine the slope and the y-intercept of the line from the equation.

7. The slope-intercept form for an equation can be found from the standard form of the equation by solving $Ax + By = C$ for the variable y, as long as $B \neq 0$. Rewrite the equation $6x + 3y = 12$ in slope-intercept form by solving the equation for y.

8. The slope-intercept form of an equation can be found from the graph of a line. Use the graph to work through the following problems.

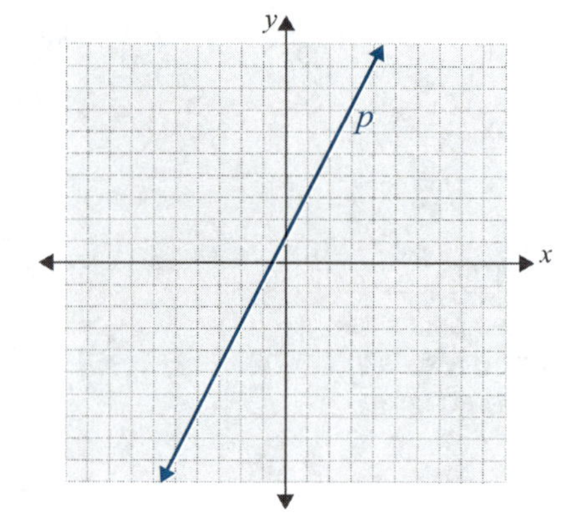

a. Determine the ordered pair that represents the y-intercept of the line.

b. Determine an ordered pair that represents a different point on the line.

Quick Tip

When given a graph of a line, the slope can be determined in two ways:

1. using rise over run between two points or
2. using the slope formula.

c. Calculate the slope using the points from parts **a.** and **b.**

d. Use parts **a.** and **c.** to write the slope-intercept form of the equation. Remember, the slope is m and the y-coordinate of the y-intercept is b.

Name: Date:

Skill Check

Go to Software Work through Practice in Lesson 9.3 of the software before attempting the following exercises.

Find the slope of the line which contains each pair of points.

9. $(3, 6)$ and $(1, -2)$

10. $\left(\frac{3}{4}, \frac{3}{2}\right)$ and $(1, 2)$

11. $(-12, 5)$ and $(24, 5)$

12. $(3, 0)$ and $(3, 7)$

The slope-intercept form of an equation can be found if the slope and y-intercept are both known. Write the slope-intercept form of the equation represented by the given slope and y-intercept.

13. $m = -2$; y-intercept $= (0, 12)$

14. $m = \frac{4}{5}$; y-intercept $= \left(0, -\frac{1}{2}\right)$

Apply Skills

Quick Tip

Writing equations from word problems is a very important skill. It can be difficult at first, but keep practicing.

Work through the problems in this section to apply the skills you have learned related to the slope-intercept form of linear equations in two variables.

15. The Millennium Force roller coaster at Cedar Point in Sandusky, Ohio, has been voted the Best Steel Coaster by Golden Ticket seven times since 2001. The Millennium Force is known for its steep slope on the first hill and speeds up to 93 miles per hour. The descent from the first hill starts at 310 feet above the ground and ends 10 feet above the ground. The descent runs approximately 53 feet. Find the slope of this hill to the nearest hundredth. (**Hint:** The slope is running downhill. Consider how that affects the slope.)

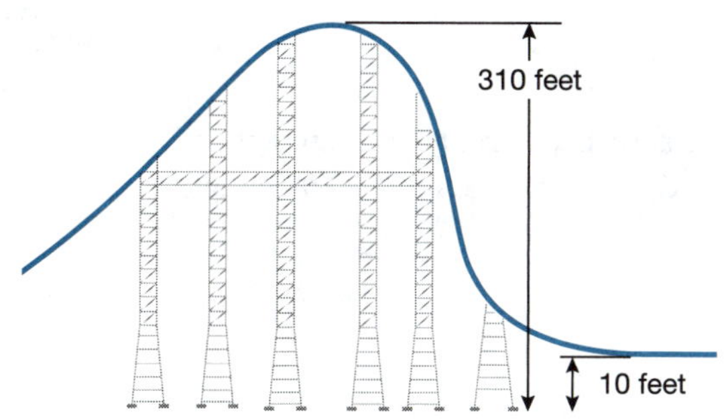

16. The grade, or slope, of a road is commonly given as a percentage. The grade can be determined by multiplying the slope by 100. Calculate the slope of each road or track and then determine its grade. Round each percent to the nearest hundredth if necessary.

   a. A road increases in height 5 feet for every 120 feet of run.

   b. The railway line with the steepest grade that does not run on a track system is the Lisbon tramway network in Portugal which has a section that increases 5 feet in height for every 37 feet of run.

   c. A road on the Route des Crêtes (Route of the Ridges) in France has an elevation that increases in height 450 feet over 1500 feet.

Quick Tip

In real-world situations, the **slope** is often referred to as the **rate of change**. It is important to note the units used for both x and y in these situations.

17. Jared sells paintings at an open-air market. He starts his work day with \$30 and sells each painting for \$15. Jared wants to create a linear equation to model this situation where y is the amount of money Jared has at the end of the work day and x is the number of paintings sold.

   a. The slope, or rate of change, is the increase in the amount of money Jared makes when he sells a painting. Determine the value of the slope and list the units for both variables.

   b. The y-coordinate of the y-intercept of this equation is the amount of money Jared has before he sells any paintings. What is the y-intercept?

   c. Write a linear equation in slope-intercept form to model this situation using the answers from parts **a.** and **b.**

   d. Graph the equation from part **c.**

   e. Are there any solutions to the equation which do not make sense in the context of the problem? Explain why.

   f. Use the graph to determine the amount of money Jared will have after selling 4 paintings.

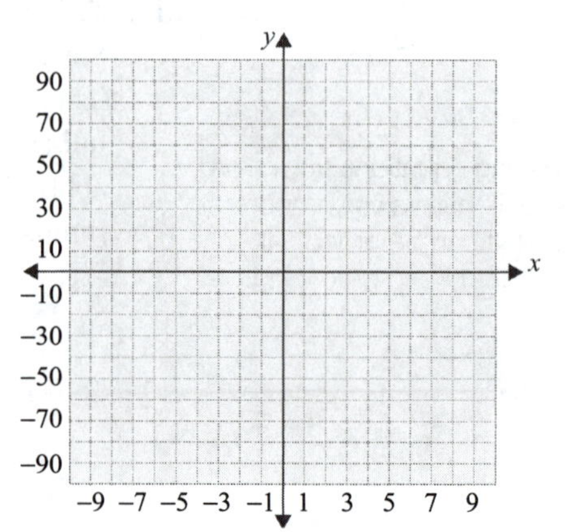

Name: Date:

9.4 The Point-Slope Form: $y - y_1 = m(x - x_1)$

Objectives

Graph a line given its slope and one point on the line.

Use point-slope form, $y - y_1 = m(x - x_1)$, to find the equation of a line given its slope and a point on the line.

Find the equation of a line given two points on the line.

Recognize and know how to find lines that are parallel and perpendicular.

Success Strategy

The point-slope form is a very useful form of a linear equation when all you are given is a point on the line, which is not the y-intercept, and the slope or if you are given two points on the line. Often you will need to rewrite an equation in this form to slope-intercept form, so do some extra practice solving equations for y.

Understand Concepts

Go to Software First, read through Learn in Lesson 9.4 of the software. Then, work through the problems in this section to expand your understanding of the concepts related to the point-slope form of linear equations in two variables.

In Section 9.2, you learned to graph an equation by finding two solutions of the equation, plotting these two points, and then drawing a line through them. You don't need an equation to graph a line. You can graph a line if you know the slope of the line and you have a point on the line.

Graph a Line Given the Slope and a Point

1. Plot the given point.
2. From that point, move up (for positive slope) or down (for negative slope) as many units as the value of the numerator of the slope.
3. From this new position, move right as many units as the value of the denominator of the slope.
4. Plot a point at the ending position.
5. Draw a line through the two points.

Quick Tip

Remember that the numerator of the slope represents the rise and the denominator of the slope represents the run.

Quick Tip

When graphing a line using a negative slope, it is helpful to write the slope with the negative sign in the numerator. This way the line can be plotted from left to right.

1. Let's graph a line using the slope $m = \frac{-1}{2}$ and the point $(-3, 2)$.

a. Plot the point on the coordinate plane.

b. Use the slope to find a second point on the line.

c. Draw a line through the two points.

d. Check: Since the slope is negative, does the line you graphed in part **c.** fall as you move left to right?

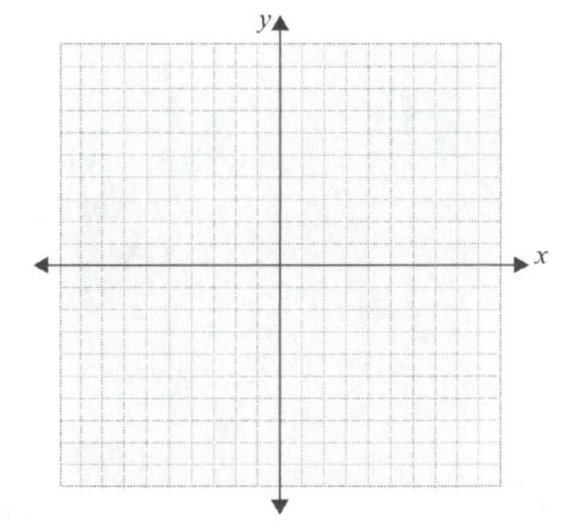

A third form of an equation of a line can be created from the slope and a point on the line. This is called the **point-slope form** of the equation. This form of an equation is useful when the slope of the line and a point on the line are important pieces of information to communicate.

Point-Slope Form

The **point-slope form** of an equation is

$$y - y_1 = m(x - x_1),$$

where x and y are variables, m is the slope, and x_1 and y_1 are the coordinates of a given point (x_1, y_1) on the line.

There are two situations where the point-slope form of an equation is useful for creating an equation that describes a specific line. When you know a point on the line and the slope of the line, or when you know two points on the line, then it is easy to write an equation using the point-slope form. Problems 2 and 3 will guide you through creating equations by using the given information.

2. Find the point-slope form of the equation that contains the point $(7, -2)$ and has slope $m = 2$.

a. Determine which part of the point is x_1 and which part is y_1.

b. Substitute the given information into the equation $y - y_1 = m(x - x_1)$.

c. Simplify both sides of the equation from part **b.**

d. Rewrite the equation from part **c.** so that it is written in standard form. That is, rearrange the equation so that the variables are on the left side of the equation in alphabetical order and the constant is on the right side of the equation.

3. A line goes through the two points $(1, 2)$ and $(4, 6)$. Find the point-slope form of the equation of the line.

a. Determine the slope of the line using the coordinates of the two points.

b. Choose one of the two points. Use this point and the slope from part **a.** to create the point-slope form of the equation.

c. Solve the equation from part **b.** for y and simplify.

d. Now, use the other point and the slope to create the point-slope form of the equation.

e. Solve the equation from part **d.** for y and simplify.

f. Are the slope-intercept forms of the equation that you found in parts **c.** and **e.** the same? If yes, why do you think this is? If no, how are they different?

True or False: Determine whether each statement is true or false. Rewrite any false statement so that it is true. (There may be more than one correct new statement.)

4. The point-slope form of an equation can be found by using only the slope.

5. The point-slope form of an equation can be found by using one point on the line and the slope of the line.

6. The point-slope form of an equation can be found by using two points from the graph of a linear equation.

7. The point-slope form of an equation can be found by using only one point on the line.

Lesson Link

Parallel and perpendicular lines were first introduced in Section 5.1.

Parallel and Perpendicular Lines

Parallel lines are lines that never intersect. Two lines that are parallel have the same slope.

Perpendicular lines are lines that intersect at a right (90°) angle. Two lines which are perpendicular have slopes that are negative reciprocals of each other.

True or False: Determine whether each statement is true or false. Rewrite any false statement so that it is true. (There may be more than one correct new statement.)

Quick Tip

The definition says that two lines are parallel if they have the same slope. It is also true that two lines that have the same slope must be parallel to one another.

8. All vertical lines are parallel to each other.

9. All horizontal lines are perpendicular to each other.

10. A horizontal line is perpendicular to a vertical line.

Quick Tip

The only unique form of a line is the **slope-intercept form**. To compare two lines, it is usually best to convert them to this form. If two equations have the exact same slope and the same y-intercept, then they are the same line.

Summary Table

Form Name	Equation Form	Tips
Standard	$Ax + Bx = C$	Useful in graphing using intercepts.
Slope-Intercept	$y = mx + b$	Useful when given the y-intercept as a point on the line. Useful when comparing equations.
Point-Slope	$y - y_1 = m(x - x_1)$	Useful when given any two points. Most general form and works with any point.
Vertical Line	$x = a$	Useful when x takes on a constant value of a for all y values.
Horizontal Line	$y = b$	Useful when y takes on a constant value of b for all x values.

Skill Check

Go to Software Work through Practice in Lesson 9.4 of the software before attempting the following exercises.

Find the negative reciprocal of each slope.

11. $m = \frac{3}{4}$

12. $m = -\frac{1}{4}$

Quick Tip

Remember that any integer can be written as a fraction by placing the integer in the numerator and 1 in the denominator. For example, $5 = \frac{5}{1}$.

13. $m = 5$

14. $m = 0$

Name: Date:

Determine if the pairs of equations represent parallel lines, perpendicular lines, or neither by finding the slope of each line, and then compare the slopes.

15. $y = \frac{1}{2}x + 7$ and $y = 2x - 15$

16. $y = -\frac{6}{7}x - 50$ and $y = \frac{7}{6}x + 2$

17. $y = \frac{2}{5}x + 5$ and $-2x + 5y = 10$

18. $2x + 7y = 14$ and $5x + 7y = 14$

Apply Skills

Work through the problems in this section to apply the skills you have learned related to the point-slope form of linear equations in two variables.

19. Natalie invested some money in a simple interest savings fund. After 2 years, she earned $120 in interest. After 5 years, she earned $300 in interest.

a. Write two ordered pairs from the information given where x represents the time in years and y represents the amount of interest earned.

b. Find the slope of the line which contains the two ordered pairs from part **a.**

c. Write the point-slope equation that models the situation.

d. Rewrite this equation in $y = mx + b$ form.

20. An archaeology crew finds the foundation of a house during a dig. The corners of the foundations are plotted on their grid map at the following points: $(1, 7)$, $(3, 2)$, $(9, 4)$, and $(7, 9)$.

a. Plot the points on the coordinate plane.

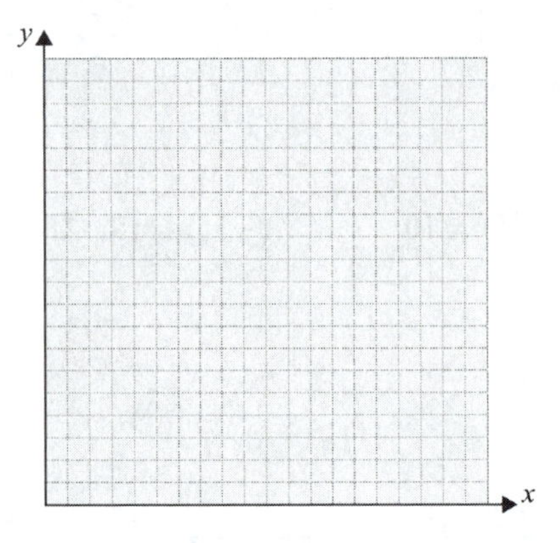

b. Find the slope of each side of the foundation.

c. Are any of the sides parallel? If so, which sides?

d. Are any of the sides perpendicular to each other? If so, which sides?

e. Is the foundation in the form of a geometric shape? If so, which shape.

Name: Date:

9.5 Introduction to Functions and Function Notation

Objectives

Learn about functions.

Find the domain and range of a relation or function.

Determine whether or not a relation is a function.

Use the vertical line test to determine whether or not a graph is the graph of a function.

Learn about linear functions.

Determine the domain of nonlinear functions.

Write a function using function notation.

Use a graphing calculator to graph functions. (Software only)

Success Strategy

Understanding function notation in this section is critical if you go on to take College Algebra or another math class that has an emphasis on modeling real-life situations. Remember that it is just special notation used for the variable y, where y is the dependent variable and x is the independent variable.

Understand Concepts

Go to Software First, read through Learn in Lesson 9.5 of the software. Then, work through the problems in this section to expand your understanding of the concepts related to functions and function notation.

Sets

A **set** is a collection or list of items. The items in a set are called **elements**. Each element should appear only once in a set.

For example: $S = \{1, 2, 3, 4, 5\}$ is a set of numbers and $T = \{\text{A, B, C}\}$ is a set of letters.

A set can also consist of ordered pairs as elements.

For example: $T = \{(1, 2), (2, 5), (3, 8), (4, 11)\}$.

Lesson Link

Sets were introduced in Section 8.7.

Quick Tip

A relation doesn't always have to map numbers to numbers. It can map numbers to variables, names to birthdays, and anything else that can be listed! However, most of the relations covered in this workbook will involve numbers.

Another name for a set of ordered pairs is **relation**. A relation connects the elements in one set to the elements in another set. If you think of the numbers on the x-axis as one set and the numbers on the y-axis as another, then a relation is a set of ordered pairs that connects a number on the x-axis to a number on the y-axis.

One way to visualize a relation is with a mapping diagram that shows the connection between elements from one set to another. A common way to visualize this is shown here for the relation $R = \{(1, b), (2, c), (3, a), (4, d)\}$.

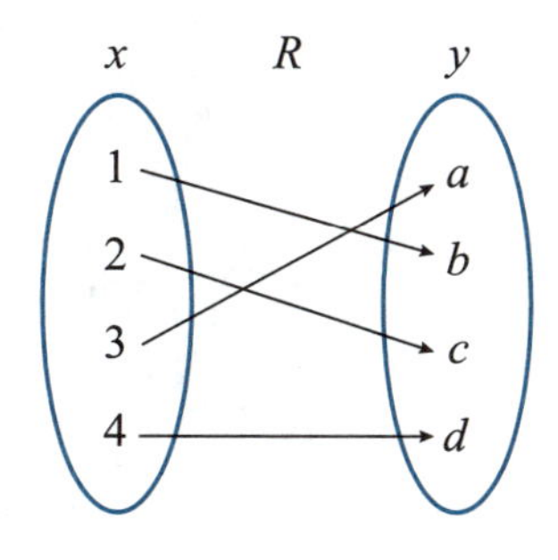

Quick Tip

The **domain** can be thought of as the input and the **range** can be thought of as the output of a relation.

Domain and Range

The **domain** D is the set of all first coordinates in a relation.

The **range** R is the set of all second coordinates in a relation.

Quick Tip

Listing the elements in the domain and range in ascending order makes the set easier to read and understand.

1. List the domain of the relation $\{(1,b),(2,c),(3,a),(4,d)\}$.

$D=\{\qquad\qquad\qquad\}$

2. List the range of the relation $\{(1,b),(2,c),(3,a),(4,d)\}$.

$R=\{\qquad\qquad\qquad\}$

Quick Tip

The range elements can appear multiple times in a function.

Functions

A **function** is a relation in which each domain element has exactly one corresponding range element. To determine if a relation is a function, check to see that each domain element appears in the relation only once.

Quick Tip

Try creating a mapping diagram to determine if each element in the domain set is mapped to only one element in the range set. If a domain element is only mapped to one range element, there will be only one arrow starting at that domain element.

3. Determine whether or not each relation is a function. If the relation is not a function, explain why.

a. $\{(0,5),(9,5),(8,4),(0,7),(6,3)\}$

b. $\{(6,2),(5,2),(4,2),(3,2),(2,2)\}$

c. $\left\{\left(\frac{1}{2},1\right),\left(\frac{3}{6},3\right),\left(\frac{6}{12},6\right),\left(\frac{7}{14},7\right)\right\}$

Read the following information about the vertical line test and work through the problems.

In situations where you are given a graph instead of a set of ordered pairs, you can use the vertical line test to determine if the graph represents a function or not.

Vertical Line Test

If a **vertical line** can be drawn that intersects the graph of a relation at more than one point, then the relation is not a function.

Name: Date:

4. Use the graphs to answer the following questions.

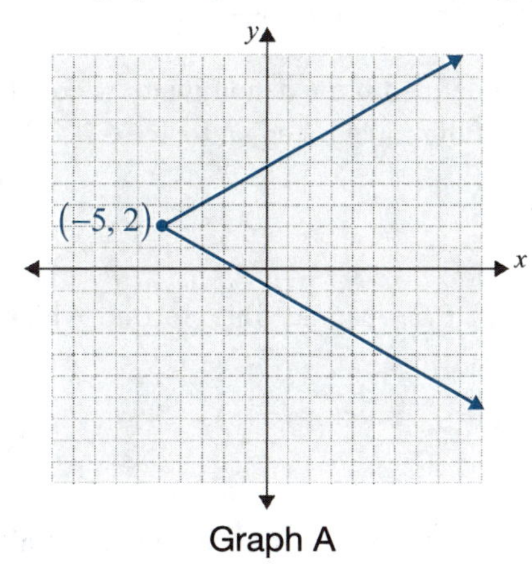

Graph A

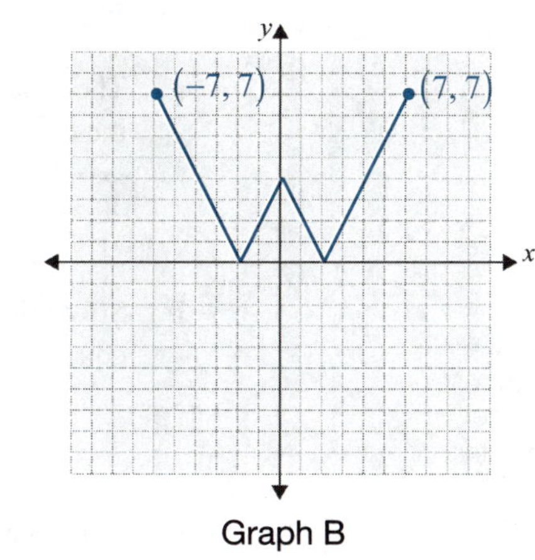

Graph B

a. Find two ordered pairs of points on graph A which have an x-coordinate of 2.

b. Does graph A represent a function? Explain why or why not.

c. Find two ordered pairs of points on graph B which have a y-coordinate of 2.

d. Does graph B represent a function? Explain why or why not.

Read the following information about the linear functions and work through the problems.

Linear Functions

A **linear function** is a function that is represented by an equation that simplifies to the form $y = mx + b$.

The domain of a linear function is the set of all real numbers; that is, $D = (-\infty, \infty)$.

Quick Tip

The domains of functions which are not linear vary from function to function.

There may be some values that need to be excluded from the domain of a function that is not a linear function. Two reasons for excluding values from the domain of a function are:

a. Division by zero is not permitted.

b. Certain measurements, such as time, length, people, etc., cannot have negative values.

In other words, to determine the domain of a function, you need to look at the denominator of the function and also consider the context of the problem.

To find the domain of a function, start with the entire set of real numbers $(-\infty, \infty)$ and then exclude any values that make the function undefined or do not make sense in the context of the problem.

5. Find the domain of the function $y = \frac{4x+1}{x-3}$.

a. The function is undefined if the denominator is equal to zero. Write an equation by setting the denominator of the function equal to zero.

b. Solve the equation from part **a.** for x to find which value of x makes the denominator equal to zero.

Quick Tip

The symbol ∪ means "union." When this symbol is used to describe a domain such as $D = (-\infty, -2) \cup (-2, \infty)$, it means that the domain consists of all numbers within the two intervals.

c. The domain of a nonlinear function is the set of all real numbers excluding the values that cause the function to be undefined. Describe the domain of $y = \frac{4x+1}{x-3}$ by writing a statement in the form $D = (-\infty, x) \cup (x, \infty)$, where x is the value from part **b.** that causes the function to be undefined.

Lesson Link

Independent and **dependent variables** were introduced in Section 9.1. Remember that independent variables are the input and dependent variables are the output.

Functions have the special notation $f(x)$ that is used in place of the variable y. This notation appears to change a two-variable equation into a one-variable equation. The variable x is still the independent variable and $f(x)$ becomes the dependent variable. The notation $f(x)$, which is read "f of x," indicates which variable the function is dependent on. In this situation, the variable is x.

When a function is to be evaluated at a specific value, that value is indicated in place of the x in $f(x)$. For instance, a function that is to be evaluated at $x = 2$ is written as $f(2)$.

Determine if any mistakes were made while evaluating each function. If any mistakes were made, describe the mistake and then correctly evaluate the function to find the actual result.

6. $f(x) = 2x+9$; Find $f(4)$.
$f(4) = 2(4)+9$
$f(4) = 8+9$
$f(4) = 17$

7. $f(x) = x^2 - 2$; Find $f(-2)$.
$f(-2) = -2^2 - 2$
$f(-2) = -4-2$
$f(-2) = -6$

Skill Check

Go to Software Work through Practice in Lesson 9.5 of the software before attempting the following exercises.

8. $T = \{(1, 2), (4, 7), (1, 7), (8, 2)\}$

a. List the domain.

b. List the range.

c. Is the relation a function? Explain why or why not.

Evaluate the functions for the given values.

9. $f(x) = x^2 - 5$

a. $f(2)$

b. $f(-3)$

10. $f(x) = \dfrac{10 - x}{3}$

a. $f(25)$

b. $f(-7)$

Apply Skills

Work through the problems in this section to apply the skills you have learned related to functions and function notation.

Quick Tip

Recall that IV is the abbreviation for intravenous. An IV bag is a bag which contains intravenous medication.

11. A nurse hangs a 1000-milliliter IV bag which is set to drip at 120 milliliters per hour. Create a model of this situation to represent the amount of IV solution left in the bag after x hours.

a. The y-intercept is the amount of IV solution in the bag initially (time = 0). What is the y-intercept?

b. The slope is equal to the rate that the IV solution is dispensed per hour. What is the slope? (**Hint:** Consider whether the amount of IV solution in the bag is increasing or decreasing and how this would affect the slope.)

Lesson Link

The slope-intercept form of equations was introduced in Section 9.3.

c. Write an equation in slope-intercept form to model this situation.

d. Write the equation from part **c.** using function notation.

e. State the domain and range of the function.

f. State any additional restrictions that should be made on the domain for it to make sense in the context of this problem.

g. How much IV solution is left in the bag after 5 hours?

12. Ariella is a full-time sales associate at a clothing store. She earns a weekly salary of $250 and earns 15% commission on all of her sales. Create a model of this situation to represent the amount of money Ariella makes after x dollars in sales.

a. What is the y-intercept and what does the y-coordinate of the y-intercept represent?

b. What is the slope and what does this value represent?

c. Write an equation in slope-intercept form to model this situation using the answers from parts **a.** and **b.**

d. Write the equation from part **c.** using function notation.

e. State the domain and range of the function.

f. State any additional restrictions that should be made on the domain for it to make sense in the context of this problem.

g. How much will Ariella make if she sells $5000 worth of merchandise?

Name: Date:

9.6 Graphing Linear Inequalities in Two Variables

Objectives

Graph linear inequalities.

Graph linear inequalities using a graphing calculator.

Success Strategy

Be sure to understand the difference between the solution sets for linear inequalities and linear equalities. Extra practice in the software will help you learn the difference.

Understand Concepts

Go to Software First, read through Learn in Lesson 9.6 of the software. Then, work through the problems in this section to expand your understanding of the concepts related to graphing linear inequalities in two variables.

Linear inequalities in two variables have a standard form of

$$Ax + By < C,$$

where x and y are variables and A and B are constants. The inequality symbol can be any of the following: $<$, $>$, $\leq$, or $\geq$.

Quick Tip

Linear inequalities in two variables can take on many forms, just like equations.

True or False: Determine whether each statement is true or false. Rewrite any false statement so that it is true. (There may be more than one correct new statement.)

1. $5x - 15y < 50$ is an example of a linear inequality in two variables.

2. $8y + 5x < 7x - 9$ is not an example of a linear inequality in two variables.

3. $10y < 5y + 2$ is an example of a linear inequality in two variables.

Read the following information about solutions of linear inequalities and work through the problems.

The solution to a linear inequality in one variable is a set of values. The solution to a linear inequality in two variables is a set of ordered pairs. A convenient and visual way to represent the solution set for a linear inequality in two variables is to use a graph.

Lesson Link

Graphing a linear inequality in one variable was covered in Section 8.7.

Terms Related to Graphing Inequalities

A line divides a plane into two **half-planes.** The points on one side of the line represent one half-plane and the points on the other side represent the other half-plane.

The line that divides a plane into half-planes is called the **boundary line**.

If the boundary line is included in the solution set, then the half-plane that represents the solution set is said to be **closed**.

If the boundary line is not included in the solution set, then the half-plane that represents the solution set is said to be **open**

4. Use the graphs to answer the following questions.

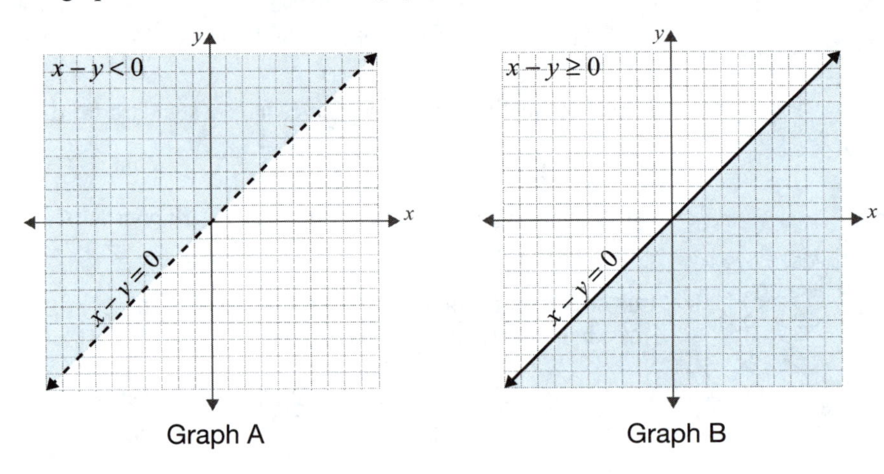

a. Which graph has a closed solution set?

b. Which graph has an open solution set?

c. Are both graphs divided into two half-planes? Explain why or why not.

d. Both graphs have the same boundary line. What is the equation of the boundary line?

Name: Date:

Quick Tip

These steps can also be used to graph linear inequalities with just one variable on the coordinate plane.

Graphing Linear Inequalities

1. First, graph the boundary line (dashed if < or >, solid if ≤ or ≥).
2. **Method 1**
   a. Pick a test point that is not on the boundary line, but lies in one of the half-planes defined by the boundary line.
   b. If the test point satisfies the inequality, then shade the half-plane that contains the point. If it does not satisfy the inequality, shade the half-plane that does not contain the point.

   Method 2
   a. Solve the inequality for y.
   b. If the inequality shows $y <$ or $y \leq$, then shade the half-plane below the line. If the inequality shows $y >$ or $y \geq$, then shade the half-plane above the line.
3. The shaded half-plane (and the line, if it is solid) represents the solution set for the inequality.

Quick Tip

Remember that when an inequality is multiplied or divided by a negative value, the inequality symbol changes direction.

The trickiest part of graphing a linear inequality is determining which half-plane is included in the solution set. Problems 5 and 6 will walk you through the process to graph the linear inequality $x - y > 3$. Step 1 of the process has been completed.

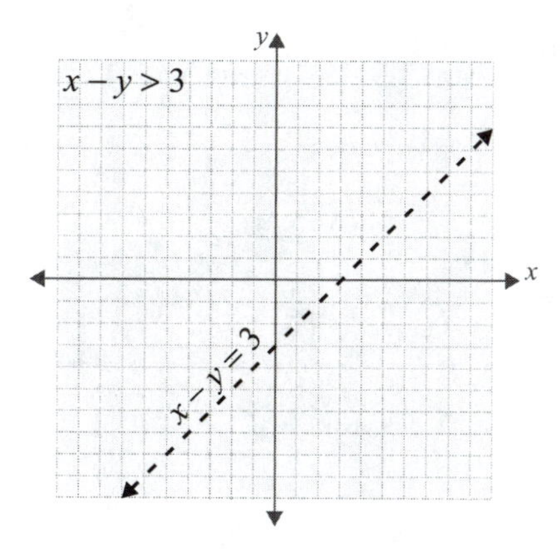

Quick Tip

Choosing (0, 0) as the test point will make the calculations easier. However, (0, 0) cannot be used as the test point if it lies on the boundary line.

5. Use Method 1 to complete the graph of this linear inequality.

a. Choose a test point in one of the half-planes.

b. Evaluate the linear inequality at the test point from part **a.**

c. Based on the results of part **b.**, which half-plane should be shaded?

6. Use Method 2 to complete the graph of this linear inequality.

   a. Solve the linear inequality for the variable y.

   b. Based on the results of part **a.**, which half-plane should be shaded?

7. Explain which method you prefer when determining which half-plane is the solution set to an inequality.

Skill Check

Go to Software Work through Practice in Lesson 9.6 of the software before attempting the following exercises.

Graph each linear inequality.

8. $2x + 4y \geq 12$

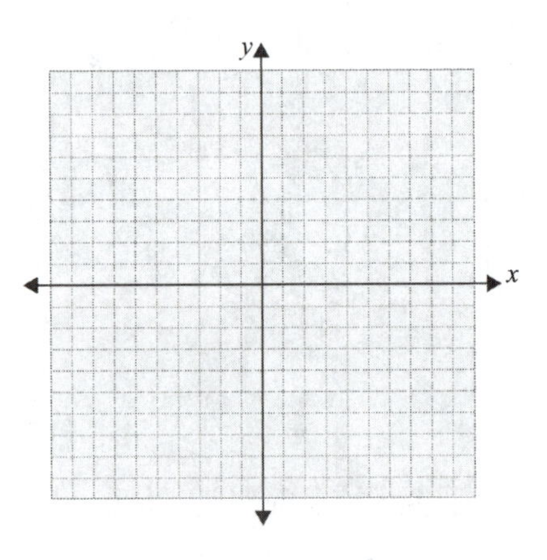

9. $4x < -3y + 9$

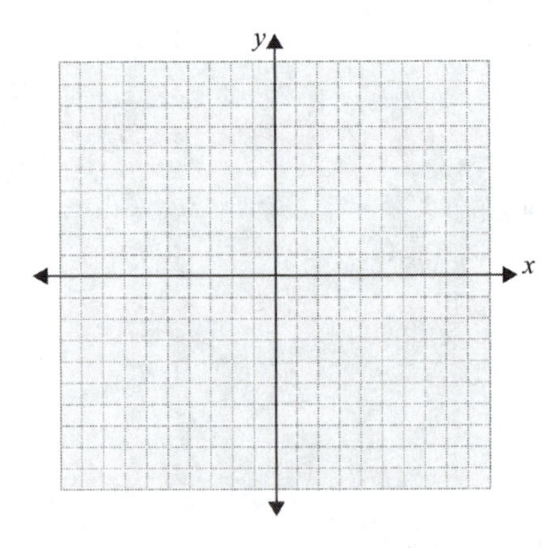

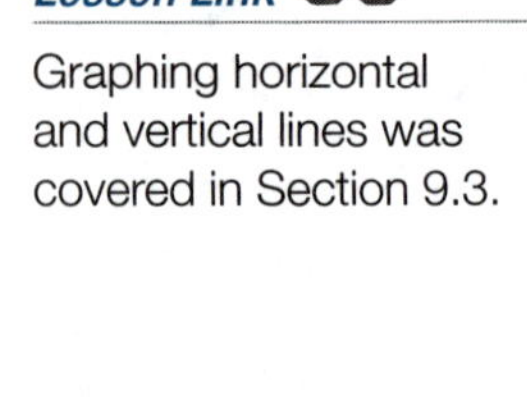

Lesson Link

Graphing horizontal and vertical lines was covered in Section 9.3.

10. $x > 0$

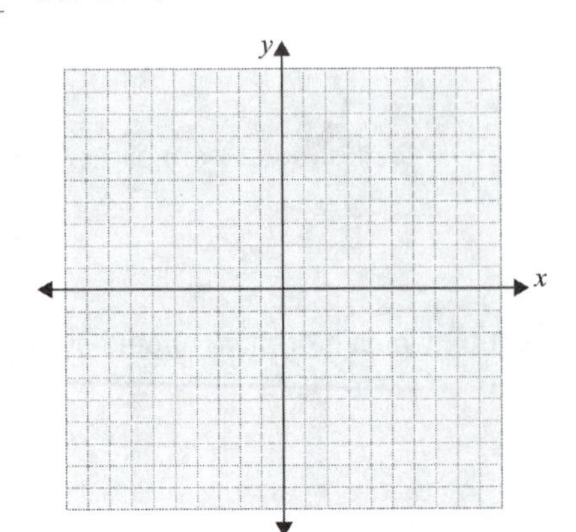

11. $y < 3$

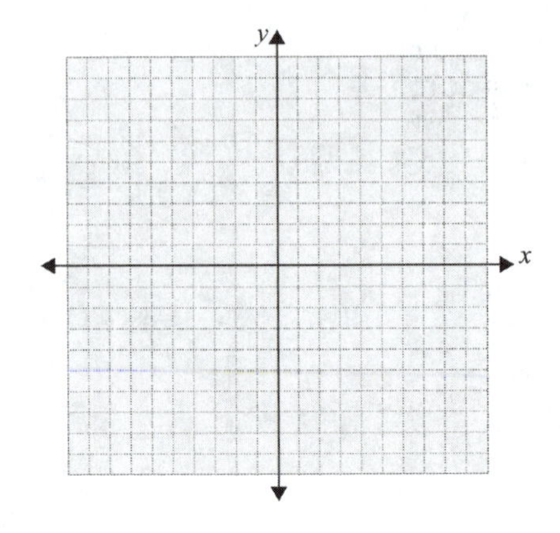

Name: Date:

Apply Skills

Work through the problems in this section to apply the skills you have learned related to graphing linear inequalities in two variables.

12. The grade for a 1-credit-hour survey class is based on an exam and a project, which are worth a maximum of 50 points each. The sum of the two scores must be at least 75 points for a student to earn a passing grade.

a. Let the amount of points earned on the exam be represented by the variable x and the amount of points earned on the project be represented by the variable y. Create a linear inequality to describe the solution set for a passing grade.

Quick Tip

Use a scale of 10 units for each increment on the graph. Or use a separate sheet of graph paper to graph the equation.

b. Graph the linear inequality from part **a.**

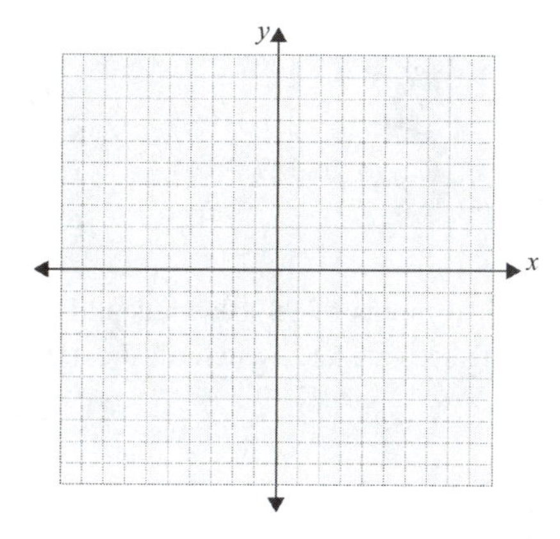

c. A student earns 45 points on their final exam and 22 points on their project. Plot this point on the graph. Did this student earn a passing grade?

d. Are there any points in the solution set which do not make sense for this situation?

Quick Tip

A **fail-safe** is a device that responds in a way that will cause little-to-no harm to other devices it is attached to in case of a failure. A commonly used fail-safe is a surge protector which protects devices from voltage spikes.

13. A fail-safe is installed on a device with two electrical inputs. If the sum of the inputs is greater than 250 kilowatts, the fail-safe will activate and cause the machine to switch off.

a. If one electrical input is represented by the variable x and the other is represented by the variable y, create a linear inequality to describe the values that will activate the fail-safe.

b. Graph the linear inequality.

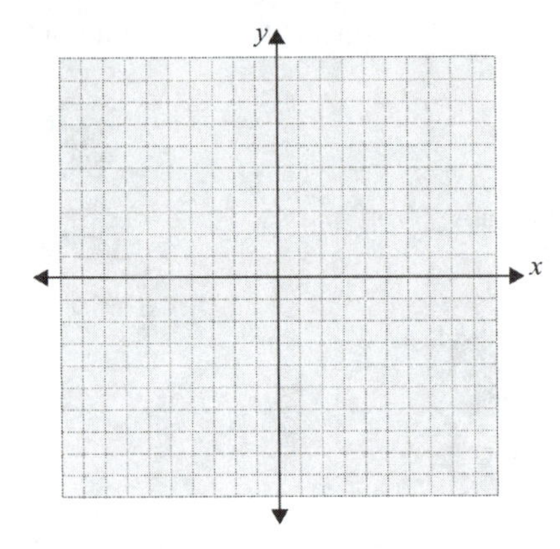

c. The device has electrical inputs of 95 kilowatts and 145 kilowatts. Plot this point on the graph. Will the fail-safe activate and switch off the device? Explain why.

Name: Date:

Chapter 9 Projects

Project A: What's Your Car Worth?

An activity to demonstrate the use of linear models in real life

When buying a new car there are a number of things to keep in mind: your monthly budget, the length of the warranty, routine maintenance costs, potential repair costs, cost of insurance, etc.

One thing you may not have considered is the *depreciation*, or reduction in value, of the car over time. If you like to purchase a new car every 3 to 5 years, then the retention value of a car, or the portion of the original price remaining, is an important factor to keep in mind. If your new car depreciates in value quickly, you may have to settle for less money if you choose to resell it later or trade it in for a new one.

Below is a table of original Manufacturer's Suggested Retail Price (MSRP) values and the anticipated retention value after 3 years for three 2012 mid-price car models.

Car Model	2012 MSRP	Expected Value in 2015	Rate of Depreciation (slope)	Linear Equation
Mini Cooper	$19,500	$13,065		
Toyota Camry	$22,500	$13,950		
Ford Taurus	$25,500	$ 7,140		

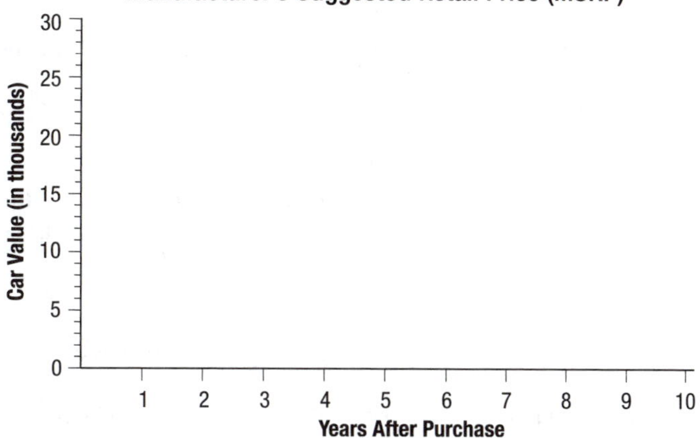

1. The x-axis of the graph is labeled "Years After Purchase." Recall that the MSRP value for each car is for the year 2012 when the car was purchased.

   a. What value on the x-axis will correspond to the year 2012?

   b. Using the value from part **a.** as the x coordinate and the MSRP values in column two as the y-coordinates, plot three points on the graph corresponding to the value of the three cars at time of purchase.

   c. What value on the x-axis will correspond to the year 2015?

   d. Using the value from part **c.** as the x-coordinate and the expected car values in column three as the y-coordinates, plot three points on the graph corresponding to the value of the three cars in 2015.

2. Draw a line segment on the graph connecting the pair of points for each car model. Label each line segment after the car model it represents and label each point with a coordinate pair (x, y). Consider using a different color when plotting each line segment to help you identify the three models.

3. Use the slope formula $m = \frac{y_2 - y_1}{x_2 - x_1}$ to answer the following questions.

a. Calculate the rate of depreciation for each model by calculating the slope (or rate of change) between each pair of corresponding points and enter it into the appropriate row of column four of the table.

b. Are the slopes calculated in part **a.** positive or negative? Explain why.

c. Interpret the meaning of the slope for the Toyota Camry making sure to include the units for the variables.

d. Which car model depreciates in value the fastest? Explain how you determined this.

4. Use the slope-intercept form of an equation $y = mx + b$ for the following problems.

a. Write an equation to model the depreciation in value over time of each car (in years). Place these in column five of the table.

b. What does the y-intercept represent for each car?

5. Let's use the equations from Problem 4 to predict a car's value over time.

a. Predict the value of the Mini Cooper 4 years after purchase.

b. Predict the value of the Ford Taurus $2\frac{1}{2}$ years after purchase.

Name: Date:

Optional Problems

6. When you plot the line segments for each car model, the line segment for the Ford Taurus crossed the other two. Can you determine from the graph how long it takes from the time of purchase until the Ford Taurus and the Mini Cooper have the same value?

(It may be difficult to read the coordinates for the point of intersection, but you can get a rough idea of the value from the graph. You can find the exact point of intersection by setting the two equations equal to one another and solving for x.)

a. After how many years are the car values for the Ford Taurus and the Mini Cooper the same? (Round to the nearest tenth.)

b. What is the approximate value of both cars at this point in time? (Round to the nearest 100 dollars.)

7. How long will it take for the Toyota Camry to fully depreciate (reach a value of zero)? This can be determined in two ways.

a. For the first method, extend the line segment between the two points plotted for the Toyota Camry until it intercepts the horizontal axis. The x-intercept is the time at which the value of the car is zero.

b. For the second method, substitute 0 for y in the equation you developed for the Toyota Camry and solve for x. (Round to the nearest year.)

c. Compare the results from parts **a.** and **b.** Are the results similar? Why or why not?

8. How long will it take for the Ford Taurus to fully depreciate? (Repeat Problem 7 for the Ford Taurus. Round to the nearest year.)

9. Why is there such a difference in depreciation for the Camry and the Taurus? Do some research on a reliable Internet site and list two reasons why cars depreciate at different rates.

10. Based on what you have learned from this activity, do you think retention value will be a significant factor when you purchase your next car? Why or why not?

Name: Date:

Project B: Game On!

An activity to demonstrate the use of linear inequalities (in two variables) in real life

Tristen has just purchased a new computer game in which he acts as the leader of an ancient civilization. In the game, he plays the role of the King of Capitol City and he must protect his land, and the villagers who live there, from invaders. Based on his skilled playing at this point, he has earned 15,000 gold coins. In order to continue fighting the invaders he needs to buy more weapons and more horses for his cavalry. The weapons cost 100 gold coins each and the price of a horse is 250 gold coins.

1. Write a **linear inequality in two variables** where x is the number of weapons he can buy and y is the number of horses he can buy with the amount of gold he has earned so far in the game.

2. Has Tristen earned enough gold to purchase 47 weapons and 41 horses? Why or why not?

3. Graph the inequality you determined in Problem 1 on the grid below. Should the graph extend outside of quadrant 1? Explain why or why not.

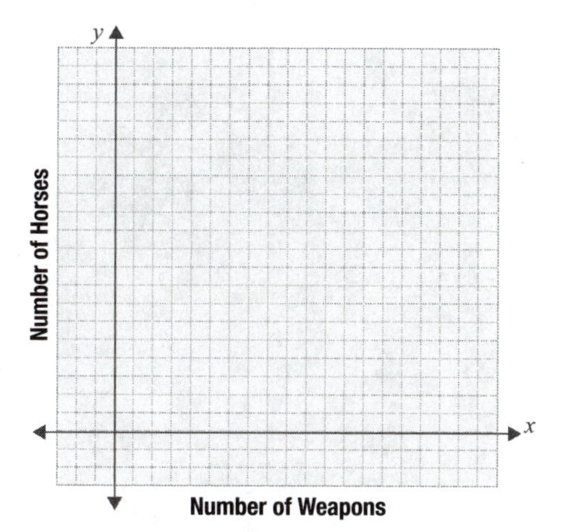

4. Suppose that Tristen has only enough property to support 80 horses at one time, and he currently has 48 horses.

a. What is the maximum number of horses that he can purchase?

b. Write a **compound inequality** involving y to express this new condition (**Hint:** You can't purchase a negative amount of horses.)

c. Graph the inequality from part **b.** on the grid in Problem 3.

5. Based on the inequalities you determined in Problems 1 and 4 and the graph of these inequalities, pick the pair of points below that you think gives Tristen the best advantage in the game and explain your reasoning. Be sure to verify that the coordinate pair is a solution by substituting the coordinates into both inequalities. Show all work.

   a. (0 weapons, 60 horses)

   b. (150 weapons, 0 horses)

   c. (150 weapons, 60 horses)

   d. (70 weapons, 32 horses)

   e. (100 weapons, 32 horses)

   f. (46 weapons, 46 horses)

Name: Date:

Math@Work

Market Research Analyst

As a market research analyst, you may work alone at a computer, collecting and analyzing data, and preparing reports. You may also work as part of a team or work directly with the public to collect information and data. Either way, a market research analyst must have strong math and analytical skills and be very detail-oriented. They must have strong critical-thinking skills to assess large amounts of information and be able to develop a marketing strategy for the company. They must also possess good communication skills in order to interpret their research findings and be able to present their results to clients.

Suppose you work for a shoe manufacturer who wants to produce a new type of lightweight basketball sneaker similar to a product a competitor recently released into the market. You have gathered some sales data on the competitor in order to determine if this venture would be worthwhile, which is shown in the table below. To begin your analysis, you create a scatter plot of the data to see the sales trend. (A scatter plot is a graph made by plotting ordered pairs in a coordinate plane in order to show the relationship between two variables.) You determine that the x-axis will represent the number of weeks after the competitors new sneaker went on the market and the y-axis will represent the amount of sales in thousands of dollars.

Number of Weeks x	Sales (in 1000s) y
3	15
6	22
9	28
12	35
15	43

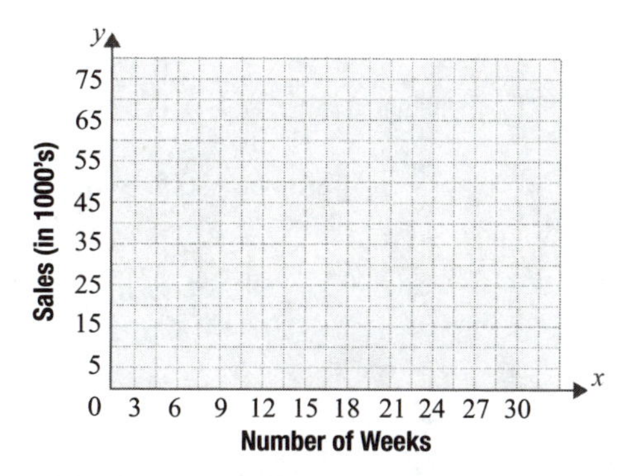

1. Create a scatter plot of the sales data by plotting the ordered pairs in the table on the coordinate plane. Does the data on the graph appear to follow a linear pattern? If so, sketch a line that you feel would "best" fit this set of data. (A market research analyst would typically use computer software to perform a technique called regression analysis to fit a "best" line to this data.)

2. Using the ordered pairs corresponding to weeks 9 and 15, find the equation of a line running through these two data points.

3. Interpret the value calculated for the slope of the equation in Problem 3 as a rate of change in the context of the problem. Write a complete sentence.

4. If you assume that the sales trend in sneaker sales follows the model determined by the linear equation in Problem 2, predict the sneaker sales in 6 months. Use the approximation that 1 month is equal to 4 weeks.

5. Give at least two reasons why the assumption made in Problem 4 may be invalid?

Foundations Skill Check for Chapter 10

This page lists several skills covered previously in the book and software that are needed to learn new skills in Chapter 10. To make sure you are prepared to learn these new skills, take the self-test below and determine if any specific skills need to be reviewed.

Each skill includes an easy (**e.**), medium (**m.**), and hard (**h.**) version. You should be able to complete each problem type at each skill level. If you are unable to complete the problems at the easy or medium level, go back to the given lesson in the software and review until you feel confident in your ability. If you are unable to complete the hard problem for a skill, or are able to complete it but with minor errors, a review of the skill may not be necessary. You can wait until the skill is needed in the chapter to decide whether or not you should work through a quick review.

7.7 Simplify the expression by combining like terms.

e. $23n - 16n$ **m.** $1.2z - 1.8 + 2.5z + 4.2$ **h.** $\frac{1}{2}x - 2\frac{3}{4}y + 3\frac{1}{5}x + y$

8.2 Solve the linear equation for the variable.

e. $9x - 5 = 13$ **m.** $4.8 - 0.5x - 0.4x = -3.3$ **h.** $2x + 4 - \frac{1}{2}x = 9$

8.3 Solve the linear equation for the variable.

e. $5x + 1 = 2x - 5$ **m.** $2(0.2x + 3) = 9 - 0.1x$ **h.** $\frac{1}{2}\left(\frac{x}{2} + 1\right) = \frac{1}{3}\left(\frac{x}{2} - 1\right)$

9.1 Determine if the given value is a solution to the equation.

e. $4x + 5y = 23$; given $(2, 3)$ **m.** $2.5x - 3y = 15$; given $(3, 4.5)$ **h.** $\frac{1}{2}x - \frac{5}{8}y = -2$; given $\left(-6, -\frac{8}{5}\right)$

9.3 Rewrite the equation in slope-intercept form.

e. $2x + 2y = 16$ **m.** $y + 5 = \frac{1}{3}(x - 6)$ **h.** $3y + 2x - 4 = 3x + 5$

Chapter 10: Systems of Linear Equations

Study Skills

10.1 Systems of Linear Equations: Solutions by Graphing

10.2 Systems of Linear Equations: Solutions by Substitution

10.3 Systems of Linear Equations: Solutions by Addition

10.4 Applications: Distance-Rate-Time, Number Problems, Amounts, and Costs

10.5 Applications: Interest and Mixture

Chapter 10 Projects

Math@Work

Foundations Skill Check for Chapter 11

Math@Work

Introduction

If you decide to go into the field of chemistry, you will have many different careers to choose from. Chemistry is linked closely to several other fields of study, such as medicine, physics, and biology, which adds variety to your list of career choices. Three main sectors that you can work in are education, industry, or government. If you choose to work in education, you can teach at the high school or college level and even continue to do research. If you choose to work in industry, you can work in pharmaceuticals creating new medications, or you can work in cosmetics creating perfumes and makeup. You can also work as an industrial chemist studying the composition of various chemicals. If you choose to work for the government, you can work in the military, with health and human services, or in agriculture. No matter which career path you choose to follow with chemistry, math will be a part of your daily work.

Suppose you decide to pursue a career as a chemist in the pharmaceuticals industry. The chemicals you create will be used by humans, so there are several things you will need to keep in mind. For example, what concentration level of a substance is toxic to humans? Which chemical is a better choice given a list of requirements to create a solution? How do you know how much of two substances or solutions to combine? Finding the answers to these questions (and many more) require several of the skills covered in this chapter and the previous chapter. At the end of the chapter, we'll come back to this topic and explore how math is used as a pharmaceutical chemist.

Study Skills

Do I Need a Math Tutor?

If you do not understand the material being presented in class, if you are struggling with completing homework assignments, or if you are doing poorly on tests, then you may need to consider getting a tutor. In college, everyone needs help at some point in time. What's important is to recognize that you need help before it's too late and you end up having to retake the class.

Alternatives to Tutoring

Before getting a tutor, you might consider setting up a meeting with your instructor during their office hours to get help. Unfortunately, you may find that your instructor's office hours don't coincide with your schedule or don't provide enough time for one-on-one help.

Another alternative is to put together a study group of classmates from your math class. Working in groups and explaining your work to others can be very beneficial to your understanding of mathematics. Study groups work best if there are three to six members. If you have too many people in the study group, you may find it hard to schedule a time for all group members to get together and it may be hard to get work done due to distractions. If you have too few people and those that attend are just as lost as you, then you aren't going to be helpful to each other.

Where to Find a Tutor

Many schools have both group and individual tutoring available and in most cases the cost of this tutoring is included in tuition costs. If your college offers tutoring through a learning lab or tutoring center, then you should take advantage of it. Often you must complete an application to be considered for tutoring, so it is highly recommended that you get the necessary paperwork at the start of each semester to increase your chances of getting an appointment time that works well with your schedule. This is especially important if you know that you struggle with math or haven't taken any math classes in a while. Math is one of those subjects that "if you don't use it, you lose it," which means that if you don't apply mathematics currently in your coursework, then you will forget it quickly.

If you find that you need more help than the tutoring center can provide or your school doesn't offer tutoring, you can hire a private tutor. The hourly cost to hire a private tutor varies significantly depending on the area you live in along with the education and experience level of the tutor. You might be able to find a tutor by asking your instructor for references or by asking friends who have taken higher level math classes than you have. You can also try researching the internet for local reputable tutoring organizations in your area.

What to Look for in a Tutor

Whether you obtain a tutor through your college or hire a personal tutor, look for someone who has experience, educational qualifications, and who is friendly and easy to work with. Many professional tutors have been trained and are certified to tutor particular subject areas. If you find that the tutor's personality or learning style isn't similar to yours, then you should look for a different tutor that matches your style. It may take some effort to find a tutor who works well with you.

How to Prepare for a Tutoring Session

To get the most out of your tutoring session, come prepared by bringing your text, class notes, and any homework or questions you need help with. If you know ahead of time what you will be working on, communicate this to the tutor so they can also come prepared. You should attempt the homework prior to the session and write notes or questions for the tutor. Do not use the tutor to do your homework for you. The tutor will explain to you how to do the work and let you work some problems on your own while he or she observes. Ask the tutor to explain the steps aloud while working through a problem. Be sure to do the same so that the tutor can correct any mistakes in your reasoning. Take notes during your tutoring session and ask the tutor if he or she has any additional resources such as websites, videos, or handouts that may help you.

Name: Date:

10.1 Systems of Linear Equations: Solutions by Graphing

Objectives

Determine whether a given ordered pair is a solution to a specified system of linear equations.

Estimate the solution, by graphing, of a system of linear equations with one solution.

Use a graphing calculator to solve a system of linear equations. (Software only)

Success Strategy

Using graph paper will allow you to graph lines by hand with more precision. More precision will enable you to determine the point of intersection of two lines, if one exists, more easily.

Understand Concepts

Go to Software First, read through Learn in Lesson 10.1 of the software. Then, work through the problems in this section to expand your understanding of the concepts related to solving systems of linear equations by graphing.

In Chapter 8, you learned how to solve linear equations in one variable. In Chapter 9, you learned how to work with linear equations in two variables. In this chapter, you will learn how to solve **systems of linear equations**. A system of linear equations is made of two or more equations that use the same set of variables. The simplest system of linear equations has two equations and two variables. The **solution of a system of linear equations** that consists of two equations in two variables is an ordered pair that satisfies both equations.

There are three main ways to solve systems of linear equations. In this section, we will discuss how to solve systems by graphing. In Section 10.2, we will discuss how to solve systems by substitution. In Section 10.3, we will discuss how to solve systems by addition. All three methods will give the same solution to the system, but one method may be easier to use when finding the solution, depending on the equations in the system.

Solving systems of linear equations by graphing is the most visual method of the three. By graphing the equations, you can see how many solutions there are. The downfall to this method is that the solutions can't always be precisely determined by looking at the graph. Graphing systems of equations provides a good visual for the different types of systems of linear equations.

Types of Systems of Linear Equations

A **consistent** system of linear equations in two variables has one or more solutions.

An **inconsistent** system of linear equations in two variables has no solution.

A **dependent** system of linear equations in two variables consists of equations which describe the same line. This type of system has an infinite number of solutions.

An **independent** system of linear equations in two variables consists of equations which describe different lines. This type of system can have one solution or no solution.

1. Indicate whether each graph represents a **consistent or inconsistent** system of equations and whether the system is **dependent or independent**.

a.

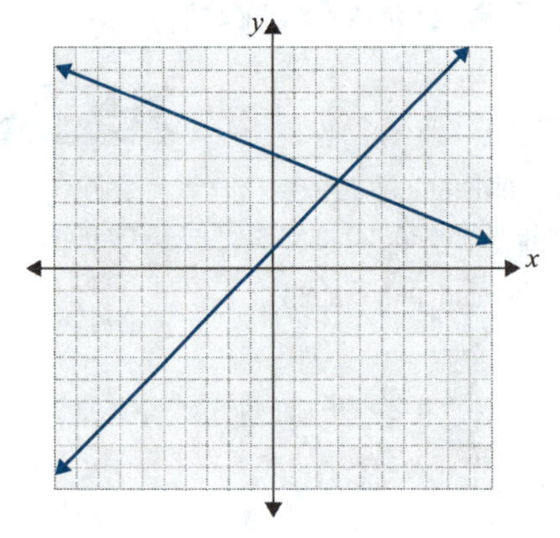

b.

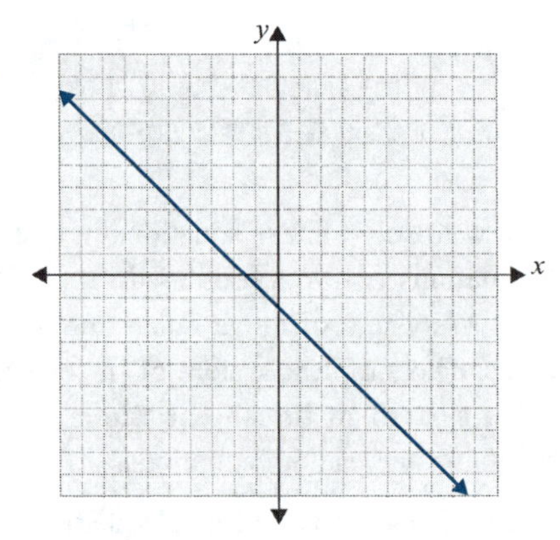

c.

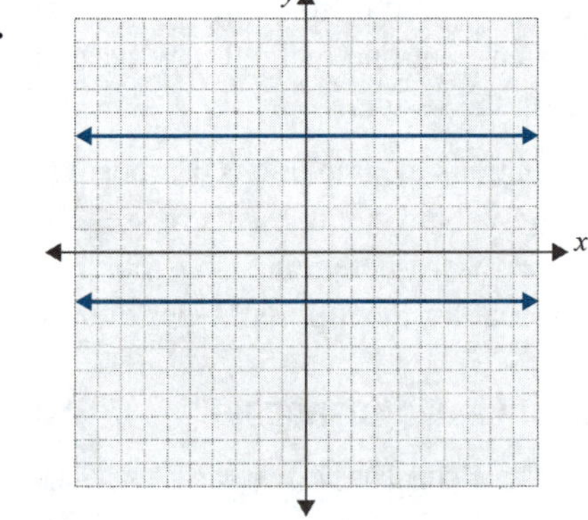

Quick Tip

Use a straightedge or ruler to draw the lines as precisely as possible. Being careless in drawing the lines could result in an incorrect solution. Be sure and check the solution from the graph in the original equations.

Solving a System of Linear Equations by Graphing

1. Graph the equations on the same coordinate plane.
2. Determine the type of system: consistent or inconsistent and dependent or independent. Then, determine the coordinates of the point of intersection, if one exists.
3. Check the solution in **both** equations.

Name: Date:

Read the following problem statement and then work through the following problems to explore how to solve a word problem that involves a system of linear equations by graphing.

Catherine is figuring out her schedule for the upcoming semester. She needs 17 credit hours for the semester to stay on track to graduate on time. The classes that she can choose from for the semester are either 3 or 4 credit hours. Catherine plans on taking 5 courses during the semester. How many 3-credit hour classes and how many 4-credit hour classes should Catherine take?

Quick Tip

Be sure to look at how the variables are used in each equation in the system when determining what the variables in an equation represent.

2. Two equations that represent this situation are $\begin{cases} x+y=5 \\ 3x+4y=17 \end{cases}$

a. What unknown value does the variable x represent in this situation?

b. What unknown value does the variable y represent in this situation?

c. What does the equation $x+y=5$ describe?

d. What does the equation $3x+4y=17$ describe?

Lesson Link

The methods covered in Sections 10.2 and 10.3 can be used to find an exact solution for any system of linear equations.

3. Graph both $x+y=5$ and $3x+4y=17$ on the given coordinate plane.

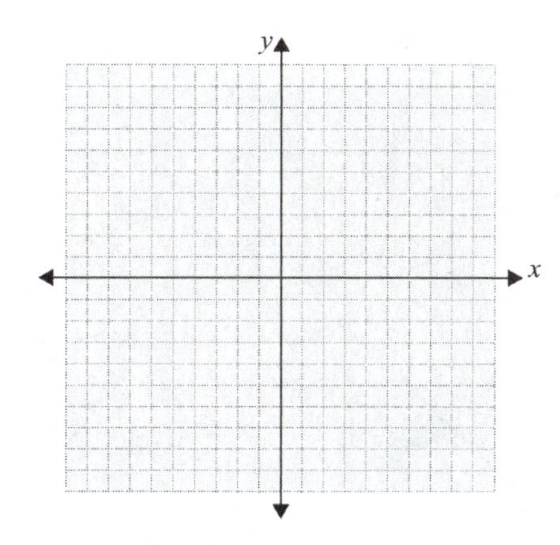

4. Is the system of linear equations **consistent or inconsistent** and is the system **dependent or independent**?

5. Use the graph created in Problem 3 to find the solution to the system of equations.

a. Determine the solution to this system of linear equations and write it as an ordered pair.

Quick Tip

The solution to a system of equations must satisfy all equations in the system. If it does not satisfy one of the equations, it is not a solution.

Quick Tip TECH

Instructions on how to find the point of intersection with a calculator is introduced in the Learn portion of the software.

b. Verify that the solution found in part **a.** is a solution to the system by substituting it into the equations from Problem 2. Remember that the solution needs to be a solution for both equations in the system.

6. Determine if an accurate solution can be found from the given graph for the systems of linear equations or if an estimate would have to be made. If there is no solution to the system of equations, write "no solution".

a.

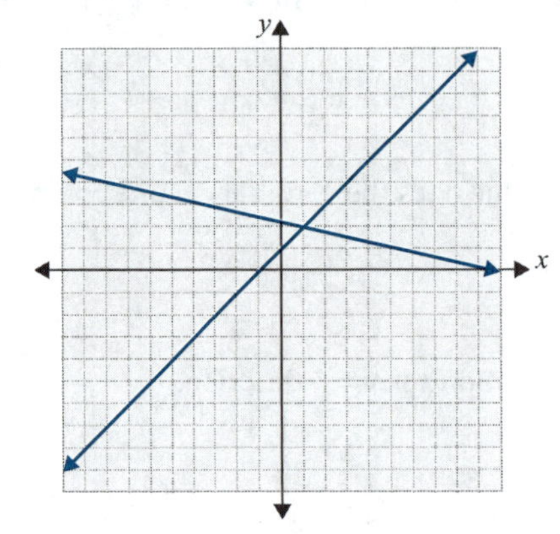

b.

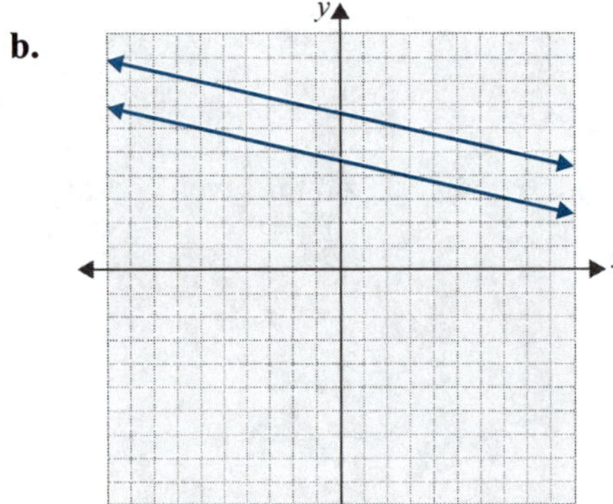

c.

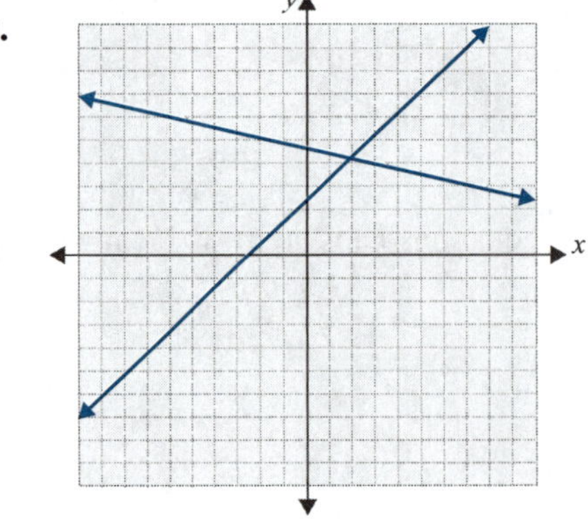

Name: Date:

Skill Check

Go to Software Work through Practice in Lesson 10.1 of the software before attempting the following exercises.

Determine if the given ordered pair is a solution to the equation.

7. $\begin{cases} x - y = 4 \\ 3x - y = 6 \end{cases}$

$(1, -3)$

8. $\begin{cases} 4x = 3 + y \\ x + 8 = 2y \end{cases}$

$(2, 6)$

Quick Tip

Recall that the lines are parallel in an **inconsistent system**. Parallel lines must have the same slope. For equations to represent different parallel lines, they must have a different y-intercept.

Show that each system of equations is inconsistent by determining the slope and the y-intercept of each line.

9. $\begin{cases} 2x + y = 3 \\ 2x + y = 5 \end{cases}$

10. $\begin{cases} 3x - y = 8 \\ x - \frac{1}{3}y = 2 \end{cases}$

Apply Skills

Work through the problems in this section to apply the skills you have learned related to solving systems of linear equations by graphing.

11. Mackenzie plans to join a gym and wants to decide between the two gyms that are close to her home, Fit4Life and Workout Nation. At Fit4Life, she would pay a \$20 per month membership fee and \$10 per class. At Workout Nation, she would pay a \$45 per month membership fee and \$5 per class. Mackenzie wants to determine how many classes she would have to take per month for both gyms to have the same cost.

    a. Write two equations to represent the situation. Use the variable y to represent the total cost per month and the variable x to represent the number of classes.

    b. Graph the two equations on the same coordinate plane.

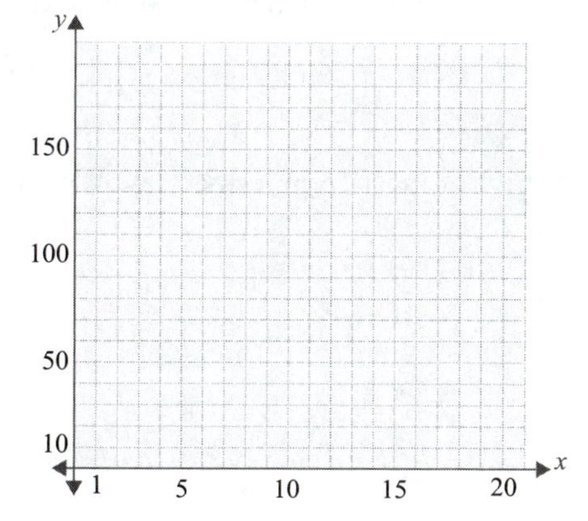

    c. Find the point of intersection.

d. What does the point of intersection mean? Write a complete sentence.

e. If Mackenzie plans to take 10 classes per month, which gym would be the better deal?

12. You are planning a vacation and would like to spend your money wisely. You decide to fly to your destination and then rent a car when you get there. Discount Car Rentals charges \$10 per day for the economy class car and \$0.10 per mile. Cars For Hire charges \$15 per day and \$0.05 per mile. You need to determine which car rental service offers the best deal.

a. Write two equations to represent the situation. Use the variable y to represent the total cost per day and the variable x to represent the number of miles driven per day.

b. Graph the two equations on the same coordinate plane.

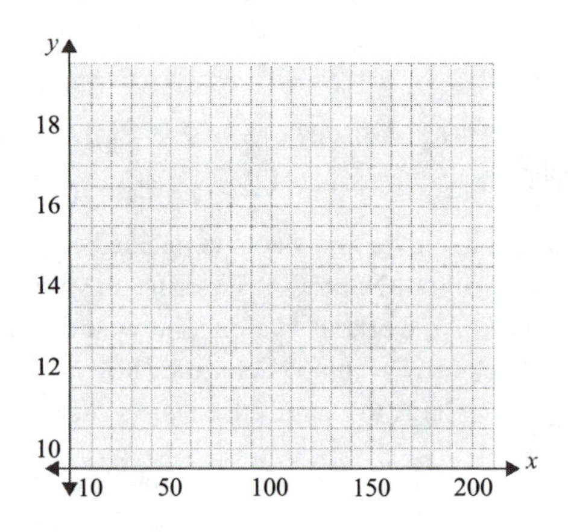

c. Find the point of intersection.

d. What does the point of intersection mean? Write a complete sentence.

e. You expect to drive at most 75 miles per day. Which car rental company should you rent the car from?

Name: Date:

10.2 Systems of Linear Equations: Solutions by Substitution

Objectives

Solve systems of linear equations by using the method of substitution.

Success Strategy

Before starting this section, practice solving linear equations in two variables for one of the variables like you did with formulas in Section 8.5.

Understand Concepts

Go to Software First, read through Learn in Lesson 10.2 of the software. Then, work through the problems in this section to expand your understanding of the concepts related to solving systems of linear equations by substitution.

The second method of solving systems of linear equations that we will discuss is called the **method of substitution**. The main idea of this method is to solve one of the equations for a variable and substitute this expression into the other equation. This process reduces the problem down to solving one linear equation in one variable. This method will give an exact answer as long as no rounding is used during the solving process. The downfall with the substitution method is that solving an equation for a variable can sometimes result in coefficients which are difficult to work with.

Solving a System of Linear Equations by Substitution

1. Solve one of the equations for one of the variables.
2. Substitute the resulting expression into the **other** equation.
3. Solve the resulting equation for the remaining variable, if possible.
4. Substitute the value of the solved variable into one of the original equations to get the value of the other variable.
5. Check the solution in **both** the original equations.

Quick Tip

Substituting the value of the solved variable into one of the original equations to find the value of the other variable is known as **back substitution**.

The trickiest part when solving linear equations by substitution is determining which variable in which equation would be the best to solve for. Ideally, you want to solve for a variable that will not produce fractional coefficients.

1. For each equation, determine which variable should be solved for so that the resulting expression doesn't involve fractions. Solve the equation for this variable.

a. $6x + 3y = 9$

b. $4y - x = 12$

c. $2x + 8y = 16$

Quick Tip

Substituting an expression for a variable is similar to substituting a numerical value for the variable. The only difference is that the distributive property may be needed to simplify the expression.

Determine if any mistakes were made while using the method of substitution. If any mistakes were made, describe the mistake and then correctly apply the method of substitution and simplify to find the actual result.

2. Given $\begin{cases} y = 3 - 2x \\ 5x - 3y = 2 \end{cases}$

$$5x - 3(3 - 2x) = 2$$
$$5x - 9 + 6x = 2$$
$$11x - 9 = 2$$

3. Given $\begin{cases} x = 4y - 10 \\ 4x + 2y = 14 \end{cases}$

$$4x + 2(4y - 10) = 14$$
$$4x + 8y - 20 = 14$$

Read the following problem statement and then work through the following problems to explore how to use substitution to solve a word problem involving a system of linear equations.

Jamie found some leftover 2-cent and 3-cent stamps from the last postage increase and would like to use them. She determines that one less than the number of 3-cent stamps is equal to twice the number of 2-cent stamps that she has. The total worth of the stamps is 43 cents. How many of each stamp does Jamie have?

4. Two equations that represent this situation are $\begin{cases} y - 1 = 2x \\ 2x + 3y = 43 \end{cases}$

Quick Tip

Be sure to look at how the variables are used in each equation in the system when determining what the variables in an equation represent.

a. What unknown value does the variable x represent in this situation?

b. What unknown value does the variable y represent in this situation?

c. What does the equation $y - 1 = 2x$ describe?

d. What does the equation $2x + 3y = 43$ describe?

Name: Date:

5. Consider the system of equations $\begin{cases} y-1=2x \\ 2x+3y=43 \end{cases}$

   a. Which variable in which equation would be easiest to solve for and substitute into the other equation? Explain why.

   b. Solve for the variable you specified in part **a.**

   c. Substitute the expression from part **b.** into the other equation and solve for the remaining variable.

   d. Determine the value of the other variable in the system using back substitution.

Lesson Link

The concept of infinity was introduced in Sections 7.1 and 8.7.

Quick Tip

A system of equations has an infinite number of solutions when the solving process results in $c = c$, where c is a real number. No variable will be present in the equation.

Occasionally a system of linear equations will have infinitely many solutions. The answer for such a system should be written in a specific format. The next problem will guide you through the process.

6. Solve the system of equations $\begin{cases} 4y-8x=20 \\ 25-5y=-10x \end{cases}$

   a. Solve the system of equations to verify that it has an infinite number of solutions.

   b. Solve both of the equations for y. Isolate y on the left side of the equation.

   c. Write the solution as an ordered pair in the form (x, y) by substituting y with the expression from part **b.**

   d. Write an explanation for why this ordered pair represents an infinite number of solutions.

Skill Check

Go to Software Work through Practice in Lesson 10.2 of the software before attempting the following exercises.

Solve each system of linear equations by using the method of substitution.

7. $\begin{cases} x+y=6 \\ x+y=7 \end{cases}$

8. $\begin{cases} 2x+5y=15 \\ x=y-3 \end{cases}$

9. $\begin{cases} x-2y=-4 \\ 3x+y=-5 \end{cases}$

10. $\begin{cases} 6x-y=15 \\ 18x-45=3y \end{cases}$

Apply Skills

In real life, the information to solve a problem won't be presented in already-assembled equations. You will need to gather the data to solve the problems and construct the system of equations yourself. Work through the problems in this section to apply the skills you have learned related to solving systems of linear equations by substitution.

Quick Tip

The **area** of a rectangle is equal to the length multiplied by the width.

11. Connor is retiling the backsplash of his kitchen counter. He plans on using square tiles that measure 2 inches by 2 inches and rectangular tiles that measure 2 inches by 4 inches. The backsplash measures 6 inches by 60 inches. He wants to use an equal number of rectangular tiles and square tiles. How many of each tile will he need to buy?

a. Find the area of each size tile and the area of the backsplash.

b. Write two equations to represent the situation. Use the variable s to represent the number of square tiles and the variable r to represent the number of rectangular tiles.

c. Solve the system of equations by substitution.

d. What does the solution mean? Write a complete sentence.

e. If the square tiles come in boxes of 30 and the rectangular tiles come in boxes of 18, how many boxes of each type of tile will Connor need to buy to retile the backsplash?

12. Paul needs to decide which cell phone plan is the best option between two providers. He wants a single line with a data plan, but will pay per text message since he doesn't text very often. The first provider charges $45 per month for a single line with 1 GB of data and an additional $0.15 per text message. The second provider charges $55 per month for a single line with 1 GB of data and an additional $0.10 per text message. When will the plans cost the same amount?

a. Write two equations to represent the situation. Use the variable c to represent the total cost of the data plan per month and the variable t to represent the number of text messages sent per month.

b. Solve the system of equations by substitution.

c. What does the solution mean? Write a complete sentence.

d. How many texts must Paul send per month for the second provider to have the less expensive option?

13. Carlos is buying supplies for the tutoring center where he works. He was given a budget of \$230 to spend. Calculators cost \$15 each and packs of paper are \$2.50 each. He would like to buy twice as many packs of paper as calculators. How many calculators and notebooks can Carlos buy and stay in budget? (**Note:** The tutoring center is non-profit organization and therefore does not have to pay sales tax.)

a. Write two equations to represent the situation. Use the variable c to represent the number of calculators and the variable p to represent the number of packs of paper.

b. Solve the system of equations by substitution. Write the solutions in decimal form.

c. What does the solution mean? Write a complete sentence.

d. Does the answer to part **c.** make sense? Explain why.

e. Based on your answer from part **d.**, how many calculators should Carlos buy to stay within budget and to buy twice as many packs of paper as calculators?

f. If the tutoring center has at most 12 students at a time, will each student have a new calculator to work with?

Name: Date:

10.3 Systems of Linear Equations: Solutions by Addition

Objectives

Solve systems of linear equations using the method of addition.

Use a system of equations to find the equation of a line through two points.

Success Strategy

Before beginning this section, review the meaning of *like terms* from Section 7.7 and *opposites* from Section 7.1.

Understand Concepts

Go to Software First, read through Learn in Lesson 10.3 of the software. Then, work through the problems in this section to expand your understanding of the concepts related to solving systems of linear equations by addition.

In this section, we will discuss the final method of solving systems of linear equations: **the method of addition**. This method is useful when none of the variables in the system are easy to solve for and the system would be difficult to graph. The downside of this method is that it might involve more steps and large numbers to work with.

Quick Tip

The method of addition is also known as the **method of elimination**.

Solving Systems of Equations by Addition

1. Write the equations in standard form $ax + by = c$, one under the other, so that like terms are aligned underneath one another and the constant terms are isolated.
2. Multiply all terms of one equation by a constant so that one pair of like terms in both equations has opposite coefficients. The second equation may need to be multiplied by a different constant to create like terms with opposite coefficients.
3. Add like terms by adding the two equations vertically.
4. Solve the resulting equation for the remaining variable.
5. Back substitute into one of the original equations to find the value of the other variable.
6. Check the solution in both equations, if there is one.

Lesson Link

Step 3 is the result of using the addition principle of equality, which was introduced in Section 8.1.

Lesson Link

Finding the least common multiple of two numbers was covered in Section 2.3. Using the least common multiple isn't necessary for this process, but it helps to keep the coefficients from getting too large.

The trickiest part of solving a system by addition is creating two equations that have like terms with opposite coefficients (Step 2). Another way to think of this is that you want to find the least common multiple (LCM) of the coefficients of a pair of like terms.

1. Determine the LCM of the coefficients of the indicated variable in the system of linear equations. Then, decide which constant you would multiply each equation by. Remember that the coefficients need to be opposites (that is, one will be positive and the other negative). There is more than one correct answer.

a. $\begin{cases} 3x+6y=14 \\ 7x-3y=26 \end{cases}$ variable: y

b. $\begin{cases} 4x+6y=-8 \\ 8x+7y=-26 \end{cases}$ variable: x

c. $\begin{cases} 2x+6y=10 \\ 3x-4y=15 \end{cases}$ variable: y

Work through the following problems to explore how to use the addition method to solve a word problem which involves a system of linear equations.

A midterm exam is worth 100 points and has 26 problems. The problems are worth either 3 points or 5 points each. How many 3-point problems are on the test? How many 5-point problems are on the test?

2. Two equations that represent this situation are $\begin{cases} x+y=26 \\ 3x+5y=100 \end{cases}$

a. What unknown value does the variable x represent in this situation?

b. What unknown value does the variable y represent in this situation?

c. What does the equation $x+y=26$ describe?

d. What does the equation $3x+5y=100$ describe?

Quick Tip

Be sure to look at how the variables are used in each equation in the system when determining what the variables in an equation represent.

3. Solve the system of linear equations from Problem 2 using the addition method.

a. Write the two equations in standard form, one under the other, so that like terms are aligned underneath one another.

b. Multiply each equation by the appropriate constant so that the terms with y have opposite coefficients. Rewrite the system here.

c. Add the equations from part **b.**

d. Solve the resulting equation from part **c.** for x.

e. Determine the value of y using back substitution.

f. What do the answers from parts **d.** and **e.** mean? Write a complete sentence.

Lesson Link

The standard form of an equation was introduced in Section 9.2.

Name: Date:

The following table is a summary of the three methods for solving linear equations. Even though you may be more comfortable with one method, it is important to practice all three methods and understand the advantages and disadvantages of each.

Pros and Cons of the Different Methods for Solving Systems of Linear Equations		
Method	**Pros**	**Cons**
Graphing	Good for determining the number of solutions. Good for visualizing the solution.	Can't always find an exact solution unless you use a graphing calculator.
Substitution	Exact solutions. Easy if one variable already has a coefficient of 1 or –1.	Coefficients aren't always easy to work with and you may end up with fractions for coefficients. May be difficult to determine which variable to solve for in the equation.
Addition	Exact solutions. Easy if coefficients are small or opposite coefficients already exist.	Coefficients may become large. May be difficult to determine which variable to eliminate.

Skill Check

Go to Software Work through Practice in Lesson 10.3 of the software before attempting the following exercises.

Solve each system of linear equations using the method of addition.

Quick Tip

Equations should be written in standard form to use the addition method.

4. $\begin{cases} 3x+y=-10 \\ 2y-1=x \end{cases}$

5. $\begin{cases} 6x-5y=-40 \\ 8x-7y=-54 \end{cases}$

Quick Tip

To make equations easier to work with, change decimal numbers into whole numbers by multiplying the entire equation by a power of 10.

6. $\begin{cases} 0.75x-0.5y=2 \\ 1.5x-0.75y=7.5 \end{cases}$

7. $\begin{cases} 10x+4y=7 \\ 5x+2y=15 \end{cases}$

Apply Skills

Work through the problems in this section to apply the skills you have learned related to solving systems of linear equations by addition.

8. You are deciding between two credit cards with similar rewards programs. The City credit card will give you 3500 points as a sign-up bonus and 1.5 points for every dollar you spend. The International credit card gives you 1000 points as a sign-up bonus and gives you 2 points for every dollar you spend. How much would you have to spend to earn the same amount of rewards on each credit card?

a. Write two equations to represent the situation. Use the variable x to represent the number of dollars spent and the variable y to represent the total number of points earned.

b. Solve the system of equations by addition.

c. What does the solution mean? Write a complete sentence.

d. If you only plan to purchase $4000 in merchandise, which credit card will give you the most points?

9. Barbara's Bombtastic Bakery uses chocolate chips in one type of cookie and in one type of muffin. The cookie recipe calls for 5 cups of chocolate chips and the muffin recipe calls for 2 cups of chocolate chips. The cookie recipe makes 30 large cookies and the muffin recipe makes 18 giant muffins. The bakery currently has 50 cups of chocolate chips and only has room in the display case for a combination of 360 cookies and muffins. The manager wants to determine how many of each item to bake to use all of the chocolate chips.

a. Write two equations to represent the situation. Use the variable x to represent the number of batches of chocolate chip cookies and the variable y to represent the number of batches of muffins.

b. Solve the system of equations by addition.

c. What does the solution mean? Write a complete sentence.

d. If the manager makes the amount of chocolate chip cookies and muffins described in part **c.**, will the bakery be able to fulfill an order for 150 chocolate chip cookies and 160 chocolate chip muffins?

Name: Date:

10.4 Applications: Distance-Rate-Time, Number Problems, Amounts, and Costs

Objectives

Use systems of linear equations to solve distance-rate-time application problems.

Use systems of linear equations to solve number application problems.

Use systems of linear equations to solve amounts and costs application problems.

Success Strategy

Regardless of the problem type, you should try using tables to organize the information when solving application problems. This will help you set up the equations correctly.

Understand Concepts

Go to Software First, read through Learn in Lesson 10.4 of the software. Then, work through the problems in this section to expand your understanding of the concepts related to using systems of linear equations to solve application problems.

Quick Tip

Remember to follow Pólya's problem-solving steps, which are located at the beginning of the workbook.

In Section 8.6, a method of using tables to solve word problems was introduced for linear equations in one variable. This section will discuss how to use a similar table to solve word problems involving systems of linear equations in two variables. Keep in mind that while using a table isn't necessary, it can be helpful to keep your work organized.

A common type of problem that occurs in real life is called a distance-rate-time problem and uses the formula $d = r \cdot t$. One example of a distance-rate-time problem involves an airplane that flies *with* the wind on one leg of the trip and *against* the wind on the return leg. To write a system of equations to describe this situation, you need to define two variables:

s = speed of the airplane in still air, or when no wind is blowing

w = speed of the wind, or wind speed

1. Using the two variables defined above, determine an expression for the speed of an airplane in the following scenarios.

a. An airplane flying in still air.

b. An airplane flying with the wind, that is, flying in the same direction that the wind is blowing. (**Hint:** The wind helps move the airplane faster since it pushes from behind.)

Quick Tip

Flying with the wind is sometimes referred to as having a **tail wind** and flying against the wind is sometimes referred to as having a **head wind**.

c. An airplane flying against the wind, that is, flying in the opposite direction of the wind. (**Hint:** The wind hinders the airplane and slows it down since it pushes from the front.)

2. A small airplane traveled 300 miles in 2 hours while flying with the wind. On the return trip, it flew against the wind and traveled only 200 miles in 2 hours. Determine the wind speed and the speed of the airplane.

a. Using your work from Problem 1, write an expression for the rate for each scenario described in column one and place in the rate column of the table.

	Rate	·	Time	=	Distance
With the Wind					
Against the Wind					

b. Pick out the time values from the problem for each scenario and place in the time column.

c. Use the formula $d \cdot r = t$ to calculate an expression for distance for each scenario and place in the distance column.

d. What do you know about the distance traveled in each scenario?

e. Create two equations for the situation using part **d.** and the information in the distance column.

f. The pair of equations from part **e.** creates a system of linear equations. Solve the system.

g. Do the answers you obtained for the airplane speed and wind speed make sense in the context of the problem? Explain your reasoning.

Another situation where systems of linear equations appear is cost problems. For these problems, a table is generally not needed to set up the equations.

3. You and your friends watch football every Sunday afternoon at your favorite local pub. Last week, you ordered 3-dozen hot wings and 8 drinks for a total cost of $32.29. This Sunday, you ordered 4-dozen hot wings and 12 drinks for a total cost of $45.44. How much does 1-dozen hot wings cost?

a. Using the second sentence of the problem, write an equation in standard form to represent the total cost of last week's outing. Let the variable h represent the cost of 1-dozen hot wings and the variable d represent the cost of each drink.

Name: Date:

Quick Tip

To rewrite the equation without decimal numbers, multiply both sides of the equation by a power of 10.

b. Using the third sentence of the problem, write an equation in standard form to represent the total cost of this Sunday's outing. Use the same variables from part **a.**

c. Which method do you think would be best to use to solve this system of equations? Explain why.

d. Solve the system of equations formed by parts **a.** and **b.**

e. Write a complete sentence to answer the question from the problem.

Skill Check

Go to Software Work through Practice in Lesson 10.4 of the software before attempting the following exercises.

Solve.

4. The sum of two numbers is 56. Their difference is 10. Find the numbers.

5. Two angles are complementary if the sum of their measures is 90°. Find two complementary angles such that one is 15° less than six times the other.

Apply Skills

Work through the problems in this section to apply the skills you have learned related to using systems of linear equations to solve application problems.

6. A petting zoo charges \$5 for children and \$10 for adults. On Tuesday, the petting zoo made \$1400 from ticket sales and sold a total of 200 tickets. How many adults and how many children visited the petting zoo on Tuesday?

   a. Write two equations to describe the situation. Use the variable c to represent the number of children and the variable a to represent the number of adults.

   b. Solve the system of linear equations.

   c. Use the solution from part **b.** to write a complete sentence to answer the question from the problem.

Quick Tip

The current of a river works *with* the boat when traveling downstream and *against* the boat when traveling upstream.

7. A boat tour travels 4 miles downstream in 20 minutes and the return trip upstream takes 30 minutes. Find the rate of the boat and the rate of the current.

   a. Change each time from minutes to hours. Write each time value as a fraction.

   b. Use the table to set up a system of linear equations. Use the variable b to represent the rate of the boat and the variable c to represent the rate of the current.

	Rate	·	Time	=	Distance
Downstream					
Upstream					

   c. Solve the system of linear equations.

   d. Use the solution from part **c.** to write a complete sentence to answer the question from the problem.

Name: Date:

10.5 Applications: Interest and Mixture

Objectives

Use systems of linear equations to solve interest problems.

Use systems of linear equations to solve mixture problems.

Success Strategy

By this point you should know how to solve systems of equations using the three different methods. For this section, focus on understanding new terminology involving interest and mixtures.

Understand Concepts

Go to Software First, read through Learn in Lesson 10.5 of the software. Then, work through the problems in this section to expand your understanding of the concepts related to using systems of linear equations to solve application problems.

This section will introduce two more types of applications for systems of linear equations in two variables. Interest problems were first introduced in Section 4.8 and again in Section 8.6. In this section, we will learn how to solve more complicated interest problems. Setting up and solving mixture problems requires organization and an understanding of what information the problem is asking for. A table can help you set up and solve both of these types of problems.

Interest problems can be set up using a table similar to the one introduced in Section 8.6. The main difference is the addition of the "totals" row to the bottom of the table. The next problem guides you through the process of solving an interest problem.

Quick Tip

In Problem 1, investing in Fund A is considered a lower risk than investing in Fund B. Investing money in a mix of low and high risk funds is called **diversification**. It helps to prevent major losses if the stock market takes a downturn.

1. Lila has \$7000 to invest and she would like to invest this amount into two different funds. Fund A has an interest rate of 7% and Fund B has an interest rate of 12%. Lila would like to earn a combined annual interest of \$690 on her investments after one year. How much money should Lila put into each fund?

   a. The total principal Lila has to invest is \$7000. The amount that she should invest in each account needs to be determined. Let the variable x represent the principal invested in Fund A and the variable y represent the principal invested in Fund B. Fill in the principal column of the table.

Fund	Principal (\$)	·	Interest Rate	=	Interest (\$)
Fund A					
Fund B					
Totals					

   b. The interest rate for each account is known. There is no total interest amount. Fill in the first two rows of the interest rate column of the table. Remember to convert the percents to decimal numbers.

   c. Lila wants to earn a total interest amount of \$690. The interest earned for each fund can be found by using the equation *Principal* · *Interest Rate* = *Interest*. Fill in the interest column of the table.

   d. Create an equation in two variables that represents the relationship between the amount invested in each fund and the total principal using the terms in the principal column.

Lesson Link

The calculation of simple interest was introduced in Section 4.8.

e. Create an equation in two variables that represents the relationship between the interest earned in each fund and the total interest Lila would like to earn using the terms in the interest column.

f. Solve the system of linear equations formed by the equations from parts **d.** and **e.**

g. Check the solution from part **f.** in both of the original equations.

h. Write a complete sentence to answer the question from the problem.

Mixture problems can be set up using a table similar to the one used for interest problems. The next problem will guide you through the process of solving a mixture problem.

2. A chemist needs 50 ounces of 12% salt solution for an experiment. The chemistry lab only has a 10% salt solution and a 15% salt solution in stock. How many ounces of the 10% salt solution and how many ounces of the 15% salt solution should be combined to create the required amount of 12% salt solution?

a. The only known amount is that 50 ounces of 12% solution are needed. The other two amounts are unknown. Let the variable x represent the amount of the 10% salt solution needed and let y represent the amount of the 15% salt solution needed. Fill in the amount of solution column of the table.

% Solution	Amount of Solution	·	Percent of Salt	=	Amount of Salt
10%					
15%					
12%					

Quick Tip

The amount of salt in each of the solutions that are mixed together has to equal the amount of salt in the final solution. This relates to a law in physics that mass cannot be created called **The Law of the Conservation of Mass.**

b. The percent of salt in each solution is given. Write the percents in decimal number form in the percent of salt column.

c. Use the equation *amount of solution* · *percent of salt* = *amount of salt* to complete the amount of salt column.

Name: Date:

d. Two amounts of two different solutions are being combined to create the 12% salt solution. Create an equation to represent the sum of the volumes of the solutions by using the terms in the amount of solution column.

e. To create a system of linear equations, we need two equations with two variables. Create another equation to represent the sum of the amounts of salt from each solution set equal to the total amount of salt found in the mixture by using the terms in the amount of salt column.

f. Solve the system of linear equations formed by the equations from parts **d.** and **e.**

g. Check the answer from part **f.** in the original equations.

h. Write a complete sentence to answer the question from the problem.

Skill Check

Go to Software Work through Practice in Lesson 10.5 of the software before attempting the following exercises.

Solve each system.

3. $\begin{cases} x + y = 100 \\ 0.20x + 0.30y = 23 \end{cases}$

4. $\begin{cases} y - x = 1000 \\ 0.10y - 0.06x = 260 \end{cases}$

Apply Skills

Work through the problems in this section to apply the skills you have learned related to using systems of linear equations to solve application problems.

5. Sanjay has $5000 to invest and he has an option to split the amount between two simple interest accounts. Account A is expected to earn 4% interest and Account B is expected to earn 9% interest. Sanjay has a goal to make $350 in interest after one year. How much should he invest in each account?

a. Use the table to set up a system of linear equations to describe the situation. Use the variable x to represent the amount invested in Account A and the variable y to represent the amount invested in Account B.

Account	Principal	·	Interest Rate	=	Interest
Account A					
Account B					
Totals					

b. Solve the system of linear equations from part **a.**

c. Write a complete sentence to answer the question from the problem.

d. Is it possible for Sanjay to earn more than $350 in interest? If yes, explain how.

Name: Date:

6. You have 3% hydrogen peroxide and 12% hydrogen peroxide. You need 20 ounces of 6% hydrogen peroxide. How many ounces of each grade of hydrogen peroxide do you need to mix together to obtain the required amount of 6% hydrogen peroxide?

a. Set up a system of linear equations to describe the situation. Use the variable x to represent the amount of 3% solution and the variable y to represent the amount of 12% solution.

% Solution	Amount of Solution	·	Percent of Peroxide	=	Amount of Peroxide
3%					
12%					
6%					

b. Solve the system of linear equations.

c. Write a complete sentence to answer the question from the problem.

7. King Nut Company sells freshly roasted nuts. A one-pound bag of broken fancy cashews costs \$5. A one-pound bag of almonds costs \$7.50. The manager wants to sell a mixture of the nuts that will cost \$6.50 for a one-pound bag. How much of each type of nut should he combine to make a one-pound bag of mixed nuts?

a. Write two equations to describe the situation. Use the variable x to represent the amount of cashews and the variable y to represent the amount of almonds.

b. Solve the system of linear equations.

c. Write a complete sentence to answer the question from the problem.

Name: Date:

Chapter 10 Projects

Project A: Don't Put all Your Eggs in One Basket!

An activity to demonstrate the use of linear systems in real life.

Have you ever heard the phrase, "Don't put all your eggs in one basket"? This is a common saying that is often quoted in the investment world—and it's true. In an ever-changing economy it is important to diversify your investments. Splitting your money up among two or more funds may keep you from losing it all if one of the funds performs poorly. You may be thinking that you are too young to consider investments and saving money for retirement, but it is never too soon—especially in today's economy where interest rates are extremely low. With low interest rates it takes even longer to build up your nest egg. So start saving now and be sure to have more than one basket to put your eggs in!

For this activity, if you need help understanding some of the investment terms, use the following website as a resource: http://www.investopedia.com/

Let's suppose that you received a total of \$5000 in cash as a graduation present from your relatives. You also have an additional \$2500 that you saved from your summer job. You are thinking about investing the \$7500 in two investment funds that have been recommended to you. One is currently earning 4% interest annually (a conservative fund) and the other is earning 8% annually (an aggressive fund). Keep in mind that interest rates fluctuate as the economy changes and there are few guarantees on the amount you will actually earn from any investment. Also, note that higher rates of interest typically indicate a higher risk on your investment.

1. If you want to earn \$400 total in interest on your investments this year, how much money will you have to invest in each fund? Let the variable x represent the amount invested in Fund 1 and the variable y represent the amount invested in Fund 2. Recall that to calculate the interest on an investment, use the formula $I = Prt$, where P is the principal or amount invested, r is the annual interest rate, and t is the amount of time invested, which for our problem will be 1 year $(t = 1)$. Use the table below to help you organize the information. Note that interest rates have to be converted to decimals before using them in an equation.

	Principal	·	Interest Rate	=	Interest
Fund 1	x		0.04		$0.04x$
Fund 2	y		0.08		$0.08y$
Total	**a.**				**b.**

   a. Fill in the total amount available for investment in the bottom row of the table.

   b. Fill in the total amount of interest desired in the bottom row of the table.

   c. What does $0.04x$ represent in the context of this problem?

   d. What does $0.08y$ represent in the context of this problem?

   e. Using the principal column of the table, write an equation in standard form involving the variables x and y to represent the total amount available for investment.

   f. Using the interest column of the table, write an equation in standard form involving the variables x and y to represent the total amount of interest desired.

g. Solve the linear system of two equations derived in parts **e.** and **f.** to determine the amount to invest in each fund to earn $400 in interest. (You may use any method you choose: substitution, addition/elimination, or graphing.)

h. Check to make sure that your solution to the system is correct by substituting the values from part **g.** for x and y into *both* equations and verify that the equations are true statements.

2. Suppose you decide that you want to earn more interest on your investment. You now want to earn $500 in interest next year instead of $400. Using a table similar to the one in Problem 1, organize the information and follow a similar format to determine the amounts to invest in each of the funds that will earn $500 in interest in a year.

3. Compare the results you obtained from Problems 1 and 2. How did the amounts in each investment change when the desired amount of interest increased by $100?

4. Suppose you decide that $500 is not enough interest and you want to earn an additional $100 on your investments for a total of $600 in interest. Using a table similar to the one in Problem 1, organize the information and follow a similar format to determine the amounts to invest in each of the funds that will earn $600 in interest in a year.

5. Compare the results from Problem 4 to the results from Problems 1 and 2.

a. How much are you investing in Fund 1 to earn $600 in interest?

b. How much are you investing in Fund 2 to earn $600 in interest?

c. How do your results contradict the advice provided to you at the start of this activity?

d. Is it possible to make more than $600 in interest on your $7500 investment using these two funds? Explain why or why not?

Name: Date:

Project B: Super Bowl Mania!

An activity to demonstrate the use of linear systems in real life.

The 2013 Super Bowl XLVII between the Baltimore Ravens and the San Francisco 49ers was held in New Orleans, Louisiana, on February 3. It was a Super Bowl history first—the head coaches of the two teams were brothers. It has been nicknamed the "Bro Bowl" and the "Harbowl" after the two coaches—Jim and John Harbaugh. It has also been called the "Blackout Bowl" due to a power outage that affected half of the stadium and caused play to be suspended for 34 minutes.

For each problem below, write a system of linear equations in two variables, x and y, based on the information given in the problem. (Be sure to label what each variable represents.) Then, solve the system and answer any questions. Be sure to show all work and check your answers.

1. Darby flew to New Orleans for Super Bowl XLVII on an airplane that was flying with a strong tail wind and made the 800 mile trip in only 2 hours. Coming back from the game the airplane was flying against the wind and had only traveled $\frac{3}{4}$ of the distance in 2 hours. Determine the wind speed and the speed of the airplane.

2. The ticket prices for the Super Bowl were $2000 for the "cheap" seats and $2400 for the remaining seats in the stadium. There were 69,700 seats available in the stadium and $159 million dollars in ticket sales. If all the seats in the stadium were filled, how many of each type of ticket was sold? (**Hint:** Let x be the number of "cheap" seats priced at $2000 a ticket and let y be the number of remaining seats priced at $2400 a ticket.)

3. The football field in the Mercedes-Benz Superdome in New Orleans has a length that is 40 feet more than twice the width of the field. If the perimeter of the field is 1040 feet, what are the dimensions of the field?

4. Jackson and Palmer are having a Super Bowl party at their apartment. Both went to the neighborhood grocery store and bought snacks, but unfortunately they bought the same thing! Jackson purchased 3 bags of chips and 4 bottles of soda for the party at a cost of $14.10 before sales tax. Palmer bought 4 bags of chips and 5 bottles of soda for a cost of $18.25 before sales tax. How much did each bag of chips and each bottle of soda cost?

5. At the end of the game, the winning team scored 3 more points than the losing team. The total number of points scored by both teams was 65. What was the final score of the game?

Name: Date:

Math@Work

Chemistry

As a pharmaceutical chemist, you will need an advanced degree in pharmaceutical chemistry, which combines biology, biochemistry, and pharmaceuticals. In this career, you will most likely spend your day in a lab setting creating new medications or researching their effectiveness. You will often work as part of a team working towards a joint goal. As a result, in addition to strong math skills and an understanding of chemistry, you will need to have good communication and leadership skills. Since you will be working directly with chemicals, you will also need to have a strong understanding of lab safety rules to ensure the safety of not only yourself, but your coworkers as well.

Suppose you work at a pharmaceutical company which creates and produces medications for various skin conditions. You are currently on a team which is developing an acne-controlling facial cleanser. Your team is working on determining the gentlest formula possible that is still effective so that the cleanser can be used on sensitive skin. Half of your team is working with salicylic acid and the other half is working with benzoyl peroxide.

As a part of your work, you will need to keep up on current research. Learning about new chemicals, new methods, and new research will be a continuous part of your life.

1. Perform an Internet search for benzoyl peroxide. How does it work to clean skin and prevent acne?

2. Perform an Internet search for salicylic acid. How does it work to clean skin and prevent acne?

3. Based on your research, which chemical seems better suited to treat acne on sensitive skin?

Another aspect of your career will involve the mixing of chemicals to create new compounds. Having the correct concentrations of chemicals is also important so the resulting solution works as you expect it to. When you don't have the correct concentration of a chemical in stock, it is possible to mix two concentrations together to obtain the desired concentration.

4. Your team wants to create a cleanser with 4% benzoyl peroxide. The lab currently has 2.5% and 10% concentrations of benzoyl peroxide in stock. To create 500 mL of 4% benzoyl peroxide, how much of each concentration should be combined?

Foundations Skill Check for Chapter 11

This page lists several skills covered previously in the book and software that are needed to learn new skills in Chapter 11. To make sure you are prepared to learn these new skills, take the self-test below and determine if any specific skills need to be reviewed.

Each skill includes an easy (**e.**), medium (**m.**), and hard (**h.**) version. You should be able to complete each problem type at each skill level. If you are unable to complete the problems at the easy or medium level, go back to the given lesson in the software and review until you feel confident in your ability. If you are unable to complete the hard problem for a skill, or are able to complete it but with minor errors, a review of the skill may not be necessary. You can wait until the skill is needed in the chapter to decide whether or not you should work through a quick review.

1.4 Divide using long division. Indicate any remainder.

e. $3\overline{)72}$ **m.** $20\overline{)305}$ **h.** $50\overline{)3065}$

1.6 Simplify the expression.

e. 10^3 **m.** 2^5 **h.** -5^4

3.3 Find the product.

e. $1.2 \cdot 100$ **m.** $0.0045 \cdot 100$ **h.** $1.0035 \cdot 1000$

7.6 Distribute.

e. $2(5+4)$ **m.** $9(x^2+4)$ **h.** $12(3x+15)$

7.7 Simplify the expression by combining like terms.

e. $14x - 23x$ **m.** $-3.5x^2 + 4.2 - 5.8x^2 + 9.3$ **h.** $2\frac{1}{3}x + 4x^3 - 2\frac{7}{8}x^3 + 1\frac{1}{6}x$

Chapter 11: Exponents and Polynomials

Study Skills

11.1 Exponents

11.2 Exponents and Scientific Notation

11.3 Introduction to Polynomials

11.4 Addition and Subtraction with Polynomials

11.5 Multiplication with Polynomials

11.6 Special Products of Binomials

11.7 Division with Polynomials

Chapter 11 Projects

Math@Work

Foundations Skill Check for Chapter 12

Math@Work

Introduction

If you plan to go into any field of science, your daily work will involve numbers, both large and small. These numbers will commonly be written in scientific notation since it is an easy and convenient way to work with very large and very small numbers. In biology, which is the study of life and living organisms, biologists have to work with very small numbers, such as the length of a bacterium, and very large numbers, such as the number of bacteria in a sample at the end of an experiment. In chemistry, which is the study of the composition and change of matter, chemists have to work with very small numbers, such as the distance between atoms in a molecule of water, and very large numbers, such as the number of water molecules in a liter. No matter which scientific career you pursue, part of your work will involve answering questions and communicating results related to numbers, both large and small.

Suppose you decide to pursue a career in astronomy, which is the study of celestial bodies such as planets, asteroids, and stars. Part of your job may require you to spend your nights in an observatory observing the universe and searching for previously undiscovered objects. One set of objects of special importance are known as near-Earth objects, which are objects that have the potential to collide with the Earth. If you identify a new object, how will you determine if it is a near-Earth object? How do you determine the potential damage that may occur if the object crashes into the Earth? Finding answers to these questions, and many more, requires several of the skills learned in this chapter and previous chapters. At the end of this chapter, we'll come back to this problem and explore how near-Earth objects are classified.

Study Skills

Tips for Improving Your Memory

Experts believe that there are three ways that we store memories: first in the sensory stage, then in short-term memory, and for some memories, the final stage of long-term memory. Because we can't retain all the information that bombards us on a daily basis, the different stages of memory act as a filter. Your sensory memory lasts only a fraction of a second and it holds your perception of a visual image, a sound, or a touch. The sensation then moves to your short-term memory, which has the limited capacity to hold about seven items for no more than 20 to 30 seconds at a time. Important information is gradually transferred over to long-term memory. The more the information is repeated or used, the greater the chance that it will end up in long-term memory. Unlike sensory and short-term memory, long-term memory can store unlimited amounts of information indefinitely. The problem is, just how exactly do you get new information into long-term memory? Here are some tips on how to improve your chances of remembering important information.

1. **To move information from short-term memory to long-term memory, you have to be attentive and focused on the information.** Study in a location that is free of distractions and avoid watching TV or listening to music with lyrics while studying.

2. **Reciting information aloud after you have read or studied it can improve your memory.** Ask yourself questions about the material to see if you can recall important facts and details. Pretend you are teaching or explaining the material to someone else. This will help you put the information into your own words.

3. **Associate the information with something you already know.** Think about how you can make the information personally meaningful—how does it relate to your life, your experiences, and your current knowledge? If you can link new information to memories already stored, you create "mental hooks" that help you recall the information. For example, when trying to remember the formula for slope using rise and run, remember that rise would come alphabetically before run so rise will be in the numerator in the slope fraction and run will be in the denominator. Also, things that rise, like balloons, go up and down, so rise is the change in the vertical direction when calculating slope.

4. **Use visual images like diagrams, charts, and pictures to help you remember.** Using index cards and the Frayer Model presented in Chapter 2, you can make your own pictures and diagrams that may help you in recalling important definitions, theorems, or concepts.

5. **Since most people can only store 4 to 7 different items in short-term memory at one time, splitting larger pieces of information into smaller "chunks" will help you to remember the information more easily.** This method is useful in remembering social security numbers and telephone numbers. Instead of remembering a sequence of digits such as 555777213 you can break it into chunks such as 555-777-213.

6. **Group long lists of information into categories that make sense.** For example, instead of remembering all the properties of real numbers individually, try grouping them into shorter lists by operation, such as addition and multiplication.

7. **Use mnemonics or memory techniques to help you remember important concepts and facts.** An important mnemonic that is commonly used to help students remember the order of operations is "**P**lease **E**xcuse **M**y **D**ear **A**unt **S**ally," which uses the first letters of the words **P**arentheses, **E**xponents, **M**ultiplication, **D**ivision, **A**ddition, and **S**ubtraction to help you remember the correct order to perform basic arithmetic calculations. To make the mnemonic more personal and possibly more memorable, make up one of your own.

8. **Use acronyms to help you remember important concepts or procedures.** An acronym is a type of mnemonic device which is a word made up by taking the first letter from each word that you want to remember and making a new word from the letters. For example, the word **HOMES** is often used to remember the five Great Lakes in North America where each letter in the word represents the first letter of one of the lakes: **H**uron, **O**ntario, **M**ichigan, **E**rie, and **S**uperior. The word **FOIL** is used in math to remember how to multiply binomials and stands for **F**irst, **O**uter, **I**nner, and **L**ast.

Source: http://science.howstuffworks.com/life/inside-the-mind/human-brain/human-memory2.htm

Name: Date:

11.1 Exponents

Objectives

Use the product rule for exponents to simplify expressions.

Use the rule for 0 as an exponent to simplify expressions.

Use the quotient rule for exponents to simplify expressions.

Use the rule for negative exponents to simplify expressions.

Use a calculator to evaluate expressions that involve exponents.

Success Strategy

The rules for exponents that were introduced in Chapter 1 still apply in this chapter. The only difference is that exponents can now have variable bases instead of only whole number bases. You will also learn how to work with negative bases and negative exponents. Take some time to review Section 1.6 before beginning this section.

Understand Concepts

Go to Software First, read through Learn in Lesson 11.1 of the software. Then, work through the problems in this section to expand your understanding of the concepts related to exponents.

Exponents were first introduced in Section 1.6 for whole numbers. This section will cover the topic of exponents in more depth and introduce several new rules related to exponents.

Product Rule for Exponents

When multiplying two powers with the same base together, the base remains the same and the exponents are added. In mathematical terms, if a is a nonzero real number and m and n are integers, then $a^m \cdot a^n = a^{m+n}$.

For example, $2^5 \cdot 2^3 = 2^{5+3} = 2^8$.

Quick Tip

Remember that the terms *base* and *exponent* refer to two different parts of an exponential expression.

base → 2^4 ← exponent

1. In Section 1.6, we learned that the exponent stands for the number of times the base is being multiplied times itself. Let's explore how this relates to the product rule for exponents by considering the expression $2^4 \cdot 2^3$.

a. Write 2^4 as a multiplication expression without using exponents.

b. Write 2^3 as a multiplication expression without using exponents.

c. Use the answers to parts **a.** and **b.** to write $2^4 \cdot 2^3$ as a multiplication expression without using exponents. Do not simplify.

d. Write the product from part **c.** in exponential form.

e. Simplify the expression $2^4 \cdot 2^3$ according to the product rule for exponents.

f. How do the answers from parts **d.** and **e.** compare?

Quick Tip

Terms with a coefficient and multiple variables are usually written with the coefficient first followed by the variables listed in alphabetical order.

Using the Product Rule for Exponents to Multiply Like Terms

1. Group the coefficients together and group each like variable together.
2. Multiply the coefficients together and apply the product rule for exponents to each group of like variables.

For example, $4xy \cdot 5x^2y = (4 \cdot 5)(x \cdot x^2)(y \cdot y) = (4 \cdot 5)(x^{1+2})(y^{1+1}) = 20x^3y^2$.

Determine if any mistakes were made while using the product rule for exponents. If any mistakes were made, describe the mistake and then correctly apply the product rule for exponents to find the actual result.

Quick Tip

Remember that a variable without an exponent is understood to have an exponent of 1. For example, $y \cdot y = y^1 \cdot y^1 = y^2$.

2.
$$\begin{aligned} 2y^2 \cdot 3y^4 &= (2 \cdot 3)(y^2 \cdot y^4) \\ &= 6 \cdot (y^{2+4}) \\ &= 6y^6 \end{aligned}$$

3.
$$\begin{aligned} 4ab^2 \cdot 12ab^2 &= (4 \cdot 12) \cdot (a \cdot b^2) \\ &= 48ab^2 \end{aligned}$$

Quotient Rule for Exponents

When dividing two powers with the same base, the base remains the same and the exponent in the denominator is subtracted from the exponent in the numerator. In mathematical terms, if a is a nonzero real number and m and n are integers, then $\frac{a^m}{a^n} = a^{m-n}$.

For example, $\frac{9^{10}}{9^4} = 9^{10-4} = 9^6$.

4. Why is it important that a is not equal to 0 in the quotient rule for exponents?

Name: Date:

5. Let's explore how the quotient rule for exponents works by considering the expression $\frac{5^7}{5^5}$.

a. Write 5^7 as a multiplication expression without using exponents.

b. Write 5^5 as a multiplication expression without using exponents.

c. Use the answers to parts **a.** and **b.** to write $\frac{5^7}{5^5}$ as a division expression without using exponents. Do not simplify.

d. Simplify the fraction from part **c.** by canceling like terms. Write the answer in exponential form.

e. Simplify the expression $\frac{5^7}{5^5}$ according to the quotient rule for exponents.

f. How do the answers from parts **d.** and **e.** compare?

Lesson Link

The reason why $a^0 = 1$ was covered in Section 1.6.

The Exponent 0

If a is a nonzero real number, then $a^0 = 1$.

The expression 0^0 is undefined.

Negative Exponents

If a term has a **negative exponent**, write the reciprocal of the entire term and remove the negative sign from the exponent. In mathematical terms, if a is a nonzero real number, then $a^{-1} = \frac{1}{a}$.

For example, $2^{-1} = \frac{1}{2}$.

If a is a nonzero real number and n is an integer, then $a^{-n} = \frac{1}{a^n}$.

For example, $3^{-5} = \frac{1}{3^5}$.

6. In Section 1.6, we discovered a pattern when exploring the reason why $a^0 = 1$. We will continue this pattern now to determine how the values of terms with negative exponents are found.

Variable Notation	$a = 3$	Pattern
a^2	$3^2 = 9$	
a^1	$3^1 = 3$	
a^0	$3^0 = 1$	
a^{-1}	$3^{-1} =$ ___	
a^{-2}	$3^{-2} =$ ___	
a^{-3}	$3^{-3} =$ ___	

a. Looking at the values in column two of the table, what is the pattern when moving from one row to the row below it? Write the pattern in column three of the table.

b. If you continue this pattern, going from 3^0 to 3^{-1}, what algebraic expression would you need to evaluate?

c. Evaluate the expression from part **b.** and place the answer in column two of the table. Does it match the rule for negative exponents?

d. Continue the pattern for 3^{-2} and 3^{-3} in the table.

e. Calculate the values of 3^{-2} and 3^{-3} using the negative exponent rule. Do the results match the values from part **d.**?

Skill Check

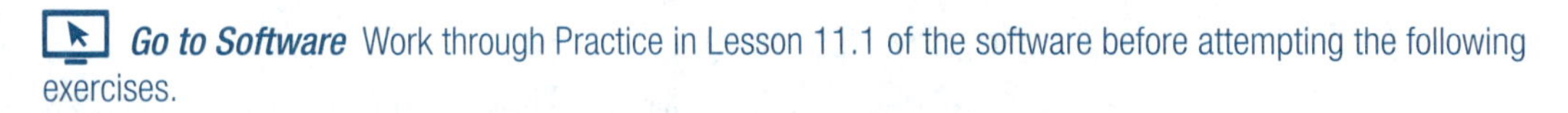

Go to Software Work through Practice in Lesson 11.1 of the software before attempting the following exercises.

Simplify each exponential expression. Keep the answer in exponential form. Make sure all exponents are positive and remove all grouping symbols.

7. $7^4 \cdot 7^2$

8. $-5\left(x^{-2}\right)$

9. $\dfrac{9^5}{9^2}$

10. $y^0 \cdot y^6 \cdot y^{-3}$

Name: Date:

Simplify each exponential expression completely. Make sure all exponents are positive.

11. $(3x^5)(4x^{-7})$

12. $(-6a^3b^4)(4a^{-2}b^8)$

13. $3x^0 + 9y^0$

14. $-4x^{-3}$

Apply Skills

Work through the problems in this section to apply the skills you have learned related to exponents.

15. Barbara's Bombtastic Bakery makes *petit four glaces*, which are small bite-sized cakes. Each cake is in the shape of a cube that has a side length of $2x$, where x is a positive length which varies depending on the cake flavor.

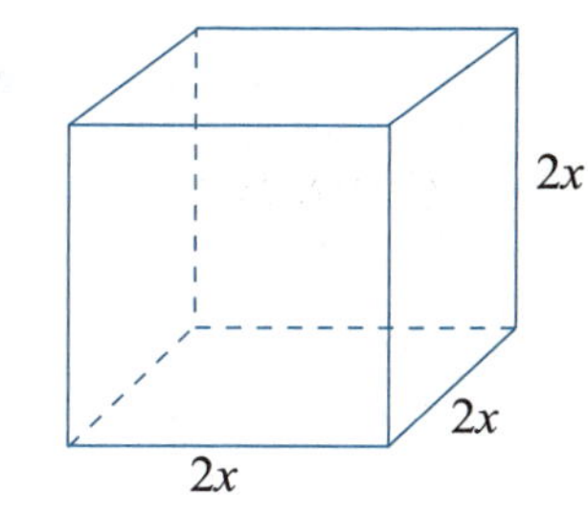

a. Write an expression using exponents to find the volume of the *petit four glaces*. Do not simplify.

b. Which exponential rule will you need to use to simplify the expression from part **a.**?

c. Simplify the expression from part **a.**

d. If $x = 2$ cm, determine the volume of the *petit four glaces* using the expression from part **c.**

Quick Tip

The ***petit four*** originated in France and the name refers to a variety of small pastries. *Petit four* means "small oven" in French, which describes the way the cakes were cooked: in a large stone oven after the fire had died out.

Lesson Link

Section 5.5 introduced formulas to find the volume of geometric figures.

Quick Tip TECH

On many calculators, the caret key [^] is used to indicate an exponent. Type the base number, the caret, the exponent, and then press enter [=] to evaluate a numerical expression with exponents.

16. A strain of influenza virus is spreading throughout a community and the number of confirmed cases of the flu doubles every day. On day 0 (the initial day) of the outbreak, 1 person has the virus. On day 1 of the outbreak, $1 \cdot 2 = 2$ people will have the virus. On day 2 of the outbreak, $1 \cdot 2 \cdot 2 = 1 \cdot 2^2 = 4$ people will have the virus.

a. Write an exponential expression to describe how many people will have influenza virus on day 5. Write in exponential form and simplify.

b. Write an exponential expression to describe how many people would have the virus on day n if 3 people had the virus on day 0 of the outbreak. Write the expression in exponential form and simplify.

c. Use the expression from part **b.** to determine the number of people that will have the virus on day 5 of the outbreak if 3 people had the virus on day 0?

17. A standard hard drive has 2^{38} bytes of data. 1 gigabyte is equivalent to 2^{30} bytes.

a. Write an exponential expression to determine how many gigabytes are equivalent to 2^{38} bytes?

b. Simplify the expression from part **a.** to determine how many gigabytes are in 2^{38} bytes.

c. What rule of exponents did you use to simplify part **b.**?

Name: Date:

11.2 Exponents and Scientific Notation

Objectives

Use the power rule for exponents to simplify expressions.

Use the rule for a power of a product to simplify expressions.

Use the rule for a power of a quotient to simplify expressions.

Use combinations of rules for exponents to simplify expressions.

Write decimal numbers in scientific notation.

Operate with decimal numbers by using scientific notation.

Success Strategy

Be sure to have a good understanding of how to use the exponent rules from Section 11.1 and how to work with negative exponents before starting this section. Work through more practice in Lesson 11.1 of the software, if needed.

Understand Concepts

Go to Software First, read through Learn in Lesson 11.2 of the software. Then, work through the problems in this section to expand your understanding of the concepts related to exponents and scientific notation.

Several rules for exponents were introduced in Section 11.1. This section introduces a few more rules.

Power Rule for Exponents

When an exponential expression is raised to a power, the power of each part of the exponential expression is multiplied by that power. In mathematical terms, if a is a nonzero real number and m and n are integers, then $\left(a^m\right)^n = a^{mn}$.

For example, $\left(4^3\right)^2 = 4^{3 \cdot 2} = 4^6$.

1. Let's explore how the power rule for exponents works by considering the expression $\left(2^3\right)^4$.

a. The exponent outside of the parentheses determines how many times the term inside the parentheses is multiplied by itself. Write $\left(2^3\right)^4$ as a multiplication expression involving only the term 2^3. Do not simplify.

b. Use the product rule for exponents to simplify the expression from part **a.**

c. Use the power rule for exponents to simplify $\left(2^3\right)^4$.

d. How do the answers from parts **b.** and **c.** compare?

2. Using the power rule for exponents, any negative exponent can be rewritten as $a^{-m} = \left(a^{m}\right)^{-1} = \left(a^{-1}\right)^{m}$. Use this fact to write a short explanation about why $a^{-n} = \frac{1}{a^{n}}$ is true for $a \neq 0$.

Rule for Power of a Product

A **power of a product** is found by raising each factor to that power. In mathematical notation, if a and b are nonzero real numbers and n is an integer, then $(ab)^{n} = a^{n}b^{n}$.
For example, $(2x)^{5} = 2^{5}x^{5}$ and $\left(-3x^{3}\right)^{2} = (-3)^{2}\left(x^{3}\right)^{2} = 9x^{6}$.

3. Let's explore how the power of a product rule works by considering the expression $(5m)^{3}$.

   a. The exponent outside of the parentheses determines how many times the term inside the parentheses is multiplied by itself. Write $(5m)^{3}$ as a multiplication expression with only the term $5m$. Do not simplify.

   b. Use the commutative and associative properties of multiplication to rewrite the expression from part **a.** so that the coefficients are grouped together and the variables are grouped together.

   c. Rewrite the expression from part **b.** using exponents and simplify.

Quick Tip

Remember that $y = y^{1}$.

   d. Simplify the expression $(5m)^{3}$ using the rule for power of a product.

   e. How do the answers from parts **c.** and **d.** compare?

Rule for Power of a Quotient

A **power of a quotient** (in fraction form) is found by raising both the numerator and denominator to the power. In mathematical notation, if a and b are nonzero real numbers and n is an integer, then $\left(\frac{a}{b}\right)^{n} = \frac{a^{n}}{b^{n}}$.

For example, $\left(\frac{3}{5}\right)^{2} = \frac{3^{2}}{5^{2}} = \frac{9}{25}$.

Name: Date:

4. Let's explore how the power of a quotient works by considering the expression $\left(\frac{x}{2}\right)^3$.

a. The exponent outside of the parentheses determines how many times the term inside the parentheses is multiplied by itself. Write $\left(\frac{x}{2}\right)^3$ as a multiplication expression with only the term $\frac{x}{2}$. Do not simplify.

Lesson Link

Multiplication of fractions was introduced in Sections 2.1 and 2.2.

b. Write the expression from part **a.** as a single fraction involving a product of factors in both the numerator and denominator. Do not simplify.

c. Rewrite the numerator and denominator in the rational expression from part **b.** using exponents and then simplify.

d. Simplify the expression $\left(\frac{x}{2}\right)^3$ using the rule for power of a quotient.

e. How do the answers from parts **c.** and **d.** compare?

Determine if any mistakes were made while simplifying the exponential expressions. If any mistakes were made, describe the mistake and then correctly simplify the exponential expression to find the actual result.

5. $$\left(\frac{2a^2b^5}{b^2}\right)^{-3} = \left(2a^2b^3\right)^{-3} = \frac{1}{\left(2a^2b^3\right)^3} = \frac{1}{8a^5b^6}$$

6. $$\frac{\left(-3x^2y^{-1}\right)^{-2}}{\left(6x^{-1}y\right)^{-3}} = \frac{\left(6x^{-1}y\right)^3}{\left(-3x^2y^{-1}\right)^2} = \frac{216x^{-3}y^3}{-9x^4y^{-2}} = -\frac{24y^5}{x^7}$$

Scientific Notation

A positive decimal number written in **scientific notation** has the form $a \times 10^n$, where $1 \le a < 10$ and n is an integer.

Writing a Number in Scientific Notation

1. Move the decimal point in the number so that the number is greater than or equal to 1 and less than 10.
2. Count the number of places the decimal point was moved in Step 1. If the decimal point moves to the right, the exponent will be negative. If the decimal point moves to the left, the exponent will be positive.
3. Multiply the result of Step 1 by 10 raised to a power equal to the result from Step 2 and with the appropriate sign.

Note: To avoid memorizing which direction makes the exponent negative when moving the decimal point, just think about the actual number. A number that is less than 1, such as 0.001, will have a negative exponent. A number that is larger than 1, such as 2000, will have a positive exponent. Numbers between 1 and 10 will have 0 as the exponent.

Quick Tip

A **tenfold error** means that the result is incorrect by a factor of ten.

7. Scientific notation is commonly used to represent extremely large or extremely small values such as the distance from the Earth to the sun (149,597,870,700 meters) or the radius of a carbon atom (0.00000000772 meters). It simplifies writing the number and avoids mistakes in placement of the decimal point. A misplacement of the decimal point by 1 place is a tenfold error in any calculations involving the number. Write both of these numbers in scientific notation.

Quick Tip

The rules for multiplying and dividing numbers in scientific notation use the product rule and the quotient rule for exponents.

Multiplying or Dividing Using Scientific Notation

1. Write both numbers in scientific notation.
2. For **multiplication**, follow the rule $(a \times 10^n) \cdot (b \times 10^m) = (a \cdot b) \times 10^{m+n}$

   For **division**, follow the rule $(a \times 10^m) \div (b \times 10^n) = \left(\frac{a}{b}\right) \times 10^{m-n}$
3. Make sure that the answer is in scientific notation.

Determine if any mistakes were made while simplifying expressions with scientific notation. If any mistakes were made, describe the mistake and then correctly simplify the expression with scientific notation to find the actual result.

Quick Tip

The multiplication symbol × is commonly used to multiply the power of 10 in scientific notation. However, the multiplication symbol · can also be used.

8. $$\begin{aligned}(2.5 \times 10^4) \cdot (5 \times 10^{-2}) &= (2.5 \cdot 5) \times (10^4 \cdot 10^{-2}) \\ &= 12.5 \times 10^2 \\ &= 1.25 \times 10^3\end{aligned}$$

9. $$\begin{aligned}(5 \times 10^{-2}) \div (4 \times 10^{-5}) &= \left(\frac{5}{4}\right) \times \left(\frac{10^{-2}}{10^{-5}}\right) \\ &= 1.25 \times 10^{-7}\end{aligned}$$

Name: Date:

Skill Check

Go to Software Work through Practice in Lesson 11.2 of the software before attempting the following exercises.

Simplify the exponential expressions completely. Make sure all exponents are positive in your answer.

10. $\left(6x^3\right)^2$

11. $-4\left(5x^{-3}y\right)^{-1}$

12. $\left(\dfrac{x}{y}\right)^{-2}$

13. $\left(\dfrac{5xy^3}{y}\right)^2$

14. $\dfrac{\left(7x^{-2}y\right)^2}{\left(xy^{-1}\right)^2}$

15. $\dfrac{\left(5x^2\right)\left(3x^{-1}\right)^2}{\left(25y^3\right)\left(6y^{-2}\right)}$

Write each number in scientific notation and then perform each operation. Make sure the answer is in scientific notation.

16. $300 \cdot 0.00015$

17. $125 \div 50{,}000$

Apply Skills

Work through the problems in this section to apply the skills you have learned related to exponents and scientific notation.

18. A scientist calculated that her experiment consumed 520,000 joules (J) of energy. She wrote the value as 52×10^4 J on a report of the experiment. Did she write the value correctly in scientific notation? If not, what should it be?

19. A molecule of table salt weighs approximately 9.704×10^{-23} grams. What would be the weight of 4,000,000 molecules of table salt?

a. Write 4,000,000 in scientific notation.

b. Write an expression to find the weight of 4,000,000 molecules of table salt.

c. Simplify the expression from part **b.**

d. What does the answer from part **c.** mean? Write a complete sentence.

Quick Tip

The orbits of the planets in the solar system are not circular. The planets follow an elliptical path.

20. An astronomical unit (AU) is defined as the furthest distance the Earth is from the Sun. The approximate value of 1 AU is 150,000,000,000 meters.

a. Write the approximate value of 1 AU in scientific notation.

b. The furthest distance that Mars's orbit takes it from the Sun is approximately equal to 200,000,000,000 meters. Write this value in scientific notation.

c. The result from part **b.** is approximately equal to how many AUs? Round your answer to the nearest hundredth.

Quick Tip

The formula relating distance, rate, and time is $d = rt$.

d. Light travels at a speed of approximately 2.0×10^{-3} AU per second. Determine approximately how long it will take light to travel to Mars when it is the furthest from the sun.

Name: Date:

11.3 Introduction to Polynomials

Objectives

Define a polynomial and learn how to classify polynomials.

Evaluate a polynomial for given values of the variable.

Success Strategy

There is a lot of new terminology in this section. Be sure to write the terms and definitions in your notebook or use the Frayer Model from Chapter 2.

Understand Concepts

Go to Software First, read through Learn in Lesson 11.3 of the software. Then, work through the problems in this section to expand your understanding of the concepts related to polynomials.

We've worked with expressions and equations in several chapters. This section will introduce some terminology related to specific types of expressions used in algebra.

Monomials

A **monomial in** x is a term of the form kx^n, where k is a real number and n is a positive integer.

The value of n is the **degree** of the term and k is the **coefficient** of the term.

Quick Tip

Monomials can have coefficients which are negative or fractional, but monomials cannot have exponents on the variables that are negative or fractional.

Some examples of monomials are $5y$, $3x^2$, and $-4a^5$. Monomials are similar to the concept of terms that we learned about in Section 7.7. The main difference is that the terms we will work with in this chapter contain only one variable.

1. Determine the coefficient and degree of each monomial.

Monomial	Coefficient	Degree
x		
$-3x^5$		
$\frac{1}{6}y^{24}$		
$4x^0$		

Quick Tip

Remember that any nonzero number or variable raised to the power 0 is equal to 1.

The following table summarizes the important terms related to polynomials.

Polynomials		
Term	**Definition**	**Example**
Polynomial	A monomial or the sum or difference of monomials.	$\frac{1}{2}x^3 - 6x^5 + 17$
Degree of a Polynomial	The largest degree of any of its terms.	The degree of $16x^4 + 2x^7$ is 7.
Leading Coefficient	The coefficient of the term with the largest degree.	The leading coefficient of $-14x + 7x^9 - 3x^{15}$ is -3.
Binomial	A polynomial with two terms.	$2x + 7$
Trinomial	A polynomial with three terms.	$1.5x^3 - 4.2x + 12$

Quick Tip

The names of polynomials have Greek roots. *Poly-* means many, *mono-* means one, *bi-* means two, and *tri-* means three.

Quick Tip

The terms in a polynomial can be written in any order. However, polynomials are commonly written with their terms ordered in descending order of degree or ascending order of degree.

2. Fill in the table by first writing each polynomial with its terms in descending order of degree (that is, with the *exponents* decreasing in order from left to right). Then, determine the degree of the polynomial and the leading coefficient.

Polynomial	Descending Order	Degree	Leading Coefficient
$6x - 12x^2 + 3x^4$			
$-14a^3 + 4 - 2a$			
$56y^2 + y^5 - 4 + 3y$			

Skill Check

Go to Software Work through Practice in Lesson 11.3 of the software before attempting the following exercises.

When evaluating polynomials, the polynomial may be written in function notation. Evaluate each polynomial at the given value.

3. Given $p(x) = x^2 - 4x + 13$, find $p(2)$.

4. Given $y = 3$, evaluate $-3y^2 - 15$.

5. Given $p(m) = 2m^4 + 3m^2 - 8m$, find $p(-1)$.

Lesson Link

Combining like terms was introduced in Section 7.7.

Simplifying the terms in a polynomial follows the same method as simplifying the terms in an expression or an equation. Simplify each polynomial by combining like terms and write it in descending order of degree.

6. $y + 3y - 12$

7. $6x^5 + 2x^4 - 5x^3 - 3x^4$

8. $3x + 7 - x - 1 - 3x$

Name: ______________________ Date: ______________________

Apply Skills

Work through the problems in this section to apply the skills you have learned related to polynomials.

9. The value of a car starts to depreciate the moment you drive it off of the car dealership's lot. After two years, the approximate value of a car which originally costs \$25,000 is calculated by $V = \$25{,}000x^2$, where x is the average percent that the car retains its value per year. Cars that are well taken care of and driven sparingly will retain more value than cars that are poorly maintained and driven frequently.

Lesson Link

Changing percents to decimal numbers was covered in Section 4.3.

a. Suppose that a car that is well maintained and rarely driven will retain an average of 90% of its value each year. What will be the approximate value of the car after 2 years?

b. Suppose that a car that is not well maintained will retain an average of 80% of its value each year. What will be the approximate value of the car after 2 years?

10. Forensic scientists can use a simple formula to approximate the time of death. This formula is based on the average body temperature of humans being 37 °C and the fact that a deceased body will lose an average of 1.5 °C per hour until the body temperature matches the temperature of the surrounding environment. The formula is $f(t) = 37 - 1.5t$, where t is the time in hours since death.

a. What is the approximate body temperature of a person that died 4 hours ago?

b. What is the approximate body temperature of a person that died 14 hours ago?

c. If the temperature of the environment in which the body was found was 20 °C, would it be reasonable for the body temperature to be the temperature from part **b.**? Explain why or why not.

11. A sled going down a hill has an initial speed of 5 feet per second and a constant acceleration of 1 foot per second squared. The distance of the sled in feet from the top of the hill can be modeled by the polynomial $d(t) = 5t + \frac{1}{2}t^2$, where t is the time in seconds after the sled leaves the top of the hill.

a. Determine the distance the sled is from the top of the hill after 2 seconds.

b. Determine the distance the sled is from the top of the hill after 4 seconds.

c. Determine the distance the sled is from the top of the hill after 8 seconds.

d. Does the distance that the sled travels double when the time doubles? Explain why or why not.

12. Camilla is creating square baby quilts. She determines that the sale price of each quilt should be $p(x) = \$1.80x^2 + 4(\$0.50)x + \$15$, where x is the side length of each square blanket in feet, \$1.80 is the cost per square foot of material, \$0.50 is the cost per foot of border material, and \$15 is the amount of profit Camilla wants to make on each blanket.

a. How much will a blanket that has a side length of 3 feet cost?

b. How much will a blanket that has a side length of 4 feet cost?

Name: Date:

11.4 Addition and Subtraction with Polynomials

Objectives

Add polynomials.

Subtract polynomials.

Simplify algebraic expressions by removing grouping symbols and combining like terms.

Success Strategy

When adding and subtracting polynomials, you should only add the coefficients of like terms. Do **not** add the exponents of like terms.

Understand Concepts

Go to Software First, read through Learn in Lesson 11.4 of the software. Then, work through the problems in this section to expand your understanding of the concepts related to adding and subtracting polynomials.

Adding Polynomials

If adding horizontally,

1. Use the associative and commutative properties to rearrange the polynomial so that like terms are grouped together.
2. Add like terms.

If adding vertically,

1. Line up like terms underneath one another.
2. Add like terms.

Lesson Link

Combining like terms was introduced in Section 7.7.

1. Adding and subtracting polynomials involves combining like terms. Determine which terms in each list are like terms. You will have multiple sets of like terms for each list.

a. $5a, 7, 2b, 6a, 140, 19a$

b. $9x, -2y, 20xy, 14y, 15, 7x$

c. $2x^3, 4x, 8x^2, 16x, 32x^3, 64x$

2. Find the sum of $(2x^2+5x-1)+(x^2+2x+3)$.

a. Rewrite the sum so that like terms are next to each other or write the sum vertically.

b. Add like terms together.

Subtracting Polynomials

1. Find the additive inverse, or opposite, of every term in the polynomial being subtracted.
2. Add like terms.

3. Let's find the difference of $(2x^2+4x+8)-(x^2+3x+2)$.

a. Distribute the negative sign over all terms of the second polynomial. Remember that a negative sign can be thought of as multiplication by -1. Rewrite as an addition problem.

b. Rearrange the terms so that like terms are grouped together or write the sum vertically.

c. Add like terms together.

When simplifying polynomial expressions, the order of operations must be followed. The order of operations when working with polynomial expressions is the same as the order of operations for real numbers. The order of operations are restated here for easy reference.

Lesson Link

The order of operations has been covered in Sections 1.6, 2.6, 3.4, and 7.5.

Order of Operations

1. Simplify within grouping symbols. If there are multiple grouping symbols, start with the innermost grouping.
2. Simplify any numbers or expressions with exponents.
3. From left to right, perform any multiplication or division in the order they appear.
4. From left to right, perform any addition or subtraction in the order they appear.

Determine if any mistakes were made while simplifying the polynomial expressions. If any mistakes were made, describe the mistake and then correctly simplify the polynomial expression to find the actual result.

4. $$\begin{aligned} 5x+2(x-3)-(3x+7) &= 5x+2x-6-3x+7 \\ &= 4x+1 \end{aligned}$$

5. $$\begin{aligned} 2+[9x-4(3x+2)] &= 2+[9x-12x-8] \\ &= 2-3x-8 \\ &= -3x-6 \end{aligned}$$

Name: Date:

Skill Check

Go to Software Work through Practice in Lesson 11.4 of the software before attempting the following exercises.

Simplify each algebraic expression.

6. $\begin{array}{r} -2x^2 - 3x + 9 \\ +\ (3x^2 - \ x + 2) \\ \hline \end{array}$

7. $\begin{array}{r} x^2 - 9x + 2 \\ -(4x^2 - 3x + 4) \\ \hline \end{array}$

8. $(-4x^2 + 2x + 1) + (2 - x + 3x^2) + (x - 8)$

9. $(x^4 + 8x^3 - 2x^2 - 5) - (11 - 2x^2 + 10x^3)$

Apply Skills

Work through the problems in this section to apply the skills you have learned related to adding and subtracting polynomials.

10. A manufacturer estimates that it costs $2x^3 + 4x^2 - 35$ dollars to create the amount of items it would take to fill a box which has a side length of x feet. The warehouse manager determines it will cost $1.50x^3 + 5$ dollars to store each box for one month.

a. Add the two polynomials to determine the manufacturing and storage costs for each box of items for one month.

b. The warehouse manager knows that each box will be stored for an average of 3 months. Determine the cost to produce a box of items and store it for 3 months.

c. If the box has a side length of 4 feet, use the expression from part **b.** to determine how much will it cost to create and store a box of items for 3 months.

11. Bernard has two loans, a loan for his car and a home equity loan that he used for home improvements. The car loan is for $15,000 and Bernard plans to make monthly payments of $500 per month. The home improvement loan is for $9000 and Bernard plans to make monthly payments of $300.

a. Write an algebraic expression to describe the value of the car loan after x months.

b. Write an algebraic expression to describe the value of the home equity loan after x months.

c. Add together the two algebraic expressions to determine the remaining loan amount to be paid after x months for both loans combined.

d. How much will Bernard still owe on both loans after 10 months?

Name: Date:

11.5 Multiplication with Polynomials

Objectives

Multiply a polynomial by a monomial.

Multiply two polynomials.

Success Strategy

When multiplying polynomials you will need to use the rules for exponents to simplify terms. Briefly review the rules in Section 11.1 before starting this section.

Understand Concepts

Go to Software First, read through Learn in Lesson 11.5 of the software. Then, work through the problems in this section to expand your understanding of the concepts related to multiplication with polynomials.

Multiplying a Monomial by a Monomial

1. Multiply the coefficients together.
2. Use the rules for exponents to multiply the variables together.

For example, $(4y^5)(3y^2) = (4 \cdot 3)(y^5 \cdot y^2) = (4 \cdot 3)(y^{5+2}) = 12y^7$.

Quick Tip

Remember that $3x$ is $3 \cdot x$ and $4x^2$ is $4 \cdot x \cdot x$. Use the associative and commutative properties of multiplication to rearrange the terms so that the coefficients are together and the variables are grouped together.

1. Find the product of $(3x)(4x^2)$.

a. Rewrite the product so that the coefficients are grouped together and the variables are grouped together. Do not simplify.

b. Simplify each product in the expression from part **a.**

Multiplying a Monomial by a Polynomial

1. Using the distributive property, multiply each term of the polynomial by the monomial in front of the parentheses.
2. Simplify each product of monomials.

For example, $2x(3x+5) = (2x)(3x) + (2x)(5) = (2 \cdot 3)(x \cdot x) + (2 \cdot 5)x = 6x^2 + 10x$.

Lesson Link

The distributive property for real numbers was introduced in Section 7.6.

2. The distributive property is used to find the product of $3x(2x^2+5x-3)$. Distributing the monomial term over the trinomial results in the sum $(3x)(2x^2)+(3x)(5x)-(3x)(3)$. Simplify each product of monomials and write the resulting polynomial in simplest form.

Multiplying a Binomial by a Binomial

1. Take the first term of the first binomial factor and multiply it by the second binomial factor. Apply the distributive property and simplify each product.
2. Take the second term of the first binomial factor and multiply by the second binomial factor. Apply the distributive property and simplify each product.
3. Combine any like terms.

For example,
$(2x-1)(4x-3)=(2x)(4x-3)+(-1)(4x-3)=8x^2-6x-4x+3=8x^2-10x+3.$

Multiplying a Polynomial by a Polynomial

1. Take the first term of the first polynomial factor and multiply it by the second polynomial factor. Apply the distributive property and simplify each product.
2. Repeat Step 1 for each term in the first polynomial.
3. Combine any like terms.

Determine if any mistakes were made while multiplying the polynomials. If any mistakes were made, describe the mistake and then correctly multiply the polynomials to find the actual result.

3.
$$\begin{aligned}(x+1)(x^2-x+1) &= x(x^2-x+1)+1(x^2-x+1)\\ &= x^3-2x+x+x^2-x+1\\ &= x^3+x^2-2x+1\end{aligned}$$

4.
$$\begin{aligned}(x^2+2x-1)(x^2-4x+3) &= x^2(x^2-4x+3)-1(x^2-4x+3)\\ &= x^4-4x^3+3x^2-x^2+4x-3\\ &= x^4-4x^3+2x^2+4x-3\end{aligned}$$

Skill Check

Go to Software Work through Practice in Lesson 11.5 of the software before attempting the following exercises.

Find each product.

5. $5x^2(-4x+6)$

6. $(t+6)(4t-7)$

Name: Date:

7. $(2x+1)(x^2-7x+2)$

Apply Skills

Work through the problems in this section to apply the skills you have learned related to multiplication with polynomials. For each problem, drawing a figure and labeling the known values may be helpful.

8. The smallest plot of land that you can rent at a community garden is 3 feet long by 4 feet wide.

a. Suppose you want to rent a plot of land which is x feet longer than the smallest available plot. What would the area of this plot of land be?

b. Suppose you want to rent a plot of land which is x feet wider than the smallest plot with a length of 3 feet. What would the area of this plot of land be?

c. Suppose you want to rent a plot of land which is x feet longer and x feet wider than the smallest plot of land. What would the area of this plot of land be?

9. Lee is making a box. He starts with a piece of cardboard that is 14 inches by 20 inches. He cuts a square with side length x from each corner of the box.

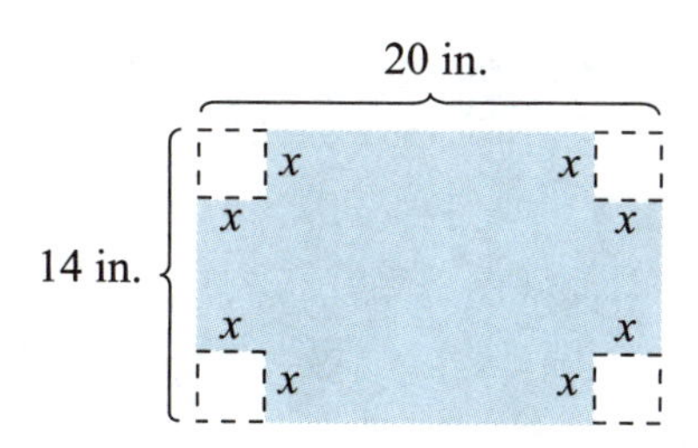

a. Write a polynomial function $A(x)$ to represent the area of the cardboard that remains after the corners are cut out.

b. When the sides of the box are folded up, what will be the side lengths of the base of the box?

c. Write a polynomial function $B(x)$ to represent the area of the base of the box when the sides are folded up.

d. The height of the box will be x inches. Write a polynomial function $V(x)$ to determine the volume of the box.

Lesson Link

Ratios were covered in Section 4.1.

10. The glass portion of a sliding glass door has a ratio of height to width of 2:1. The framework around the window adds 8 inches to the width of the door and 10 inches to the height.

a. Write a polynomial expression to represent the width of the door, including the framework. Use the variable x to represent the width of the door.

b. Write a polynomial expression to represent the height of the door, including the framework.

c. Write a polynomial expression for the total area of the window, including the framework.

Name: Date:

11.6 Special Products of Binomials

Objectives

Multiply binomials using the FOIL method.

Multiply binomials, finding products that are the difference of squares.

Multiply binomials, finding products that are perfect square trinomials.

Success Strategy

There is a lot of new terminology in this section. Be sure to write the terms and definitions in your notebook or use the Frayer Model from Chapter 2.

Understand Concepts

Go to Software First, read through Learn in Lesson 11.6 of the software. Then, work through the problems in this section to expand your understanding of the concepts related to special products of binomials.

The FOIL method for multiplying binomials is useful to ensure you include all terms in the multiplication process.

The FOIL method

1. Multiply the first terms of each binomial together.
2. Multiply the "outer" terms of each binomial together. These are the first term of the first binomial and the second term of the second binomial.
3. Multiply the "inner" terms of each binomial together. These are the second term of the first binomial and the first term of the second binomial.
4. Multiply the last terms of each binomial together. These are the second term in each binomial.
5. Combine any like terms.

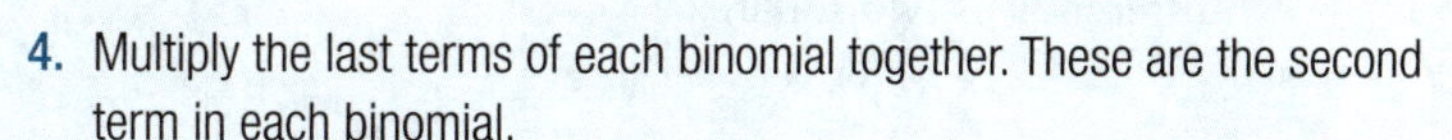

First
Last
$(2x+5)(3x-7)$
Inner
Outer

Quick Tip

The term **FOIL** is an acronym that is used to help you remember the steps for multiplying two binomials:
First, **O**uter, **I**nner, **L**ast.

1. Find the product of $(x+5)(2x+1)$ by using the FOIL method.

a. Find the product of the first terms of each binomial.

b. Find the product of the outer terms of each binomial.

c. Find the product of the inner terms of each binomial.

d. Find the product of the last terms of each binomial.

e. Write an expression to represent the sum of the terms from parts. **a.** through **d.** Simplify by combining any like terms.

2. How does the FOIL method compare to using the distributive method to multiply binomials?

Difference of Two Squares

The product of the sum and the difference of two terms is equal to the **difference of the squares** of the terms. In mathematical notation, $(a+b)(a-b)=a^2-b^2$.

3. Use the FOIL method to find the product of $(a+b)(a-b)$. Does it match the formula for the difference of two squares?

4. The difference of two squares can be visualized geometrically. Suppose you have a square with side length x that has a smaller square with side length a cut out.

Lesson Link

Determining the area of a figure with a cut out was covered in Section 5.3.

a. Use the area formula for a square to find the area of the shaded region of Figure 1.

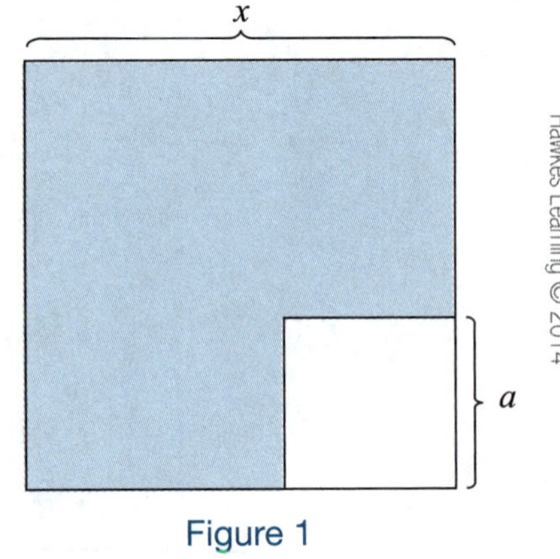

Figure 1

b. What is the length of the side marked with the question mark in Figure 2?

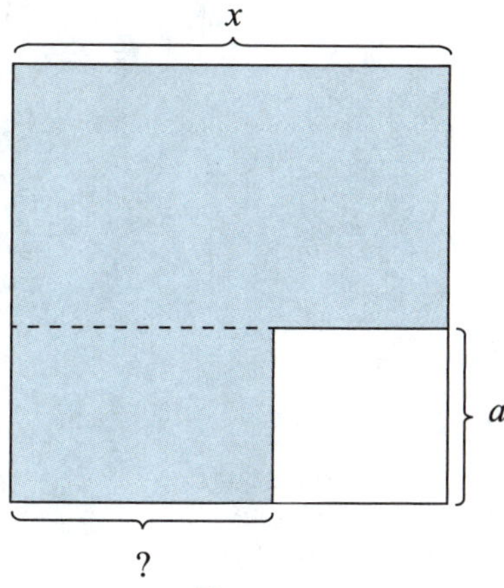

Figure 2

c. The shaded portion of the square in Figure 2 is divided into two rectangles. Find the area of each rectangle.

d. What is the sum of the areas of the two rectangles from Figure 2?

e. Rotating and relocating the smaller rectangle allows us to redraw Figure 2 as a rectangle. Determine the length and width of the new rectangle in Figure 3.

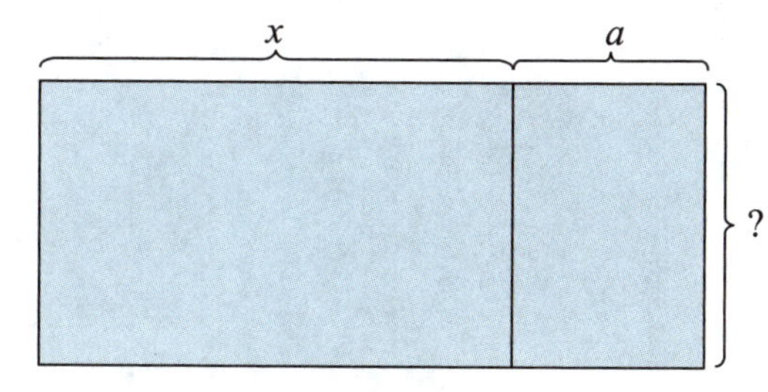

Figure 3

f. Find the area of the rectangle in Figure 3.

g. How does the result from part **f.** compare to the result from part **a.** and part **d.**?

Squares of Binomials (Perfect Square Trinomials)

To **square a binomial** you square both terms individually and then the middle term is twice the product of the two terms.

Square of a binomial sum: $(x+a)^2 = x^2 + 2ax + a^2$

Square of a binomial difference: $(x-a)^2 = x^2 - 2ax + a^2$

Quick Tip

Remember that **squaring** a quantity means to multiply it by itself.

5. Use the FOIL method to find the product of $(x+a)^2$ to verify that the formula for the square of a binomial sum is true.

6. Use the FOIL method to find the product of $(x-a)^2$ to verify that the formula for the square of a binomial difference is true.

Determine if any mistakes were made while using special products. If any mistakes were made, describe the mistake and then correctly use the special products to find the actual result.

7. $(x+2)^2 = x^2 + 4$

8. $(4x-3)^2 = 4x^2 - 24x + 9$

Name: Date:

Skill Check

Go to Software Work through Practice in Lesson 11.6 of the software before attempting the following exercises.

Find each product.

9. $(x+2)(5x+1)$

10. $(x^2+1)(x^2-1)$

11. $\left(x+\frac{3}{4}\right)\left(x-\frac{3}{4}\right)$

12. $(2x+0.5)^2$

Apply Skills

Work through the problems in this section to apply the skills you have learned related to special products of binomials.

13. A rug manufacturer puts a 1-inch trim on every rug they make to prevent the rugs from unravelling. The area of each rug, without the trim, needs to be treated with a color guard before it can be sold. The rugs vary in size. One style of rug is in the shape of a square where the total side length is x inches, including the trim.

a. Write a polynomial to represent the side length of a square rug without the trim. (**Hint:** The trim is on every side of the rug. Try drawing a figure.)

b. Write a polynomial to represent the area of the rug without the trim.

c. If the total length of the rug is 60 inches, find the area of the rug without the trim by evaluating the polynomial in part **b.** with $x = 60$.

d. Calculate the side length of the rug without the trim using the polynomial in **a.**

e. Find the area of the rug without the trim by squaring the value from part **d.**

Quick Tip

Remember that the **area of a square** is s^2, where s is the side length of the square.

f. How does the answer from part **e.** compare to the answer you obtained in part **c.**?

Quick Tip

There are two possible outcomes when tossing a fair coin. As a result, the probability of each outcome is ½ or 0.5.

14. When performing an experiment with two outcomes, the probability of a success is x and the probability of failure is $1 - x$, where $0 \le x \le 1$. The equation $p(x) = 15x^4(1-x)^2$ represents the probability of 4 successes in 6 trials of the experiment.

a. Rewrite the right side the equation as a polynomial by squaring the binomial $1 - x$ and then distributing the monomial $15x^4$ to all of the terms.

b. When a fair coin is tossed, the probability of the outcome being heads is 0.5. If heads is considered to be a success, then $x = 0.5$. Use the equation from part **a.** to determine the probability of 4 heads occurring during 6 tosses of a fair coin. Round your answer to the nearest hundredth.

15. Lee is making a box. He starts with a piece of cardboard that is 20 inches by 20 inches. He cuts a square with side length x from each corner of the box.

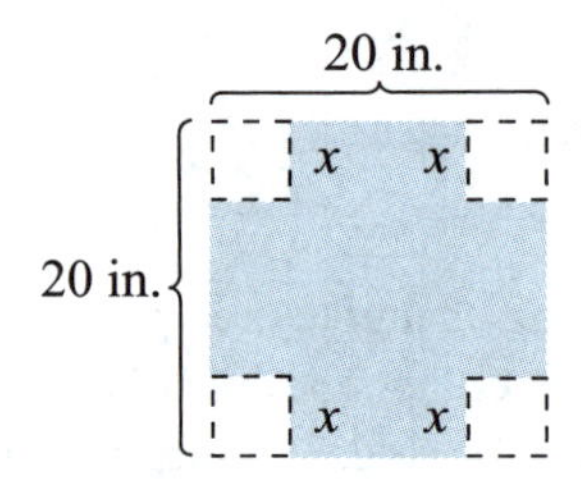

a. Write a polynomial function $A(x)$ to represent the area of the cardboard that remains after the corners are cut out.

b. When the sides of the box are folded up, what will be the side lengths of the base of the box?

c. Write a polynomial function $B(x)$ to represent the area of the base of the box when the sides are folded up.

d. The height of the box will be x inches. Write a polynomial function $V(x)$ to determine the volume of the box.

Name: Date:

11.7 Division with Polynomials

Objectives

Divide a polynomial by a monomial.

Divide polynomials by using the division algorithm.

Success Strategy

Review the quotient rule for exponents in Section 11.1 before starting this section.

Understand Concepts

Go to Software First, read through Learn in Lesson 11.7 of the software. Then, work through the problems in this section to expand your understanding of the concepts related to division with polynomials.

This section will use a form of a fraction called a **rational expression**. A fraction in which the numerator and denominator are polynomials is called a rational expression. As with all fractions, the denominator of a rational expression cannot be equal to 0.

Lesson Link Dividing by a monomial uses the techniques of reducing fractions that were introduced in Section 2.1.

Dividing By a Monomial

1. Break the rational expression into separate rational expressions where each consists of a term from the numerator over the denominator.

For example, $\frac{2x+1}{x}=\frac{2x}{x}+\frac{1}{x}$.

2. Simplify each rational expression.

For example, $\frac{2x}{x}+\frac{1}{x}=2+\frac{1}{x}$.

1. Simplify $\frac{24x^3+4x^2-8x+5}{4x}$.

a. Rewrite the rational expression as a sum of four separate rational expressions.

b. Simplify each separate rational expression in the sum from part **a.**

Lesson Link The division algorithm is similar to the method for long division, which was introduced in Section 1.4.

The Division Algorithm

For polynomials P and D, where $D \neq 0$, the division algorithm gives

$$\frac{P}{D}=Q+\frac{R}{D},$$

where Q and R are polynomials and the degree of R is less than the degree of Q.

P is the original polynomial or dividend.

D is the divisor polynomial.

Q is the quotient polynomial.

R is the remainder polynomial.

Quick Tip

Both the dividend and the divisor must be written in descending order. Missing terms in either the divisor or dividend should be represented by a 0 coefficient.

2. Use the division algorithm to find the quotient of $\frac{x^2-12x+35}{x-5}$.

a. Set up the long division problem.

The divisor has two terms, so start by looking at the first two terms of the dividend. To determine the number of times the divisor will divide into the first two terms of the dividend, we divide the first term of the dividend by the first term of the divisor.

$$\frac{x^2}{x}=x$$

The result is placed over the second term in the dividend since these are like terms.

$$x-5\overline{\smash{)}x^2-12x+35}\quad\text{(quotient } x \text{ written above } -12x\text{)}$$

Quick Tip

Try to rewrite the division problem from Problem 1 in the margin to keep the work all together.

b. Find the product of the first term of the quotient (the value just found) and all terms of the divisor. Place this expression under the like terms of the dividend.

c. Subtract the answer from part **b.** from the first two terms of the dividend. Be sure that the terms are aligned vertically according to their degree and that you *subtract* both terms.

d. Bring down the next term in the dividend.

e. The new dividend to work with should be $-7x+35$. Determine the number of times the divisor will divide into $-7x+35$. Place this result over the third term of the dividend in the long division problem.

f. Find the product of the second term of the quotient and all terms of the divisor.

g. Subtract the answer from part **f.** from the terms of the remaining dividend. Be sure that the terms are aligned vertically according to their degree.

h. Check that the quotient of the long division problem is correct by multiplying the quotient by the divisor. Their product should equal the dividend.

Skill Check

Go to Software Work through Practice in Lesson 11.7 of the software before attempting the following exercises.

Find each quotient.

3. $\dfrac{8y^3 - 16y^2 + 24y}{8y}$

4. $\dfrac{110x^4 - 121x^3 + 11x^2}{-11x}$

Lesson Link

An alternative method for dividing polynomials by a binomial of the form $x - c$ is introduced in Section A.5. This method is called synthetic division.

5. $\dfrac{x^2 - 12x + 27}{x - 3}$

6. $\dfrac{6x^2 - 11x - 3}{2x - 1}$

7. $\dfrac{3x^3 + x^2 - 7x - 5}{x^2 + 2x + 1}$

Apply Skills

Work through the problems in this section to apply the skills you have learned related to division with polynomials.

Quick Tip

Since the formula for area is $A = l \cdot w$ and the formula for volume of a rectangular solid is $V = l \cdot w \cdot h$, an alternate formula for volume is $V = A \cdot h$, where A is the area of the base of the rectangular solid.

8. A moving company uses a box that has a volume of $x^3 - 2x^2 - 13x - 10$ cubic inches.

a. If the height of the box is $x + 2$, what is the area of the base of the box?

b. If the height of the box is $x + 1$, what is the area of the base of the box?

9. A rectangular garden requires a volume of top soil modeled by the equation $200x^3 + 350x^2 + 150x$, where x is the depth in inches of top soil needed.

a. Determine an expression for the area of the garden.(**Hint:** Volume = length · width · height and Area = length · width.)

b. If the width of the garden is $10x + 10$ inches, use the expression from part **a.** to find an expression for the length of the garden.

c. If the depth of soil needed is 3 inches, find the volume of top soil that needs to be purchased for the garden.

d. Determine how many cubic feet of top soil is needed by dividing the answer from part **c.** by 1728 in.^{3}. Round your answer to the nearest tenth.
(**Note:** 1 cubic foot $= 12$ in. · 12 in. · 12 in. = 1728 in.^{3})

e. If the top soil comes in bags which contain 0.75 cubic feet of soil, how many bags will need to be purchased?

f. If the cost of the top soil is $2.10 per bag (including tax), what will be the total cost of the top soil needed for the garden?

Name: Date:

Chapter 11 Projects

Project A: There's gold in them there ... asteroids?

An activity to demonstrate the use of scientific notation in real life

According to a recent news article, the future of mining for precious metals may lie in space. A group of wealthy entrepreneurs, called Planetary Resources, Inc., plans to send robots into space to mine precious metals such as gold and platinum from asteroids. It is estimated that this would add trillions of dollars to the global GDP (Gross Domestic Product).

There are nearly 9000 asteroids larger than 150 feet in diameter that orbit close to the Earth. Some of these asteroids could contain as much platinum as is mined in an entire year on Earth. The platinum group metals—platinum, palladium, osmium, and iridium—are used in medical devices, renewable energy products, catalytic converters, and can possibly be used in automotive fuel cells.

These asteroids, called near-Earth asteroids, are defined as those that pass within 0.983 to 1.3 Astronomical Units (AU) from the Sun during their orbit. One AU is Earth's furthest distance from the Sun and is approximately equal to 150 million kilometers or 93 million miles.

1. If 150 million kilometers is equal to 93 million miles, how many kilometers are there in one mile? (Round your answer to thousandths.)

2. Write the distance from the Earth to the Sun (1 AU) in scientific notation.

a. In miles:

b. In kilometers:

3. To be called a "near-Earth asteroid," an asteroid must be within a certain distance of Earth during it's orbit.

a. What is the closest distance in miles a near-Earth asteroid can come to Earth during its orbit?

b. What is the furthest distance in miles a near-Earth asteroid can be from Earth during its orbit?

4. Write the distances from Problem 3 in scientific notation.

Two near-Earth asteroids have been visited by robotic spacecraft: the asteroid called 433 Eros by NASA's NEAR mission in 2000, and 25143 Itokawa by Japan's Hayabusa mission in 2005. The closest distance that Eros has come to the Earth during its orbit was 1.66×10^7 miles.

5. Suppose a spacecraft were to travel to 433 Eros and return to Earth.

a. What is the shortest distance it would have to travel? Write your answer in scientific notation.

b. Write the number from part **a.** in decimal notation.

6. The closest distance between the Earth and the 25143 Itokawa asteroid has been 0.013 AU. Write the value 0.013 in scientific notation.

7. In September 2011, the 25143 Itokawa asteroid was 0.177 AUs from Earth. In 2037, it is projected to be 0.099 AUs from Earth.

a. What is the difference between these two measurements in AUs?

b. What is this difference in miles?

c. Write the differences from parts **a.** and **b.** in scientific notation.

NASA is currently working on the OSIRIS-REx mission to visit the asteroid 1999 RQ36 in 2016. This asteroid will use a robotic arm to take samples of the asteroid and bring them back to Earth. The samples will allow scientists to investigate how planets form and the origin of life.

8. Every 6 years, 1999 RQ36's orbit brings it close to earth, which is within 448,794 km.

a. Convert this measurement to miles and AUs. (Use the conversion factor you derived in Problem 1 and round your answers to thousandths.)

b. Write all three measurements from part **a.** in scientific notation.

Name: Date:

Project B: Math in a Box

An activity to demonstrate the use of polynomials in real life

Suppose you have a piece of cardboard with length 32 inches and width 20 inches and you want to use it to create a box. You would need to cut a square out of each corner of the cardboard, so that you can fold the edges up. But what size square should you cut? Cutting a small square will make a shorter box. Cutting a large square will make a taller box. Look at the diagram below.

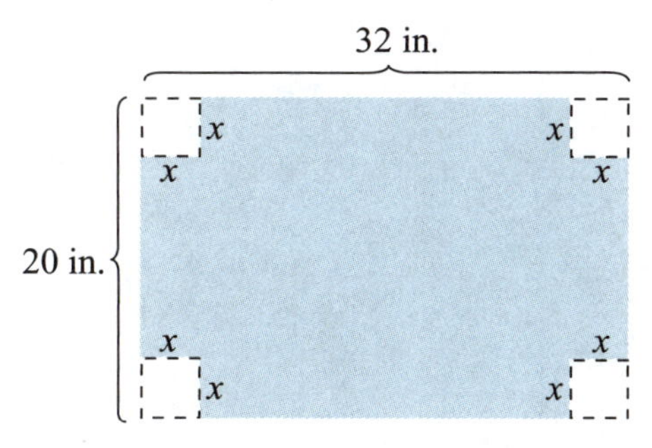

1. Since we haven't determined the size of the square to cut from each corner, let the side length of the square be represented by the variable x. Write a simplified polynomial expression in x and note the degree of the polynomial for each of the following geometric concepts:

   a. The length of the base of the box once the corners are cut out.

   b. The width of the base of the box once the corners are cut out.

   c. The height of the box.

   d. The perimeter of the base of the box.

   e. The area of the base of the box.

   f. The volume of the box.

2. Evaluate the volume expression for the following values of x. (Be sure to include the units of measurement.)

   a. $x = 1$ in.

b. $x = 2$ in.

c. $x = 3$ in.

d. $x = 3.5$ in.

e. $x = 6$ in.

f. $x = 7$ in.

3. Based on your volume calculations for the different values of x in Problem 2, if you were trying to maximize the volume of the box, between what two values of x do you think the maximum will be?

4. Using trial and error, see if you can determine the side length x of the square that maximizes the volume of the box. (**Hint:** It will be a value in the interval from Problem 3.)

5. Using the value you found for x in Problem 4, determine the dimensions of the box that maximize its volume.

Name: Date:

Math@Work

Astronomy

Astronomy is the study of celestial bodies, such as planets, asteroids, and stars. While you work in the field of astronomy, you will use knowledge and skills from several other fields, such as mathematics, physics, and chemistry. An important tool of astronomers is the telescope. Several powerful telescopes are housed in observatories around the world. One of the many things astronomers use observatories for is discovering new celestial objects such as a near-Earth object (NEO). NEOs are comets, asteroids, and meteoroids that orbit the sun and cross the orbital path of Earth. The danger presented by NEOs is that they may strike the Earth and result in global catastrophic damage. (**Note:** The National Aeronautics and Space Administration (NASA) keeps track of all NEOs which are a potential threat at the website http://neo.jpl.nasa.gov/risk/)

For an asteroid to be classified as an NEO, the asteroid must have an orbit that partially lies within 0.983 and 1.3 astronomical units (AU) from the sun, where 1 AU is the furthest distance from the Earth to the sun, approximately 9.3×10^7 miles.

Near-Earth Object Distance			
	Minimum		**Maximum**
Distance in AU	0.983 AU	1 AU	1.3 AU
Distance in Miles		9.3×10^7 miles	

Suppose you discover three asteroids that you suspect may be NEOs. You perform some calculations and come up with the following facts. The furthest that Asteroid A is ever from the sun is 81,958,000 miles. The closest Asteroid B is ever to the sun is 12,529,000 miles. The closest Asteroid C is ever to the sun is 92,595,000 miles.

1. To determine if any of the asteroids pass within the range to be classified as an NEO, fill in the missing values from the table.

2. Based on the measurements from Problem 1, do any of the three asteroids qualify as an NEO?

There are two scales that astronomers use to explain the potential danger of NEOs. The Torino Scale is a scale from 0 to 10 that indicates the chance that an object will collide with the Earth. A rating of 0 means there is an extremely small chance of a collision and a 10 indicates that a collision is certain to happen. The Palermo Technical Impact Hazard Scale is used to rate the potential impact hazard of an NEO. If the rating is less than –2, the object poses a very minor threat with no drastic consequences if the object hits the Earth. If the rating is between –2 and 0, then the object should be closely monitored as it could cause serious damage.

Go to the NASA website http://neo.jpl.nasa.gov/risk/ to answer the following questions.

3. Does any NEO have a Torino Scale rating higher than 0? If so, what is the object's designation (or name) and during which year range could a potential impact occur?

4. Which NEO has the highest Palermo Scale rating? During which year range could a potential impact occur?

Foundations Skill Check for Chapter 12

This page lists several skills covered previously in the book and software that are needed to learn new skills in Chapter 12. To make sure you are prepared to learn these new skills, take the self-test below and determine if any specific skills need to be reviewed.

Each skill includes an easy (**e.**), medium (**m.**), and hard (**h.**) version. You should be able to complete each problem type at each skill level. If you are unable to complete the problems at the easy or medium level, go back to the given lesson in the software and review until you feel confident in your ability. If you are unable to complete the hard problem for a skill, or are able to complete it but with minor errors, a review of the skill may not be necessary. You can wait until the skill is needed in the chapter to decide whether or not you should work through a quick review.

1.9 Find all factors of the given number.

e. 12 **m.** 51 **h.** 275

11.5 Find the product of the polynomial and simplify the expression.

e. $-4x\left(x^2-3x+5\right)$ **m.** $(2x+5)(x-7)$ **h.** $\left(m^2-2m+1\right)\left(2m^2-3m+4\right)$

11.6 Find the product of the polynomial and simplify the expression.

e. $(x+a)(x-a)$ **m.** $(x-a)^2$ **h.** $(3x+7)^2$

11.7 Find the quotient.

e. $\dfrac{5x^4-13x^2-23x}{x}$ **m.** $\dfrac{12x^3+9x^2-39x}{3x}$ **h.** $\dfrac{16y^6-56y^5-120y^4+64y^3}{16y^3}$

Chapter 12: Factoring Polynomials and Solving Quadratic Equations

Study Skills

12.1 Greatest Common Factor and Factoring by Grouping

12.2 Factoring Trinomials with Leading Coefficient 1

12.3 Factoring Trinomials with Leading Coefficient Not 1

12.4 Special Factoring Techniques

12.5 Additional Factoring Practice

12.6 Solving Quadratic Equations by Factoring

12.7 Applications of Quadratic Equations

Chapter 12 Projects

Math@Work

Foundations Skill Check for Chapter 13

Math@Work

Introduction

If you plan to go into math education, mathematics will be a part of your everyday work. With a degree in math education, you can work in a variety of settings. You can earn a bachelor's degree and teach primary or secondary students or you can earn a master's degree or PhD and teach at the college level. You could even work as a tutor or education coach. Besides learning essential teaching skills and methods, a degree in math education will provide you with a broad understanding of the type of math you will be teaching. As a math instructor, it is very important to have a deeper understanding of the subject matter than your students will need to learn. This understanding will allow you to easily answer any questions that students ask. If you are unable to answer the question, you will have an idea of where to look to find the answer for the student.

Suppose you decide to use your degree in math education to teach math at a high school. Teaching requires preparation outside of the classroom and this preparation will require you to answer several questions. How will you set up a lecture to cover a specific topic, such as factoring? What will you say to motivate and encourage students to learn the material? Finding the answers to these questions (and many more) requires several of the skills covered in this chapter and the previous chapters. At the end of the chapter, we'll come back to this topic and explore how math is used when working as a math instructor.

Study Skills

Working Effectively in a Group

Learning how to work with others in a group, also known as collaborative learning, should be an important part of everyone's college education. In college, students are often put into groups for discussion or to complete an assignment or project. Learning in a group setting prepares you to work effectively with groups or teams in the workplace. Employers need people who can work productively and successfully with suppliers, customers, and other employees.

What are the personal benefits of working in groups?

1. Being in a group will make it easier to get to know people in your class. You may find someone that you can study with or that has similar interests as you.
2. You will learn how to work with others who may have a different learning style than you. This may offer you some insight on other ways to be a successful student.
3. Working in a group can be a more relaxed environment than the classroom. You may find it easier to ask questions in a group setting than to ask questions in front of the entire class.
4. Your group members may be able to help you understand a concept or topic that you have been struggling with and provide help on material you miss when you are unable to attend class.

Tips for working in groups:

1. If the project or assignment will take several meetings to complete, set up a regular time and place for the group to meet and make sure that everyone has a copy of the contact information for all members of the group.
2. Be on time to group meetings and come prepared with your assigned tasks or reading completed. Coming unprepared wastes everyone's time.
3. Be considerate when others are talking and be sure to listen to their ideas, even if they are different than yours.
4. Pay attention and do not waste the group's time by checking your phone messages or e-mail during the group meeting time.
5. Encourage less talkative members of the group to share their ideas. Provide a supportive environment so that everyone will feel comfortable participating.
6. Stay away from side conversations that stray from the task at hand. Help the group focus on the work that needs to be done.
7. Be active in the group by expressing your ideas and participate in making decisions without dominating the group discussion.
8. Plan out tasks for the assignment or project based on the strengths and weaknesses of the group members and make sure everyone feels comfortable with the task given to them.
9. Make sure that everyone in the group understands the assignment directions and requirements. Be sure to ask questions if you don't understand something, so that you can make good decisions as to the direction of the project or assignment.
10. If the assignment or project has a due date, have the group create a timeline and discuss what needs to be done, when it needs to be done, and who will do each task.
11. At the end of each group meeting, make sure that someone summarizes the results of the meeting and that everyone knows what they are to do before the next meeting.
12. Keep in mind that conflicts of opinion often arise and that it is a natural part of group discussions. Agree as a group to resolve conflicts quickly and fairly so as not to hinder the group's progress on the assignment.

Name: Date:

12.1 Greatest Common Factor and Factoring by Grouping

Objectives

Find the greatest common factor of a set of terms.

Factor polynomials by factoring out the greatest common monomial factor.

Factor polynomials by grouping.

Success Strategy

Review the method of finding factors of a number in Section 1.4 and the exponent rules for variables in Section 11.1 to help you get off to a good start with factoring.

Understand Concepts

Go to Software First, read through Learn in Lesson 12.1 of the software. Then, work through the problems in this section to expand your understanding of the concepts related to greatest common factors and factoring by grouping.

If you plan on going into a field of study which requires you to take college algebra and higher level math courses, knowing how to factor will be a critical skill for success in these math classes. While nice polynomials such as those presented in this chapter don't often occur in nature, knowing how to work with them will build your critical thinking skills and allow you to be more comfortable with polynomials in general.

The first type of factoring we will cover is factoring out the greatest common factor of a set of terms. We will start with finding the greatest common factor (GCF) of a set of integers and then move on to finding the GCF of a set of terms.

Greatest Common Factor of Integers

The **greatest common factor** (GCF) of two or more integers is the largest integer that is a factor, or divisor, of all of the integers.

Lesson Link

Finding the prime factorization of a number was covered in Section 1.9.

The greatest common factor of a set of integers can be found by listing out all of the factors of each integer, using a Venn diagram, or using a method which involves prime factorization. It's important to try both methods and see which one you prefer. You may find that one method is preferable over the other in certain situations.

1. Find the GCF of 12, 18, and 36.

a. List out the factors of 12, 18, and 36.

12:

18:

36:

b. Find the GCF of 12, 18, and 36 by finding the largest factor in the lists of factors that is common to all of the integers in the set.

Lesson Link

Venn diagrams were introduced in Section 2.3 to find the LCM of a set of numbers.

2. A Venn diagram can be used to find the LCM and GCF of two numbers using their prime factorizations. The product of the prime factors in the overlapping sections of the circles is the GCF of the numbers. Find the LCM and GCF of 45 and 75 by using a Venn diagram.

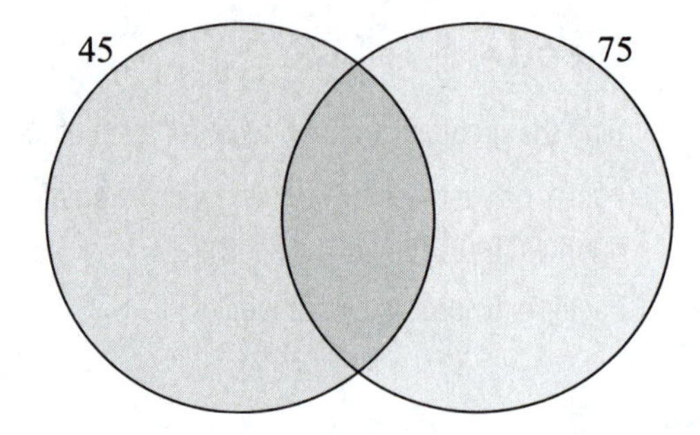

Finding the GCF of a Set of Terms

1. Find the GCF of the coefficients and any constant terms.
2. For each variable, find the largest exponent that is common to all terms. Use these exponents to create the largest variable factor that is common to all terms.
3. The product of the results from Steps 1 and 2 is the GCF of the set of terms.

Note: Finding the GCF of a set of terms involves one more step compared to finding the GCF of a set of integers because the variables need to be considered.

3. Find the GCF of $20x^4$ and $15x^3y^2$.

a. Find the prime factorizations of each coefficient.

20:

15:

b. Find the GCF of the coefficients.

c. Find the largest exponent of each variable that is common to both terms and create the largest variable factor that is common to all terms.

d. The product of the results from parts **b.** and **c.** is the GCF of the set of terms. What is the GCF of $20x^4$ and $15x^3y^2$?

Name: Date:

Read the following paragraph about factoring polynomials and work through the problems.

Factoring the GCF out of the terms of a polynomial is similar to using the distributive property in reverse. The GCF of the polynomial $5x^3 + 10x^2 - 15x$ is $5x$. Factoring out the $5x$ results in the expression $5x(x^2+2x-3)$. Distributing the factored out GCF to the terms inside the parentheses will result in the original polynomial. Keep in mind that factoring out a GCF is equivalent to multiplying and dividing the original polynomial by the GCF.

For example, $\frac{5x}{5x}(5x^3+10x^2-15x)=5x\left(\frac{5x^3}{5x}+\frac{10x^2}{5x}-\frac{15x}{5x}\right)=5x(x^2+2x-3)$.

Quick Tip

Remember that $\frac{5x}{5x}=1$, so the original polynomial is being multiplied by 1.

Factoring the GCF Out of a Polynomial

1. Find the GCF of all of the terms of the polynomial.
2. Divide each term of the polynomial by the GCF and simplify using the rules for exponents.
3. Write the factored polynomial as the product of the GCF and the polynomial resulting from Step 2.

Quick Tip

It is very important to keep track of the GCF when factoring polynomials. The factored out GCF should always be part of the final factored expression.

4. Factor the polynomial $6y^5 - 12y^3 + 24y^2$ by finding a GCF.

a. Find the GCF of all terms of the polynomial.

b. Divide each term of the polynomial by the GCF from part **a.** and simplify.

c. Write the factored polynomial as the product of the answers from parts **a.** and **b.**

Sometimes, the GCF of a set of terms is a binomial. This binomial is factored out of the polynomial in the same way as a monomial. The next problem will guide you through the process of factoring out a binomial.

Quick Tip

Remember that a **binomial** is a polynomial with exactly two terms.

5. Factor the common binomial out of $3x^2(5x+1)-2(5x+1)$.

a. What is the common binomial factor?

b. Divide each term by the common binomial factor and simplify.

c. Write the factored polynomial as a product of the answers from part **a.** and part **b.**

Read the following information about factoring by grouping and work through the problems.

Factoring by grouping is a method of factoring a polynomial with four terms. The four terms are paired off, then the GCF of each pair is found and factored out. If the polynomial can be factored, a common binomial can be factored out.

Factoring by Grouping

1. Group the terms into two pairs. Each pair should have a common factor.
2. Factor the GCF out of the first two terms and the GCF out of the last two terms. The binomial that results from factoring out each GCF should be the same in each pair. If it is not the same, go back to Step 1 or try to factor out -1 along with one of the GCFs.
3. Factor out the common binomial. The GCF from each pair will form a binomial factor.

For example:
$$\begin{aligned} xy + x - y - 1 &= (xy + x) + (-y - 1) \\ &= x(y+1) + (-1)(y+1) \\ &= (y+1)(x+(-1)) \\ &= (y+1)(x-1) \end{aligned}$$

Note: The common binomial factor for this example is $y + 1$.

If the terms are unable to be paired in a way that results in a common binomial factor, then the polynomial is **not factorable**.

Quick Tip

When polynomials are not factorable, they are also referred to as **prime polynomials**. The meaning of prime for polynomials is the same as the meaning of prime for numbers.

6. Factor $xy + 5x + 3y + 15$ by using factoring by grouping.

a. Pair the first two terms together and the last two terms together by placing the pairs in parentheses.

b. Factor the GCF out of the first pair of terms. Rewrite the polynomial.

c. Factor the GCF out of the second pair of terms. Rewrite the polynomial.

d. What is the common binomial factor in the polynomial from part **c.**?

e. Divide each term in part **c.** by the common binomial factor from part **d.**

f. Write the factored polynomial as a product of the answers from parts **d.** and **e.**

Name: Date:

g. Check your factoring by multiplying the two factors in part **f.** together. You should obtain the original polynomial.

Skill Check

Go to Software Work through Practice in Lesson 12.1 of the software before attempting the following exercises.

Factor each polynomial by finding the GCF so that the leading coefficient of the remaining binomial or trinomial factor is positive.

7. $11x - 121$

8. $-2x - 14$

9. $3x^2 - 6x$

10. $8m^2x^3 - 12m^2y + 4m^2z$

Factor each polynomial by grouping.

11. $x^2 + x + 5x + 5$

12. $2z^3 - 14z^2 + 3z - 21$

13. $2ac - 3bc + 6ad - 9bd$

14. $x^2 - 5 + x^2y - 5y$

Apply Skills

Work through the problems in this section to apply the skills you have learned related to greatest common factors and factoring by grouping.

15. The area of a rectangular photo can be represented by the polynomial $15x^2 + 5x$.

a. If $x = 2$ inches, find the area of the photo.

Quick Tip

The area of a rectangle is found by multiplying the length by the width.

b. Factor the polynomial to find a variable expression for the length and width of the photo.

c. If $x = 2$ inches, use the answer from part **b.** to find the length and the width of the photo.

d. Find the area of the photo by multiplying the length and width values from part **c.**

e. Are the answers from parts **a.** and **d.** the same? Explain why or why not.

16. A circus performer is shot vertically into the air with an initial velocity of 48 feet per second. The height of the performer above the ground in feet can be described by the polynomial $48x - 16x^2$ after x seconds.

a. Find the height of the circus performer after 2 seconds.

b. Factor the polynomial $48x - 16x^2$.

c. Use the factored form of the polynomial from part **b.** to find the height of the circus performer after 2 seconds.

d. Are the answers from parts **a.** and **c.** the same? Explain why or why not.

Name: Date:

12.2 Factoring Trinomials with Leading Coefficient 1

Objectives

Factor trinomials with leading coefficient 1 (of the form $x^2 + bx + c$).

Factor trinomials with leading coefficient not 1 by first factoring out the GCF.

Success Strategy

A lot of mathematics is about patterns and recognition of patterns between numbers. When factoring trinomials with leading coefficient 1, you are looking for a particular pair of factors of c that have a sum equal to b. Do extra practice in the software to make spotting these patterns easier.

Understand Concepts

Go to Software First, read through Learn in Lesson 12.2 of the software. Then, work through the problems in this section to expand your understanding of the concepts related to factoring trinomials with leading coefficient 1.

The method of factoring by grouping introduced in Section 12.1 cannot be used on a trinomial since it only has three terms. In this section, we will discuss how to factor a trinomial with leading coefficient 1. In Section 12.3, we will discuss how to factor any trinomial.

This section introduces the **trial-and-error method**, which uses your knowledge of the distributive property or the FOIL process when multiplying two binomials to determine the two binomial factors. Recall the FOIL process with the following equation, where x and y are variables and u and v are constants.

$$(x+u)(x+v) = x^2 + ux + vx + uv = x^2 + (u+v)x + uv$$

If the trinomial is written in descending order, then the sum of the two factors of the constant term, u and v, is equal to the coefficient of the middle term.

To factor a trinomial with coefficient 1, we will essentially be reversing the FOIL process to find the factors of the constant term whose sum is the coefficient of the middle term. Remember that a trinomial in standard form is written as $ax^2 + bx + c$. Comparing this to the equation above, we have $a = 1$, $b = u + v$, and $c = uv$.

A table can be used to organize your work while using the trial-and-error method. The next problem will guide you through this process.

1. Factor the trinomial $x^2 + 8x + 15$ by using the trial-and-error method.

a. Fill in columns one and two of the table by finding the two pairs of factors for the constant term 15.

Factors of c		Sum of Factors	Factored Form of the Trinomial
u	v	$u+v$	$(x+u)(x+v)$

b. Fill in column three of the table by finding the sum of each pair of factors.

c. Fill in column four of the table by substituting the factors from columns one and two in for the variables u and v in the factored form $(x+u)(x+v)$.

d. To factor this trinomial, the sum of the factors of 15 will be equal to the coefficient of the middle term, which is 8. Which pair of factors of 15 from the table gives you a sum of 8?

e. Verify that the answer to part **d.** is correct by finding the product of the two binomials.

2. Factor the trinomial $x^2 - 5x - 14$ by using the trial-and-error method.

Quick Tip

A negative integer has twice as many factors as a positive integer. For example, 5 has the factor pair 1 and 5 while –5 has the factor pairs 1 and –5, and –1 and 5.

a. Fill in columns one and two of the table by finding all pairs of factors for the constant term –14. Don't forget that one of the factors will need to be negative.

Factors of *c*		Sum of Factors	Factored Form of the Trinomial
u	v	$u+v$	$(x+u)(x+v)$

b. Fill in column three of the table by finding the sum of each pair of factors.

c. Fill in column four of the table by substituting the factors from columns one and two in for the variables u and v in the factored form $(x+u)(x+v)$.

d. For this trinomial, the sum of the factors of –14 will be equal to the coefficient of the middle term, which is –5. Which pair of factors of –14 from the table gives you a sum of –5?

e. Verify that the answer to part **d.** is correct by finding the product of the two binomials.

The sign of each term in a polynomial will not always be positive. The following table will help you determine which final form the factored polynomial will take based on the signs of the terms.

In the following table, u and v are factors of c.

Quick Tip

The order of the binomials in the factored form doesn't matter. This is because of the commutative property of multiplication. As a result $(x-2)(x+5) = (x+5)(x-2)$.

Factoring Patterns			
General Form	**Factored Form**	**Notes**	**Example**
$x^2 + bx + c$	$(x+u)(x+v)$	It doesn't matter which factor is the larger factor.	$x^2+10x+24=(x+4)(x+6)$
$x^2 + bx - c$	$(x-u)(x+v)$	v is the larger of the two factors so that the middle term is positive.	$x^2+2x-24=(x-4)(x+6)$
$x^2 - bx + c$	$(x-u)(x-v)$	It doesn't matter which factor is the larger factor.	$x^2-10x+24=(x-4)(x-6)$
$x^2 - bx - c$	$(x+u)(x-v)$	v is the larger of the two factors so that the middle term is negative.	$x^2-2x-24=(x+4)(x-6)$

Name: Date:

True or False: Determine whether each statement is true or false. Rewrite any false statement so that the factored form of the polynomial is correct. (There may be more than one correct new statement.)

3. The factored form of $x^2 - 6x - 16$ is $(x+2)(x-8)$.

4. The factored form of $x^2 - 8x - 20$ is $(x-2)(x+10)$.

5. The factored form of $x^2 + 5x + 6$ is $(x-2)(x-3)$.

6. The factored form of $x^2 - 9x + 20$ is $(x-4)(x-5)$.

The first step when factoring any polynomial should be to factor out the GCF of all of the terms. After factoring the GCF out of a polynomial in this section, you will be left with the product of the GCF and a trinomial with leading coefficient 1. A polynomial is considered to be completely factored when all of its factors cannot be factored any further.

Determine if any mistakes were made while factoring the trinomial. If any mistakes were made, describe the mistake and then correctly factor the trinomial to find the actual result.

7. $5x^3 - 15x^2 + 10x = 5x(x^2 - 3x + 2)$
$= 5x(x-1)(x+2)$

8. $2x^4 - 14x^3 - 36x^2 = 2x^2(x^2 - 7x - 18)$
$= 2x^2(x-9)(x+2)$

Skill Check

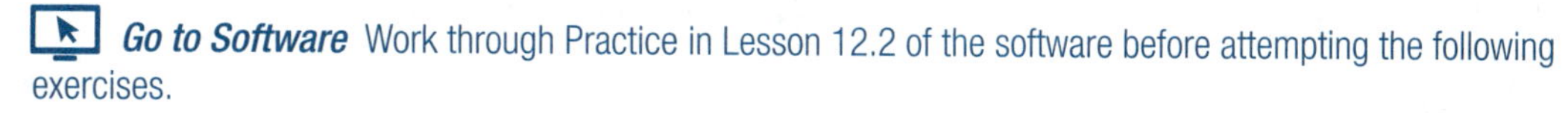

Go to Software Work through Practice in Lesson 12.2 of the software before attempting the following exercises.

Factor each trinomial completely.

9. $x^2 - x - 12$

10. $y^2 - 14y + 24$

11. $4x^5 + 28x^4 + 24x^3$

12. $-4p^3 - 36p^2 - 32p$

Apply Skills

Work through the problems in this section to apply the skills you have learned related to factoring trinomials with leading coefficient 1.

13. A ball is thrown upward from an initial height of 96 feet with an initial velocity of 16 feet per second. After t seconds, the height of the ball can be described by the polynomial $-16t^2 + 16t + 96$.

a. What is the height of the ball after 3 seconds?

b. Completely factor the polynomial $-16t^2 + 16t + 96$.

c. Use the factored form of the polynomial from part **b.** to find the height of the ball after 3 seconds.

d. Are the answers from parts **a.** and **c.** the same? Why do you think this is?

14. A large call center determines that the average number of calls they receive per hour of the day can be modeled by the polynomial $-x^2 + 25x - 100$, where x is the hour of the day, 1 through 24.

a. Factor the polynomial completely.

b. If the average number of calls at a certain time of day equals 26, write an equation using the polynomial given to demonstrate this fact.

c. Rewrite the equation in part **b.** so that all terms are on the left side of the equation and zero is on the right.

d. Factor the expression on the left side of the equation from part **c.**

Lesson Link

The process for solving equations by factoring will be introduced in Section 12.6.

Name: Date:

12.3 Factoring Trinomials with Leading Coefficient Not 1

Objectives

Factor trinomials using the trial-and-error method.

Factor trinomials using the *ac*-method

Success Strategy

Learning to factor takes a lot of practice. Be sure to spend extra time in the software and ask for help when you need it. Knowing the multiplication tables up to $12 \cdot 12$ also helps in finding factors of the coefficients.

Understand Concepts

Go to Software First, read through Learn in Lesson 12.3 of the software. Then, work through the problems in this section to expand your understanding of the concepts related to factoring trinomials.

There are two methods that can be used to factor trinomials which do not have a leading coefficient of 1. The first method is similar to the trial-and-error method introduced in Section 12.2. The second method is a procedure called the *ac*-method. Be sure to practice both methods and determine which one you prefer. In some situations the trial-and-error method may be easier while in other situations the *ac*-method may be easier.

The trial-and-error method of factoring trinomials that do not have a leading coefficient of 1 involves listing out all possible factorizations.

Trial-and-Error Method for Factoring $ax^2 + bx + c$

1. List out all possible combinations of the factors of ax^2 and c in their respective F and L positions of the FOIL process.
2. Check the sum of the products in the O and I positions until you find a sum equal to bx.
3. If none of these sums are equal to bx, then the trinomial is not factorable.

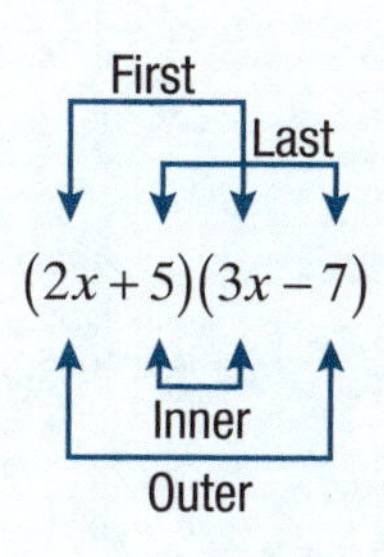

Lesson Link The FOIL method was introduced in Section 11.6.

The maximum number of possible factorizations depends on the number of factor pairs of a and c. In Problem 1, a has 1 factor pair and c has 2 factor pairs. The maximum number of possible factorizations is twice the product of the number of factor pairs. That is, there are a maximum of $2(1 \cdot 2) = 4$ possible factorizations.

Quick Tip When factoring a binomial, each possible factor of the term ax^2 should have a variable in it.

1. When factoring $3x^2 + 19x + 26$, the value of ax^2 is $3x^2$ and the value of c is 16. The possible factors of $3x^2$ are $3x$ and x and the possible factors of 26 are 1 and 26, and 2 and 13. Find all possible factorizations of $3x^2 + 19x + 26$ by filling in the parentheses below. The first two have been filled out for you.

$(3x+1)(x+26)$
$(3x+26)(x+1)$
$(\quad\quad)(\quad\quad)$
$(\quad\quad)(\quad\quad)$

2. Factor $2x^2 + 8x + 8$ using the trial-and-error method.

a. The term $2x^2$ is the product of two terms. What are the possible pairs of factors that have a product $2x^2$? (In this situation, each factor needs to contain a variable.)

b. The term 8 is the product of two terms. What are the possible pairs of factors that have a product of 8?

c. Fill in the parentheses with the possible combinations.

( )( )
( )( )
( )( )
( )( )

d. The correct factorization depends on the sum of the O and I steps of the FOIL process. Find the sum of the inner product and outer product to determine which combination will result in $8x$ as the middle term. What is the factorization of $2x^2 + 8x + 8$?

As you know, the sign of each term in a polynomial will not always be positive. The following table will help you determine which final form the factored polynomial will take based on the signs of the terms.

In the following table, n and m are factors of a and u and v are factors of c.

Factoring Patterns			
General Form	**Factored Form**	**Notes**	**Example**
$ax^2 + bx + c$	$(mx+u)(nx+v)$	It doesn't matter which factor is the larger factor.	$6x^2 + 11x + 3 = (3x+1)(2x+3)$
$ax^2 + bx - c$	$(mx+u)(nx-v)$	$u \cdot n$ is the larger of the two products so that b is positive.	$20x^2 + 7x - 6 = (4x+3)(5x-2)$
$ax^2 - bx + c$	$(mx-u)(nx-v)$	It doesn't matter which factor is the larger factor.	$12x^2 - 14x + 2 = (6x-1)(2x-2)$
$ax^2 - bx - c$	$(mx-u)(nx+v)$	$u \cdot n$ is the larger of the two products so that b is negative.	$20x^2 - 7x - 6 = (4x-3)(5x+2)$

The *ac*-Method for Factoring Trinomials of the Form $ax^2 + bx + c$

1. Find the product of a and c. Be sure to keep track of the sign of the product ac.
2. Find two factors, u and v, of ac whose sum is b. That is, $uv = ac$ and $u + v = b$. If this is not possible, then the trinomial is **not factorable**.
3. Rewrite the middle term of the trinomial as two terms using the factors found in Step 2 as coefficients. That is, $bx = ux + vx$. Be sure to keep track of the signs of the factors.
4. Factor the four-term polynomial using the method of factoring by grouping.
5. (Optional) Check that the factored polynomial is correct by finding the product of all its factors.

Note: The *ac*-method can also be used with a leading coefficient of 1 (that is, $a = 1$).

Quick Tip

Before using the *ac*-method, be sure that the trinomial is written in descending order and any GCF is factored out.

Lesson Link

Factoring by grouping was introduced in Section 12.1.

Name: Date:

3. Factor the trinomial $8x^2 - 10x - 3$ using the *ac*-method.

a. Find the product of 8 and -3.

b. Find two factors of the product in part **a.** that have a sum of -10. (Notice that one of the factors in each pair will be negative because of the negative sign on the product.)

c. Rewrite the trinomial as a polynomial with four terms by rewriting $-10x$ as a sum of two terms using the factors found in part **b.**

d. Factor the four term polynomial from part **c.** using the method of factoring by grouping.

e. Find the product of the factored form from part **d.** and verify that it is equal to the original trinomial.

Quick Tip

When factoring, the leading coefficient should be positive. If the leading coefficient of a trinomial is negative, factor -1 out of all terms of the polynomial.

The first step when factoring any polynomial should be to factor out the GCF of all of the terms. After factoring the GCF out of a polynomial in this section, you will be left with the product of the GCF and a trinomial. A polynomial is considered to be completely factored when all of its factors cannot be factored further.

Determine if any mistakes were made while factoring the trinomial. If any mistakes were made, describe the mistake and then correctly factor the trinomial to find the actual result.

4.
$$\begin{aligned} -2x^3 + x^2 + x &= -x(2x^2 - x - 1) \\ &= -x(2x^2 - 2x + x - 1) \\ &= -x(2x(x-1) + 1(x-1)) \\ &= -x(2x+1)(x-1) \end{aligned}$$

5.
$$\begin{aligned} 7z^2 - 11z - 6 &= 7z^2 + 14z - 3z - 6 \\ &= 7z(z+2) - 3(z+2) \\ &= (7z-3)(z+2) \end{aligned}$$

Skill Check

Go to Software Work through Practice in Lesson 12.3 of the software before attempting the following exercises.

Factor each trinomial completely.

6. $2x^2 - 3x - 5$

7. $-2y^3 - 3y^2 - y$

8. $12b^2 - 12b + 3$

9. $16x^2 - 8x + 1$

Apply Skills

Work through the problems in this section to apply the skills you have learned related to factoring trinomials.

10. Edith is participating in the annual cheese hurling competition in her county. She throws a block of cheese upward over the edge of a building towards a target on the ground. The cheese is thrown from an initial height of 120 feet with an initial velocity of 8 feet per second. After t seconds, the height of the cheese can be described by the polynomial $-16t^2 + 8t + 120$.

a. What is the height of the cheese after 2 seconds?

b. Completely factor the polynomial $-16t^2 + 8t + 120$.

c. Use the factored form of the polynomial from part **b.** to find the height of the cheese after 2 seconds.

d. Are the answers from parts **a.** and **c.** the same? Why do you think this is?

e. Describe what the answers from parts **a.** and **c.** mean.

Name: Date:

12.4 Special Factoring Techniques

Objectives

Factor the difference of two squares.

Factor perfect square trinomials.

Factor the sum and differences of two cubes.

Success Strategy

As previously mentioned, a lot of math is about patterns and recognizing patterns in numbers. This section requires you to recognize perfect squares and perfect cubes of numbers and terms in order to factor special forms of polynomials.

Understand Concepts

Lesson Link

Section 11.6 introduced several special products of binomials. Some of these products are used in this section.

Go to Software First, read through Learn in Lesson 12.4 of the software. Then, work through the problems in this section to expand your understanding of the concepts related to special factoring techniques.

The special factoring techniques introduced in this section can be considered "shortcuts" to factoring. With the exception of the difference or sum of cubes, the factorizations can also be found using the factoring methods introduced in Section 12.3.

Difference of Two Squares

The difference of two terms which are perfect squares can be factored using the following relationship.

$$x^2 - a^2 = (x+a)(x-a)$$

Variables x and a in the relationship can stand for constants, variables, or a combination of both.

1. Factor $m^2 - 400$.

a. Are m^2 and 400 both perfect squares?

b. Find the square root of m^2.

c. Find the square root of 400.

d. Substitute the answer from part **b.** for x and the answer from part **c.** for a into the relationship $x^2 - a^2 = (x+a)(x-a)$.

e. Simplify the right side of the equation from part **d.** to verify that the factorization is correct.

Sum of Two Squares

The sum of two squares $x^2 + a^2$ is **not factorable** over the set of real numbers.

Perfect Square Trinomials

A perfect square trinomial can be factored using one of the following relationships.

Square of a binomial sum: $x^2 + 2ax + a^2 = (x + a)^2$

Square of a binomial difference: $x^2 - 2ax + a^2 = (x - a)^2$

Variables x and a in the relationships can stand for constants, variables, or a combination of both.

2. Factor $z^2 - 12z + 36$.

a. Are the first term and the last term of the polynomial both perfect squares?

b. Find the square root of the first term.

c. Find the square root of the last term.

d. Ignoring the sign, is the value of the second term equal to two times the product of the results from parts **a.** and **b.**?

e. The sign of the second term of the polynomial determines the sign that is present in the factored form. To factor the polynomial $z^2 - 12z + 36$, which relationship should be used, $x^2 + 2ax + a^2 = (x+a)^2$ or $x^2 - 2ax + a^2 = (x-a)^2$?

f. Substitute the answer from part **b.** for x and the answer from part **c.** for a into the relationship from part **e.**

g. Simplify the right side of the equation from part **f.** to verify that the factorization is correct.

Name: Date:

Quick Tip

One way to remember the correct signs when factoring the sum and difference of two cubes is the SOAP method. S stands for **Same** because the sign between the two cubes determines the sign in the binomial. The first sign in the trinomial is the **Opposite** of the sign in the binomial. The sign of the last term is **Always Positive**.

Sum and Difference of Two Cubes

A difference of two cubed terms can be factored using one of the following relationships.

Sum of two cubes: $x^3 + a^3 = (x+a)(x^2 - ax + a^2)$

Difference of two cubes: $x^3 - a^3 = (x-a)(x^2 + ax + a^2)$

Variables x and a in the relationships can stand for constants, variables, or a combination of both.

3. Factor $x^6 - 27$.

a. Are x^6 and 27 both perfect cubes?

b. Find the cube root of x^6.

c. Find the cube root of 27.

d. The sign of the second term of the polynomial determines the signs that are used in the factored form. To factor the polynomial $x^6 - 27$, which relationship should be used, $x^3 + a^3 = (x+a)(x^2 - ax + a^2)$ or $x^3 - a^3 = (x-a)(x^2 + ax + a^2)$?

e. Substitute the answer from part **b.** for x and the answer from part **c.** for a into the equation from part **d.** Simplify the terms of the trinomial.

f. Simplify the right side of the equation from part **e.** to verify that the factorization is correct.

Skill Check

Go to Software Work through Practice in Lesson 12.4 of the software before attempting the following exercises.

Completely factor each polynomial.

4. $x^2 - 49$

5. $z^2 + 18z + 81$

6. $x^3 - 64$

7. $x^4 - 16$

8. $16x^4 + 8x^2 + 1$

9. $x^2 + 64y^2$

Apply Skills

Work through the problems in this section to apply the skills you have learned related to special factoring techniques.

10. A landscaper plans to put a square fountain in the middle of a square fish pond. The fountain has a side length of 4 feet and the fish pond will have a side length of x feet.

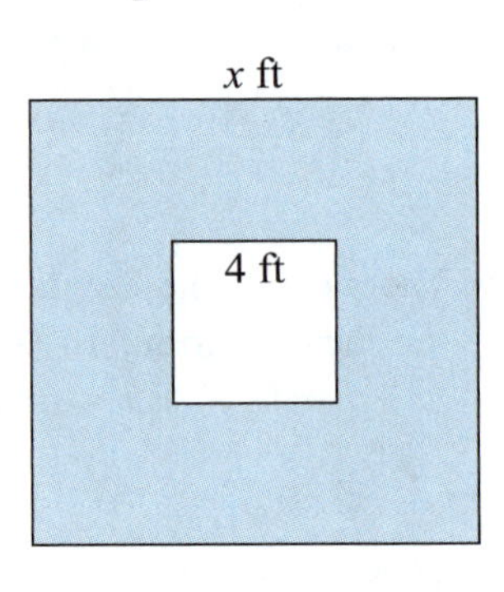

Lesson Link

Finding the area of figures with cut outs was discussed in Section 5.3.

a. Write a polynomial to describe the water surface area of the fish pond (the shaded region).

b. Use the polynomial from part **a.** to determine the water surface area of the fish pond if $x = 10$ feet.

c. Factor the polynomial from part **a.**

d. The landscaper wants to put a rectangular fish pond with the same water surface area at another location of the property. What should the length and width of this fish pond be? (**Hint:** Use $x = 10$ and the factors from part **c.**)

Name: Date:

12.5 Additional Factoring Practice

Objectives

Use the guidelines for deciding which method to use when factoring a polynomial.

Success Strategy

When factoring polynomials, create a flow chart or table to help you remember all of the steps in the guidelines presented in this section. A sample table for this purpose is provided below.

Understand Concepts

Go to Software First, read through Learn in Lesson 12.5 of the software. Then, work through the problems in this section to expand your understanding of the concepts related to factoring polynomials.

Learning to factor takes a lot of practice. This section will introduce some helpful guidelines to reference while factoring polynomials. These guidelines should be followed in the order that they are presented in this section. As you work through this section, fill out the following table with the guidelines for factoring.

Summary Table for Factoring Polynomials		
Factor Out a GCF First		
2 Terms	3 Terms	4 Terms

Factor Out a GFC

The first step in the guidelines for factoring polynomials is to factor out the GFC, if possible. Remember that if the leading coefficient is negative, factor out a negative coefficient as part of the GCF.

1. Find the GCF of the polynomial.

a. $3n^3 + 15n^2 + 18n$

b. $x^4y^3 - x^4$

c. $-4x^2 + 16x - 20$

d. $64 + 49t^2$

The remaining steps of the guidelines depends on the number of terms in the polynomial.

Count the Number of Terms in the Polynomial

Check the number of terms in the polynomial. If the polynomial has two terms, consider the following.

a. The difference of two squares is factorable.

b. The sum of two squares is **not** factorable.

c. The difference of two cubes is factorable.

d. The sum of two cubes is factorable.

If the polynomial has three terms, consider the following.

a. A perfect square trinomial is factorable.

b. Try the trial-and-error method.

c. Try the *ac*-method.

If the polynomial has four terms, consider the following.

a. Try to factor by grouping.

2. Determine which factoring technique you would use to factor each polynomial.

a. $a^9 + 64b^3$

b. $x^2 + 12x + 35$

c. $x^2 + 2xy - 6x - 12y$

d. $2x^2 + 7x + 3$

Make Sure the Polynomial Is Completely Factored

Check to see if any factors can be factored further. If any of the factors have a degree higher than 1, check to make sure it will not factor again.

Note: This does not apply when factoring the sum or difference of cubes since the trinomial that results will not factor any further.

3. Determine if the polynomial is completely factored. If it is not completely factored, state which factor can be factored further.

a. $5(24m^2 + 2m + 15)$

b. $(x+5)(x^4 - 4)$

c. $(4x - 10)(x - 2)$

d. $3x(4x - 3)(4x + 3)$

Name: Date:

Check Your Answer

Check your answer by multiplying the factors together. Their product should be equal to the original polynomial.

4. Determine if each polynomial was factored properly by multiplying out the factored forms.

a. The trinomial $a^2 - 4a + 3$ factors to $(a-3)(a-1)$.

b. The polynomial $-150x^2 + 96$ factors to $6(5x-4)(5x+4)$.

Skill Check

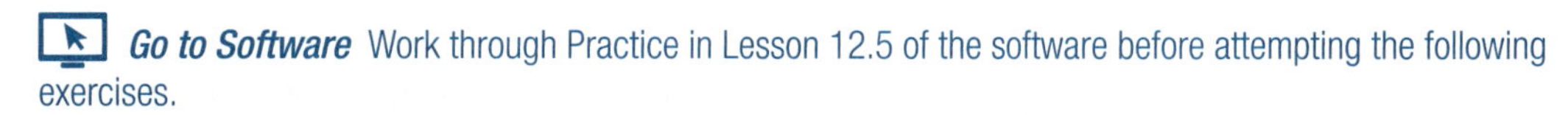

Go to Software Work through Practice in Lesson 12.5 of the software before attempting the following exercises.

Completely factor the polynomial.

5. $x^2 + 11x + 18$

6. $-x^2 - 12x - 35$

7. $x^2 + 3x - 10$

8. $6x^2 - 11x + 4$

9. $6t^2 + t - 35$

10. $x^3 + 125$

11. $9y^2 + 24y + 16$

12. $16x^3 - 100x$

Apply Skills

Work through the problems in this section to apply the skills you have learned related to factoring polynomials.

13. The manager at Barbara's Bombtastic Bakery has determined that the revenue function for cupcakes is $R(p) = 1500p - 5p^2$ when the price is p cents per cupcake.

a. Factor the expression on the right side of the function.

b. If the price is \$1.25 per cupcake, what is the revenue made by the bakery? (Remember that the value of p is in cents.)

c. If the price is \$3 per cupcake, what is the revenue made by the bakery?

d. Write an equation to represent the situation where the revenue made by the bakery is \$1000.

e. Rearrange the equation in part **d.** so that all terms are on the left side of the equation and zero is on the right.

f. Factor the expression on the left side of the equation from part **e.** (**Hint:** Remember to factor out the GCF first. This will make the coefficients easier to work with.)

Name: Date:

12.6 Solving Quadratic Equations by Factoring

Objectives

Solve quadratic equations by factoring.

Write equations given the roots.

Success Strategy

Use the table or flowchart you created in Section 12.5 to help you factor the polynomials in this section.

Understand Concepts

Go to Software First, read through Learn in Lesson 12.6 of the software. Then, work through the problems in this section to expand your understanding of the concepts related to solving quadratic equations by factoring.

Many of the trinomials that we have worked with so far in this chapter have been **quadratic polynomials**. A quadratic polynomial is another name for a second degree polynomial. A **quadratic equation** is an equation with the general form of

$$ax^2 + bx + c = 0,$$

where x is the variable and a, b, and c are constants and $a \neq 0$. Keep in mind that not every quadratic equation will look like this. All quadratic equations, however, can be rewritten in this form.

Zero-Factor Property

If the product of two (or more) factors is 0, then at least one of the factors must be 0. That is, for real numbers a and b, if $a \cdot b = 0$, then $a = 0$, $b = 0$, or both a and b are equal to 0.

The zero-factor property can be used to solve quadratic equations that can be factored. The property allows us to take each factor of a quadratic equation and set it equal to zero.

Solving Quadratic Equations by Factoring

1. Rewrite the quadratic equation in standard form $ax^2 + bx + c = 0$.
2. Completely factor the left side of the equation.
3. Set each factor equal to 0 and solve for the variable.
4. Check each solution in the original equation.

Quick Tip

Remember that a **solution** to an equation is a value of the variable which makes the equation true.

1. Solve $x^2 - x = 12$.

a. Rewrite the equation so that it is in the standard form $ax^2 + bx + c = 0$.

b. Completely factor the left side of the equation.

c. Take each factor of the left side of the equation and set it equal to zero.

Quick Tip

The zero-factor property can only be applied to set each factor of the equation equal to zero if the entire equation is equal to zero. That is, one side of the equation *must* be zero.

d. Solve each of the equations from part **c.** for the variable. You should end up with two solutions.

Quick Tip

The reason for checking solutions in the original equation is to verify that the factorization is correct.

e. Verify the solutions by substituting them, one at a time, into the original equation and then simplifying.

Factor Theorem

If $x = c$ is a root of a polynomial equation in the form $P(x) = 0$, then $x - c$ is a factor of the polynomial $P(x)$.

For example, since $x = 2$ is a solution of $x^2 - 5x + 6 = 0$, then $x - 2$ is a factor of $x^2 - 5x + 6$.

Quick Tip

The solutions to an equation are also called **zeros** or **roots**.

The factor theorem can be used to find an equation that has certain given solutions. The next problem will walk you through this process.

2. Find an equation that has $x = 5$ and $x = -3$ as solutions.

a. Use the addition principle of equality to rewrite $x = 5$ as an equation that is equal to 0. Write the equation so 0 is on the right side.

b. Use the addition principle of equality to rewrite $x = -3$ as an equation that is equal to 0. Write the equation so 0 is on the right side.

c. Write an equation that has both $x = 5$ and $x = -3$ as solutions by finding the product of the binomials on the left side of the equations in parts **a.** and **b.** and setting this product equal to zero.

d. Verify that $x = 5$ and $x = -3$ are solutions to the equation from part **c.**

e. Multiply the entire equation from part **c.** by 2.

f. Check if $x = -3$ and $x = 5$ are solutions to the new equation from part **e.**

g. Does multiplying an equation by a constant number change its solutions? Explain your answer.

Name: Date:

h. If multiplying an equation by a nonzero constant doesn't change the solution, then how many polynomials exist that have the two solutions $x = -3$ and $x = 5$?

Skill Check

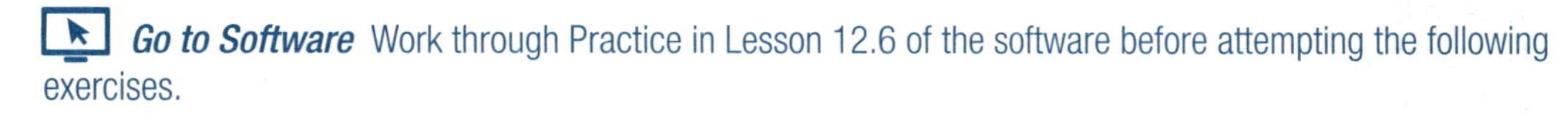

Go to Software Work through Practice in Lesson 12.6 of the software before attempting the following exercises.

Solve each equation by factoring.

3. $x^2 - 3x - 4 = 0$

4. $4x^2 - 12x + 9 = 0$

5. $x^2 = x + 30$

6. $9x^2 = 36$

Apply Skills

Work through the problems in this section to apply the skills you have learned related to solving quadratic equations by factoring.

7. A ball is thrown upward from an initial height of 96 feet with an initial velocity of 16 feet per second. After t seconds, the height of the ball can be described by the equation $h = -16t^2 + 16t + 96$.

a. What happens when $h = 0$?

b. Rewrite the equation with $h = 0$.

c. Solve the equation by factoring.

d. What does the answer to part **c.** mean?

e. Do both solutions from part **c.** make sense in the context of the problem? Explain why or why not.

Quick Tip

Drawing a diagram or figure may be useful for solving Problem 8.

8. Robin is putting the finishing touches on a quilt. The quilt is currently 80 inches long by 60 inches wide and she plans to add a border around the quilt. The width of the border on the sides will be twice the width of the border on the top and bottom of the quilt.

a. If x is the width of the border in inches that will be added to the top and bottom of the quilt, write an expression for the length and width of the quilt with the border added.

b. Write a simplified expression to find the area of the quilt with the border added.

c. Robin has a total of 5712 square inches of fabric to use for the back of the quilt. Use the expression from part **b.** to write an equation to describe the total area of the back of the quilt.

d. Solve the quadratic equation from part **c.** by factoring.

e. Do both of the solutions from part **d.** make sense in the context of the problem?

f. What is the total length and width of the quilt?

Name: Date:

12.7 Applications of Quadratic Equations

Objectives

Use quadratic equations to solve problems related to numbers and geometry.

Use quadratic equations to solve problems involving consecutive integers.

Use quadratic equations to solve problems related to the Pythagorean Theorem.

Success Strategy

In this section you will be introduced to some new problem-solving strategies. Try each strategy and determine which one you like the best. Learn this strategy and use it as you work through the problems in this section and the remainder of the workbook.

Understand Concepts

Go to Software First, read through Learn in Lesson 12.7 of the software. Then, work through the problems in this section to expand your understanding of the concepts related to applications of quadratic equations.

We have previously discussed Pólya's problem-solving process for solving applications or word problems, which is a very simple four-step plan that has been around since the 1960's. Since that time, there have been some other strategies developed which may appeal to you more. These strategies may be easier to remember since the name of each of them is an acronym for the steps involved in the strategy. A few of these problem-solving strategies are presented here and even more can be found with a little bit of research.

Quick Tip

These problem-solving strategies come from TeachaPedia http://teachapedia.org

RIDGES Strategy

This strategy was developed by Kathleen Snyder in 1988 to help students solve word problems.

1. **Read** the problem. If you don't understand it completely the first time then re-read it.
2. **Identify** all the information given in the problem. List the information separately and circle the pieces of information needed to solve the problem.
3. **Draw** a picture or diagram of the situation to help pick out relevant information.
4. **Goal** statement. Express the problem being asked in your own words.
5. **Equation or expression** development. Write a mathematical equation or expression for the problem.
6. **Solve** the equation by substituting the given information into the equation, or **simplify** the expression.

SOLVE Strategy

1. **Study** the problem: Highlight or underline the question. What am I to find?
2. **Organize** the facts: Identify each fact. Eliminate unnecessary facts by drawing a line through them.
3. **Line** up a plan: Decide (in words) how to solve the problem and choose the operation(s) needed.
4. **Verify** your plan with action: Estimate your answer and carry out the plan.
5. **Examine** the results: Evaluate your answer. Does it make sense? Is it reasonable and accurate?

S.T.A.R. Strategy

This strategy was developed in 1998 by Paula Maccini.

1. **Search** the word problem: Read the problem carefully. What do you know? What do you need to find? Write down the facts.
2. **Translate** the problem: Translate the problem into a picture and/or equation.
3. **Answer** the problem being asked (which may be different than the solution to your equation).
4. **Review** your answer in the context of the problem. Reread the problem. Does the answer make sense? Why? Check the answer.

1. Do any of the three problem-solving strategies introduced in this section appeal to you more than Pólya's problem-solving method? Write an explanation for your answer.

Skill Check

Go to Software Work through Practice in Lesson 12.7 of the software before attempting the following exercises.

Solve.

2. One number is eight more than another. Their product is negative sixteen. What are the numbers?

Quick Tip

When creating expressions and equations, it is important to determine and label what each variable stands for.

3. Find two consecutive positive integers whose product is one hundred ten.

4. Find three consecutive negative odd integers such that the product of the first and third is seventy-one more than ten times the second.

Name: ______________________ Date: ______________________

Apply Skills

Work through the problems in this section to apply the skills you have learned related to applications of quadratic equations.

5. A support wire is attached x feet from the top of a 17-foot pole to protect the pole during a blizzard. The other end of each wire is attached to a stake x feet from the base of the pole. The wire used is 13 feet long.

a. Draw a diagram to describe the situation. Be sure to label the figure with the known information.

Lesson Link

The Pythagorean Theorem was introduced in Section 5.7.

b. Use the Pythagorean Theorem to write an equation that describes the situation. Do not simplify.

c. Simplify the equation from part **b.** and solve for x.

d. Do both solutions from part **b.** make sense in this situation? That is, do they both result in a positive distance on the pole and a positive distance from the pole?

e. What do the answers from part **b.** mean? (You should have two answers.)

f. Which answer from part **d.** seems like the better option? Write an explanation for your choice.

6. A family wants to fence in a rectangular area of their yard next to the house so their dog can play outside without being on a leash. One side of the fenced-in area will be along the side of the house, so they will only need to fence in three sides. The family decides to fence in an area of 4000 square feet and they purchase 180 feet of fencing. What are the dimensions of the fenced in area?

   a. Draw a diagram to represent the situation. Use the variable x to label the two sides of the fence which will have the same length.

Quick Tip

To answer part **b.**, use the fact that the perimeter is 180 feet and two sides have a length of x feet.

   b. Write an expression involving x to represent the length of the third side of the fence.

   c. Write an equation to represent the area of the fenced-in yard.

   d. Solve the equation from part **c.**

   e. Do both solutions make sense in the context of the problem?

   f. What are the possible dimensions of the fenced-in yard?

Name: Date:

Chapter 12 Projects

Project A: Building a Dog Pen

An activity to demonstrate the use of the Pythagorean Theorem and quadratic equations in real life

Justin's house sits on a lot that is shaped like a trapezoid. He decides to make the right side of the lot usable by building a triangular dog pen for his dog Blackjack. On his lunch break at work, he decides to order the materials but realizes that he forgot to write down the actual dimensions of that area of the lot. He wants to get the pen done this weekend because his friends are coming over to help him. As a result, he has to order the materials today so that they will arrive on time. Justin remembers that one of the sides is 5 feet more than the length of the shortest side and the longest side is 10 feet more than the shortest side. Can he figure out the dimensions of the area so that he can order the materials today?

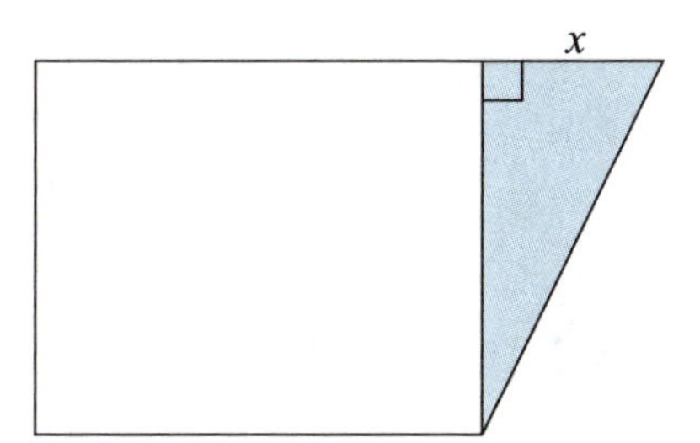

1. Using the diagram of the lot and the variable x for the length of the shortest side of the dog pen, write an expression for the other two sides of the triangular pen and label them on the diagram.

2. Using the Pythagorean Theorem, substitute the three expressions for the sides of the triangle into the formula and simplify the resulting polynomial. Be sure to move all terms to one side of the equation with the other side equal to zero. (Remember that the longest side is the hypotenuse in the formula. Make sure your leading term has a positive coefficient and that your squared binomials result in a trinomial.)

3. Use the equation from Problem 2.

a. Factor the resulting quadratic equation into two linear binomial factors.

b. Find the two solutions to the equation from part **a.** using the zero-factor law.

c. Do both of these solutions make sense? Explain your reasoning.

d. Using the solution that makes sense, substitute this value for x and determine the dimensions of the dog pen.

4. To fence in the dog pen, Justin plans to purchase chain link fencing at a cost of $1.90 per foot.

a. How much fencing will he need?

b. How much will the fencing cost?

5. How much area will the dog pen have?

6. Justin decides to put a dog house in the pen to protect Blackjack in bad weather. The dog house is rectangular in shape and measures 2.5 feet by 3 feet. Once the dog house is in the pen, how much area will Blackjack have to run in?

7. The sides of the pen form a right triangle and the measurements of the sides of the pen were found using the Pythagorean Theorem. Any 3 positive integers that satisfy the Pythagorean Theorem are called a **Pythagorean triple**. There are an infinite number of these triples and numerous formulas that can be used to generate them. Do some research on the Internet to find one of these formulas and use the formula to generate three more sets of Pythagorean triples. Verify that they are Pythagorean triples by substituting them into the Pythagorean Theorem.

a. (, ,)

b. (, ,)

c. (, ,)

8. Another way to generate a Pythagorean triple is to take an existing triple and multiply each integer by a constant. Take the Pythagorean triple $(3, 4, 5)$ and multiply each integer in the triple by the factors below and verify that the result is also a Pythagorean triple by substituting into the Pythagorean Theorem.

a. Multiply by 3:

b. Multiply by 5:

c. Multiply by 8:

Name: Date:

Project B: A Little Support, Please!

An activity to demonstrate the use of quadratic equations in real life

You may have noticed that most utility poles have a wire that runs diagonally from the top of the pole to an anchor on the ground. This wire is called a guy-wire and the tension placed on the wire acts as a stabilizer against the lateral pull of the electrical wires. In this project you will be calculating the length of wire needed to support a typical utility pole that is used to carry electrical power lines.

A utility pole is being installed in a new subdivision and will need to be supported by a guy-wire that attaches to the top of the pole and to a point on the ground as shown in the diagram. The wire will need to be 12 feet less than twice the height of the pole and will need to be anchored at a distance from the base of the pole that is 12 feet longer than the height of the pole.

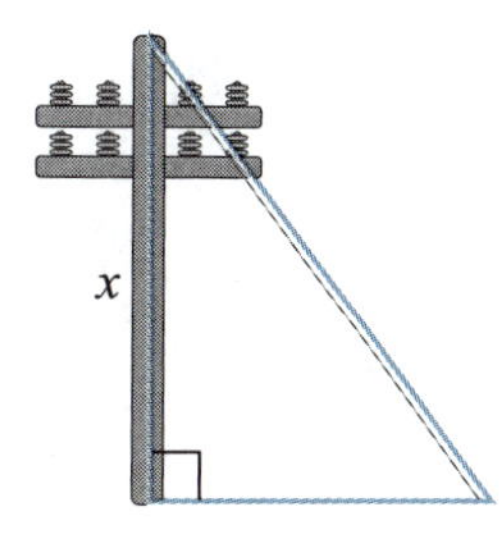

1. If x represents the height of the pole, write an expression using the variable x that represents the length of the guy-wire.

2. Write an expression in x for the distance from the base where the pole is anchored.

3. Using the Pythagorean Theorem, write a quadratic equation in x that expresses the relationship of the height of the pole, the length of the wire, and the distance from the base of the pole where the wire is anchored.

4. Solve the equation above for x. Since this is a quadratic equation, you will get two solutions. Which solution does not make sense? Why?

5. Determine the height of the utility pole.

6. How long is the guy-wire that is needed to support the utility pole?

7. How far away from the base will the wire need to be anchored in order for it to be taut and, therefore, support the pole properly?

Name: Date:

Math@Work

Math Education

As a math instructor at a public high school, your day will be spent preparing class lectures, grading assignments and tests, and teaching students with a wide variety of backgrounds. While teaching math, it is your job to explain the concepts and skills of math in a variety of ways to help students learn and understand the material. As a result, a solid understanding of math and strong communication skills are very important. Teaching math is a challenge and being able to understand the reasons that students struggle with math and empathize with these students is a critical aspect of the job.

Suppose that the next topics you plan to teach to your algebra students involve finding the greatest common factor and factoring by grouping. To teach these skills, you will need to plan how much material to cover each day, choose examples to walk through during the lecture, and assign in-class work and homework. You decide to spend the first day on this topic explaining how to find the greatest common factor of a list of integers.

1. It is usually easier to teach a group of students a new topic by initially showing them a single method. If a student has difficulty with that method, then showing the student an alternative method can be helpful. Which method for finding the greatest common factor would you teach to the class during the class lecture?

2. On a separate piece of paper, sketch out a short lecture on finding the greatest common factor of a list of integers. Be sure to include examples that range from easy to difficult.

3. While the class is working on an in-class assignment, you find that a student is having trouble following the method that you taught to the entire class. Describe an alternative method that you could show the student.

4. From your experience with learning how to find the greatest common factor of a list of integers, what do you think are some areas that might confuse students and cause them to struggle while learning this topic? Explain how understanding the areas that might cause confusion can help you become a better teacher.

Foundations Skill Check for Chapter 13

This page lists several skills covered previously in the book and software that are needed to learn new skills in Chapter 13. To make sure you are prepared to learn these new skills, take the self-test below and determine if any specific skills need to be reviewed.

Each skill includes an easy (**e.**), medium (**m.**), and hard (**h.**) version. You should be able to complete each problem type at each skill level. If you are unable to complete the problems at the easy or medium level, go back to the given lesson in the software and review until you feel confident in your ability. If you are unable to complete the hard problem for a skill, or are able to complete it but with minor errors, a review of the skill may not be necessary. You can wait until the skill is needed in the chapter to decide whether or not you should work through a quick review.

2.1 Reduce the fraction to lowest terms.

e. $\frac{3}{9}$ **m.** $\frac{24}{100}$ **h.** $\frac{34}{51}$

2.2 Find the product and reduce to lowest terms.

e. $\frac{2}{3} \cdot \frac{1}{5}$ **m.** $\frac{11}{8} \cdot \frac{6}{22}$ **h.** $\frac{32}{20} \cdot \frac{13}{9} \cdot \frac{7}{26}$

2.2 Find the quotient and reduce to lowest terms.

e. $\frac{2}{5} \div \frac{5}{4}$ **m.** $\frac{5}{16} \div \frac{15}{16}$ **h.** $\frac{26}{35} \div \frac{39}{40}$

2.4 Find the sum and reduce to lowest terms.

e. $\frac{4}{9} + \frac{1}{9}$ **m.** $\frac{3}{10} + \frac{1}{100} + \frac{9}{1000}$ **h.** $\frac{1}{8} + \frac{1}{12} + \frac{1}{9}$

2.4 Find the difference and reduce to lowest terms.

e. $\frac{7}{11} - \frac{4}{11}$ **m.** $\frac{5}{4} - \frac{3}{5}$ **h.** $\frac{8}{45} - \frac{11}{72}$

Chapter 13: Rational Expressions

Study Skills

13.1 Multiplication and Division with Rational Expressions

13.2 Addition and Subtraction with Rational Expressions

13.3 Complex Fractions

13.4 Solving Equations with Rational Expressions

13.5 Applications of Rational Expressions

13.6 Variation

Chapter 13 Projects

Math@Work

Foundations Skill Check for Chapter 14

Math@Work

Introduction

If you plan to study physics, you will gain a broad educational background that you can use to pursue a variety of careers. In general, physicists study matter and its motion through space and time, along with studying energy and force. Some career options that a degree in physics could lead to include the field of education as an instructor, the medical field as a researcher creating cures for cancer, or in electronics creating parts which run anything from a mobile phone to a spacecraft. No matter which career path you choose, you can count on math being a part of your daily work.

Suppose you decide to focus your physics education on electronics and you start your career at a company that creates and manufactures circuit boards. While creating circuit boards, you will be required to answer a variety of questions that involve math. For example, what is the total resistance of the circuit? What is the resistance of an unlabeled resistor in the circuit board? Finding the answers to these questions (and many more) requires several of the skills covered in this chapter and previous chapters. At the end of the chapter, we'll come back to this topic and explore how math is used in electronics.

Study Skills

Overcoming Anxiety

People who are anxious about math are often just not good at taking math tests. If you understand the math you are learning but don't do well on math tests, you may be in the same situation. If there are other subject areas in which you also perform poorly on tests, then you may be experiencing test anxiety.

How to Reduce Math Anxiety

1. Learn effective math study skills. Sit near the front of your class and take notes. Ask questions when you don't understand the material. Review your notes after class and preread material before it is covered in class. Keep up with your assignments and do a lot of practice problems.
2. Don't accept negative self-talk such as "I am not good at math" or "I just don't get it and never will." Maintain a positive attitude and set small math achievement goals to keep you positively moving toward bigger goals.
3. Visualize yourself doing well in math, whether it is on a quiz or test, or passing a math class. Rehearse how you will feel and perform on an upcoming math test. It may also help to visualize how you will celebrate your success after doing well on the test.
4. Form a math study group. Working with others may help you feel more relaxed about math in general and you may find that other people have the same fears.
5. If you panic or freeze during a math test, try to work around the panic by finding something on the math test that you can do. Once you gain confidence, go back and work through all the problems you know how to do. Then, try completing the harder problems, knowing that you have a large part of the test completed already.
6. If you have trouble remembering important math formulas or procedures for tests, do what is called a "brain drain" and write down all the formulas and important facts that you have studied on your test or scratch paper as soon as you are given the test. Do this before you look at any questions on the test so that you won't panic and forget everything. Having this information available to you should help boost your confidence and reduce your anxiety.

How to Reduce Test Anxiety

1. Be prepared. Knowing you have prepared well for the test will make you more confident in your abilities and less anxious.
2. Get plenty of sleep the night before a big test and be sure to eat nutritious meals on the day of the test. It is helpful to exercise regularly and establish a set routine for test days. For example, your routine might include eating your favorite food, putting on your lucky shirt, and packing yourself a special treat for after the test.
3. Talk to your instructor about your anxiety. Your instructor may be able to make some accommodations for you when taking tests that may make you feel more relaxed, such as extra time or a more calming testing place.
4. Learn how to manage your anxiety by taking deep, slow breaths and thinking about places or people who make you happy and peaceful.
5. When you receive a low score on a test, take time to analyze the reasons why you performed poorly. Did you prepare enough? Did you study the right material? Did you get enough rest the night before? Resolve to change those things that may have negatively affected your performance in the past before the next test.
6. Learn effective test-taking strategies. (See the Chapter 5 Study Skills section.)

Name: Date:

13.1 Multiplication and Division with Rational Expressions

Objectives

Become familiar with rational expressions.

Reduce rational expressions to lowest terms.

Multiply rational expressions.

Divide rational expressions.

Success Strategy

Rational expressions are similar to fractions, with the exception that they contain variables. Be sure to review how to add, subtract, multiply, and divide fractions before starting this section.

Understand Concepts

Go to Software First, read through Learn in Lesson 13.1 of the software. Then, work through the problems in this section to expand your understanding of the concepts related to multiplication and division with rational expressions.

Fractions were introduced in Chapter 2 and have been used throughout the workbook. Polynomials were introduced in Chapter 11 and further explained in Chapter 12. This section will use your knowledge of fractions and polynomials to introduce you to rational expressions.

Lesson Link

Polynomials were introduced in Section 11.3.

Rational Expressions

A **rational expression** is an algebraic expression that can be written in the form $\frac{P}{Q}$, where P and Q are polynomials and $Q \neq 0$.

1. Determine which expressions are rational expressions. If the expression is not a rational expression, briefly explain why.

a. $4x^2 - 3x$

b. $\frac{\sqrt{2x}}{3x+5}$

c. $\frac{19x^2 + 2x - 3}{7x}$

d. $\frac{(11x-7)(14x+3)}{5x^3 - 7x + 2}$

e. $\frac{x^{\frac{1}{2}} - 3}{8x+5}$

Quick Tip

A **constant monomial** is a monomial which consists of only a number. For example, 5, 8, and −13 are constant monomials.

Fractions are a type of rational expression. In a fraction, P and Q are constant monomials. The rational expressions that we deal with in this section will contain variables in the polynomial P, the polynomial Q, or both polynomials. One thing to keep in mind is that the denominator of a rational expression cannot equal 0. A value of the variable that causes the denominator to equal 0 is called a **restriction** on the domain, or **excluded value**.

Finding Restrictions of Rational Expressions

1. Set the expression in the denominator of the rational expression equal to 0.
2. Solve the equation for the variable.

2. Let's find the restrictions for the rational expression $\frac{7x}{2x-5}$.

a. Set the denominator of the rational expression equal to 0.

b. Solve the equation from part **a.** for the variable.

c. What does the answer to part **b.** mean? Write a complete sentence.

Read the following paragraph about fractions and rational expressions then work through the following problem.

Quick Tip

The fundamental principle of rational expressions *divides out* common factors from the numerator and denominator of an expression.

Rational expressions follow the same arithmetic rules as fractions. The following table contains a summary of the rules for fractions that were covered in Chapter 2. This section and the next will reintroduce these rules for rational expressions. First recall that a **fraction** (or **rational number**) is a number that can be written in the form $\frac{a}{b}$, where a and b are integers and $b \neq 0$.

3. In the example column, fill in the boxes with the missing numbers.

Arithmetic Rules for Rational Numbers

Rule	Example
The Fundamental Principle: $\frac{a}{b}=\frac{a\cdot k}{b\cdot k}$, where $b \neq 0$, $k \neq 0$.	$\frac{1}{2}=\frac{1\cdot 5}{2\cdot 5}=\frac{\square}{\square}$.
The **reciprocal** of $\frac{a}{b}$ is $\frac{b}{a}$ and $\frac{a}{b}\cdot\frac{b}{a}=1$, where $a \neq 0$, $b \neq 0$.	The reciprocal of $\frac{3}{4}$ is $\frac{\square}{\square}$ and $\frac{3}{4}\cdot\frac{\square}{\square}=\square$.
Multiplication: $\frac{a}{b}\cdot\frac{c}{d}=\frac{a\cdot c}{b\cdot d}$, where $b \neq 0$, $d \neq 0$.	$\frac{3}{5}\cdot\frac{2}{7}=\frac{\square\cdot\square}{\square\cdot\square}=\frac{\square}{\square}$.
Division: $\frac{a}{b}\div\frac{c}{d}=\frac{a}{b}\cdot\frac{d}{c}=\frac{a\cdot d}{b\cdot c}$, where $b \neq 0$, $c \neq 0$, $d \neq 0$.	$\frac{1}{2}\div\frac{3}{5}=\frac{1}{2}\cdot\frac{\square}{\square}=\frac{1\cdot\square}{2\cdot\square}=\frac{\square}{\square}$.
Addition: $\frac{a}{b}+\frac{c}{b}=\frac{a+c}{b}$, where $b \neq 0$.	$\frac{4}{13}+\frac{7}{13}=\frac{\square+\square}{\square}=\frac{\square}{\square}$.
Subtraction: $\frac{a}{b}-\frac{c}{b}=\frac{a-c}{b}$, where $b \neq 0$.	$\frac{15}{23}-\frac{7}{23}=\frac{\square-\square}{\square}=\frac{\square}{\square}$.

Name: Date:

The Fundamental Principle of Rational Expressions

If $\frac{P}{Q}$ is a rational expression and P, Q, and K are polynomials where $Q \neq 0$ and $K \neq 0$,

then $\frac{P}{Q} = \frac{P \cdot K}{Q \cdot K}$.

Lesson Link

Simplifying rational expressions requires the ability to factor expressions. Be sure to review the factoring process, which was introduced in Chapter 12.

The fundamental principle of rational expressions can be used to simplify rational expressions or build rational expression to a higher term. The next problem will walk you through the process of simplifying rational expressions.

4. Use the fundamental principle of rational expressions to simplify $\frac{2x-10}{3x-15}$.

a. Factor the numerator of the rational expression.

b. Factor the denominator of the rational expression.

c. Rewrite the rational expression by using the factored numerator and denominator.

d. Use the fundamental principle of rational expressions to simplify the factored rational expression. Remember that $\frac{K}{K} = 1$.

Read the following information about simplifying rational expressions then work through the problems.

Quick Tip

The sum of opposite polynomials is 0. This is similar to adding opposites or additive inverses.

The rational expression $\frac{2-x}{x-2}$ appears as though it cannot be simplified. However, the numerator and denominator are **opposite polynomials**. With a little work, this rational expression can be simplified.

Opposites in Rational Expressions

For a polynomial P, $\frac{-P}{P} = -1$, where $P \neq 0$.

In general, $\frac{a-x}{x-a} = \frac{-(x-a)}{x-a} = -1$, where $x \neq a$.

For example, $\frac{1-x}{x-1} = \frac{-1(x-1)}{x-1} = -1$, where $x \neq 1$.

The step $\frac{a-x}{x-a} = \frac{-(x-a)}{x-a}$ can be thought of as either multiplying $x - a$ by -1 or factoring -1 out of $a - x$. This process is similar to finding the additive inverse of a number.

Determine if any mistakes were made while simplifying the rational expression. If any mistakes were made, describe the mistake and then correctly simplify the rational expression to find the actual result.

5.
$$\begin{aligned}\frac{2x-8}{16-4x} &= \frac{2(x-4)}{4(4-x)} \\ &= \frac{2(x-4)}{-4(x-4)} \\ &= -\frac{1}{2}\end{aligned}$$

6.
$$\begin{aligned}\frac{3x-15}{-20-4x} &= \frac{3(x-5)}{4(-5-x)} \\ &= \frac{3(x-5)}{-4(x-5)} \\ &= -\frac{3}{4}\end{aligned}$$

Multiplication with Rational Expressions

To multiply two rational expressions, multiply the numerators together and multiply the denominators together. In mathematical notation, if P, Q, R, and S are polynomials and $Q \neq 0$ and $S \neq 0$, then

$$\frac{P}{Q} \cdot \frac{R}{S} = \frac{P \cdot R}{Q \cdot S}.$$

To multiply rational expressions,

1. Determine any restrictions on the variable.
2. Completely factor each numerator and denominator.
3. Multiply the numerators and multiply the denominators, keeping the expressions in factored form.
4. Use the fundamental principle of rational expressions to divide out any common factors from the numerator and denominator.

7. Simplify $\frac{x^2-9}{x^2+2x} \cdot \frac{x+2}{x-3}$.

a. Find any restrictions on the variable.

b. Factor each numerator and denominator and rewrite the product of the two rational expressions.

c. Multiply the numerators and multiply the denominators. Try to place common factors in the numerator above the common factors in the denominator. Do not distribute or simplify.

Quick Tip

The associative and commutative properties of multiplication allow factors to be rearranged in any order.

Name: Date:

d. Use the fundamental principle of rational expressions to simplify the rational expression from part **b.**

Division with Rational Expressions

To divide two rational expressions, multiply the first rational expression by the reciprocal of the second rational expression. In mathematical notation, if P, Q, R, and S are polynomials and $Q \neq 0$, $R \neq 0$ and $S \neq 0$, then

$$\frac{P}{Q} \div \frac{R}{S} = \frac{P}{Q} \cdot \frac{S}{R} = \frac{P \cdot S}{Q \cdot R}.$$

Notice that division by a rational expression is similar to division by a fraction: multiply the first rational expression by the reciprocal of the second rational expression.

Quick Tip

When dividing rational expressions, always change the operation to multiplication before performing any simplification.

8. Simplify $\frac{7x-14}{x^2} \div \frac{x^2-4}{x^3}$.

a. Rewrite the division problem as a multiplication problem.

b. Factor each numerator and denominator of the expression from part **a.** and rewrite the product of the two rational expressions.

c. Multiply the numerators and multiply the denominators. Try to place common factors in the numerator above the common factors in the denominator. Do not distribute or simplify.

d. Simplify the rational expression from part **c.**

Skill Check

Quick Tip

Restrictions on variables should be found before any simplification takes place.

Go to Software Work through Practice in Lesson 13.1 of the software before attempting the following exercises.

State any restrictions on the variable(s) and reduce to lowest terms.

9. $\frac{14x^3}{42x^5}$

10. $\frac{9x^2y^3}{12xy^4}$

11. $\frac{1+3y}{4x+12xy}$

12. $\frac{x^2-y^2}{3x^2+3xy}$

State any restrictions on the variable(s). Simplify the expression and reduce to lowest terms.

13. $\frac{24x^3}{25y^2} \cdot \frac{10y^5}{18x}$

14. $\frac{x^2+6x-16}{x^2-64} \cdot \frac{1}{2-x}$

15. $\frac{6x^2-54}{x^4} \div \frac{x-3}{x^2}$

16. $\frac{2x+1}{4x-x^2} \div \frac{4x^2-1}{x^2-16}$

Apply Skills

Quick Tip

Compound interest was introduced in Section 4.8. Compound interest earns interest on both the principal and the interest earned.

Work through the problems in this section to apply the skills you have learned related to applications of quadratic equations.

17. An annuity is a type of savings account that you put money into after equal periods of time to reach a goal amount. Annuities are a type of investment that is generally used to meet long-term savings goals such as college funds or retirement funds. The future value of an annuity is determined by the equation

$$FV = P\left[\frac{(1+r)^n - 1}{r}\right],$$

where FV is the future value of the annuity, P is the size of the periodic payment, r is the interest rate, and n is the number of payments or times the interest is compounded.

a. Determine the future value of an annuity if the monthly payment is \$100, the interest rate is 3%, and the payments are made for 60 months. Round your answer to the nearest cent.

b. Determine the future value of an annuity if the monthly payment is \$200, the interest rate is 3%, and the payments are made for 60 months. Round your answer to the nearest cent.

c. What was the total amount of money paid into the annuity from part **a.**?

d. What was the total amount of money paid into the annuity from part **b.**?

e. The regular payment in part **b.** is double the regular payment in part **a.** Is the future value from part **b.** double the future value from part **a.**? Why do you think this is?

Name: Date:

13.2 Addition and Subtraction with Rational Expressions

Objectives

Add rational expressions.

Subtract rational expressions.

Success Strategy

Remember that rational expressions are like fractions and you must have the same denominator to add or subtract them. Review how to find the LCD of two fractions in Section 2.4 before starting this section

Understand Concepts

Go to Software First, read through Learn in Lesson 13.2 of the software. Then, work through the problems in this section to expand your understanding of the concepts related to addition and subtraction with rational expressions.

The operations of addition and subtraction with rational expressions follow similar rules as those used for addition and subtraction with fractions. To add or subtract, the denominators need to be equal. If the denominators are not the same, you will need to find the least common denominator (LCD).

Adding and Subtracting Rational Expressions with the Same Denominator

When **adding (or subtracting) rational expressions** with the same denominator, add (or subtract) the numerators together and place that sum (or difference) over the common denominator. In mathematical notation, for polynomials P, Q, and R, with $Q \neq 0$,

$$\frac{P}{Q}+\frac{R}{Q}=\frac{P+R}{Q} \quad \text{or} \quad \frac{P}{Q}-\frac{R}{Q}=\frac{P-R}{Q}.$$

1. Simplify $\dfrac{x}{x^2-1}+\dfrac{1}{x^2-1}$.

a. Find any restrictions on the variable of the rational expression.

b. Add the rational expressions.

c. Factor the numerator and denominator of the rational expression.

Quick Tip

Dividing out common factors from an expression is the same process as **reducing** the expression.

d. Divide out the common factors to simplify the rational expression from part **c.**

e. Find any restrictions on the variable of the rational expression from part **d.** How does this compare to the restrictions from part **a.**?

2. Simplify $\frac{x}{x-5} - \frac{3}{5-x}$.

Lesson Link

Opposites in rational expressions were introduced in Section 13.1.

a. The denominators are opposites. Rewrite the second fraction so that its denominator matches the first fraction's denominator. Be sure to keep track of the sign of the second fraction.

b. Simplify the expression from part **a.**

c. Can the simplified expression from part **b.** be reduced? If yes, reduce it. If not, explain why.

Lesson Link

LCM stands for least common multiple. Methods for determining the LCM were introduced in Section 2.3.

Finding the LCM of a Set of Polynomials

1. Completely factor each polynomial (including prime factors for numerical factors).
2. Form the product of all factors that appear, using each factor the most number of times it appears in any one polynomial.

3. Find the LCM of the polynomials $3x^2 - 12$ and $x^2 + x - 6$.

a. Factor each polynomial completely.

$3x^2 - 12 =$

$x^2 + x - 6 =$

b. Which factors are common to both polynomials?

c. Determine the most number of times each factor from part **b.** appears in each polynomial.

d. Determine the LCM by forming a product of the common factors from part **b.** along with their associated powers from part **c.** and multiplying by the remaining factors that were unique in each polynomial. Do not distribute.

Name: Date:

Adding or Subtracting Rational Expressions with Different Denominators

1. Find the LCD, which is the LCM of all denominators in the rational expressions.
2. Rewrite each rational expression as an equivalent expression with the LCD as the denominator.
3. Add or subtract the numerators and keep the common denominator the same.
4. Reduce if possible.

Lesson Link

Adding and subtracting rational expressions with different denominators follows the same steps as adding and subtracting fractions, which was introduced in Section 2.4.

4. Simplify $\frac{y}{y-3}+\frac{6}{y+4}$.

a. Find the LCD of the rational expressions.

Lesson Link

The method for rewriting a fraction as an equivalent fraction was introduced in Section 2.1. This same method can be used to rewrite rational expressions.

b. Rewrite each fraction as an equivalent fraction with the LCD as the denominator. Simplify the numerator but keep the denominator factored.

c. Add the fractions from part **b.** Simplify the numerator by combining any like terms.

d. Factor the numerator in the expression from part **c.**, if possible.

e. Reduce the rational expression from part **d.**, if possible.

Determine if any mistakes were made while simplifying the rational expressions. If any mistakes were made, describe the mistake and then correctly simplify the rational expression to find the actual result.

5. $$\frac{6x}{x-6}+\frac{36}{6-x}=\frac{6x}{x-6}+\frac{-36}{(x-6)}$$
$$=\frac{6x+36}{x-6}$$
$$=\frac{6(x+6)}{x-6}$$

6. $$\frac{3x-4}{x^2-x-20}-\frac{2}{5-x}=\frac{3x-4}{(x-5)(x+4)}-\frac{2}{(5-x)}$$
$$=\frac{3x-4}{(x-5)(x+4)}+\frac{2}{x-5}$$
$$=\frac{3x-4}{(x-5)(x+4)}+\frac{2}{x-5}\cdot\frac{x+4}{x+4}$$
$$=\frac{3x-4}{(x-5)(x+4)}+\frac{2x+8}{(x-5)(x+4)}$$
$$=\frac{5x+4}{(x-5)(x+4)}$$

7. $$\frac{x^2+2}{x^2-4}-\frac{4x-2}{x^2-4}=\frac{x^2-4x}{x^2-4}$$
$$=\frac{x(x-4)}{(x-2)(x+2)}$$

Skill Check

Go to Software Work through Practice in Lesson 13.2 of the software before attempting the following exercises.

Simplify. Reduce if possible.

8. $\frac{4}{x}+\frac{3}{x}$

9. $\frac{3x}{x+4}+\frac{12}{x+4}$

10. $\frac{x+3}{7x-2}+\frac{2x-1}{14x-4}$

11. $\frac{2}{x}+\frac{1}{3}$

Name: Date:

12. $\dfrac{x}{x-1}-\dfrac{4}{x+2}$

13. $\dfrac{4}{x+5}-\dfrac{2x+3}{x^2+4x-5}$

14. $\dfrac{4x-1}{x^2-5x+4}+\dfrac{2x+7}{x^2-11x+28}$

15. $\dfrac{3x}{4-x}+\dfrac{7x}{x+4}-\dfrac{x-3}{x^2-16}$

Apply Skills

Work through the problems in this section to apply the skills you have learned related to addition and subtraction with rational expressions.

16. During Expedition 34 to the International Space Station, three crew members are tasked with unloading supplies from the SpaceX Dragon spacecraft. In one hour, Chris Hadfield can unload $\dfrac{1}{x}$ of the supplies, Thomas Marshburn can unload $\dfrac{1}{x+4}$ of the supplies, and Oleg Novitskiy can unload $\dfrac{1}{x-2}$ of the supplies. If they work together, what portion of the supplies will they unload in one hour?

Lesson Link

Work problems use the fraction of a job that can be completed in one hour by each person to determine how long it will take a group of people to complete the entire job. The method for setting up and solving these problems will be introduced in Section 13.5.

a. Find the sum of the fractions of the supplies each crew member can unload in one hour.

b. If $x = 10$, what fraction of the supplies will be unloaded after one hour?

c. When $x = 10$, will more than half of the supplies be unloaded in one hour? Explain your answer.

17. Barbara's Bombtastic Bakery was a cupcake shop when it first opened up. The bakery space that Barbara rented came with most of the equipment that was needed, such as a commercial oven and a display case. This meant that she only had to buy a mixer for \$5000, cupcake pans for \$250, and various utensils for \$500. Barbara estimated that the ingredients for each cupcake would cost \$0.35.

a. What was the total amount that Barbara spent on equipment to bake the cupcakes?

b. The total cost to bake the cupcakes is equal to the sum of the total amount spent on equipment plus the total amount spent on ingredients. Create a function $C(x)$ to describe the total cost to bake the cupcakes and use the variable x to represent the number of cupcakes baked.

c. The average cost to bake each cupcake is calculated by dividing the total cost by the number of cupcakes baked. Create a formula $A(x)$ to describe the average cost to create each cupcake.

d. After Barbara's Bombtastic Bakery's first week of business, Barbara baked a total of 950 cupcakes. What was the average cost to bake each cupcake after the first week? Round your answer to the nearest cent.

e. After one month of business, Barbara baked a total of 3500 cupcakes. What was the average cost to bake each cupcake after the first month? Round your answer to the nearest cent.

Name: Date:

13.3 Complex Fractions

Objectives

Simplify complex fractions.

Simplify complex algebraic expressions.

Success Strategy

Working with complex fractions requires diligence, patience, and organization of your work. Do extra practice in the software to ensure success.

Understand Concepts

Go to Software First, read through Learn in Lesson 13.3 of the software. Then, work through the problems in this section to expand your understanding of the concepts related to complex fractions.

A **complex fraction** is a fraction in which the numerator, the denominator, or both contain fractions. When complex fractions contain variables, they are called **complex rational expressions**. This section will introduce two methods for simplifying complex rational expressions. The first method involves simplifying the numerator and denominator before dividing. The second method involves multiplying the numerator and denominator of the rational expression by the LCD before reducing.

Lesson Link

Complex fractions were first introduced in Section 2.6 along with this method to simplify them.

Simplifying Complex Rational Expressions: Method 1

1. Simplify the numerator so that it is a single rational expression.
2. Simplify the denominator so that it is a single rational expression.
3. Divide the numerator by the denominator and reduce to lowest terms.

Quick Tip

Complex fractions are a form of complex rational expressions. The only difference is that a complex fraction contains no variables.

1. Simplify the complex fraction $\dfrac{\frac{1}{2}+\frac{1}{3}}{\frac{1}{2}-\frac{1}{3}}$.

a. Simplify the numerator so that it is a single fraction.

b. Simplify the denominator so that it is a single fraction.

c. Divide the numerator by the denominator and reduce if possible. Remember that division by a fraction is the same as multiplying by the reciprocal of the fraction.

2. Let's simplify the complex rational expression $\dfrac{\frac{1}{3}+\frac{1}{x}}{\frac{1}{2}-\frac{1}{x}}$.

a. Simplify the numerator so that it is a single rational expression.

b. Simplify the denominator so that it is a single rational expression.

c. Divide the numerator by the denominator and reduce if possible. Remember that division by a fraction is the same as multiplying by the reciprocal of the fraction.

Simplifying Complex Rational Expressions: Method 2

1. Find the LCD of all the denominators in the numerator and the denominator of the complex fraction.
2. Multiply both the numerator and denominator of the complex rational expression by the LCD from Step 1.
3. Simplify both the numerator and denominator and reduce the complex rational expression to lowest terms.

3. Simplify $\dfrac{\frac{4}{3x}-\frac{5}{y}}{\frac{1}{3}+\frac{3}{y}}$.

a. List the four denominators in the complex rational expression.

b. Find the LCD of the four denominators from part **a.**

Quick Tip

Remember to write the LCD as a fraction over 1 to ensure it is multiplied correctly.

c. Multiply the numerator and the denominator of the complex fraction by the LCD found in part **b.** Simplify.

d. Can the simplified complex fraction from part **d.** be reduced? If yes, reduce it. If not, explain why.

4. Which method of simplifying complex fractions do you prefer? Write an explanation for your choice.

Lesson Link

The order of operations was given in Sections 1.6, 2.6, 3.4, and 7.5.

Read the following information about complex algebraic expressions then work through the problems.

A **complex algebraic expression** is an expression that involves rational expressions and more than one operation. To simplify these expressions, follow the order of operations and simplify any resulting complex fractions by using the methods presented in this section.

Name: Date:

Determine if any mistakes were made while simplifying complex rational expressions. If any mistakes were made, describe the mistake and then correctly simplify the complex rational expression to find the actual result.

5. $$\frac{1}{x+1}-\frac{3}{2x}\cdot\frac{4x}{x+1}=\frac{1}{x+1}-\frac{12x}{2x(x+1)}$$
$$=\frac{2x}{2x(x+1)}-\frac{12x}{2x(x+1)}$$
$$=\frac{-10x}{2x(x+1)}$$
$$=\frac{-5}{x+1}$$

6. $$\frac{x}{x-1}-\frac{3}{x-1}\cdot\frac{x+2}{x}=\frac{x-3}{x-1}\cdot\frac{x+2}{x}$$
$$=\frac{x^2-x-6}{x(x-1)}$$

Skill Check

Go to Software Work through Practice in Lesson 13.3 of the software before attempting the following exercises.

Simplify.

7. $$\frac{\frac{2x}{3y^2}}{\frac{5x^2}{6}}$$

8. $$\frac{\frac{x+3}{2x}}{\frac{2x-1}{4x^2}}$$

9. $$\frac{\frac{1}{x}+\frac{1}{3x}}{\frac{x+6}{x^2}}$$

10. $$\frac{1+\frac{4}{2x-3}}{1+\frac{x}{x+1}}$$

Apply Skills

Work through the problems in this section to apply the skills you have learned related to complex fractions.

11. To calculate the average rate of a two-part commute, where each part is the same distance, the following formula is used.

$$\frac{2d}{\frac{d}{r_1}+\frac{d}{r_2}}$$

In the formula, d is the commute distance traveled one way, r_1 is the rate, or speed, during the first part of the trip, and r_2 is the rate during the second part of the trip.

a. Simplify the expression.

b. Calculate the average rate of the trip if you can travel 35 miles per hour during the first part of the trip and 60 miles per hour during the second part of the trip. Round your answer to the nearest tenth.

c. Use the answer from part **b.** and the formula $d = rt$ to calculate how long the commute took if the total distance of the trip was 80 miles. Round your answer to the nearest tenth.

Quick Tip

In real-life problems, a variable can be represented by more than one letter. For example, in this problem the unknown value is represented by the three letters APY.

12. The average percent yield (APY) of an annuity is the annual interest rate earned in a given year that accounts for the effects of compounding. The APY acts as the interest rate for a simple interest account and is larger than the stated interest rate on the compound interest account. The formula to calculate the APY on an annuity after 2 years is

$$\text{APY} = \left(1 + \frac{r}{2}\right)^2 - 1,$$

where r is the stated interest rate.

a. Simplify the expression for APY and write as a single rational expression.

b. Using the original formula, calculate the APY for an annuity whose interest rate is 6%. Do not round.

c. Using the expression in part **a.**, calculate the APY for an annuity whose interest rate is 6%. Do not round.

d. Does the result from part **c.** match the result from part **b.**? Explain why or why not.

e. How much larger is the APY than the interest rate?

f. Why do you think the APY is larger than the interest rate? Write a complete sentence.

Name: Date:

13.4 Solving Equations with Rational Expressions

Objectives

Solve equations that involve rational expressions.

Use the process of solving equations to manipulate formulas.

Use proportions to relate similar triangles.

Success Strategy

Solving equations with rational expressions uses the same principles of equality that were introduced in Section 8.1. Review solving equations involving fractions from Sections 8.1, 8.2, and 8.3 before starting this section.

Understand Concepts

Go to Software First, read through Learn in Lesson 13.4 of the software. Then, work through the problems in this section to expand your understanding of the concepts related to solving equations with rational expressions.

When solving equations with rational expressions, it is important to remember that they are not the same as proportions and extra care must be taken. Multiplying an equation by a variable can introduce false solutions, which are called **extraneous solutions** or **extraneous roots**. Because of this, it is important to **always** check your solutions.

Solving Equations with Rational Expressions

1. Find any restrictions on the variable.
2. Find the LCD of all denominators in the equation.
3. Multiply all terms on both sides of the equation by the LCD.
4. Solve the resulting polynomial equation for the variable.
5. If a solution is equal to a restriction, discard it. Check the remaining solutions in the original equation.

1. Solve $\frac{x-5}{2x}=\frac{6}{3x}$ for x.

a. Find any restrictions on the variable of the rational equation.

b. Find the LCD of all fractions in the equation. Write the LCD in factored form.

c. Multiply both sides of the equation by the LCD from part **b.** and simplify.

d. Solve the equation from part **c.** for the variable.

e. Are any of the solutions from part **d.** equal to any of the restrictions from part **a.**?

f. If any of the solutions from part **d.** are not equal to the restrictions from part **a.**, check them in the original equation and list the solutions. Otherwise, write "no solution".

2. Solve $\frac{2}{x^2-9}=\frac{1}{x^2}+\frac{1}{x^2-3x}$ for x.

a. Find any restrictions on the variable of the rational equation.

b. Find the LCD of all fractions in the equation. Write the LCD in factored form.

c. Multiply both sides of the equation by the LCD from part **b.** and simplify.

d. Solve the equation from part **c.** for the variable.

e. Are any of the solutions from part **d.** equal to any of the restrictions from part **a.**?

f. If any of the solutions from part **d.** are not equal to the restrictions from part **a.**, check them in the original equation and list the solutions. Otherwise, write "no solution".

3. Why is it important to check the solutions of a rational expression equation against the restrictions on the variable?

In Section 8.5, you learned how to solve formulas for different variables. Sometimes the resulting formula is in the form of a rational expression, as in Problem 4.

4. The formula $S=2\pi r^2+2\pi rh$ is used to find the surface area of a circular cylinder. If the surface area and radius of the circular cylinder are known, then the formula can be solved for h to determine the height. Solve the formula for h.

Name: Date:

Similar triangles were introduced in Section 5.6. If the side lengths of the triangles are defined with variable expressions, then the equations to solve for proportional sides are rational expressions.

5. In the figure, $\triangle ABC \sim \triangle PQR$. Use the figure to work through the following problems.

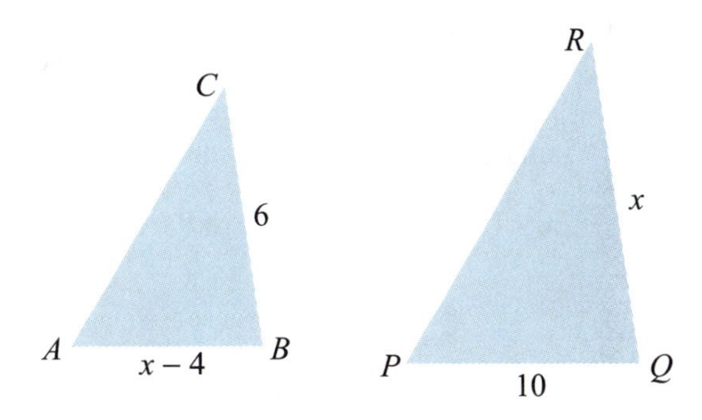

a. Set up an equation with rational expressions involving the corresponding sides. This will be similar to the proportion $\frac{AB}{PQ} = \frac{BC}{QR}$.

b. Determine any restrictions on the variable of the equation from part **a.**

c. Solve the equation from part **a.**

d. Check the solutions against the restrictions from part **b.** and also remove any solutions that do not fit the context of the problem.

e. Determine the lengths of sides AB and QR.

Skill Check

Go to Software Work through Practice in Lesson 13.4 of the software before attempting the following exercises.

Determine any restrictions on the variable and then solve the equation.

6. $\frac{4x}{7} = \frac{x+5}{3}$

7. $\frac{6}{x-2} = \frac{3}{x+5}$

8. $\frac{5x}{4} - \frac{1}{2} = -\frac{3}{16}$

9. $\frac{x}{x+3} + \frac{1}{x+2} = 1$

Apply Skills

Work through the problems in this section to apply the skills you have learned related to solving equations with rational expressions.

Lesson Link

Setting up and solving distance-rate-time problems using complex rational expressions will be discussed in Section 13.5.

10. Terrence and Alicia are competing in a marathon where the average running speed is x kilometers per hour. Terrence is running 2 kilometers per hour slower than the average running speed. Alicia is running 2 kilometers per hour faster than the average running speed. After a certain amount of time, Terrence ran 4 kilometers and Alicia ran 6 kilometers.

a. Determine the speed of the average runner by solving the equation $\frac{4}{x-2}=\frac{6}{x+2}$ for x.

b. What was Terrence's average running speed?

c. What was Alicia's average running speed?

d. How long did it take Terrence to run 4 kilometers and Alicia to run 6 kilometers?

Quick Tip

An **isosceles triangle** is a triangle that has at least two equal sides. If all three sides are equal, the triangle is also called an **equilateral triangle**.

11. A team of gardeners is making two flower beds that are in the shape of similar triangles outside of an art museum. The apprentice gardener wasn't completely paying attention to the instructions given by the master gardener. All that he can remember is that the flower beds are isosceles triangles, the base of the small triangle is 3 feet wide, one side of the larger triangle is 16 feet long, and the base of the large triangle is three times the side length of the small triangle. The apprentice gardener needs to determine the unknown dimensions of the triangles.

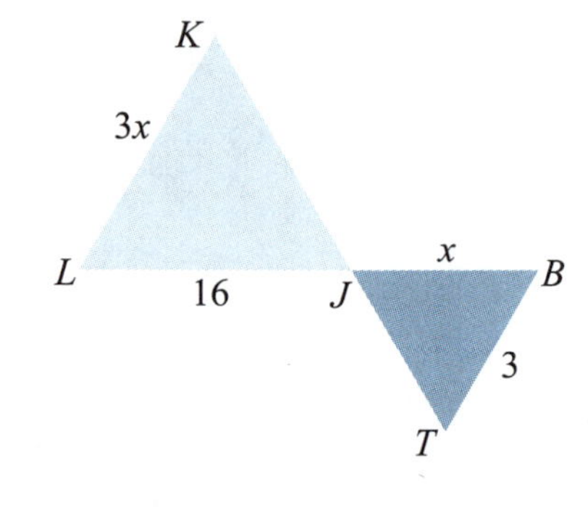

a. Use the figure to write an equation to show that the side lengths are proportional.

b. Solve the equation from part **a.** for x.

c. Do any of the solutions from part **b.** not make sense in the context of the problem? If yes, explain why.

d. What are the lengths of the unknown sides of the triangles?

Name: Date:

13.5 Applications of Rational Expressions

Objectives

Solve applied problems related to fractions.

Solve applications involving distance, rate, and time.

Solve applied problems related to work.

Success Strategy

While working through application problems similar to the ones in this section, be sure to organize your work. Try using a table to help you set up the problem and determine the equation that needs to be solved.

Understand Concepts

Go to Software First, read through Learn in Lesson 13.5 of the software. Then, work through the problems in this section to expand your understanding of the concepts related to applications of rational expressions.

Number problems were introduced in Section 8.4. This section expands the methods of solving number problems to include problems involving rational expressions. The next problem will guide you through solving one of these problems.

Quick Tip

When solving word problems, remember to follow one of the problem-solving strategies introduced in Section 12.7, or use Pólya's problem-solving strategy at the beginning of the workbook.

1. The denominator of a fraction is six more than the numerator. If both the numerator and denominator are increased by five, the new fraction is equal to one half. Find the original fraction.

a. Let the variable n represent the original numerator. Write an algebraic expression to describe the original denominator.

b. Use the information from part **a.** to write a rational expression to represent the original fraction.

c. Write a rational equation to show that when the numerator and denominator of the original fraction are both increased by five, the fraction is equal to one half.

d. Find any restrictions on the variable of the rational equation from part **c.**

e. Solve the rational equation from part **c.** for the variable.

f. Is the solution from part **e.** equal to any of the restrictions from part **d.**? If not, verify the solution in the original equation. If yes, write "no solution".

g. What was the original fraction?

A method for solving distance-rate-time problems using a table was introduced in Section 8.6. This section expands these problems to include equations containing rational expressions. The next problem will guide you through solving one of these problems.

2. In still water, a man can row his boat 5 miles per hour. On a river, it takes him the same time to row 5 miles downstream (with the current) as it does to row 3 miles upstream (against the current). What is the rate of the river's current in miles per hour?

a. Let the variable c represent the rate of the current. Fill in the distance and rate columns of the table.

	Distance (d)	÷	Rate (r)	=	Time $\left(t = \frac{d}{r}\right)$
Downstream					
Upstream					

b. Use the information from the distance and the rate columns to fill in the time column.

c. The time it takes for the man to row upstream is equal to the time it takes for him to row downstream. Since the times are equal, we can set the two rational expressions equal to each other. Write an equation with rational expressions using the information in the time column.

d. Find any restrictions on the variable of the rational equation from part **c.**

e. Solve the equation from part **c.** for the variable.

f. Is the solution from part **e.** equal to one of the restrictions from part **d.**? If not, verify the solution in the original equation. If yes, write "no solution".

g. Use the solution from part **f.** to answer the question in the problem statement.

Read the following paragraph about work problems and work through Problem 3.

Problems involving work can be solved using a table similar to the one used to solve distance-rate-time problems. The basic idea with work problems is to represent the amount of work that is done in one unit of time, typically an hour. The next problem will introduce you to work problems and guide you through the solving process.

3. A carpenter can build a certain type of patio cover in 6 hours. His partner takes 8 hours to build the same cover. How long would it take build this type of patio cover if they worked together?

a. Let the variable x represent the amount of time in hours it will take the two carpenters to complete the patio cover when working together. Fill in the time of work column.

Name: Date:

Person(s)	Time of Work (in Hours)	Part of Work Done in 1 Hour
Carpenter		
Partner		
Together		

Quick Tip

The part of work done in 1 hour can also be thought of as the reciprocal of the time of work.

b. The part of work done in 1 hour is the ratio of 1 to the total time it takes to complete the patio cover. For instance, it takes the carpenter 6 hours to complete the job. This means that he will complete $\frac{1}{6}$ of the job in 1 hour. Fill in the last column of the table.

c. The part of work done in 1 hour by both people working together is equal to the sum of the part of work done in 1 hour when they work separately. Write an equation to describe this situation.

d. Find any restrictions on the variable of the rational equation from part **c.**

e. Solve the rational equation from part **c.** for the variable.

f. Is the solution from part **e.** equal to one of the restrictions? If not, verify the solution in the original equation. If yes, write "no solution".

g. Use the solution from part **e.** to answer the question in the problem statement.

Skill Check

 *Go to Software* Work through Practice in Lesson 13.5 of the software before attempting the following exercises.

Determine any restrictions on the variable and then solve the equation.

4. $\frac{x+9}{3x+2}=\frac{5}{8}$

5. $\frac{2}{3x}=\frac{1}{4}-\frac{1}{6x}$

6. $\frac{3x-2}{x+4}+\frac{2x+5}{x-1}=5$

Apply Skills

Work through the problems in this section to apply the skills you have learned related to applications of rational expressions.

7. A local print shop has a big order of pamphlets to print, so they decide to use two of their printers for the one job. The newest printer can print the pamphlets four times as fast as the old printer. Working together the printers can complete the job in 4 hours. How many hours would it take each printer to print all of the pamphlets by itself?

a. Use the table to set up a rational equation to describe the situation. Use the variable x to represent the time it takes the newest printer to complete the job.

Printer	Time of Work (in Hours)	Part of Work Done in 1 Hour
Newest		
Old		
Together		

b. Solve the equation from part **a.** for x.

c. Use the solution from part **b.** to answer the question in the problem statement.

Quick Tip

Remember that **speed** is another word for **rate**.

8. The Winston family is moving to another state. The family is driving to their new house in a car and all of their belongings are in a moving truck. The car is traveling at speed that is 9 miles per hour faster than the speed of the truck. After a certain amount of time, the family's car traveled 350 miles and the moving truck traveled 300 miles. What are the speeds of the car and the truck?

a. Use the table to set up a rational equation to describe the situation. Use the variable x to represent the rate of the truck.

	Distance (d)	÷	Rate (r)	=	Time $\left(t = \frac{d}{r}\right)$
Car					
Truck					

b. Solve the equation from part **a.** for x.

c. Use the solution from part **b.** to answer the question in the problem statement.

d. If the Winston family's new home is 378 miles away, how long will it take the car and the truck to make the trip?

Name: Date:

13.6 Variation

Objectives

Solve problems related to direct variation.

Solve problems related to inverse variation.

Solve problems involving combined variation.

Success Strategy

Variation problems use equations that involve two or more variables which are related in a certain way. Extra practice in the software will help you learn the steps to solve these problems.

Understand Concepts

Go to Software First, read through Learn in Lesson 13.6 of the software. Then, work through the problems in this section to expand your understanding of the concepts related to variation.

We have worked with variation throughout this workbook, we just haven't used the term variation yet. Variation comes in two forms: direct variation and inverse variation. When the two types of variation occur at the same time with multiple variables, it is called combined variation.

Quick Tip

The distance formula $d = rt$ is a form of direct variation when r is a constant value.

Direct Variation

A variable quantity y **varies directly as** (or **directly proportional to**) a variable x if there is a constant k such that

$$y = kx.$$

The constant k is called the **constant of variation**.

The process of solving variation problems when the constant of variation is unknown tends to follow the method outlined in the following box.

Solving Variation Problems

1. Determine the general formula that describes the relationship between the variables.
2. Use the information given about the initial scenario to solve for the constant k.
3. Replace k in the general formula from Step 1 with the value found in Step 2.
4. Answer the question by substituting the values for the new scenario into the formula from Step 3.

1. The value of y varies directly as x. When $x = 2$, $y = 6$. Find the value of y if $x = 6$.

a. Substitute the pair of known values of x and y into the direct variation formula.

b. Solve the equation from part **a.** for the constant of variation k.

c. Substitute the constant of variation from part **b.** into the direct variation equation.

d. Substitute $x = 6$ into the equation from part **c.** and solve for y.

Inverse Variation

A variable quantity y **varies inversely as** (or **inversely proportional to**) a variable x if there is a constant k such that

$$y = \frac{k}{x}.$$

The constant k is called the **constant of variation.**

The variation equation isn't always a linear equation. It can involve powers or roots. The next problem illustrates such a relationship.

2. The value of y varies inversely as the cube of x. When $x = 3$, $y = -1$. Find the value of y when $x = -3$.

a. Write the inverse variation equation. Remember to cube x.

b. Substitute the pair of known values of x and y into the equation from part **a.**

c. Solve the equation from part **b.** for the constant of variation k.

d. Substitute the constant of variation from part **c.** into the inverse variation equation from part **a.**

e. Substitute $x = -3$ into the equation from part **d.** and solve for y.

Combined and Join Variation

If a variable varies either directly or indirectly with more than one other variable, then it is called **combined variation**.

If the combined variation is all direct variation, then it is called **joint variation**.

While both combined and joint variation use three or more variables, they only have one constant of variation k.

Name: Date:

True or False: Determine whether each statement is true or false. Rewrite any false statement so that it is true. (There may be more than one correct new statement.)

3. The equation $z = kx^2y$ represents joint variation.

4. The equation $z = \frac{7x}{y^3}$ represents only inverse variation.

5. The equation $y = 2.25x$ represents joint variation.

Skill Check

Go to Software Work through Practice in Lesson 13.6 of the software before attempting the following exercises.

Use the information given to find the unknown value.

6. If y varies directly as x, and $y = 3$ when $x = 9$, what is y if $x = 7$?

7. If y is inversely proportional to the cube of x, and $y = 40$ when $x = \frac{1}{2}$, what is y when $x = \frac{1}{3}$?

8. The variable z varies directly as the cube of x and inversely as the square of y. If $z = 24$ when $x = 2$ and $y = 2$, find z if $x = 3$ and $y = 2$.

Apply Skills

Work through the problems in this section to apply the skills you have learned related to variation.

9. The distance that an object falls is directly proportional to the square of the time that has passed since the object started to fall. A rock falls a distance of 64 feet in 2 seconds. How long will it take the rock to fall a distance of 100 feet?

Quick Tip

Notice that both a lower case *l* and a capital *L* are used as variables in this equation. Keep in mind that they stand for different measurements.

10. For a certain type of wooden beam that carries a load at its center, the safe load *SL* varies jointly as the width w and the cube of the depth d and inversely as the square of the length l. A wooden beam that is 4 inches wide, 6 inches deep, and 12 feet long can safely support a load of 2400 pounds.

a. Set up the variation equation.

b. Determine the constant of variation.

c. How much weight can a wooden beam that is 5 inches wide, 6 inches deep, and 10 feet long safely support?

Name: Date:

Chapter 13 Projects

Project A: Let's Be Rational Here!

An activity to demonstrate the use of rational expressions in real life

You may be surprised to learn how many different situations in real life involve working with rational expressions and rational equations. Hopefully, after spending the day with Meghan and her friends you'll be convinced of their importance.

It's Saturday and Meghan has a list of things to accomplish today: revise her budget, paint the walls in the spare bedroom, travel to the lake with her friends, and then ski on the lake for the rest of the day, provided the weather stays nice.

1. Meghan decides to tackle the budget first. After reviewing her budget, she decides that she really needs to get a part-time job to earn some extra money. She is remodeling the living room and would like to buy some new furniture. Letting x represent her new monthly combined salary, she estimates that $\frac{1}{4}$ of her new salary will be used for bills and approximately $\frac{1}{5}$ for her car payment. She would like to have \$1100 left over each month of which \$100 will be saved for the new furniture. What must her new monthly salary be?

a. Let the variable x represent Meghan's new monthly salary. Write an expression to represent the amount of her new salary used for bills. (Remember that the word **of** implies multiplication.)

b. Write an expression to represent the amount of Meghan's new monthly salary used for her car payment.

c. Write an equation that sums the expenditures and leftover balance and set this sum equal to the new monthly salary of x.

d. Find the LCD of the rational expressions from part **c.** and multiply it times each term in the equation to remove the fractions.

e. Solve the equation from part **d.** to determine what Meghan's new monthly salary needs to be.

2. With the budget done, Meghan prepares to paint the spare bedroom. She just finished painting her bedroom last week and it took her about 4 hours. Her roommate Ashley painted her bedroom a couple of weeks ago and it took her 6 hours. All the bedrooms are similar in size. Meghan realizes that if she gets Ashley to help her with the painting, it will take them less time and they can get to the lake sooner. How long will it take Meghan and Ashley working together to paint the spare bedroom? Use the table below to help you set up the problem.

Person	Time (in hours)	Part of Work Done in 1 hour
Meghan	4	
Ashley	6	
Together	x	$\frac{1}{x}$

a. Fill in the missing information in column three of the table.

b. Use the entries in the last column of the table to set up an equation to represent the sum of the amount of work done by both Meghan and Ashley in an hour.

c. Solve the equation to determine how long it will take the two girls to paint the spare bedroom when working together. Express the result as a decimal to the nearest tenth. Convert this measurement to hours and minutes.

3. With the spare bedroom painted, the girls call Lucas to let him know they are ready to head to the lake. While he is preparing the boat for the lake, Lucas is trying to decide which route they should travel to get there. If he travels the highway, he can travel 20 mph faster than the scenic route. However, the highway is 30 miles longer than the scenic route, which is 60 miles long. Lucas thinks it should take him the same amount of time to get there using either route. Use the following table to help you organize the information for this problem.

	Distance (miles)	÷	Rate (mph)	=	Time (hours)
Highway			$x+20$		
Scenic Route	60		x		

a. Fill in the missing information in the table. (**Hint:** Recall that the formula that involves distance, rate, and time is $d = r \cdot t$.)

b. Since Lucas expects the time to be the same for each route, create an equation from the time column and solve for x.

c. How fast must Lucas travel on the highway to get to the lake in the same amount of time as traveling the scenic route?

Name: Date:

Project B: Looking at Lifting Force in the Wright Way

An activity to demonstrate the use of variation models in real life

You may be thinking that there is a mistake in the spelling of the title of this activity, but the title is referring to some very famous brothers, Orville and Wilbur Wright, who dreamed of inventing a machine that could fly like a bird. Their creativity and ingenuity led to one of the greatest inventions of the 20th century—the airplane.

The force that keeps an airplane in the air is called **lift**. Although lift can be generated by any part of an airplane, it is mostly the result of the wing design. For more information on lift and how it is generated, go to the following website: http://wright.nasa.gov/airplane/lift1.html. Be sure to view the short movie of "Orville and Wilbur Wright" at the bottom of the web page, as they discuss lift and its importance for flight.

For the purposes of this activity we will use a simpler version of the formula for lift which only uses one coefficient k, called the **constant of variation** or **constant of proportionality**.

1. Write the mathematical model (or formula) for lifting force based on the following description: The lifting force L in pounds exerted on the wings of an airplane varies **jointly** as the surface area of the wings A in square feet and the square of the speed (or velocity) v of the plane in miles per hour. (Don't forget to use the constant k in your formula.)

2. Calculate the constant of variation for the lifting force formula given the following information: A plane with a wing area of 170 ft^2 and flying at a speed of 100 mph generates a lift force of 5560 pounds on its wings. Round to the nearest hundred-thousandth.

3. Calculate the lifting force for the Wright's first flight in 1903. The wing area was 510 ft^2 and the plane traveled at a velocity of 30 mph. Use the constant of variation that you determined from the previous question. Do not round your answer.

4. In order to fly, the lifting force of an airplane must be more than its weight. Do some research on the web to determine the weight of the Wright's airplane on the famous flight of December 17, 1903, in Kitty Hawk, North Carolina. Assume that the pilot weighed around 150 pounds. How does the lifting force of the plane calculated in the previous question compare to the weight of the plane including the pilot?

5. Solve the lifting force formula that you created for the wing area A.

6. If the wing area is doubled while the speed of the plane remains constant, what effect does this have on the lifting force?

7. Solve the lifting force formula for velocity v.

8. If the speed of the plane is doubled and the wing area remains constant, what effect does this have on the lifting force?

9. To increase the lifting force of an airplane, which variable has the greatest impact, wing area or velocity? Explain why.

Name: Date:

Math@Work

Physics

As an employee of a company that creates circuit boards, your job may vary from designing new circuit boards, setting up machines to mass produce the circuit boards, to testing the finished circuit boards as part of quality control. Depending on your position, you may work alone or as part of a team. Regardless of who you work with, you will need strong math skills to be able to create new circuit board designs and strong communication skills to describe the specifications for a new circuit board design, describe how to set up the production line, or explain why a part is faulty.

Suppose your job requires you to create new circuit boards for a variety of electronic equipment. The latest circuit board that you are designing is a small part of a complicated device. The circuit board you create has three resistors which run in parallel, as shown in the diagram.

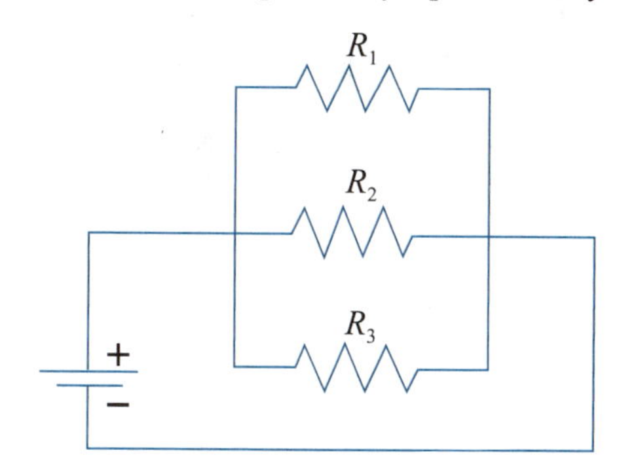

Two of the resistors were properly labeled with their correct resistance, which is measured in ohms. The first resistor has a rating of 2 ohms. The second resistor has a rating of 3 ohms. The third resistor was taken from the supply shelf for resistors of a certain rating, but the resistor was unlabeled. As a result, you are unsure if it has the correct resistance for the current you want to produce. You use an ohmmeter, a device that measures resistance in a circuit, to determine that the total resistance of the circuit you created is $\frac{30}{31}$ ohms.

You know that the equation to determine the total resistance R_t is $\frac{1}{R_t} = \frac{1}{R_1} + \frac{1}{R_2} + \frac{1}{R_3}$, where R_1 is the resistance of the first resistor, R_2 is the resistance of the second resistor, and R_3 is the resistance of the third resistor.

1. Use the formula to determine the resistance of the third resistor given that the total resistance of the circuit is $\frac{30}{31}$ ohms.

2. Was the third resistor on the correct shelf if you took it from the supply shelf that holds resistors with a rating of 7 ohms?

3. What would be the total resistance of the circuit if the third resistor had a rating of 7 ohms?

4. What do you think would happen if the resistance of the unlabeled resistor wasn't determined and the circuit board was sent to the production line to be mass produced?

Foundations Skill Check for Chapter 14

This page lists several skills covered previously in the book and software that are needed to learn new skills in Chapter 14. To make sure you are prepared to learn these new skills, take the self-test below and determine if any specific skills need to be reviewed.

Each skill includes an easy (**e.**), medium (**m.**), and hard (**h.**) version. You should be able to complete each problem type at each skill level. If you are unable to complete the problems at the easy or medium level, go back to the given lesson in the software and review until you feel confident in your ability. If you are unable to complete the hard problem for a skill, or are able to complete it but with minor errors, a review of the skill may not be necessary. You can wait until the skill is needed in the chapter to decide whether or not you should work through a quick review.

5.7 Simplify the radical expression.

e. $\sqrt{9}$ **m.** $\sqrt{169}$ **h.** $\sqrt{324}$

7.7 Simplify the expression by combining like terms.

e. $5n + 12n$ **m.** $5x^2 + 2x - 15x^2 + 9x$ **h.** $5a - 7b + 3(a + 2b)$

8.3 Solve the equation for the variable.

e. $4n - 5 = n + 4$ **m.** $2(b + 3) = 3b + 9$ **h.** $4(5 - x) = 8(3x + 10)$

11.1 Simplify the expression using the rules of exponents.

e. $\frac{x^4}{x^2}$ **m.** $\frac{-8z^5}{2z^3}$ **h.** $\frac{15x^5y^2}{-3x^2y^6}$

11.6 Find the product using the difference of two squares formula.

e. $(x+1)(x-1)$ **m.** $\left(x+\frac{3}{4}\right)\left(x-\frac{3}{4}\right)$ **h.** $(5x-8)(5x+8)$

Chapter 14: Radicals

Study Skills

14.1 Roots and Radicals

14.2 Simplifying Radicals

14.3 Addition, Subtraction, and Multiplication with Radicals

14.4 Rationalizing Denominators

14.5 Equations with Radicals

14.6 Rational Exponents

14.7 Functions with Radicals

Chapter 14 Projects

Math@Work

Foundations Skill Check for Chapter 15

Math@Work

Introduction

If you plan to study biology, you will have a variety of career options to choose from. In general, biologists study life and living organisms and apply this knowledge in their daily work. With an education in biology, you can be a researcher, work in environmental conservation, work in education, or work in health care. As in most fields of science, some career options require additional specialization, so be sure to look into specialization requirements before deciding on a particular career path. Regardless of the particular area of biology that you choose, math and strong communication skills will be an important part of your daily work.

Suppose you decide to combine your knowledge of biology with criminal law and go into the forensic sciences. As a forensic scientist, you will work as part of a team to determine what happened at a crime scene, even when there are no witnesses to provide any factual information. You will use your analysis of the known information to answer a variety of questions. Does any of the evidence found at the crime scene place any of the potential suspects at the scene of the crime? What was the actual cause of death? How long ago did the victim die? Finding the answers to these questions (and many more) requires several of the skills covered in this chapter. At the end of the chapter, we'll come back to this problem and explore how a forensic scientist uses math to solve a crime.

Study Skills

Online Resources

With the invention of the Internet, there are numerous resources available to students who need help with mathematics. Here are some quality online resources that we recommend.

HawkesTV

Web Link: http://tv.hawkeslearning.com/

If you are looking for instructional videos on a particular topic, then start with HawkesTV. There are hundreds of videos which can be found by looking under a particular math subject area such as Introductory Algebra, Precalculus, or Statistics. You can also find videos on Study Skills.

Khan Academy

Web Link: http://www.khanacademy.org/

Khan Academy is a nonprofit educational website that was created in 2006 by Salman Khan, a graduate of MIT and Harvard Business School. Khan Academy started as a small collection of math videos that Khan posted on YouTube to tutor his cousins who lived in a different state. There are now thousands of videos on the website, which means you may have to sort through a few videos to find the exact one(s) that you need.

YouTube

Web Link: http://www.youtube.com/

You can also find math instructional videos on YouTube, but you have to search for videos by topic or key words. You may have to use various combinations of key words to find the particular topic you are looking for. Keep in mind that the quality of the videos varies considerably depending on who produces them.

Google Hangouts

Web Link: https://plus.google.com/hangouts

You can organize a virtual study group of up to 10 people using Google Hangouts. This is a terrific tool when schedules are hectic and it avoids everyone having to travel to a central location. You do have to set up a Google+ profile to use Hangouts. In addition to video chat the group members can share documents using Google Docs. This is a great tool for group projects!

Wolfram|Alpha

Web Link: http://www.wolframalpha.com/

Wolfram Alpha is a computational knowledge engine developed by Wolfram Research that answers questions posed to it by computing the answer from "curated data." Typical search engines search all of the data on the Internet based on the key words given and then provide a list of documents or web pages that might contain relevant information. The data used by Wolfram|Alpha is said to be "curated" because someone has to verify its integrity before it can be added to the database, therefore ensuring that the data is of high quality. Users can submit questions and request calculations or graphs by typing their request into a text field. Wolfram|Alpha then computes the answers and related graphics from data gathered from both academic and commercial websites such as the CIA's World Factbook, the United States Geological Survey, financial data from Dow Jones, etc. Wolfram Alpha uses the basic features of Mathematica, which is a computational toolkit designed earlier by Wolfram Research that includes computer algebra, symbol and number computation, graphics, and statistical capabilities.

Name: Date:

14.1 Roots and Radicals

Objectives

Evaluate square roots.

Evaluate cube roots.

Use a calculator to evaluate square and cube roots. (Software only)

Success Strategy

It is important to understand the difference between roots and powers of a number. Pay close attention to the terminology in this section and be sure to include any new terminology in your notebook or your collection of index cards.

Understand Concepts

Quick Tip

An **exponent** is also referred to as a ***power***. For example, 5^4 is 5 raised to the fourth power.

Go to Software First, read through Learn in Lesson 14.1 of the software. Then, work through the problems in this section to expand your understanding of the concepts related to roots and radicals.

We have previously worked with powers in Sections 1.6 and 11.1 and roots in Section 5.7. The following table is a summary of the terminology related to powers and roots with examples.

Terminology for Powers and Roots		
Term	**Definition**	**Example**
Squared Number	A number used as a factor two times.	x^2, 4^2
Cubed Number	A number used as a factor three times.	y^3, 5^3
Square Root	If $b^2 = a$, then b is a square root of a. Written $b = \sqrt{a}$.	$\sqrt{25} = 5$
Cube Root	If $b^3 = a$, then b is a cube root of a. Written $b = \sqrt[3]{a}$.	$\sqrt[3]{27} = 3$

Parts of a Radical Expression

A **radical expression** has the form $\sqrt[n]{a}$, where n is called the **index** of the radical and a is the **radicand**. The $\sqrt{}$ symbol is called a **radical**.

When the index of a radical expression is equal to 2, it is often not written. For example, $\sqrt{5} = \sqrt[2]{5}$.

1. Identify the parts of the radical expression.

Radical Expression	Index	Radicand
$\sqrt{13}$		
$\sqrt[3]{-8}$		
$\sqrt[6]{64}$		
$\sqrt[2]{18}$		

Lesson Link

The square root of a negative number is called an **imaginary number**. Imaginary numbers are introduced in Section A.8.

The square root of a negative number is not a real number. The cube root of a negative number is a real number. While square roots have a **principal square root** and a **negative square root**, there is only one cube root of a number. The cube root of any number will have the same sign as the number itself.

Square Roots

If a is a nonnegative real number, then

$\sqrt{a}$ is the **principal square root** of a, and

$-\sqrt{a}$ is the **negative square root** of a.

True or False: Determine whether each statement is true or false. Rewrite any false statement so that it is true. (There may be more than one correct new statement.)

2. The principal square root of –25 is –5.

3. The cube root of –27 is –3.

Quick Tip

Only radicals that have an even index, such as 2, 4, and 6, simplify to have a principle and a negative root.

4. The square roots of 16 are –4 and 4.

Skill Check

Go to Software Work through Practice in Lesson 14.1 of the software before attempting the following exercises.

Simplify. If the result is not a real number, write "not a real number".

5. $\sqrt{36}$

6. $\sqrt{\frac{1}{4}}$

7. $\sqrt[3]{-27}$

8. $\sqrt{\frac{25}{81}}$

9. $\sqrt{-49}$

10. $\sqrt[3]{\frac{8}{125}}$

Name: ______________________________ Date: ______________

Quick Tip TECH

Use the square root function on the calculator to determine the approximate value of a square root that is not a perfect square. This function is often the 2nd function of the x^2 button. To find the value of $\sqrt{3}$, type 2ND x^2 3) ENTER. The display will read "1.732050808...".

Evaluate with a calculator. Round your answer to the nearest thousandth.

11. $\sqrt{39}$

12. $4\sqrt{5}$

13. $\sqrt{\frac{1}{5}}$

14. $-3\sqrt{6}$

Apply Skills

Work through the problems in this section to apply the skills you have learned related to roots and radicals.

15. A square flower garden covers an area of 68 square feet.

a. What is the approximate length of each side of the square? Round your answer to the nearest tenth.

b. Use the answer from part **a.** to determine the amount of edging material needed to create a border around the flower garden.

c. The edging material costs $1.39 per foot. How much will the amount of edging material from part **b.** cost? Round your answer to the nearest cent.

Lesson Link

The Pythagorean Theorem was introduced in Sections 5.7 and 12.7.

16. Barbara's Bombtastic Bakery is installing a corner display stand for custom decorated cakes. The top of the display stand is designed in the shape of a right triangle, as shown.

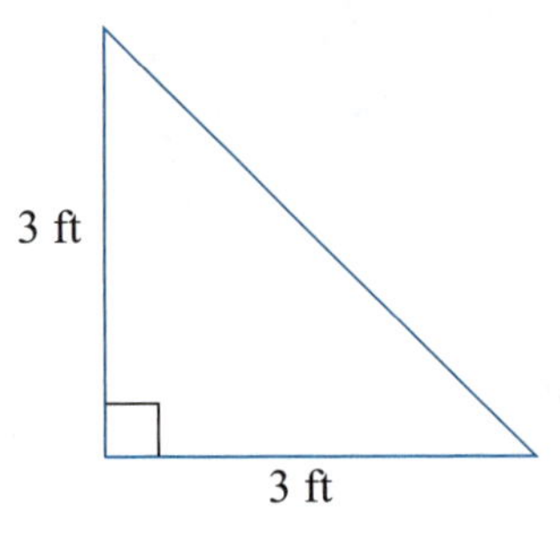

a. What is the length of the longest side of the display stand? Round your answer to the nearest tenth.

b. The top of the display has a decorative edge around all three sides to prevent the cakes from falling off. How much edging is required for the display?

c. If the top of the display stand is to be covered by square tiles that have a side length of 6 inches, how many tiles will be needed?

17. A glass company makes a paperweight in the shape of a cube that has a volume of 91.125 cubic inches.

a. What is the length of each side of the cube?

b. What is the area of the base of the cube?

Lesson Link

Surface area was covered in Section 5.5.

c. What is the surface area of the cube?

Name: Date:

14.2 Simplifying Radicals

Objectives

Simplify algebraic expressions that contain square roots.

Simplify radical expressions that contain variables.

Simplify algebraic expressions that contain cube roots.

Success Strategy

It is important to know how to apply the properties of radicals to simplify radical expressions. Be sure to do extra practice in the software.

Understand Concepts

Go to Software First, read through Learn in Lesson 14.2 of the software. Then, work through the problems in this section to expand your understanding of the concepts related to simplifying radicals.

It is sometimes necessary to work with square roots of nonperfect squares in radical form instead of changing them to a decimal number. When working with square roots in radical form, it is best to work with them in their **simplest form**. A square root is considered to be in simplest form when the radicand has no perfect square as a factor. To simplify square roots, the properties in the following box will be used.

Properties of Roots

If a and b are nonnegative real numbers and $b \neq 0$, then

1. **Product Property of Radicals:** $\sqrt{a} \cdot \sqrt{b} = \sqrt{a \cdot b}$
2. **Quotient Property of Radicals:** $\dfrac{\sqrt{a}}{\sqrt{b}} = \sqrt{\dfrac{a}{b}}$
3. **Simplifying Square Roots:** $\sqrt{a} \cdot \sqrt{a} = \sqrt{a \cdot a} = \sqrt{a^2} = |a|$.

Note: If the problem states that $a \geq 0$, then we can write $\sqrt{a^2} = a$.

Quick Tip

The product property of radicals and the quotient property of radicals are true for any pair of radical expressions as long as they have the same index.

True or False: Determine whether each statement is true or false. Rewrite any false statement so that it is true. (There may be more than one correct new statement.)

1. The expression $\dfrac{\sqrt{15}}{\sqrt{3}}$ simplifies to $\sqrt{5}$.

2. The expression $\sqrt{10} \cdot \sqrt{10}$ simplifies to 10.

3. The expression $\sqrt{5} \cdot \sqrt{7}$ simplifies to $\sqrt{12}$.

There are several approaches that can be used to simplify square roots. One method is to find perfect squares in the radicand and pull them out. The next problem will guide you through this process.

4. Write $\sqrt{200}$ in simplest form.

a. 200 is equal to $2 \cdot 100$. Use the product property of radicals to rewrite the radical expression.

b. Is either of 2 or 100 a perfect square? If so, which one?

c. Simplify the perfect square factor and rewrite the radical expression.

Quick Tip

It's possible to look at the radicand and determine what the largest square factor is. It may take a lot of practice to develop this insight.

Using prime factors can help you find all perfect squares in a number if you are unable to immediately determine them. The next problem will guide you through this process.

5. Write $\sqrt{80}$ in simplest form.

Lesson Link

Prime factorization was introduced in Section 1.9.

a. Write $\sqrt{80}$ as the square root of the prime factorization of 80. That is, write the prime factorization of 80 inside of the radical.

b. Circle the pairs of prime factors that are alike in the answer from part **a.**

c. Use the product property of radicals to rewrite the radical expression from part **b.** as a product of square roots, one containing all pairs of like prime factors and the other containing all unpaired prime factors.

Quick Tip

When simplifying square roots or any even index radical, the principal square root is typically used, especially in word problems.

d. Use the idea that $\sqrt{a \cdot a} = \sqrt{a^2} = a$ to simplify the square root that has pairs of like prime factors.

e. Simplify any factors without radicals by multiplying them together and using their result as the coefficient of the remaining radical term.

Name: Date:

6. Simplify $\sqrt{72x^2}$, where $x \geq 0$.

a. Write $\sqrt{72x^2}$ as the square root of the prime factorization of $72x^2$. That is, write the prime factorization of $72x^2$ inside of the radical.

b. Circle the pairs of prime factors that are alike in the answer from part **a.**

c. Use the product property of radicals to rewrite the radical expression from part **b.** as a product of square roots, one containing all pairs of like prime factors and the other containing all unpaired prime factors.

d. Use the idea that $\sqrt{a \cdot a} = \sqrt{a^2} = a$ to simplify the square root that has pairs of like prime factors.

e. Simplify any factors without radicals by multiplying them together and using their result as the coefficient of the remaining radical term.

Read the following information about the simplest form of cube roots then work through the problem.

A cube root is considered to be in simplest form when the radicand has no perfect cube as a factor. The key to keep in mind when simplifying cube roots is $\sqrt[3]{a} \cdot \sqrt[3]{a} \cdot \sqrt[3]{a} = \sqrt[3]{a \cdot a \cdot a} = \sqrt[3]{a^3} = a$.

7. Simplify $\sqrt[3]{108y^6}$.

a. Write $\sqrt[3]{108y^6}$ as the cube root of the prime factorization of $108y^6$. That is, write the prime factorization of $108y^6$ inside of the radical.

b. Circle the sets of three prime factors that are alike in the answer from part **a.**

c. Use the product property of radicals to rewrite the radical expression from part **b.** as a product of cube roots, one containing all sets of like prime factors and the other containing all remaining prime factors.

d. Use the idea that $\sqrt[3]{a \cdot a \cdot a} = \sqrt[3]{a^3} = a$ to simplify the cube root that has like triples of prime factors.

e. Simplify any factors without radicals by multiplying them together and using their result as the coefficient of the remaining radical term.

Skill Check

Go to Software Work through Practice in Lesson 14.2 of the software before attempting the following exercises.

Simplify the expressions, assuming that $x \geq 0$ and $y \geq 0$.

8. $-\sqrt{45}$

9. $\sqrt{8x^3}$

10. $-\sqrt{9x^2y^2}$

11. $\sqrt{\dfrac{5x^4}{9}}$

12. $\sqrt[3]{-64a^5}$

13. $\sqrt{\dfrac{196x^{15}y^{10}}{49}}$

Name: Date:

Apply Skills

Work through the problems in this section to apply the skills you have learned related to simplifying radicals.

14. A nut company is determining how to package their new type of party mix. The marketing department is experimenting with different-sized cans for the party mix packaging. The designers use the equation $V = \pi r^2 h$ to determine the radius of the can for a certain height h and volume V. The company decides they want the can to have a volume of 1200π cm³. Keep your answers in simplified radical form.

a. Solve the formula for r.

b. Find the radius of the can if the height is 12 cm.

c. Find the radius of the can if the height is 10 cm. Leave your answer in radical form.

d. Looking at the results from parts **b.** and **c.**, what relationship do you notice between the height of the can and the radius of the can? That is, as the height decreases, what happens to the radius when the volume is kept constant at 1200π cm³?

15. The amount of power that flows through a circuit depends on the current and the amount of resistance in the circuit. The power P in watts is calculated by the equation $P = I^2R$, where I is the current in amperes and R is the resistance in ohms.

a. Solve the equation for I.

b. What is the current that runs through a 90 watt light bulb with a resistance of 30 ohms? Keep the answer in simplified radical notation.

c. What is the approximate amount of amperes running through the light bulb in part **b.**, to the nearest hundredth?

d. What is the current that runs through a 90 watt light bulb with a resistance of 15 ohms? Keep the answer in simplified radical notation.

e. What is the approximate amount of amperes running through the light bulb in part **d.**, to the nearest hundredth?

f. Looking at the results from parts **c.** and **e.**, what relationship do you notice between the current and the resistance of the light bulb? That is, as the resistance decreases, what happens to the current when the power is kept constant at 90 watts?

Name: Date:

14.3 Addition, Subtraction, and Multiplication with Radicals

Objectives

Perform addition and subtraction involving radical expressions.

Perform multiplication involving radical expressions.

Use a graphing calculator to evaluate radical expressions. (Software only)

Success Strategy

In order to be prepared for the material in this section, review the meaning of like terms and the method for combining like terms that was introduced in Section 7.7.

Understand Concepts

Go to Software First, read through Learn in Lesson 14.3 of the software. Then, work through the problems in this section to expand your understanding of the concepts related to addition, subtraction, and multiplication with radicals.

In Section 7.7, **like terms** were introduced as terms that are constants or terms that contain the same variables raised to the same powers. In this section, we are going to work with **like radicals** to add or subtract expressions with radicals.

Like Radicals

Like radicals are radicals that have the same index and radicand or can be simplified so that they have the same index and radicand.

Combining radicals follows a similar method as combining like terms. Keep in mind that all radicals should be completely simplified before trying to combine them.

Lesson Link 

Recall that the distributive property allows us to combine coefficients of like terms. The distributive property was introduced in Section 7.6.

Combining Like Radicals

1. Simplify all radical terms.
2. Determine which radicals are alike and group them together.
3. Find the sum of the coefficients of the like radicals (be sure to keep track of negative coefficients).
4. Attach the common radical expression to the sum of the coefficients.

For example, $\sqrt{63}+5\sqrt{7}=3\sqrt{7}+5\sqrt{7}=(3+5)\sqrt{7}=8\sqrt{7}$.

1. Determine whether or not each pair of radical expressions are like radicals or can be simplified to like radicals.

a. $4\sqrt{5}, -2\sqrt{5}$

b. $5\sqrt{3}, 5\sqrt{6}$

c. $\sqrt{2}, \sqrt{8}$

d. $12\sqrt{3}, 4\sqrt{9}$

2. Simplify $3\sqrt{6}-7\sqrt{2}+5\sqrt{6}+\sqrt{8}$.

a. Simplify the radicands of each radical expression and rewrite the expression.

Quick Tip

One way to work with radicals is to treat them like variables. Only the coefficients of like terms are combined.

b. Rearrange the terms of the expression from part **a.** to group together like radicals.

c. Find the sum of the coefficients of the like radicals from part **b.** Be sure to keep track of any negative coefficients.

Determine if any mistakes were made while combining radical expressions. If any mistakes were made, describe the mistake and then correctly combine the radical expressions to find the actual result.

3. $$\begin{aligned}4\sqrt{3}+3\sqrt{5}+\sqrt{125} &= 4\sqrt{3}+3\sqrt{5}+5\sqrt{5}\\ &= 4\sqrt{3}+8\sqrt{5}\\ &= 12\sqrt{8}\end{aligned}$$

4. $$\begin{aligned}10\sqrt{8}-\sqrt{2}+2\sqrt{32} &= 10\cdot 2\sqrt{2}-\sqrt{2}+2\cdot 4\sqrt{2}\\ &= 20\sqrt{2}-\sqrt{2}+8\sqrt{2}\\ &= 28\sqrt{2}\end{aligned}$$

5. $$\begin{aligned}5\sqrt[3]{16}-4\sqrt[3]{24}+\sqrt[3]{-250} &= 5\cdot 2\sqrt[3]{2}-4\cdot 2\sqrt[3]{3}+(-5)\sqrt[3]{2}\\ &= 10\sqrt[3]{2}-8\sqrt[3]{3}-5\sqrt[3]{2}\\ &= 5\sqrt[3]{2}-8\sqrt[3]{3}\end{aligned}$$

Name: Date:

Read the following paragraph about simplifying algebraic expressions and work through the problems.

When multiplying algebraic expressions with radical terms, the product property of radicals, $\sqrt{a} \cdot \sqrt{b} = \sqrt{ab}$, is used. At some point in the multiplication process, you should simplify all radicals that can be simplified. You can simplify the terms before you multiply them and then again at the end after finding the product, or all simplification can be done at the very end.

6. Simplify $\sqrt{6} \cdot \sqrt{8}$.

a. Use the product property of radicals to rewrite the product $\sqrt{6} \cdot \sqrt{8}$.

b. Rewrite the radical expression from part **a.** in simplest form.

c. Rewrite each radical in the expression $\sqrt{6} \cdot \sqrt{8}$ in simplest form.

d. Use the product property to rewrite the product from part **c.**

e. Rewrite the radical expression from part **d.** in simplest form.

f. How do the results from part **b.** and part **e.** compare?

g. Do you prefer to simplify the radical expression as the first step and again after finding the product or just simplifying it once at the end? Write an explanation for your choice.

7. Simplify $\sqrt{7}\left(\sqrt{7} - \sqrt{14}\right)$.

a. Use the distributive property to rewrite the radical expression. Do not simplify.

b. Simplify the two terms in the expression from part **a.** Combine any like terms.

Determine if any mistakes were made while simplifying the radical expressions. If any mistakes were made, describe the mistake and then correctly simplify the radical expression to find the actual result.

8. $$\begin{aligned}(\sqrt{6}+2)(\sqrt{6}-2) &= \sqrt{6}\cdot\sqrt{6}-2\sqrt{6}+2\sqrt{6}-4\\ &= \sqrt{36}-2\sqrt{6}+2\sqrt{6}-4\\ &= 6-2\sqrt{6}+2\sqrt{6}-4\\ &= 6-4\\ &= 2\end{aligned}$$

9. $$\begin{aligned}(4-2\sqrt{6})(\sqrt{3}+1) &= 4\sqrt{3}+4-2\sqrt{3}\cdot\sqrt{6}-2\sqrt{6}\\ &= 4\sqrt{3}+4-2\sqrt{18}-2\sqrt{6}\\ &= 4\sqrt{3}+4-3\sqrt{2}-2\sqrt{6}\end{aligned}$$

Skill Check

Go to Software Work through Practice in Lesson 14.3 of the software before attempting the following exercises.

Simplify.

10. $4\sqrt{11}-3\sqrt{11}$

11. $\sqrt{a}+4\sqrt{a}-2\sqrt{a}$

12. $\sqrt{80}+\sqrt{8}-\sqrt{45}+\sqrt{50}$

13. $2\sqrt[3]{128}+5\sqrt[3]{-54}$

14. $\sqrt{2}(\sqrt{3}-\sqrt{6})$

15. $(3+\sqrt{2})(5-\sqrt{2})$

Name: Date:

Apply Skills

Work through the problems in this section to apply the skills you have learned related to addition, subtraction, and multiplication with radicals.

16. A simple diamond-shaped kite is created from wooden dowel rods and nylon cloth. A diagram for the kite is shown.

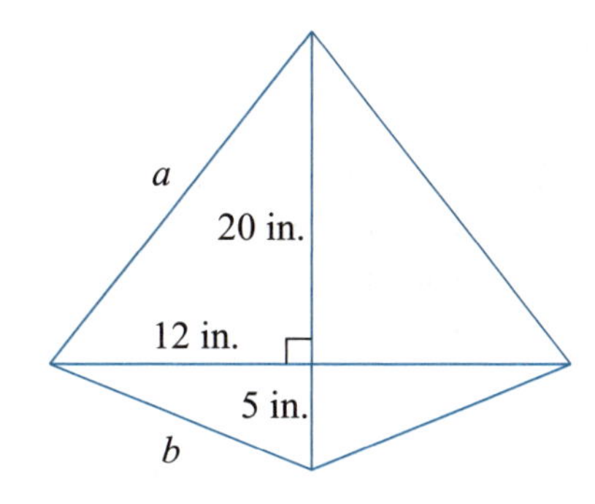

a. Determine the length of side a. Write your answer in simplified radical form.

b. Determine the length of side b. Write your answer in simplified radical form.

c. Calculate the perimeter of the kite. Write your answer in simplified radical form.

d. Use part **c.** to determine the approximate perimeter of the kite to the nearest tenth of an inch.

Quick Tip TECH

Use the square root function on the calculator to determine the approximate value of a square root that is not a perfect square. This function is often the 2nd function of the x^2 button. For example, to find the value of $\sqrt{3}$, type 2ND x^2 3) ENTER. The display will read "1.732050808…".

e. Determine the area of the kite in square inches. Round to the nearest tenth of a square inch. (**Hint:** Divide the kite into sections.)

f. Determine the area of the kite in square feet by dividing the answer from part **e.** by 144 square inches. Round your answer to the nearest tenth.
(**Note:** $(1 \text{ ft})^2 = (12 \text{ in.})^2 = 144 \text{ in.}^2$)

g. If the fabric for the kite costs $2 per square foot, use the answer from part **f.** to determine how much the fabric to create the kite will cost.

Name: Date:

14.4 Rationalizing Denominators

Objectives

Rationalize denominators with one term in the denominator.

Rationalizing denominators with a sum or difference in the denominator.

Success Strategy

Review the use of equivalent fractions in Section 2.1 to help you rewrite the quotients of radical expressions in this section.

Understand Concepts

Go to Software First, read through Learn in Lesson 14.4 of the software. Then, work through the problems in this section to expand your understanding of the concepts related to rationalizing denominators.

The main focus of this section is rationalizing the denominator of a rational expression that has an irrational number in the denominator. In Section 5.7, we discussed the connection between square roots and irrational numbers. In Section 14.1, we expanded upon this concept to include cube roots.

Perfect Squares, Perfect Cubes, and Irrational Numbers

A **perfect square** is the square of an integer. The square root of a perfect square is an integer.

A **perfect cube** is the cube of an integer. The cube root of a perfect cube is an integer.

The square root of a number that is not a perfect square or the cube root of a number that is not a perfect cube is an **irrational number**. An irrational number is an infinite nonrepeating decimal.

During the process of rationalizing a denominator, a fraction with a radical expression in the denominator, such as $\frac{1}{\sqrt{2}}$, will be rewritten so that the denominator no longer has a radical in it. The reason for rationalizing denominators is to make the expressions easier to work with and easier to understand.

Rationalizing a Denominator Containing a Single Term with a Square Root

1. Multiply the numerator and denominator of the rational expression by the radical part of the term in the denominator.
2. Simplify the numerator and denominator and reduce the resulting rational expression to lowest terms.

For example, $\frac{5}{3\sqrt{2}} = \frac{5}{3\sqrt{2}} \cdot \frac{\sqrt{2}}{\sqrt{2}} = \frac{5\sqrt{2}}{3 \cdot 2} = \frac{5\sqrt{2}}{6}$.

Quick Tip

The expression $\frac{\sqrt{a}}{\sqrt{a}} = 1$. Remember that multiplying a term by 1 does not change its value.

1. Rationalize the denominator of $\frac{4}{7\sqrt{3x}}$.

a. What is the radical part of the term in the denominator?

b. Multiply the numerator and denominator of the rational expression by the radical part of the denominator from part **a.** to get a perfect square in the radical of the denominator.

c. Check to make sure there are no radical terms remaining in the denominator of part **b.**

d. Reduce the rational expression from part **b.** to lowest terms, if possible.

Determine if any mistakes were made while rationalizing the denominator. If any mistakes were made, describe the mistake and then correctly rationalize the denominator to find the actual result.

2. $$\sqrt{\frac{3}{5}} = \frac{\sqrt{3}}{\sqrt{5}} = \frac{\sqrt{3}}{\sqrt{5}} \cdot \frac{\sqrt{5}}{\sqrt{5}} = \frac{\sqrt{15}}{\sqrt{25}} = \frac{\sqrt{15}}{5} = \frac{\sqrt{3}}{1}$$

3. $$\frac{-2y}{5\sqrt{2y}} = \frac{-2y}{5\sqrt{2y}} \cdot \frac{\sqrt{2y}}{\sqrt{2y}} = \frac{-2y\sqrt{2y}}{5\left(\sqrt{2y}\right)^2} = \frac{-2y\sqrt{2y}}{5(2y)} = \frac{-\sqrt{2y}}{5}$$

Lesson Link

Conjugates have a special relationship. Remember that the result when finding the product of $(a+b)(a-b)$ is $a^2 - b^2$. Special products of polynomials were introduced in Section 11.6.

Conjugates

The **conjugate** of the expression $a + b$ is $a - b$. Notice that the two terms remain the same, only the sign of the second term changes.

For example, the conjugate of $2+\sqrt{3}$ is $2-\sqrt{3}$.

True or False: Determine whether each statement is true or false. Rewrite any false statement so that it is true. (There may be more than one correct new statement.)

4. The conjugate of $1-3\sqrt{4}$ is $-1+3\sqrt{4}$.

5. The conjugate of $2\sqrt{5}+7$ is $2\sqrt{5}-7$.

Name: Date:

Rationalizing a Denominator Containing a Binomial

1. Multiply the numerator and denominator by the conjugate of the denominator of the rational term.
2. Simplify the numerator and the denominator.
3. Reduce, if possible.

6. This problem will explore the method of rationalizing the binomial denominator of a rational expression and why this method works.

a. Suppose $7-\sqrt{5}$ is the denominator of a rational expression. Find the conjugate of the expression.

b. Write an expression to describe the product of $7-\sqrt{5}$ and its conjugate. Do not simplify.

c. The expression from part **b.** is what type of special product?

d. Simplify the expression from part **b.**

e. Does the answer from part **d.** have any terms containing radicals, which are irrational numbers?

7. Rationalize the denominator of the expression $\dfrac{3}{2x-\sqrt{6}}$.

a. Find the conjugate of the denominator of the expression.

b. Multiply the numerator and denominator of the expression by the conjugate of the denominator and simplify.

Determine if any mistakes were made while rationalizing the denominator. If any mistakes were made, describe the mistake and then correctly rationalize the denominator to find the actual result.

8. $$\frac{x-y}{\sqrt{x}-\sqrt{y}}=\frac{x-y}{\sqrt{x}-\sqrt{y}}\cdot\frac{\sqrt{x}+\sqrt{y}}{\sqrt{x}+\sqrt{y}}$$
$$=\frac{(x-y)\left(\sqrt{x}+\sqrt{y}\right)}{x-y}$$
$$=\sqrt{x}+\sqrt{y}$$

9. $$\frac{2}{\sqrt{6}-2}=\frac{2}{\sqrt{6}-2}\cdot\frac{\sqrt{6}-2}{\sqrt{6}-2}$$
$$=\frac{2\left(\sqrt{6}-2\right)}{6-4\sqrt{6}+4}$$
$$=\frac{2\left(\sqrt{6}-2\right)}{2-4\sqrt{6}}$$
$$=\frac{\sqrt{6}-2}{1-2\sqrt{6}}$$

Skill Check

Go to Software Work through Practice in Lesson 14.4 of the software before attempting the following exercises.

Rationalize the denominators. Reduce if possible.

10. $\frac{5}{\sqrt{2}}$

11. $\sqrt{\frac{3}{8}}$

12. $\frac{-10}{3\sqrt{5}}$

13. $\frac{3}{1+\sqrt{2}}$

14. $\frac{8}{2\sqrt{x}-3}$

15. $\frac{\sqrt{2}+4}{5-\sqrt{2}}$

Name: ______________________ Date: ______________________

Apply Skills

Work through the problems in this section to apply the skills you have learned related to rationalizing denominators.

16. An intern at NASA needs to construct a cylinder to be used as a fuel cell for a scale model of a rocket. Her instructions are to make a fuel cell with a volume of 200π cm^3.

a. The equation for the volume of a circular cylinder is $V = \pi r^2 h$, where r is the radius and h is the height. Solve this equation for r.

b. Use the formula from part **a.** to determine the radius of a fuel cell that has a height of 12 cm. Write the answer in simplified radical form.

c. Find the decimal number equivalent of the answer from part **b.** Round your answer to the nearest hundredth.

d. Use the formula from part **a.** to determine the radius of a fuel cell that has a height of 24 cm. Write the answer in simplified radical form.

e. Find the decimal number equivalent of the answer from part **d.** Round your answer to the nearest hundredth.

f. Explain how the increase in height of the cylinder affects the radius of the cylinder.

Quick Tip

People go to financial consultants for advice about finances, such as creating budgets and developing savings plans.

17. A client tells his financial consultant that he has $6000 to invest and would like to earn $615 on his investment after 2 years. The client needs to know what the average interest rate of the investment will need to be to meet his expectations. The financial consultant can use the formula $A = P(r+1)^2$ to find the future amount A of an investment with a starting principle P and interest rate r after 2 years.

a. Solve the equation for r. Be sure to rationalize any denominators. (**Hint:** Divide both sides by P first and then take the square root of each side.)

b. Determine the interest rate that the $6000 would need to be invested at to meet the client's expectations. Be sure to express the rate as a percent.

Name: Date:

14.5 Equations with Radicals

Objectives

Solve equations that contain one or more radical expressions.

Success Strategy

You will need to exercise perseverance and diligence when solving radical equations, as they often involve many steps to solve. You should also review the process of squaring binomials found in Section 11.6.

Understand Concepts

Go to Software First, read through Learn in Lesson 14.5 of the software. Then, work through the problems in this section to expand your understanding of the concepts related to equations with radicals.

Solving Equations with Radicals

1. Isolate the radical on one side of the equation. (If the equation has more than one radical, isolate one of the radicals on one side of the equation.)
2. Raise both sides of the equation to the power corresponding to the index of the isolated radical and simplify.
3. If the equation still contains a radical, repeat Steps 1 and 2 until no radicals remain.
4. Solve the equation for the variable after all the radicals have been eliminated.
5. Be sure to check all possible solutions in the original equation and eliminate any extraneous solutions.

Lesson Link

Extraneous solutions are false solutions. They were discussed in Section 13.4.

When both sides of an equation are raised to a power, an extraneous solution may be introduced. Therefore, it is very important to always check your solutions when solving equations with radicals.

1. Solve the equation $\sqrt[3]{4+3x}+2=0$. (**Note:** The "+ 2" is on the outside of the radical.)

a. Isolate the radical on one side of the equation.

b. Raise each side of the equation to the power corresponding to the index of the isolated radical. Do not simplify.

c. Simplify both sides of the equation from part **b.** No radicals should remain.

d. Solve the resulting equation from part **c.** for the variable.

e. Check all solutions in the original equation. Which solutions are actual solutions?

2. Solve the equation $\sqrt{x+1}+3=0$. (**Note:** The "+ 3" is on the outside of the radical.)

a. Isolate the radical on one side of the equation.

Quick Tip

Remember that if no index is specified, the index is understood to be 2.

b. Raise each side of the equation to the power corresponding to the index of the isolated radical. Do not simplify.

c. Simplify both sides of the equation from part **b.** No radicals should remain.

d. Solve the resulting equation from part **c.** for the variable.

e. Check all solutions in the original equation. Which solutions are actual solutions?

f. Can an equation with radicals have the square root of a number equal to a negative number? Explain your reasoning.

Name: Date:

3. Solve the equation $\sqrt{x}-\sqrt{2x-14}=1$.

a. Isolate one of the radicals on one side of the equation.

b. Raise each side of the equation to the power corresponding to the index of the isolated radical. Do not simplify.

Lesson Link

Squaring a binomial was covered in Section 11.6. Remember that the square of a binomial will simplify to a trinomial.

c. Simplify both sides of the equation from part **b.** A radical should remain.

d. Isolate the remaining radical on one side of the equation.

e. Raise each side of the equation to the power corresponding to the index of the isolated radical. Do not simplify.

f. Simplify both sides of the equation from part **e.** No radicals should remain.

Lesson Link

Solving quadratic equations by factoring was covered in Section 12.6.

g. Solve the resulting quadratic equation from part **f.** for the variable.

h. Check all solutions in the original equation. Which solutions are actual solutions?

Skill Check

Go to Software Work through Practice in Lesson 14.5 of the software before attempting the following exercises.

Solve the equations. Be sure to check the solutions.

4. $\sqrt{8x+1}=5$

5. $\sqrt{x-4}+6=2$

6. $\sqrt{x+6}=x+4$

7. $\sqrt{3x+2}=\sqrt{x+4}$

8. $\sqrt{x}=\sqrt{x+16}-2$

9. $\sqrt[3]{5x+4}=4$

Apply Skills

Work through the problems in this section to apply the skills you have learned related to equations with radicals.

Quick Tip

Hang time is the amount of time that something spends in the air. The hang time for an athlete jumping is the amount of time they remain in the air before their feet return to the ground.

10. The hang time of an athlete can be represented by the formula $t=2\sqrt{\frac{2h}{g}}$, where t is the hang time in seconds, h is the vertical height of the jump in feet, and g is the acceleration due to gravity, 32 ft/sec^2.

a. If Michael Jordan had a vertical jump of 4 feet, how long would he be in the air?

b. A volleyball player has a hang time of 0.866 seconds. How high is this players vertical jump? Round your answer to the nearest foot.

Name: Date:

14.6 Rational Exponents

Objectives

Learn about n^{th} root.

Translate expressions using radicals into expressions using rational exponents.

Translate expressions using rational exponents into expressions using radicals.

Simplify expressions using the properties of rationale exponents.

Evaluate expressions of the form $a^{\frac{m}{n}}$ with a calculator. (Software only)

Success Strategy

Review the exponent rules introduced in Sections 11.1 and 11.2. Make sure you understand the rules for using integer exponents before learning about and applying the exponential rules using rational exponents.

Understand Concepts

Go to Software First, read through Learn in Lesson 14.6 of the software. Then, work through the problems in this section to expand your understanding of the concepts related to rational exponents.

Every radical expression can be rewritten to use rational exponents instead of radicals. This can be useful for problem solving and working with radical expressions. Some people also prefer to use rational exponents instead of radical symbols since the rules of exponents apply to rational exponents just as they do for integer exponents.

Quick Tip

A **rational exponent** is an exponent that can be written in fraction form.

n^{th} roots

If n is a positive integer, then $\sqrt[n]{a} = a^{\frac{1}{n}}$ (assuming $\sqrt[n]{a}$ is a real number).

Note: For $\sqrt[n]{a} = a^{\frac{1}{n}}$ to be a real number:

1. When a is nonnegative, n can be any positive integer.
2. When a is negative, n must be an odd positive integer.

Quick Tip

Remember that if no index is specified, the index is understood to be 2.

1. Fill in the missing pieces in the table. Not all expressions can be simplified.

Radical Notation	Exponential Notation	Simplified Form
$\sqrt{x}$		
$\sqrt[3]{8}$		
$\sqrt[4]{81}$		
$\sqrt[3]{64}$		

General Form of Rational Exponents

If m and n are positive integers greater than 1 with $\frac{m}{n}$ in simplest form, then

$$a^{\frac{m}{n}} = \left(a^{\frac{1}{n}}\right)^m = \left(a^m\right)^{\frac{1}{n}} = \sqrt[n]{a^m} = \left(\sqrt[n]{a}\right)^m.$$

When evaluating radical expressions such as $\sqrt{4^3}$, it doesn't matter if the square root or the power is simplified first. This is because $\sqrt{4^3} = \left(\sqrt{4}\right)^3 = \left(4^{\frac{1}{2}}\right)^3 = \left(4^3\right)^{\frac{1}{2}}$.

2. Fill in the missing pieces in the table. Not all expressions can be simplified.

Radical Notation	Exponential Notation	Simplified Form
$\sqrt[3]{x^2}$		
$\sqrt[2]{16^3}$		
$-\sqrt[3]{27^2}$		

3. The rules for exponents can be used to simplify expressions with rational exponents. The rules are summarized in the table. Fill in the boxes in the example column to correctly complete the example.

Name	Rule	Example
The Exponent 1	$a = a^1$	$2 = 2^{\square}$
The Exponent 0	$a^0 = 1$	$5^0 = \square$
The Product Rule	$a^m \times a^n = a^{m+n}$	$4^2 \cdot 4^6 = 4^{\square + \square} = 4^{\square}$
The Quotient Rule	$\frac{a^m}{a^n} = a^{m-n}$	$\frac{x^5}{x^2} = x^{\square - \square} = x^{\square}$
Negative Exponents	$a^{-n} = \frac{1}{a^n}, \frac{1}{a^{-n}} = a^n$	$5^{-1} = \frac{1}{\square}, \frac{1}{y^{-4}} = y^{\square}$
Power Rule	$\left(a^m\right)^n = a^{mn}$	$\left(2^{\frac{1}{2}}\right)^5 = 2^{\frac{\square}{\square}}$
Power of a Product	$(ab)^n = a^n b^n$	$(2x)^8 = 2^{\square} x^{\square}$
Power of a Quotient	$\left(\frac{a}{b}\right)^n = \frac{a^n}{b^n}$	$\left(\frac{x}{3}\right)^2 = \frac{x^{\square}}{3^{\square}}$

Skill Check

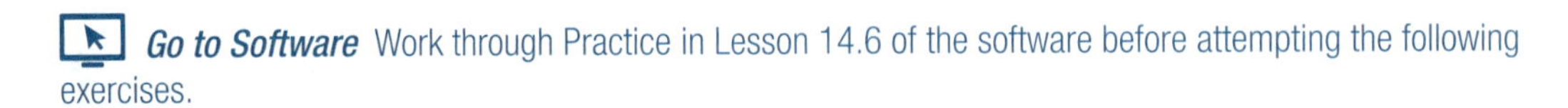

Go to Software Work through Practice in Lesson 14.6 of the software before attempting the following exercises.

Simplify.

4. $\left(2x^{\frac{1}{3}}\right)^3$

5. $\frac{a^{\frac{2}{3}}}{a^{\frac{1}{9}}}$

Name: Date:

6. $\dfrac{x^{\frac{2}{3}} \cdot x^{\frac{4}{3}}}{x^2}$

7. $\left(-x^3y^6z^{-6}\right)^{\frac{2}{3}}$

Quick Tip TECH

To evaluate expressions with rational exponents on a calculator, use the caret button and enter the exponent inside of parentheses. For example, to evaluate $2^{\frac{1}{3}}$, type the following key sequence into the calculator: [2] [^] [(] [1] [÷] [3] [)] [ENTER]. The screen should read 1.25992105.

Simplify. Round to the nearest ten-thousandth, if necessary.

8. $25^{\frac{2}{3}}$

9. $\sqrt[8]{63}$

Apply Skills

Work through the problems in this section to apply the skills you have learned related to rational exponents.

10. An amusement park is creating signs to indicate the velocity of the roller coaster car on certain hills of the most popular rides. A roller coaster car gains kinetic energy as it goes down a hill. The velocity, or speed, of an object in kilometers per hour (kph) can be determined by $V = \left(\dfrac{2k}{m}\right)^{\frac{1}{2}}$, where k is the kinetic energy of the object in joules (J) and m is the mass of the object in kilograms (kg).

Quick Tip

The **kinetic energy** of an object is the amount of energy the object has due to its motion.

a. For the most popular roller coaster, the car has a mass of 300 kg and the car has a kinetic energy of 375,000 J on the first hill. What velocity does the car obtain on the first hill?

b. For the second most popular roller coaster, the car has a mass of 350 kg and the car has a kinetic energy of 70,000 on the first hill. What velocity does the car obtain on the first hill?

Quick Tip

Despite popular belief, Newton did not discover gravity when he got hit in the head by an apple. However, acquaintances of Newton did confirm that his concept of gravity was inspired by observing apples falling and wondering why they always fell down, and not sideways or upwards.

11. Isaac Newton sat under an apple tree to drink some tea and think. While he was thinking, an apple fell off of a branch of the tree. The equation $v = (19.8d)^{\frac{1}{2}}$ can be used to find the velocity v, in meters per second, of the apple after dropping a distance d, in meters.

a. If the apple was connected to a branch 2 meters above the ground, what was the velocity of the apple, to the nearest hundredth, when it hit the ground?

b. If the apple was connected to a branch 4 meters above the ground, what was the velocity of the apple, to the nearest hundredth, when it hit the ground?

c. Compare the values from parts **a.** and **b.** The apple in part **b.** is twice as far from the ground as the apple in part **a.** Is the velocity of the apple in part **b.** twice that of the velocity in part **a.**? Why do you think this is?

Name: Date:

14.7 Functions with Radicals

Objectives

Review the concepts of functions and function notation.

Find the domain and range of radical functions.

Evaluate radical functions.

Graph radical functions.

Use a graphing calculator to graph radical functions. (Software only)

Success Strategy

Some of the information that you learned about functions from Section 9.5 is repeated at the beginning of this section as a brief review. Go back to Section 9.5 and review if you need more details or examples to refresh your memory.

Understand Concepts

Go to Software First, read through Learn in Lesson 14.7 of the software. Then, work through the problems in this section to expand your understanding of the concepts related to functions with radicals.

The following box reviews the basics of functions.

Function Review

A **relation** is a set of ordered pairs of real numbers.

The **domain** D of a relation is the set of all first coordinates in the relation.

The **range** R of a relation is the set of all second coordinates in the relation.

A **function** is a relation in which each domain element has exactly one corresponding range element.

The **vertical line test** states that if any vertical line intersects the graph of a relation at more than one point, then the relation is not a function.

1. Use the vertical line test to determine if each graph represents a function.

a.

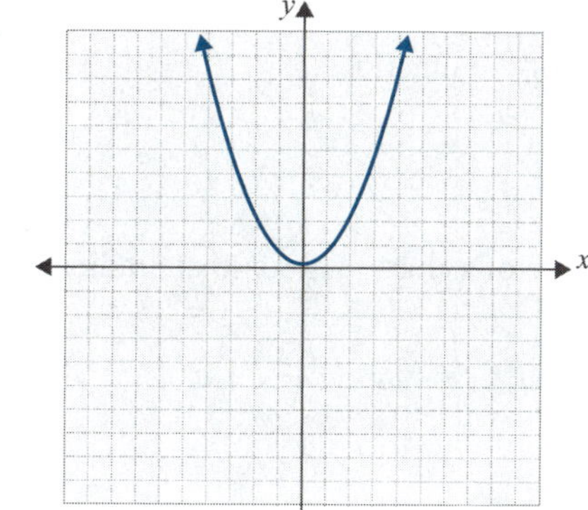

b.

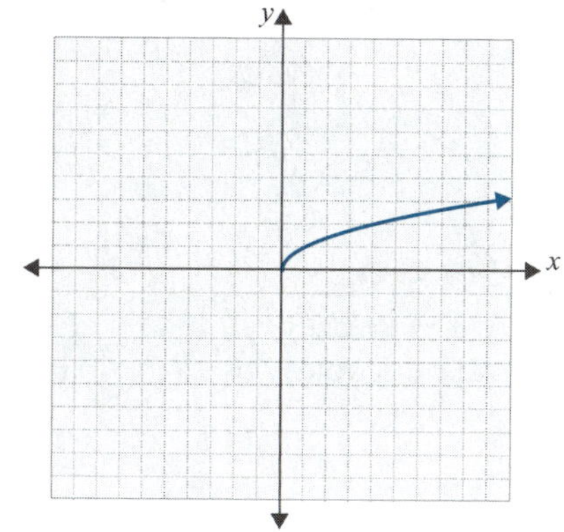

Radical Functions

A **radical function** is a function of the form $y = \sqrt[n]{g(x)}$, where $g(x)$ is a polynomial.

The domain of such a function depends on the index, *n*:

1. If *n* is an even number, the domain is the set of all *x* such that $g(x) \geq 0$.
2. If *n* is an odd number, the domain is the set of all real numbers, $(-\infty, \infty)$.

2. Find the domain of $f(x) = \sqrt{2x+3}$.

a. Since the index is even, the domain of the radical function is restricted. Write an inequality of the form $g(x) \geq 0$, where $g(x)$ is the polynomial inside of the radical.

b. Solve the inequality from part **a.** for the variable.

c. Write the domain of the function in interval notation.

Lesson Link

Interval notation was introduced in Section 8.7.

Graphing Radical Functions

1. Determine the domain of the function.
2. Create a table of ordered pairs using values of *x* that are in the domain of the function.
3. Plot the ordered pairs on a coordinate plane.
4. Draw a curve through the plotted points.

3. Graph the equation $y = \sqrt{x+5}$.

a. Find the domain of the function.

b. Complete the table of values. Approximate radical values with a calculator to the nearest hundredth.

x	*y*
−5	
−4	
−3	
0	
4	

Lesson Link

Approximating radical values with a calculator was introduced in Section 5.7.

Name: Date:

c. Plot the values from part **b.** on the coordinate plane. The approximate values may be difficult to plot by hand, so just do your best.

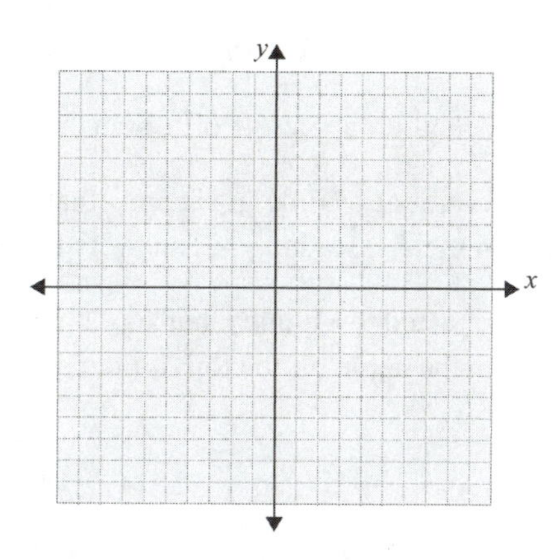

d. Draw a curve through the points. The curve will start at the point with x-coordinate -5, pass through the remaining plotted points, and then continue towards infinity.

Skill Check

Go to Software Work through Practice in Lesson 14.7 of the software before attempting the following exercises.

Complete each table by finding the corresponding $f(x)$values for the given values of x. Approximate radical values with a calculator to the nearest hundredth.

4. $f(x)=\sqrt{2x+1}$

x	y
0	
1	
2	
3	
4	

5. $f(x)=\sqrt{5-3x}$

x	y
−3	
−2	
−1	
0	
1	

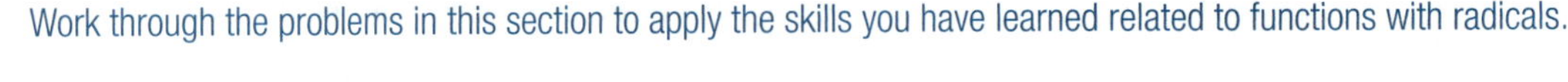

Apply Skills

Work through the problems in this section to apply the skills you have learned related to functions with radicals.

6. A farmer fell asleep under a tree in his apple orchard while thinking about pie. While he was sleeping, a squirrel knocked an apple off of a branch of the tree. The function $f(d) = \sqrt{\frac{2d}{9.8}}$ can be used to find the amount of time in seconds that it takes for the apple to drop a certain distance d, where d is in meters. Round your answers to the nearest hundredth.

   a. If the apple was connected to a branch that was 2 meters above the farmer's head, how long would it take before the apple hit the top of the farmer's head?

   b. If the squirrel knocked a second apple off of a branch that was 5 meters above the farmer's head, how long would it take before the apple hit the top of the farmer's head?

   c. Suppose the second apple missed the farmer's head and landed on the ground instead. If the farmer's head was 0.8 meters above the ground, how long did it take for the apple to hit the ground?

7. A person's Body Mass Index (BMI) is determined by the formula $B = \frac{703w}{h^2}$, where B is the BMI, w is the person's weight in pounds, and h is the person's height in inches. Having a BMI between 18.5 and 24.9 is considered optimal. A BMI between 25 and 29.9 is considered overweight and a BMI over 30 is considered obese. A BMI below 18.5 is considered underweight.

   Lesson Link

   Solving formulas for different variables was introduced in Section 8.5.

   a. Solve the BMI formula for the variable h.

   b. How tall is a person who has a BMI of 20 and a weight of 120 pounds? Round to the nearest inch.

   c. To be in the optimal BMI range with a weight of 200 pounds, what range in height should a person be? Round to the nearest inch. (**Hint:** Calculate the heights for the endpoints of the BMI range given.)

   d. How tall is a person whose BMI is 30 and who weighs 150 pounds? Round to the nearest inch.

Name: Date:

Chapter 14 Projects

Project A: Let's Get Radical!

An activity to demonstrate the use of radical expressions in real life

There are many different situations in real life that require working with radicals, such as solving right triangle problems, working with the laws of physics, calculating volumes, and even solving investment problems. Let's take a look at a simple investment problem to see how radicals are involved.

The formula for computing compound interest for a principal P that is invested at an annual rate r and compounded annually is given by $A = P(1+r)^n$, where A is the accumulated amount in the account after n years.

1. Let's suppose that you have \$5000 to invest for a term of 2 years. If you want to be sure and make at least \$600 in interest, then at what interest rate should you invest the money?

a. One way to approach this problem would be through trial-and-error by substituting various rates for r into the formula. This approach might take a while. Using the table below to organize your work, try substituting three different values for r. Remember that rates are percentages and need to be converted to decimals before using in the formula. Did you get close to \$5600 for the accumulated amount in the account after 2 years?

Annual Rate (r)	Principal (P)	Number of Years (n)	Amount, $A = P(1+r)^n$
	\$5000	2	
	\$5000	2	
	\$5000	2	

b. Let's try a different approach. Substitute the value of 2 for n and solve this formula for r. Verify that you get the following result: $r = \sqrt{\frac{A}{P}} - 1$. (**Hint:** First solve for $(1+r)^2$ and then take the square root of both sides of the equation.) Notice that you now have a radical expression to work with. Substitute \$5000 for P and \$5600 for A (which is the principal plus \$600 in interest) to see what the interest rate must be. Round your answer to the nearest percent.

2. Now, let's suppose that you won't need the money for 3 years.

a. Use $n = 3$ years and solve the compound interest formula for r.

b. What interest rate will you need to invest the principal of $5000 at in order to have at least $5600 at the end of 3 years? (To evaluate a cube root you may have to use the rational exponent of $\frac{1}{3}$ on your calculator.) Round your answer to the nearest percent.

c. Compare the rates needed to earn at least $600 when $n = 2$ years and $n = 3$ years. What did you learn from this comparison? Write a complete sentence.

3. Using the above formulas for compound interest when $n = 2$ years and $n = 3$ years, write the general formula for r for any value of n.

4. Using the formula from Problem 3, compute the interest rate needed to earn at least $3000 in interest on a $5000 investment in 7 years. Round to the nearest percent.

5. Do an Internet search on a local bank or financial institution to determine if the interest rate from Problem 4 is reasonable in the current economy. Using three to five sentences, briefly explain why or why not.

Name: Date:

Project B: Skid Marks Can Be Revealing

An activity to demonstrate the use of radical expressions in real life

Have you ever driven by an accident where police officers are measuring the skid marks left on the road? Do you know why they are doing this? The reason is that the length of the skid marks that a car makes when it suddenly brakes tells the police officers approximately how fast the car was traveling. A big factor in this determination is the condition of the road. It takes longer to stop on wet pavement than it does on dry pavement. There are other factors that affect stopping distance such as road composition—that is, whether the road is concrete, asphalt, or dirt and gravel. In this activity we are only going to compare the stopping distance on wet vs. dry roads. Hopefully this activity will convince you to slow down when traveling in rainy weather and to always turn on your headlights for better visibility!

1. The formula that is used to relate stopping distance d to velocity (or speed) v on wet pavement is

$$\frac{v^2}{12} = d$$

Solve this formula for velocity and simplify the resulting radical expression.

2. If the skid marks on a wet road are 102 feet long, then how fast was the driver going? Round your answer to the nearest whole number.

3. The formula that police officers use to relate stopping distance d to velocity v on dry pavement is

$$\frac{v^2}{24} = d$$

Solve this formula for velocity and simplify the resulting radical expression.

4. If the skid marks on a dry road are 102 feet long, then how fast was the driver going? Round your answer to the nearest whole number.

5. If a car is traveling at 60 mph, how much farther in distance will it take the car to stop on wet pavement than dry pavement?

6. What is the ratio of the two distances (wet to dry) from Problem 5? Simplify the ratio, if possible, and interpret what it means. Use a complete sentence.

7. Take the ratio of the two formulas for d (wet to dry) and simplify it to a number. How does this number compare to the ratio in Problem 6?

8. If you decrease your speed by half, how does this affect your stopping distance? (You can use either the wet or the dry formula since the comparison will be the same.) Show all work and interpret your result using complete sentences.

Name: Date:

Math@Work

Forensic Scientist

As a forensic scientist, you will work as part of a team to investigate the evidence from a crime scene. Every case you encounter will be unique and the work may be intense. Communication is especially important because you will need to be clear and honest about your findings and your conclusions. A suspect's freedom may depend on the conclusions your team draws from the evidence.

Suppose the most recent case that you are involved in is a hit-and-run accident. A body was found at the side of the road with skid marks nearby. The police are unsure if the cause of death of the victim was vehicular homicide. Among the case description, the following information is provided to you.

Accident Report	
Date:	June 14
Time:	9:30 pm
Climate:	55 degrees Fahrenheit, partly cloudy, dry
Description of crime scene:	
Victim was found at the side of a road. Body temperature upon arrival is 84.9 °F. Posted speed limit is 30 mph. Road is concrete. Conditions are dry. Skid marks near the body are 88 feet in length.	

Known formulas and data:

A body will cool at a rate of 2.7 °F per hour until the body temperature matches the temperature of the environment. Average human body temperature is 98.6 °F.

Impact Speed and Risk of Death	
Impact Speed	**Risk of Death**
23 mph	10%
32 mph	25%
42 mph	50%
58 mph	90%
Source: 2011 AAA Foundation for Traffic Safety "Impact Speed and Pedestrian's Risk of Severe Injury or Death"	

Braking distance is calculated using the formula $\frac{s}{\sqrt{l}} = k$, where s is the initial speed of the vehicle in mph, l is the length of the skid marks in feet, and k is a constant that depends on driving conditions. Based on the driving conditions on that road for the last 12 hours, $k = \sqrt{20}$.

1. Based on the length of the skid marks, how fast was the car traveling before it attempted to stop? Round to the nearest whole number.

2. Based on the table, what percent of pedestrians die after being hit by a car moving at that speed?

3. Based on the cooling of the body, if the victim died instantly, how long ago did the accident occur? Round to the nearest hour.

4. Can you think of any other factors that should be taken into consideration before determining whether the impact of the car was the cause of death?

Foundations Skill Check for Chapter 15

This page lists several skills covered previously in the book and software that are needed to learn new skills in Chapter 15. To make sure you are prepared to learn these new skills, take the self-test below and determine if any specific skills need to be reviewed.

Each skill includes an easy (**e.**), medium (**m.**), and hard (**h.**) version. You should be able to complete each problem type at each skill level. If you are unable to complete the problems at the easy or medium level, go back to the given lesson in the software and review until you feel confident in your ability. If you are unable to complete the hard problem for a skill, or are able to complete it but with minor errors, a review of the skill may not be necessary. You can wait until the skill is needed in the chapter to decide whether or not you should work through a quick review.

5.7 Simplify the radical expression.

e. $\sqrt{36}$ **m.** $\sqrt{196}$ **h.** $\sqrt{441}$

11.3 Evaluate the expression for $x = 3$.

e. $x^2 + 2x + 1$ **m.** $3x^2 - 8x - 7$ **h.** $2x^2 + 3x - \frac{3}{4}$

12.2 Factor the trinomial completely.

e. $y^2 + y - 30$ **m.** $m^2 + 3m - 1$ **h.** $20a^2 + 40a + 20$

12.3 Factor the trinomial completely.

e. $3x^2 - 4x - 7$ **m.** $8x^2 - 10x - 3$ **h.** $24y^2 + 4y - 4$

12.4 Factor the trinomial completely.

e. $x^2 + 4x + 4$ **m.** $a^2 - 6a + 9$ **h.** $x^2 + 5x + \frac{25}{4}$

Chapter 15: Quadratic Equations

Study Skills

15.1 Quadratic Equations: The Square Root Method

15.2 Quadratic Equations: Completing the Square

15.3 Quadratic Equations: The Quadratic Formula

15.4 Applications

15.5 Quadratic Functions

Chapter 15 Projects

Math@Work

Math@Work

Introduction

Chapters 1 through 14 have introduced a variety of different career paths that regularly use math. There are many more careers that use math which haven't been covered in this workbook. At the end of this chapter, we'll provide you with information about a few more careers, along with resources to learn more about them and others. No matter which career path you decide to pursue with your college education, you will most likely need to use some form of math daily in your job, whether it is simple arithmetic or more advanced problem-solving skills.

Study Skills

The POWER Strategy for Writing

The POWER strategy involves three stages of writing a paper. The first stage is the prewriting stage that takes place before you start writing and includes the **p**lanning and **o**rganizing steps. The second stage is actually when the **w**riting takes place. The third stage is the postwriting stage and includes the **e**diting and **r**evising steps.

Plan

1. Start with a well-defined topic. Know exactly what you intend to write about.
2. Brainstorm and list all you know about the chosen topic.
3. Make a list of items for which you need to gather more information.
4. Gather all the information you will need for your paper, using a variety of sources (Internet, library, etc.).
5. As you gather your information, write down complete references in the required format you will be using (APA, MLA, etc.) to include in the paper.
6. Take notes of all the information you want to include in your paper. Write down phrases to represent ideas instead of copying complete sentences from your sources. Rewrite these phrases later in your own words to avoid plagiarism.

Organize

1. Review the notes that you took in the planning step.
2. Organize your notes into an outline using the main ideas of your paper as the major headings.
3. Add subheadings and details to the outline.

Write

1. Use your outline as a guide for writing your paper.
2. Fill in any necessary details from your notes, using complete sentences.
3. Include all the ideas from your outline, stating them clearly and in the order they appear in the outline.
4. If you include any diagrams, tables, or graphs, make sure they are clearly labeled or titled.
5. If you use any math formulas, make sure to define all of the variables.

Edit

1. Use a review tool in the software you are using to type your paper to check spelling and grammar.
2. Check your capitalization and punctuation.
3. Check for errors by reading your paper aloud.
4. Have someone else proof your paper, if there is time.
5. Make sure you have cited *all* of the references to other people's works or ideas.
6. Check your formatting to make sure you have adhered to the required style (APA, MLA, etc.).
7. If you have included any mathematical statements or formulas, make sure they are correct.
8. If the purpose of the paper was to solve a problem or answer a question, make sure you have done so and have included any conclusions made.

Revise

1. Make any edits from the editing step.
2. Reread the paper one last time after the changes are made, or have someone proof it for you.
3. Make any final edits.

Name: Date:

15.1 Quadratic Equations: The Square Root Method

Objectives

Review solving quadratic equations by factoring.

Solve quadratic equations using the definition of square root.

Solve problems related to right triangles and the Pythagorean Theorem.

Success Strategy

Be sure to review the information in Section 12.6 on solving quadratic equations by factoring using the zero-factor property before starting this section.

Understand Concepts

Go to Software First, read through Learn in Lesson 15.1 of the software. Then, work through the problems in this section to expand your understanding of the concepts related to the square root method.

Solving a quadratic equation may require using the factoring techniques learned in previous chapters. The *ac*-method is restated here for convenience.

The *ac*-Method for Factoring Trinomials of the Form $ax^2 + bx + c$

1. Find the product of a and c. Be sure to keep track of the sign of the product ac.
2. Find two factors, u and v, of ac whose sum is b. That is, $uv = ac$ and $u + v = b$. If this is not possible, then the trinomial is **not factorable**.
3. Rewrite the middle term of the trinomial as two terms using the two factors found in Step 2 as coefficients. That is, $bx = ux + vx$. Be sure to keep track of the signs of the factors.
4. Factor the four-term polynomial using the method of factoring by grouping.
5. (Optional) Check that the factored polynomial is correct by finding the product of all its factors.

Lesson Link

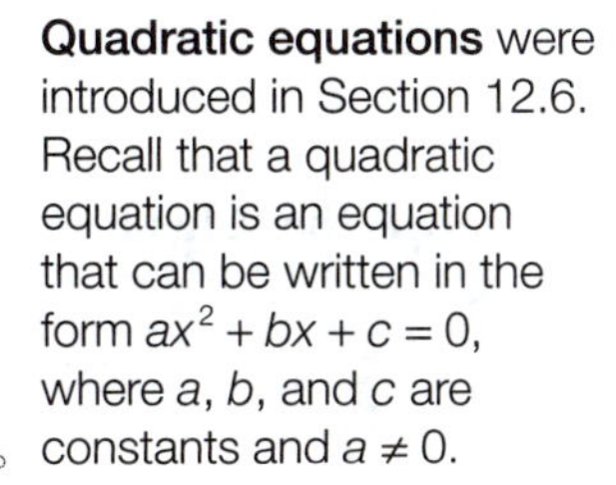

Quadratic equations were introduced in Section 12.6. Recall that a quadratic equation is an equation that can be written in the form $ax^2 + bx + c = 0$, where a, b, and c are constants and $a \neq 0$.

Not every quadratic equation can be solved by factoring. This chapter will introduce several methods of solving quadratic equations. The method introduced in this section is the square root method.

The Square Root Method

For a quadratic equation in the form $x^2 = c$, where c is a nonnegative real number,

$$x = \sqrt{c} \text{ or } x = -\sqrt{c}.$$

This can be written as $x = \pm\sqrt{c}$.

1. In the square root method, why must the value of c be nonnegative?

Solving Quadratic Equations Using the Square Root Method

1. Write the equation in the form $x^2 = c$.
2. If $c < 0$, then the equation has no real solutions. If $c > 0$, continue to the next step.
3. Take the square root of both sides of the equation. Remember that there is a positive and a negative solution.
4. Simplify any radical expressions.
5. (Optional) Check the answers in the original equation.

2. Use the square root method to solve the quadratic equation $3x^2 = 51$.

a. Write the quadratic equation in the form $x^2 = c$.

b. Will the equation have real solutions? Explain why or why not. (**Hint:** Is the value of c nonnegative?)

c. Take the square root of both sides of the equation. Remember to use the symbol $\pm$ for the sign of the constant on the right side of the equation since there are two possible square roots—one positive root and one negative root.

Quick Tip

Remember that the variable x can represent an expression, such as $x = y - 2$.

3. The square root method can be used on more complicated quadratic equations like $(x+4)^2 = 21$.

a. Is the equation written in the form $x^2 = c$? If not, rewrite the equation to match this form.

b. Will the equation have real solutions? Explain why or why not. (**Hint:** Is the value of c nonnegative?)

c. Take the square root of both sides of the equation.

d. Solve the equation for x.

Name: Date:

Problems that involve the Pythagorean Theorem can often be solved using the square root method. The following box reviews the Pythagorean Theorem, which will be used in the Apply Skills portion of this section.

Quick Tip

The **legs of a right triangle** are represented by the variables *a* and *b*. The **hypotenuse**, which is opposite the right angle and is always the longest side, is represented by the variable *c*.

The Pythagorean Theorem

In a right triangle, the square of the length of the hypotenuse is equal to the sum of the squares of the lengths of the two legs.

$$c^2 = a^2 + b^2$$

c b 90° a

4. Can the Pythagorean Theorem be used to find the missing sides of any triangle?

Skill Check

Go to Software Work through Practice in Lesson 15.1 of the software before attempting the following exercises.

Solve for the variable. Simplify any radical expressions.

5. $x^2 - 81 = 0$

6. $5x^2 = 60$

7. $(x+2)^2 = -25$

8. $(x-2)^2 = \frac{1}{16}$

9. $2(x-7)^2 = 24$

10. $(2x+1)^2 = 48$

Apply Skills

Work through the problems in this section to apply the skills you have learned related to the square root method.

11. A ball is dropped from the top of a building that is 144 feet tall. The formula for finding the height of the ball at any time is $h = 144 - 16t^2$, where t is the number of seconds since the ball was dropped.

a. What will the height of the ball be when it hits the ground?

b. Replace h in the equation with your answer from part **a.** and solve to determine how long it will take the ball to hit the ground.

c. Are there any solutions to the equation that do not make sense in this context? If so, why do they not make sense?

Quick Tip

Remember that drawing a figure of the situation can make the problem easier to solve.

12. A 38-foot ladder is leaning against a building that is 34 feet high. The top of the ladder extends 3 feet beyond the point where it touches the building.

a. How far is the base of the ladder from the base of the building? Write your answer as a decimal number rounded to the nearest tenth.

Quick Tip

The Occupational Safety and Health Administration (OSHA) is the main federal agency in charge of the enforcement of health and safety laws. See www.osha.gov for more information.

b. According to OSHA guidelines, the distance from the base of the ladder to the building should be $\frac{1}{4}$ the distance from the ground to the point of contact. Does the distance of the base of the ladder to the building meet the safety guidelines? Explain your answer.

Quick Tip

The **future value** A of an investment is equal to the amount in the fund at the end of the investment period. It is equal to the original principal plus the interest.

13. A financial consultant is asked for advice about finances and savings plans. When a client invests money, they need to know which interest rate will meet their financial goals based on the amount invested. The financial consultant can use the formula $A = P(r+1)^n$ to find the future amount A, after n years, of an investment with a starting principal P invested at an interest rate of r.

a. A client has \$3000 to invest and would like to earn \$300 on his investment after 2 years. At what interest rate will the client need to invest his money? Round to the nearest hundredth of a percent.

b. Another client has \$5000 to invest and would like to earn \$750 on her investment after 2 years. At what interest rate will the client need to invest her money? Round to the nearest hundredth of a percent.

Name: Date:

15.2 Quadratic Equations: Completing the Square

Objectives

Create perfect square trinomials by completing the square.

Solve quadratic equations by completing the square.

Success Strategy

Review Section 12.4 on factoring perfect square trinomials before starting this section.

Understand Concepts

Go to Software First, read through Learn in Lesson 15.2 of the software. Then, work through the problems in this section to expand your understanding of the concepts related to completing the square.

The method of solving quadratic equations by completing the square is introduced in this section. This method is based on the forms of perfect square trinomials, $(x-a)^2 = x^2 - 2ax + a$ and $(x+a)^2 = x^2 + 2ax + a$. The final step of completing the square uses the square root method that was introduced in the previous section. The method of completing the square can be used for quadratic functions that have no constant term or have a constant term that is not the correct square.

Completing the Square

If the leading coefficient is 1,

1. Write the terms of the quadratic polynomial in decreasing order.
2. Find half of the coefficient of the first degree term.
3. Square the result from Step 2 and add it to the polynomial.
4. Factor the resulting perfect square trinomial.

If the leading coefficient is not 1,

1. Factor the leading coefficient out of the polynomial.
2. Proceed as if the leading coefficient is 1 for the polynomial inside the parentheses.

Quick Tip

The leading coefficient can be divided out or factored out.

1. Complete the square: $2x^2 + 20x$.

a. The leading coefficient is not 1, so 2 needs to be factored out. Fill in the parentheses in the expression below.

$2(\qquad\qquad)$

b. The coefficient of x is 10. Find half of this value and then square it.

c. Add the result from part **b.** to the expression inside of the parentheses from part **a.** Rewrite the entire expression, including the 2. Do not simplify.

d. Factor the trinomial inside of the parentheses from part **c.** Rewrite the entire factored expression.

Determine if any mistakes were made while completing the square. If any mistakes were made, describe the mistake and then correctly complete the square to find the actual result.

2. $x^2 + 14x + \underline{\quad} = (x + \underline{\quad})^2$

The coefficient of x is 14; $14 \div 2 = 7$

$7^2 = 49$

$x^2 + 14x + \underline{\ 49\ } = (x + \underline{\ 7\ })^2$

3. $x^2 - 16x + \underline{\quad} = (x - \underline{\quad})^2$

The coefficient of x is 16; $16 \div 2 = 9$

$9^2 = 81$

$x^2 - 16x + \underline{\ 81\ } = (x - \underline{\ 9\ })^2$

4. $3x^2 + 18x + \underline{\quad} = 3(x + \underline{\quad})^2$

$3(x^2 + 6x + \underline{\quad}) = 3(x + \underline{\quad})^2$

The coefficient of x is 6; $6 \div 2 = 3$

$3^2 = 9$

$3(x^2 + 6x + \underline{\ 9\ }) = 3(x + \underline{\ 3\ })^2$

Solving Quadratic Equations by Completing the Square

Steps	Example: $2x^2 - 12x + 8 = 0$
1. Write the equation in the form $ax^2 + bx = c$. That is, write the equation so that all variables are on one side of the equation and the constant is on the other side.	$2x^2 - 12x = -8$
2. Divide both sides of the equation by a, the leading coefficient.	$x^2 - 6x = -4$
3. Find half of the coefficient of the first degree term. Square the result and add it to both sides.	The coefficient of x is -6, so $-6 \div 2 = -3$. $(-3)^2 = 9$. So, $x^2 - 6x + 9 = -4 + 9$.
4. Factor the resulting perfect square trinomial. Simplify the constant side of the equation.	$(x-3)^2 = 5$
5. Use the square root method to solve the equation.	$x - 3 = \pm\sqrt{5}$ $x = 3 \pm \sqrt{5}$

5. What is the solution for the equation if the constant side of the equation is negative during Step 4, such as $(x-4)^2 = -10$?

Name: Date:

Determine if any mistakes were made while solving the equation by completing the square. If any mistakes were made, describe the mistake and then correctly solve the equation by completing the square to find the actual result.

6. $$\begin{aligned} x^2+6x-13&=0\\ x^2+6x&=13\\ x^2+6x-9&=13-9\\ (x-3)^2&=4\\ x-3&=\pm 2\\ x&=3\pm 2\\ x&=1,\ 5 \end{aligned}$$

7. $$\begin{aligned} x^2-5x+1&=0\\ x^2-5x&=-1\\ x^2-5x+\left(\frac{5}{2}\right)^2&=-1+\left(\frac{5}{2}\right)^2\\ \left(x-\frac{5}{2}\right)^2&=\frac{21}{4}\\ x-\frac{5}{2}&=\pm\sqrt{\frac{21}{4}}\\ x&=\frac{5}{2}\pm\frac{\sqrt{21}}{2}=\frac{5\pm\sqrt{21}}{2} \end{aligned}$$

8. $$\begin{aligned} 2x^2+8x-6&=0\\ 2x^2+8x&=6\\ 2(x^2+4x)&=6\\ x^2+4x&=3\\ x^2+4x+4&=3\\ (x+2)^2&=3\\ x+2&=\pm\sqrt{3}\\ x&=-2\pm\sqrt{3} \end{aligned}$$

Skill Check

Go to Software Work through Practice in Lesson 15.2 of the software before attempting the following exercises.

Complete the square.

9. $x^2+12x+\_\_\_\_\_=(x+\_\_\_\_)^2$

10. $x^2-7x+\_\_\_\_\_=(x-\_\_\_\_)^2$

11. $5x^2+10x$

12. $2x^2-16x$

Solve for the variable by completing the square.

13. $x^2 + 8x + 2 = 0$

14. $2x^2 + 5x + 2 = 0$

15. $x^2 + x - 3 = 0$

Apply Skills

Work through the problems in this section to apply the skills you have learned related to completing the square.

16. A local frame shop determines that the revenue function for their custom framing service is $R(p) = 360p - 4p^2$, where p is the base price in dollars for each custom framing job.

a. Set the function equal to 0 and solve for p using the method of completing the square.

b. What do the solutions from part **a.** mean?

17. The height of a golf ball that is hit from the ground at a speed of 128 feet per second can be modeled with the expression $h(t) = -16t^2 + 128t$, where t is the time in seconds after the ball is hit.

a. Set the function equal to 0 and solve for t using the method of completing the square.

b. What do the solutions from part **a.** mean?

Name: Date:

15.3 Quadratic Equations: The Quadratic Formula

Objectives

Write quadratic equations in standard form $ax^2 + bx + c = 0$.

Solve quadratic equations using the quadratic formula.

Success Strategy

Review the order of operations before starting this section to ensure that you evaluate the quadratic formula correctly.

Understand Concepts

Go to Software First, read through Learn in Lesson 15.3 of the software. Then, work through the problems in this section to expand your understanding of the concepts related to the quadratic formula.

The quadratic formula can be used to solve any quadratic equation that is written in standard form.

Quadratic Formula

The solutions of the general quadratic equation $ax^2 + bx + c = 0$, where $a \neq 0$, are

$$x = \frac{-b \pm \sqrt{b^2 - 4ac}}{2a}.$$

1. Why can't the value of a be equal to 0 when using the quadratic formula?

2. The expression inside the radical in the quadratic formula is called the **discriminant**. When the value of the discriminant is positive, the quadratic formula will give two real solutions to the quadratic equation.

a. What is the expression inside the radical of the quadratic formula?

b. If the discriminant is equal to 0, what will the quadratic formula simplify to?

c. Use the answer from part **b.** to determine how many real solutions a quadratic equation will have if the discriminant is 0 and $a \neq 0$.

d. If the discriminant is equal to a negative value, what will the radical expression evaluate to?

e. How many real solutions will a quadratic equation have if the discriminant is a negative value?

3. Solve the equation $2x^2 = -x + 2$ using the quadratic formula.

a. First write the equation in standard form $ax^2 + bx + c = 0$.

b. Determine the values of a, b, and c. Remember to keep track of the signs of the coefficients.

c. Substitute the values from part **b.** into the quadratic formula. Do not simplify. Remember to place any negative values in parentheses.

d. Simplify the formula from part **c.**

e. How many real solutions are there to the equation $2x^2 = -x + 2$?

4. Some equations will need to be simplified before you can determine the values of a, b, and c.

a. Write the equation $(3x-1)(x+2) = 4x$ in standard form $ax^2 + bx + c = 0$.

b. Determine the values of a, b, and c. Remember to keep track of the signs of the coefficients.

c. Substitute the values from part **b.** into the quadratic formula. Do not simplify. Remember to place negative values in parentheses.

d. Simplify the formula from part **c.**

Quick Tip

Substitute all values into the quadratic formula before performing any simplifications. This will help prevent any mistakes or the loss of any negative signs.

Name: Date:

e. How many real solutions are there to the equation $(3x-1)(x+2)=4x$?

Skill Check

Go to Software Work through Practice in Lesson 15.3 of the software before attempting the following exercises.

Solve for the variable using the quadratic formula.

5. $x^2 - 4x - 1 = 0$

6. $9x^2 + 12x - 2 = -6$

7. $x^2 - 2x + 1 = 2 - 3x^2$

8. $(2x+1)(x+3) = 2x+6$

Apply Skills

Work through the problems in this section to apply the skills you have learned related to the quadratic formula.

9. An orange is thrown down from the top of a building that is 300 feet tall with an initial velocity of 6 feet per second. The distance of the object from the ground can be calculated using the equation $d = 300 - 6t - 16t^2$, where t is the time in seconds after the orange is thrown.

a. On a balcony, a cup is sitting on a table located 100 feet from the ground. If the orange is thrown with the right aim to fall into the cup, how long will the orange fall? Round to the nearest hundredth. (**Hint:** The distance is 100 feet.)

Quick Tip

Remember to make sure the solution makes sense in the context of the problem.

b. If the orange misses the cup and falls to the ground, how long will it take for the orange to splatter on the sidewalk? (**Hint:** What is the height of the orange when it hits the ground?)

c. Approximately how much longer would it take for the orange to fall to the sidewalk than it would for the orange to fall into the cup?

Quick Tip

Velocity can be split into horizontal and vertical components which depend on the angle of motion. When the angle is 45°, the horizontal and vertical components are equal.

10. Merida is practicing archery with her recurve bow. Her target is the top of a 3-foot tall bale of hay that is 400 feet away. She aims at a 45° angle and shoots the arrow with an initial velocity of 140 feet per second. The height of the arrow can be described by $h = 99t - 16t^2 + 5$, where 99 is the vertical velocity of the arrow, h is the height of the arrow, and t is the time in seconds that passes after the arrow leaves the bow.

a. Solve the equation $3 = 99t - 16t^2 + 5$ to determine the time in seconds when the height of the arrow will be 3 feet. Round your answer to the nearest hundredth.

b. When shot at a 45° angle, the horizontal velocity of the arrow is also 99 feet per second. Use this velocity to determine how long will it take the arrow to reach the bale of hay? Round your answer to the nearest hundredth. (**Hint:** Use the $d = rt$ formula.)

c. Did Merida hit the target, undershoot the target, or overshoot the target? (**Hint:** Compare the answers from part **a.** and part **b.**)

Name: Date:

15.4 Applications

Objectives

Solve applied problems using quadratic equations.

Success Strategy

Be sure to review the four methods of solving quadratic equations presented in the first three sections of this chapter before starting this section.

Understand Concepts

Go to Software First, read through Learn in Lesson 15.4 of the software. Then, work through the problems in this section to expand your understanding of the concepts related to applications of quadratic equations.

Many types of application problems can be solved with quadratic equations. For example, problems that use geometry and the Pythagorean Theorem often involve quadratic equations. Other types of problems that involve quadratic equations are distance-rate-time problems, which have been used throughout the workbook, and work problems, which were introduced in Section 13.5.

When solving application problems that involve quadratic equations, it is important to verify that the solutions you obtain make sense in the context of the problem. Work through the following true-or-false problems to test your understanding of when solutions fit the context.

True or False: Determine whether each statement is true or false. Rewrite any false statement so that it is true. (There may be more than one correct new statement.)

1. When determining the length of the side of a pool, the length can be a negative value.

2. When calculating the amount of money in a checking account, the amount of money can be negative.

3. When determining the amount of time it takes two people to complete a task together, the time should be a negative value.

4. When calculating the price of a concert ticket, the price should be positive.

Skill Check

Go to Software Work through Practice in Lesson 15.4 of the software before attempting the following exercises.

Solve for the variable.

5. $\dfrac{960}{200-x}-\dfrac{960}{200+x}=2$

6. $\dfrac{1}{x}+\dfrac{1}{x+24}=\dfrac{1}{5}$

7. $(x+7)^2+x^2=13^2$

8. $2(x-4)(x-4)=5000$

Apply Skills

Work through the problems in this section to apply the skills you have learned related to applications of quadratic equations.

9. Jack and Diane are decorating a nursery room for their baby, who will be born in a few months. Working together, they can completely decorate the nursery in 4 hours. Working alone, it would take Diane 6 hours longer to decorate the nursery than it would take Jack. How long would it take Jack and Diane to decorate the nursery by themselves?

a. Use the table to set up a rational equation to describe the situation. Use the variable x to represent the time it takes Jack to decorate the nursery by himself.

Person(s)	Time of Work (in Hours)	Part of Work Done in 1 Hour
Jack		
Diane		
Together		

b. Solve the equation from part **a.**

c. Which solution from part **b.** makes sense in the context of the situation? Explain your reasoning.

d. Use the answer from part **c.** to answer the question from the problem statement.

Name: Date:

10. Lisa traveled to a college that is located 200 miles from the city where she works to train customers how to use the software that her company sells. Due to a traffic jam, her average speed returning was 10 miles per hour less than her average speed going to the college. The total travel time to and from the college was 9 hours. What was Lisa's average speed going to the college?

Quick Tip

Remember that the terms **speed**, **velocity**, and **rate** refer to the same thing.

a. Use the table to set up a rational equation to describe the situation. Use the variable x to represent the average speed going to the college. (**Hint:** The sum of the times that it took Lisa to travel to and from the college is 9 hours.)

	Distance (d)	÷	Rate (r)	=	Time $\left(t = \frac{d}{r}\right)$
Going					
Returning					

b. Solve the equation from part **a.** Round your answer to the nearest tenth.

c. Which solution from part **b.** makes sense in the context of the situation? Explain your reasoning.

d. Use the answer from part **c.** to answer the question from the problem statement.

11. A landscaper was given the task to create a triangular flower garden in the corner of an office building. The landscaper has 12 feet of low fencing to use as a border along one side of the garden. The final garden will have the shape shown in the figure. The landscaper needs to know the remaining side lengths of the triangle to determine the area he will need to cover with fresh topsoil.

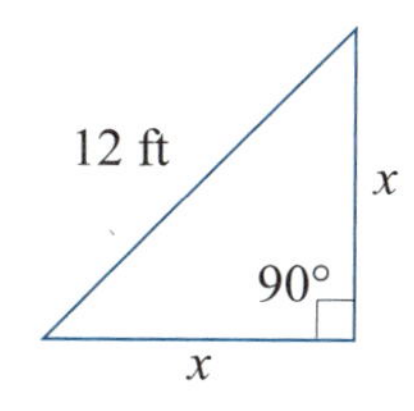

Lesson Link

The Pythagorean Theorem was introduced in Sections 5.7 and 15.1.

a. Use the Pythagorean Theorem to set up an equation which describes the relationship between the side lengths of the flower garden.

b. Solve the equation from part **a.** for the variable. Round your answer(s) to the nearest tenth.

c. Which solution from part **b.** makes sense in the context of the situation? Explain your reasoning.

d. Use the answer from part **c.** to determine the area that the landscaper will need to cover with topsoil.

12. A farmer fenced in a 198-square-meter portion of his field with 58 meters of fencing. What are the length and width of the field? (**Hint:** The length plus the width is equal to half of the perimeter.)

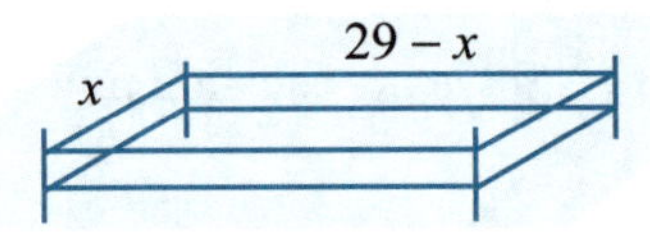

a. Write an equation to express the area of the field.

b. Solve the equation from part **a.** for the variable.

c. Use the answer from part **b.** to determine the length and width of the field.

Name: Date:

15.5 Quadratic Functions

Objectives

Learn about quadratic functions.

Graph quadratic functions.

Find maximum and minimum values of quadratic functions.

Graph quadratic functions using a graphing calculator. (Software only)

Success Strategy

Be sure to do a lot of practice graphing quadratic equations in this section, either by hand or with a graphing calculator. This will help you see the effects that changes in the values of the coefficients *a*, *b*, and *c* have on the location, orientation, and size of the parabola.

Understand Concepts

Go to Software First, read through Learn in Lesson 15.5 of the software. Then, work through the problems in this section to expand your understanding of the concepts related to applications of quadratic functions.

Linear functions were first introduced in Section 9.5. Radical functions were introduced in Section 14.7. Another type of function is the **quadratic function**. A quadratic function has the general form

$$f(x) = ax^2 + bx + c,$$

where a, b, and c are real numbers and $a \neq 0$. The shape of the graph of a quadratic function is called a **parabola**. The box and example graphs below introduce some of the basics of quadratic functions.

Quadratic Functions

1. If $a > 0$, the parabola "opens up."
2. If $a < 0$, the parabola "opens down."
3. The **line of symmetry** of a parabola is a vertical line defined by $x = -\frac{b}{2a}$.
4. The **vertex** is the turning point of the parabola. The *x*-coordinate of the vertex is $x = -\frac{b}{2a}$. To find the *y*-coordinate of the vertex, substitute the *x*-coordinate into the function and solve. The vertex is the lowest point on the parabola when it opens up and the highest point when it opens down.
5. A quadratic function can have 0, 1, or 2 real-number *x*-intercepts. To find the *x*-intercepts, set the function equal to 0 and solve the quadratic equation for *x*.
6. A quadratic function has only one *y*-intercept.

Quick Tip

The **line of symmetry** is also referred to as the **axis of symmetry**.

Lesson Link

Quadratic equations typically have two solutions, but sometimes the solutions are complex numbers. For more information on complex numbers, see Section A.8.

Quick Tip

In the graphs, the dotted line is the line of symmetry. The point where the line of symmetry intersects the parabola is the vertex.

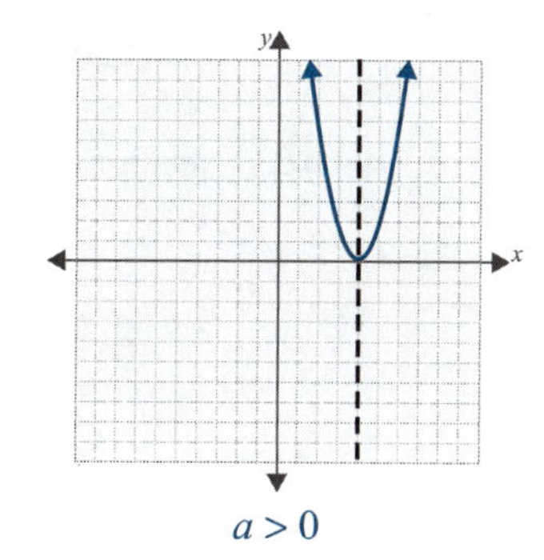

$a > 0$

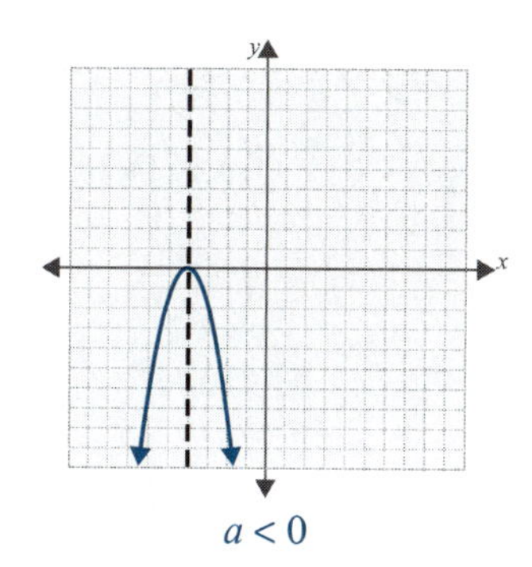

$a < 0$

1. Use the graph to answer the questions.

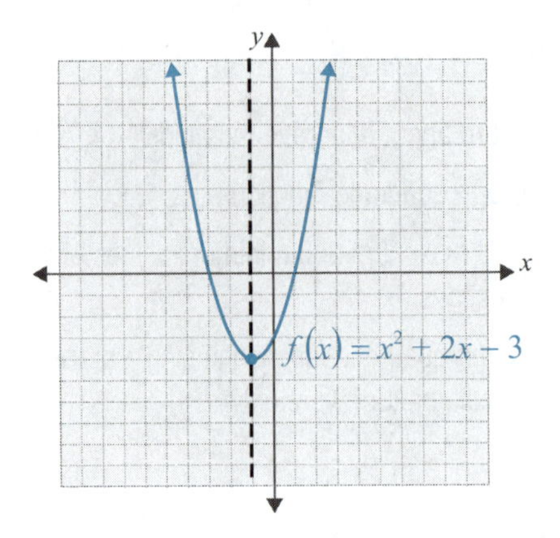

a. Does the graph open up or down?

b. What are the coordinates of the vertex of the parabola?

c. Write the equation for the line of symmetry? (**Hint:** Remember that it is a vertical line.)

d. What relationship do you notice between the vertex and the line of symmetry?

2. Use graphs A, B, and C to work through the following problems.

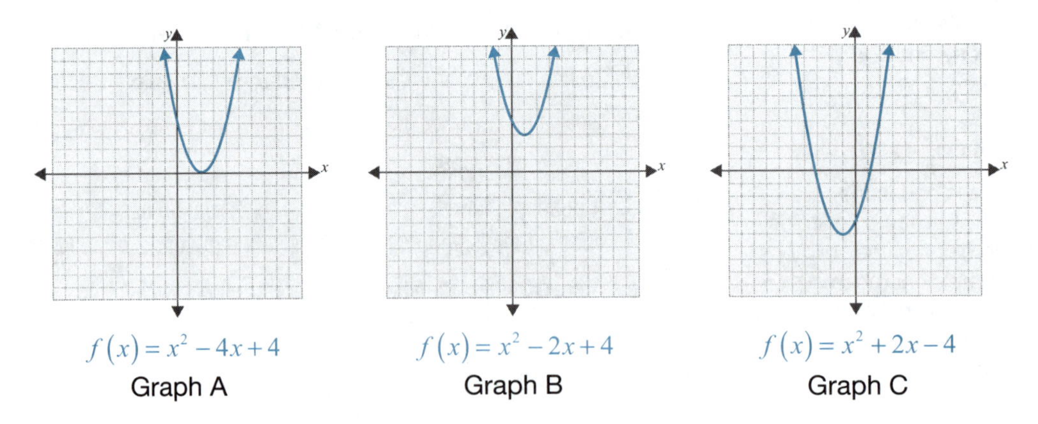

a. Determine how many x-intercepts each function has.

Name: Date:

b. Determine the value of the discriminant for each function.

Lesson Link

The discriminant of the function is the expression $b^2 - 4ac$ from the quadratic formula, which was introduced in Section 15.3.

c. There is a relationship between the discriminant of a quadratic function and the number of x-intercepts of the graph. Fill in the blanks below based on the graphs above and your results from parts **a.** and **b.**

When the discriminant is negative, there are _____ real x-intercept(s).

When the discriminant is zero, there are _____ real x-intercept(s).

When the discriminant is positive, there are _____ real x-intercept(s).

To graph a quadratic function, several solutions to the equation need to be found. A table with randomly chosen x-values can be used to find ordered pairs that satisfy the function. However, the guideline given here will typically make the parabola easier to draw by giving the vertex of the parabola and at least one solution on each side of the vertex. If a more accurate graph of the parabola is needed then you should find the vertex and at least two points on each side of the vertex.

Graphing a Quadratic Function

1. Use the value of the leading coefficient to determine if the parabola opens up or down.
2. Find the coordinates of the vertex.
3. Find any x-intercepts by substituting $y = 0$ and solving for x.
4. Find the y-intercept by substituting $x = 0$ and solving for y.
5. Plot the vertex, any x-intercepts, and the y-intercept. Draw a parabola through the points.

Note: If the quadratic function has no x-intercepts, this process may only result in 1 or 2 ordered pairs to plot. To make the parabola more accurate, find a minimum of 2 ordered pairs that lie on each side of the vertex that satisfy the equation. Find these ordered pairs by substituting values for x into the equation and solving for y.

3. Graph the equation $y = -x^2 - 3x + 4$.

a. Will the parabola open up or down?

b. What are the coordinates of the vertex of the parabola?

c. Find any x-intercepts. If there are no x-intercepts, write "no x-intercepts".

d. Find the y-intercept.

e. Plot the ordered pairs from parts **b.**, **c.**, and **d.** on the graph and draw a parabola through the points.

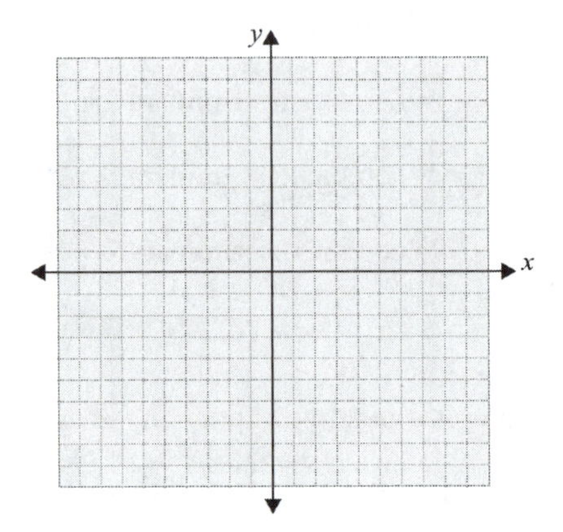

f. Does the parabola you drew open up or down? Does this match the answer to part **a.**?

Minimum and Maximum Values

A parabola that opens up will have a **minimum value** at the vertex.

A parabola that opens down will have a **maximum value** at the vertex.

4. Use the following graphs to work through the following problems.

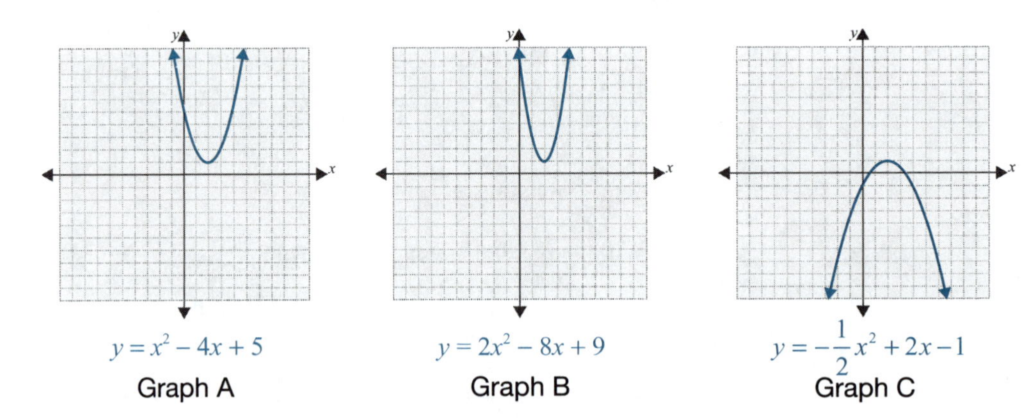

$y = x^2 - 4x + 5$ Graph A

$y = 2x^2 - 8x + 9$ Graph B

$y = -\frac{1}{2}x^2 + 2x - 1$ Graph C

a. Find the vertex of each quadratic equation.

b. Compare the width of the three graphs.

c. Which graphs have a minimum and which graphs have a maximum?

Name: Date:

Skill Check

Go to Software Work through Practice in Lesson 15.5 of the software before attempting the following exercises.

Find the x-intercept(s), y-intercept, and vertex of each quadratic equation.

5. $y = x^2$

6. $y = x^2 - 4x + 5$

7. $2x^2 - 2x - 12 = y$

8. $y + 5 = 2x^2 - 12x$

Apply Skills

Work through the problems in this section to apply the skills you have learned related to applications of quadratic equations.

9. The manager at Barbara's Bombtastic Bakery determined that the demand function for cookies is $q = 4250 - 25p$, where q is the number of cookies sold each week when the price is p cents per cookie. What price should the manager charge per cookie to maximize revenue?

a. The revenue function is defined by $R(p) = q \cdot p$. Find the revenue function for the cookies in terms of the variable p.

b. Find the x-coordinate of the minimum value for the quadratic function from part **a.**

c. Use the answer from part **b.** to calculate the maximum revenue the manager can make on cookies. (**Hint:** The price p is in cents.)

d. Answer the question in the problem statement using a complete sentence.

10. A farmer plans to use 100 feet of fencing to create a temporary rectangular-shaped corral for his goats while their barn is being repaired. The farmer would like to create a corral that has the maximum area possible.

a. Draw a figure to represent the corral. Let x represent the length of one side of the corral. (**Note:** Since parallel sides of a rectangle have the same length, two of the sides will have length x.)

Quick Tip

The **perimeter** of a geometric figure is the sum of all the side lengths.

b. Write an expression to represent the length of the side of the corral that is not x feet long. Remember that there are two sides that do not have a length of x feet.

c. Write a function $A(x)$ to represent the area of the corral.

d. Find the maximum value of the function from part **c.**

e. What are the dimensions of the corral that maximize the area?

Quick Tip

In Problem 11, the marginal cost is determined when the total number of calculators produced is increased by one thousand units.

11. Marginal cost is the change in total cost that occurs when the total number of products produced increases by one unit or a set number of units. The financial manager at a company that produces calculators determines that the marginal cost of manufacturing one type of calculator is $C(x) = 6x^2 - 300x + 12000$, where $C(x)$ is the marginal cost and x is the number of calculators manufactured in thousands.

a. How can you verify that this function has a minimum?

b. How many thousands of calculators should be manufactured to minimize the marginal cost?

c. Use part **b.** to determine the minimum marginal cost to produce this type of calculator.

Name: Date:

Chapter 15 Projects

Project A: Plowing into Math

An activity to demonstrate the use of quadratic equations in real life

Robert and his two sons, Jim and John, currently own and operate a 1000-acre farm. During the morning news, the weatherman forecasted a bright sunny day with a high temperature of 72 degrees—a perfect day to do some plowing. Robert is trying to determine how long it will take him and his sons to get the cornfield ready for planting.

1. Robert decides that Jim will work in the chicken houses today and he and John will plow the cornfield. The last time the field was plowed, it took Robert and John about 4 hours to do the job together. If Robert plows the field by himself it will take him 6 hours less than it would take John. Using your knowledge of how to set up an equation involving work, determine how long it would take each man working by himself to plow the field. Use the variable x to represent the amount of time it takes John to plow the field by himself.

a. Use the table below to organize the information and set up the equation to be solved.

Person(s)	Hours to Complete	Part of Job Completed in 1 Hour
Robert		
John	x	$\frac{1}{x}$
Together	4	$\frac{1}{4}$

b. What is the LCD of all the rational expressions in the equation from part **a.**?

c. Clear the equation from part **a.** of fractions by multiplying both sides of the equation by the LCD.

d. What is the degree of the resulting equation from part **c.**?

e. How many solutions should you have after you solve the equation from part **c.**?

f. Solve the equation from part **c.**

g. Are there any solutions to the equation that do not make sense in the context of the problem? If so, explain.

h. Fill in the actual number of hours it takes each man to do the job alone in the table (column two). Note that the total time it takes them both to do the job working together (4 hours) is less than the time it takes either man working alone to do the job.

2. While his sons are finishing breakfast, Robert changes his mind about the day's plans. It is turning out to be a really beautiful day. He decides that three of them will work together so they can get the field plowed faster, and then go fishing. He knows that the last time they all plowed the cornfield together, it only took them $2\frac{2}{3}$ hours. Determine how much time it takes Jim to plow the field alone.

 a. Use the table to organize the information and set up the equation to be solved. Use the hours you determined in Problem 1 for Robert and John. You should convert the time working together to an improper fraction before putting it in the equation. Use the variable x to represent the amount of time it takes Jim to plow the field by himself.

Person(s)	Hours to Complete	Part of Job Completed in 1 Hour
Robert		
John		
Jim	x	
Together	$2\frac{2}{3}$	

 b. What is the LCD of all the rational expressions in this equation?

 c. Clear the equation of fractions by multiplying both sides of the equation by the LCD.

 d. What is the degree of your resulting equation from part **c.**?

 e. How many solutions should you have after you solve the equation from part **c.**?

 f. Solve the equation from part **c.**

 g. Do you have any solutions to the equation that do not make sense in the context of the problem? If so, explain.

3. As they are heading to the cornfield, Robert's neighbor Eli comes by on his new 100-horsepower enclosed-cab tractor to ask them if they want to go fishing. Robert tells him that they had already planned to go after they finished plowing the cornfield. Eli offers to help and says he could plow a field this size in about 4 hours by himself. How long will it take the 4 men working together to plow the cornfield? (Let x be the time it takes for all 4 men working together to plow the field.) Express your answer as both a fraction in reduced form and a decimal.

Name: Date:

Project B: Gateway to the West

An activity to demonstrate the use of quadratic equations in real life

The Gateway Arch on the St. Louis, Missouri, riverfront serves as an iconic monument symbolizing the westward expansion of American pioneers, such as Lewis and Clark. A nationwide competition was held to choose an architect to design the monument and the winner was Eero Saarinen, a Finnish-American who immigrated to the United States with his parents when he was 13 years old. Construction began in 1962 and the monument was completed in 1965. The Gateway Arch is the tallest monument in the United States. It is constructed of stainless steel and weighs more than 43,000 tons. Although the arch is heavy, it was built to sway with the wind to prevent it from being damaged. In a 20 mph wind, the arch can move up to 1 inch. In a 150 mph wind, the arch can move up to 18 inches.

1. If you were to place the Gateway Arch on a coordinate plane centered around the y-axis, then the equation $y = -0.00635x^2 + 630$ could be used to model the height of the arch in feet.

a. The general form for a quadratic function is $y = ax^2 + bx + c$. Identify the values for a, b, and c from the Gateway Arch equation.

b. Find the vertex of the Gateway Arch equation.

c. Does the vertex represent a maximum or a minimum? Explain your answer based on the coefficients of the equation.

d. What is the height of the arch at its peak?

e. Write the equation for the axis of symmetry of the Gateway Arch.

f. Find the x-intercepts of the Gateway Arch equation using the square root method that was introduced in Section 15.1. Round your answer to the nearest foot.

2. Using the coordinate plane below and the information from Problem 1, draw a sketch of the graph of the Gateway Arch quadratic equation.

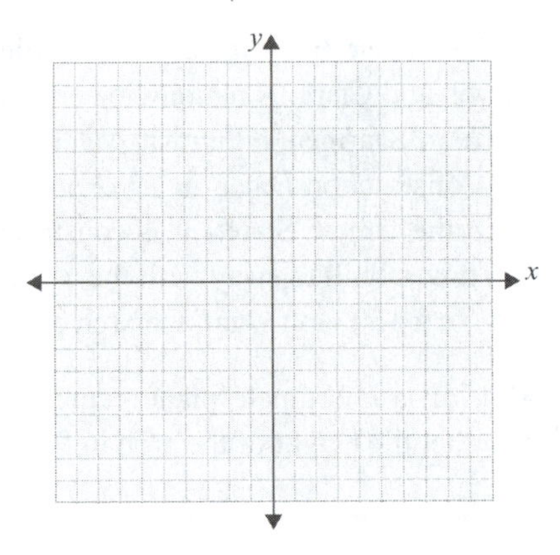

3. How far apart are the legs of the arch at its base?

4. The Gateway Arch equation is a mathematical model. Look up the actual values for the height of the arch and the distance between the legs of the arch at its base on the Internet and describe how they compare to the values calculated using the equation.

Name: Date:

Math@Work

Other Careers in Mathematics

Earning a degree in mathematics or minoring in mathematics can open many career pathways. While a degree in mathematics or a field which uses a lot of mathematics may seem like a difficult path, it is something anyone can achieve with practice, patience, and persistence. Three growing fields of study which rely on mathematics are actuarial science, computer science, and operations research. While each of these fields involves mathematics, they require special training or additional education outside of a math degree. A brief description of each career is provided below along with a source to find more information about these careers.

Growing Fields of Study

Actuarial Science: The field of actuarial science uses methods of mathematics and statistics to evaluate risk in industries such as finance and insurance. Visit www.beanactuary.org for more information

Computer Science: From creating web pages and computer programs to designing artificial intelligence, computer science uses a variety of mathematics. Visit computingcareers.acm.org for more information.

Operations Research: The discipline of operations research uses techniques from mathematical modeling, statistical analysis, and mathematical optimization to make better decisions, such as maximizing revenue or minimizing costs for a business. Visit www.informs.org for more information.

There are numerous careers that have not been discussed in this workbook. Exploring career options before choosing a major is a very important step in your academic career. Learning about the career you are interested in before completing your degree can help you choose courses that will align with your career goals. You should also explore the availability of jobs in your chosen career and whether you will have to relocate to another area to be hired. The following web sites will help you find information related to different careers that use mathematics. Another great resource is the mathematics department at your college.

The **Mathematical Association of America** has a website with information about several careers in mathematics. Visit www.maa.org/careers to learn more.

The **Society for Industrial and Applied Mathematics** also has a webpage dedicated to careers in mathematics. Visit www.siam.org/careers to learn more.

The **Occupational Outlook Handbook** is a good source for information on educational requirements, salary ranges, and employability of many careers, not just those that involve mathematics. Visit http://www.bls.gov/ooh/ to learn more.

Appendix

Study Skills

A.1 US Measurements

A.2 The Metric System

A.3 US to Metric Conversions

A.4 Absolute Value Equations and Inequalities

A.5 Synthetic Division and the Remainder Theorem

A.6 Graphing Systems of Linear Inequalities

A.7 Systems of Linear Equations in Three Variables

A.8 Introduction to Complex Numbers

A.9 Multiplication and Division with Complex Numbers

A.10 Standard Deviation and z-Scores

A.11 Mathematical Modeling

Appendix Projects

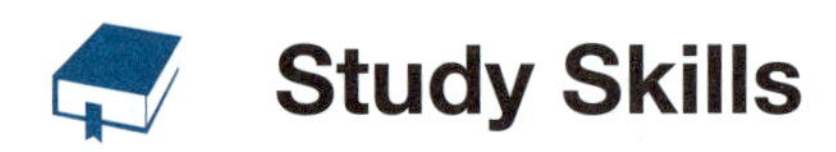

Study Skills

Preparing for a Final Math Exam

Hopefully, throughout the semester you have been attending class, taking good notes, doing your homework, and asking questions in and out of class. If you haven't, then you may have reason to be concerned. Since math concepts build on one another, a final exam in math is not one you can study for in a night or even a day or two. To pull all the concepts together for the semester you should plan to start at the very least, one week ahead of time, or even better, two weeks ahead of time. Being comfortable with the material is key to going into the exam with confidence and lowering your anxiety.

Before you start preparing for the exam, you should ask your professor some questions:

1. What is the date and time of the exam, and will the exam be given in the regular classroom or some other location? Often final exams in college are longer than regular tests. They may be given at times that differ from your regular classroom hours and possibly given at a different location.

2. How many questions will be on the exam and what are their point values? What format will the questions have: multiple choice, free response, essay, applications, true/false, fill-in-the-blanks, etc.? Knowing the types of questions to expect is essential information so that you are prepared. The questions on final exams are often presented in a different format than the ones given in class throughout the semester.

3. Is there a time limit on the exam? If you experience test anxiety on timed tests, be sure to speak to your professor about it and see if you can receive accommodations that will help reduce your anxiety, such as extended time or an alternate testing location. Your professor may be willing to let you start the exam earlier or break-up your testing session into two or more sessions.

4. Will you be able to use a formula sheet, calculator, and/or scrap paper on the exam? If you are not allowed to use a formula sheet, you should write down important formulas and memorize them. Most of the time math professors will advise you of the formulas you need to know for an exam, but in case they don't, it doesn't hurt to ask. If you cannot use a calculator on the exam, be sure to practice doing calculations by hand when you are preparing for the exam and go back and check them using the calculator. If you can use scrap paper on the exam, then be sure to number the problems as you work through the exam and circle your answers in case the professor gives partial credit. This will also make it easier for you to go back and check over the exam if you have time at the end.

A Week Before the Exam

1. Decide where to study for the exam and with whom. Make sure it's a comfortable study environment with few outside distractions. If you are studying with others, make sure the group is small and that the people in the group are motivated to study and do well on the exam. Plan to have snacks and water with you for energy and to avoid having to delay studying to go get something to eat or drink. Be sure and take small breaks every hour or two to keep focused and minimize frustration.
2. Organize your class notes and any flash cards with vocabulary or formulas. Hopefully, you have been taking notes throughout the semester and creating flash cards as suggested in Chapter 2 for vocabulary, formulas, and theorems. If you haven't used flash cards for vocabulary, go back through your notes and highlight the vocabulary with a particular-colored highlighter. Create a formula sheet to use on the exam, if the professor allows. If not, then you can use the formula sheet to memorize the formulas that will be on the exam.
3. Start studying for the exam. Studying a week before the exam gives you time to ask your instructor questions as you go over the material. Don't spend a lot of time reviewing material you already know. Go over the most difficult material or material that you don't understand so you can ask questions about it. Be sure to review old exams, and work through any questions you missed.

3 Days Before the Exam

1. Make yourself a practice test consisting of the problem types that the instructor has told you will be on the exam. Use the same number and format(s) of questions as will be on the actual exam. Be sure to mix the questions up and don't go in the order that the professor covered them in class.
2. Ask your instructor or classmates any questions that you have about the practice test so that you have time to go back and review the material you are having difficulty with before the exam.

The Night Before the Exam

1. Make sure you have all the supplies you will need to take the exam: formula sheet and calculator, if allowed, scratch paper, plain and colored pencils, highlighter, erasers, graph paper, extra batteries, etc.
2. If you won't be allowed to use your formula sheet, review it to make sure you know all the formulas. Right before going to bed, review your notes and study materials, but do not stay up all night to "cram". You will do better if you are rested and alert.
3. Go to bed early and get a good night's sleep.

The Day of the Exam

1. Get up in plenty of time to get to your exam without rushing. Eat a good breakfast and don't drink too much caffeine, which can make you anxious.
2. Review your notes, flash cards, and formula sheet again, if you have time.
3. Gather up all your materials and get to class early to get a good seat and get yourself organized and mentally prepared.

During the Exam

1. Put your name at the top of your exam immediately. If you are not allowed to use a formula sheet, before you even look at the exam, do what is called a "brain drain" or "data dump." Recall as much of the information on your formula sheet as you possibly can and write it either on the scratch paper or in the exam margins if scratch paper is not allowed. You have now transferred over everything on your "mental cheat sheet" to the exam to help yourself as you work through the exam. This should boost your confidence and make you less anxious!
2. Scan the questions on the exam, marking the ones you know how to do immediately. These are the problems you will do first. Also note any questions that have a higher point value. You should try to work these next or be sure to leave yourself plenty of time to do them later on.
3. Read the directions carefully as you go through the exam and make sure you have answered the questions being asked. Also, check your solutions as you go. If you do any work on scratch paper, write down the number of the problem on the paper and highlight or circle your answer. This will save you time when you review the exam. The instructor may also give you partial credit for showing your work.
4. As you go through the exam, if you recall other formulas or important procedures, stop and write them down before you forget them.
5. Work the harder or longer problems that you know how to do next. Show all work and check your solutions.
6. If you get to a problem you don't know how to do, skip it and come back after you finish all the ones you know how to do. You shouldn't always start at the beginning of the exam and work problems consecutively. This can waste time. Also, a problem you do later may jog your memory on how to do the problem you skipped.
7. For multiple choice questions, be sure to work the problem first before looking at the answer choices. If your answer is one of the choices, then go ahead and mark it. If not, then review your math work. Try starting with the answer choices and working backwards to see if any of them work in the problem. If this doesn't work, see if you can eliminate any of the answer choices and make an educated guess from the remaining ones. Mark the problem to come back to later when you review the exam.

8. Once you have an answer for all the problems, review the entire exam. Try working the problems differently and comparing the results or substitute the answers into the equation to verify they are correct. Do not think that those people who finish the exam early are necessarily smarter. Do not worry about what others think. You are in control of your own time—and your own success!
9. Staple your scratch paper to your exam before turning it in. Most math instructors give partial credit if you show your work or may give you the extra point or two needed to pass as a result of your efforts. Even if partial credit isn't given, it will allow you to figure out what you did wrong and correct it before your next math test or exam.

Checklist for the Exam

1. Date of the Exam: ______________________
2. Time of the Exam: ______________________
3. Location of the Exam: ____________________
4. There are _____ questions on the exam worth ______ points each.
5. The question types on the exam include: ____.
6. Items to bring to the exam:

   ___ calculator and extra batteries

   ___ formula sheet

   ___ scratch paper

   ___ graph paper

   ___ pencils

   ___ eraser

   ___ colored pencils or highlighter

   ___ ruler or straightedge

   ___ water and light snacks (if allowed)
7. Notes or other things to remember for exam day:

Name: Date:

A.1 US Measurements

Objectives

Recognize the basic units of measure in the US customary system.

Convert from one US measurement to another by using multiplication or division.

Convert from one US measurement to another by using unit fractions.

Success Strategy

If you are planning to pursue a career in the health sciences or pharmaceuticals, you will definitely need to know how to do conversions between different units of measurement. Pay close attention to the method of dimensional analysis presented in this section and do extra practice in the software.

Understand Concepts

Quick Tip

Many measurements have a common abbreviation.

inch = in.
foot = ft
yard = yd
mile = mi
ounce = oz
pound = lb
ton = T
fluid ounce = fl oz
cup = c
pint = pt
quart = qt
gallon = gal
second = sec
minute = min
hour = hr

Quick Tip

Another term for **capacity** is **volume**. Volume is a way to describe how much space an object takes up or how much it can hold.

Go to Software First, read through Learn in Lesson A.1 of the software. Then, work through the problems in this section to expand your understanding of the concepts related to US measurements.

The following table shows some common equivalent measures used in the US customary system.

Equivalent Measurements in the US Customary System

Length			Weight		
12 inches	=	1 foot	16 ounces	=	1 pound
36 inches	=	1 yard	2000 pounds	=	1 ton
3 feet	=	1 yard			
5280 feet	=	1 mile			
Capacity			**Time**		
8 fluid ounces	=	1 cup	60 seconds	=	1 minute
2 cups	=	1 pint	60 minutes	=	1 hour
2 pints	=	1 quart	24 hours	=	1 day
4 quarts	=	1 gallon	7 days	=	1 week

1. Perform an Internet search using the key words "US customary system history" to find the answers to the following questions.

a. The US customary system is similar to which system of measurement?

b. What are two interesting facts you learned about the US customary system's history?

True or False: Determine whether each statement is true or false. Rewrite any false statement so that it is true. (There may be more than one correct new statement.)

2. The units *ounce* and *fluid ounce* are both used for weight.

3. Both 36 inches and 3 feet are equivalent to 1 yard.

4. A pint is smaller than a quart.

Converting Measurements Using Multiplication and Division

1. Determine which pair of equivalent measurements has units that match the units in the original form and the desired form.
2. Determine if the conversion is from smaller units to larger units or larger units to smaller units.
3. Multiply to convert to smaller units. Divide to convert to larger units. Always multiply or divide by the number in the equivalent measurement that is not equal to one.

5. Convert 3 cups into an equivalent measure in fluid ounces.

 a. Which equivalent measurement has both cups and fluid ounces?

 b. Which is larger, 1 fluid ounce or 1 cup?

 c. When converting from cups to fluid ounces, are you converting from smaller units to larger units or larger units to smaller units?

 d. According to the steps in the box, will you multiply or divide by 8?

 e. Calculate the number of fluid ounces that are in 3 cups.

Read the following paragraph about another method of conversion using unit fractions, then work through the problems.

Quick Tip

The reciprocal of each unit fraction is also a unit fraction. The value that goes in the numerator or denominator depends on how the unit fraction will be used.

Converting values using **unit fractions** is very common in the sciences. A unit fraction is a fraction that is equivalent to 1. Unit fractions can be made from equivalent measurements because each part of the equivalent measurement represents the same amount. For instance, $\frac{1 \text{ foot}}{12 \text{ inches}}$ and $\frac{1 \text{ minute}}{60 \text{ seconds}}$ are unit fractions.

Since the value of a unit fraction is 1, multiplying an expression by a unit fraction does not change the value of the expression. Another term for this process is **dimensional analysis**. The main goal of dimensional analysis is to rewrite the original measurement so that it has a different unit of measurement.

Name: Date:

Quick Tip

Multiple unit fractions can be used in an expression to convert measurements. This method is described in Problem 7.

Converting Measurements Using Unit Fractions

1. Determine which unit fraction has units that match the original form and the desired form.
2. Write the unit fraction so that the numerator matches the desired unit and the denominator matches the original units.
3. Multiply the original measurement by the unit fraction and simplify.

For example, to convert 2 feet to inches, $\frac{2 \text{ \cancel{feet}}}{1} \cdot \frac{12 \text{ inches}}{1 \text{ \cancel{foot}}} = 24$ inches.

6. Convert 5 days into an equivalent measure in hours.

a. Which equivalent measurement has both days and hours?

b. Write the equivalent measurement as a unit fraction. Make sure the units in the denominator match the original measurement and the units in the numerator match the desired measurement.

Lesson Link

Multiplication with fractions was introduced in Sections 2.1 and 2.2.

c. Multiply 5 days by the unit fraction and simplify.

7. Convert 3 tons into an equivalent measure in ounces.

a. Notice that there isn't an equivalent measurement that gives the amount of ounces in a ton. For this conversion, you will need to use more than one equivalent measurement. Which equivalent measurement has tons?

b. Is there an equivalent measurement that connects the smaller unit from the answer of part **a.** to ounces? If so, what is it?

Quick Tip

When converting measurements with the correct unit fractions, all units should cancel out except the desired unit, which will appear in the numerator of one of the unit fractions.

c. The conversion expression has been set up for you. Use the equivalent measurement from parts **a.** and **b.** to write unit fractions using the fraction bars provided. Remember that any unit in the numerator needs to have a corresponding unit in the denominator, with the exception of the desired unit.

$\frac{3 \text{ tons}}{1} \cdot \frac{\quad}{\quad} \cdot \frac{\quad}{\quad}$

d. Simplify the expression from part **c.** to determine how many ounces are in 3 tons.

Skill Check

Go to Software Work through Practice in Lesson A.1 of the software before attempting the following exercises.

Convert the units of measurement as indicated.

8. 5 miles = _______ feet

9. 294 days = _______ weeks

10. 880 ounces = _______ pounds

11. 10 cups = ______ fluid ounces

Apply Skills

Work through the problems in this section to apply the skills you have learned related to US measurements.

12. Barbara's Bombtastic Bakery uses 1 pint of milk to make each batch of milk chocolate cupcakes. How many batches of milk chocolate cupcakes can be made from 1 gallon of milk?

Quick Tip

These conversions can be done all in one step or in multiple steps. To use multiple steps for Problem 12, convert gallons to quarts first and then convert quarts to pints.

13. On your next birthday, how old will you be in days? How old will you be in seconds? (We will assume there are 365 days in a year and ignore leap years. As a result, these answers will be an approximate amount of time.)

a. How many years old will you be on your next birthday?

b. Use your age from part **a.** to determine how many days old you will be on your next birthday.

c. Use your age from part **a.** to determine how many seconds old you will be on your next birthday.

Name: Date:

A.2 The Metric System

Objectives

Learn the metric units of measurement for length.

Learn the metric units of measurement for mass.

Learn the metric units of measurement for volume.

Learn the metric units of measurement for liquid volume.

Success Strategy

The metric system is the standard measurement system for most countries outside of the United States. If you plan to travel internationally, it would be helpful to learn this system. Create your own mnemonic device to help you remember the prefixes of the system.

Understand Concepts

Go to Software First, read through Learn in Lesson A.2 of the software. Then, work through the problems in this section to expand your understanding of the concepts related to the metric system.

1. Perform an Internet search using the keywords "metric system history" to answer the following questions.

a. When was the metric system developed?

b. The metric system was designed with several key features in mind. What were these features?

Lesson Link

Mass is different than **weight**. Mass is the amount of material in an object. The mass of an object does not change depending on location, unlike weight. This is discussed more in Section A.3.

Quick Tip

The prefix *deka-* is also written as *deca-*. They are pronounced the same way and mean the same thing.

The basic units of measurement in the metric system are **meters** (m) for length, **grams** (g) for mass, **cubic meters** (m^3) for volume (or capacity), and **liters** (L) for liquid volume. The following table shows the equivalent measurements between *units of length* in the metric system. Unlike the seemingly arbitrary equivalent measurements in the US customary system, the metric system scales units of measurement by powers of 10. The prefixes are the key difference between the measurement sizes. The equivalent measurements for weight, volume, and liquid volume follow the same pattern.

Metric Measures of Length		
1 **milli**meter (mm)	=	0.001 meter
1 **centi**meter (cm)	=	0.01 meter
1 **deci**meter (dm)	=	0.1 meter
1 meter (m)	=	1.0 meter (the basic unit)
1 **deka**meter (dam)	=	10 meters
1 **hecto**meter (hm)	=	100 meters
1 **kilo**meter (km)	=	1000 meters

Note: A mnemonic that can be used to remember the order of the prefixes in the metric system, from largest to smallest, is **K**ing **H**enry **D**ied **B**y **D**rinking **C**hocolate **M**ilk. This corresponds to kilo-, hecto-, deka-, base, deci-, centi-, and milli-.

2. The metric measurements for mass and liquid volume follow the same pattern as shown in the Metric Measures of Length table. The prefixes of milli-, centi-, deci-, deka-, hecto-, and kilo- remain the same. The only part that will change is the base measurement of meter, gram, and liter. Create a table similar to the one in Problem 1 using either gram or liter as the base unit of measurement.

Metric Measures of ________
=
=
=
=
=
=
=

True or False: Determine whether each statement is true or false. Rewrite any false statement so that it is true. (There may be more than one correct new statement.)

3. A dekameter is larger than a decimeter.

4. 1000 kilograms is equal to 1 gram.

5. 100 centimeters is equal to 1 meter.

Converting Measurements within the Metric System Using Powers of 10

To change a measurement to one that is	Example
one unit smaller, multiply by 10	6 cm = 60 mm
two units smaller, multiply by 100	3 m = 300 cm
and so on.	
To change a measurement to one that is	
one unit larger, divide by 10	35 mm = 3.5 cm
two units larger, divide by 100	250 cm = 2.5 m
and so on.	

Quick Tip

Within the metric system, dimensional analysis can be used to convert between units. However, since each equivalence relation is a power of 10, the conversions can be performed more easily by multiplying or dividing by powers of 10, as shown here.

Name: ______________________________ Date: ______________________________

Converting Measurements within the Metric System Using a Number Line

1. Create a number line with the prefixes of the metric system, ordered from greatest to least, like the one shown here.

km hm dam m dm cm mm

2. Write the measurement so that the decimal point is above the correct prefix for the original units and each digit is written in the gaps between the prefixes.
3. Move the decimal point to the prefix of the desired measurement. Fill in any place values without digits with zeros.

For example, to convert 13.5 meters to millimeters:

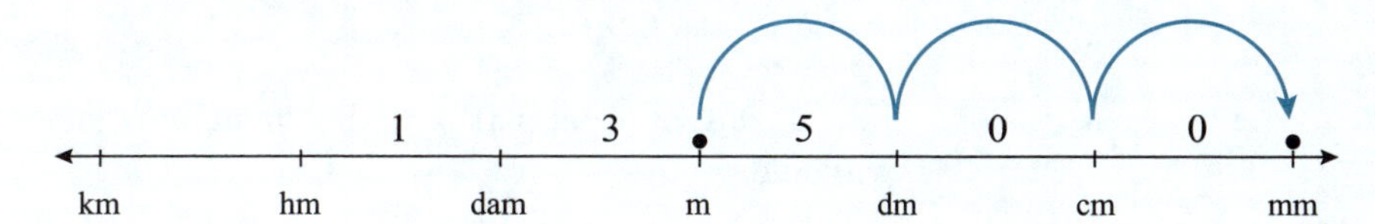

Skill Check

Go to Software Work through Practice in Lesson A.2 of the software before attempting the following exercises.

Convert the units of measurement as indicated.

6. 3 meters = _______ mm

7. 5600 grams = _________ kilograms

8. 569 mL = _______ L

Quick Tip

When the units are squared, the power of 10 that the original unit is multiplied by will be squared. When the units are cubed, the power of 10 that the original unit is multiplied by will be cubed.

9. 73 m^3 = _______ dm^3 (**Hint:** 1 m = 10 dm. Raise both sides of this equivalent measurement to a power of 3.)

Apply Skills

Work through the problems in this section to apply the skills you have learned related to the metric system.

10. An Olympic-sized swimming pool is 50 m long, 25 m wide, and 3 m deep. How many liters of water can the pool hold?

a. What is the volume of the pool in cubic meters?

b. Use the answer from part **a.** to determine how many kiloliters of water the swimming pool will hold.

Quick Tip

1 cubic meter is equal to 1 kiloliter.

c. Use the answer from part **b.** to determine how many liters of water the swimming pool will hold.

Lesson Link

A **metric tonne** is not the same as a **ton** in the US system of measurement. The difference between the two measurements is discussed in Section A.3.

11. The average weight of a female Asian elephant is 2.71 metric tonnes. A metric tonne is equal to 1000 kilograms.

a. Determine the average weight of a female Asian elephant in kilograms.

b. Determine the average weight of a female Asian elephant in grams.

c. Determine the average weight of a female Asian elephant in milligrams.

Name: Date:

A.3 US to Metric Conversions

Objectives

Use the US customary and metric equivalents for measures of temperature.

Use the US customary and metric equivalents for measures of length.

Use the US customary and metric equivalents for measures of area.

Use the US customary and metric equivalents for measures of volume.

Use the US customary and metric equivalents for measures of mass.

Success Strategy

You should focus on performing the conversions in this section using dimensional analysis instead of memorizing the equivalent measurements between the two systems, unless required by your instructor.

Understand Concepts

Go to Software First, read through Learn in Lesson A.3 of the software. Then, work through the problems in this section to expand your understanding of the concepts related to US to metric conversions.

Temperature

Temperature in the US customary system is measured in **degrees Fahrenheit**. Temperature in the metric system is measured in **degrees Celsius**.

To convert between degrees Fahrenheit and degrees Celsius, use the formula $C = \frac{5}{9}(F - 32)$, where C is degrees Celsius and F is degrees Fahrenheit.

Lesson Link

This formula can be solved for the variable F to convert from Celsius to Fahrenheit. Solving a formula for a variable was introduced in Section 8.5.

True or False: Determine whether each statement is true or false. Rewrite any false statement so that it is true. (There may be more than one correct new statement.)

1. Water boils at 212 degrees Fahrenheit. This is equivalent to 100 degrees Celsius.

2. Water freezes at 0 degrees Celsius. This is equivalent to 0 degrees Fahrenheit.

3. When the temperature is –40° F, it is also –40° C.

Quick Tip

Each equivalent measurement presented in this section is an approximation except for 1 in. = 2.54 cm.

Lesson Link

Each conversion can be done with multiplication or division using equivalent measurements or using dimensional analysis. Both of these methods were covered in Section A.1.

Length Equivalent Measurements						
US to Metric				**Metric to US**		
1 in.	=	2.54 cm		1 cm	=	0.394 in.
1 ft	=	0.305 m		1 m	=	3.28 ft
1 yd	=	0.914 m		1 m	=	1.09 yd
1 mi	=	1.61 km		1 km	=	0.62 mi

4. How many meters are in 13 feet? Round to the nearest thousandth.

a. Use the equivalent measurement 1 ft = 0.305 m.

b. Use the equivalent measurement 1 m = 3.28 ft.

c. Are the values from parts **a.** and **b.** exactly the same? Why do you think this happens?

Quick Tip

A **hectare** (ha) is equivalent to 10,000 square meters. An **acre** is equivalent to 4840 square yards.

Area Equivalent Measurements						
US to Metric				**Metric to US**		
1 $in.^2$	=	6.45 cm^2		1 cm^2	=	0.155 $in.^2$
1 ft^2	=	0.093 m^2		1 m^2	=	10.764 ft^2
1 yd^2	=	0.836 m^2		1 m^2	=	1.196 yd^2
1 acre	=	0.405 ha		1 ha	=	2.47 acres

Volume Equivalent Measurements						
US to Metric				**Metric to US**		
1 $in.^3$	=	16.387 cm^3		1 cm^3	=	0.06 $in.^3$
1 ft^3	=	0.028 m^3		1 m^3	=	35.315 ft^3
1 qt	=	0.946 L		1 L	=	1.06 qt
1 gal	=	3.785 L		1 L	=	0.264 gal

Name: Date:

Quick Tip

Knowing how to find equivalent measurements will reduce the amount of factors that need to be memorized.

5. The volume equivalents for cubic inches and cubic feet are found by using the length equivalent measurements and cubing them. This problem will guide you through finding the volume equivalent measurements between cubic inches and cubic centimeters.

   a. Start with the equivalent measurement 1 in. = 2.54 cm. Cube each side of the equation. Round your answers to the nearest thousandth.

   b. Does the answer from part **a.** match the equivalent measurement in the volume equivalent measurements table?

   c. What is the other equivalent measurement that involves inches and centimeters?

   d. Cube each side of the equation from part **c.** Round your answers to the nearest hundredth.

   e. Does the answer from part **d.** match the equivalent measurement in the volume equivalents table?

Read the following paragraph about mass and weight then work through the problems.

The conversion between mass and weight is not exact. Weight is equal to the force of gravity multiplied by the mass of an object. The force of gravity on an object changes depending on the objects location. On Earth, the force of gravity is weaker at the top of a mountain than it is at sea level. This means that your weight will vary depending on location. However, your mass stays the same no matter what your location is. The equivalent measurements given in the table are for measurements taken at sea level.

Quick Tip

A **metric tonne** is equal to 1000 kilograms. The abbreviation for tonne is T. This can be confused with the abbreviation for the US ton, so make sure the context is clear.

Mass Equivalent Measurements						
US to Metric				Metric to US		
1 oz	=	28.35 g		1 g	=	0.035 oz
1 lb	=	0.454 kg		1 kg	=	2.205 lb

Determine if any mistakes were made while converting between units. If any mistakes were made, describe the mistake and then correctly convert between the units to find the actual result. Round to the nearest tenth, when necessary.

6. 2 pounds = ________ grams

$$\frac{2\text{ pounds}}{1}\cdot\frac{16\text{ ounces}}{1\text{ pound}}\cdot\frac{1\text{ gram}}{0.035\text{ ounces}} = 914.3\text{ grams}$$

7. 5 tonne = _______ pounds

$$\frac{5\text{ tonne}}{1}\cdot\frac{2000\text{ kg}}{1\text{ tonne}}\cdot\frac{2.205\text{ lb}}{1\text{ kg}}=22,050\text{ lb}$$

8. Use the information in the tables for US Customary to metric equivalent measurements to answer the following questions.

a. Which is larger, a metric tonne or a US ton?

b. Which is longer, an inch or a centimeter?

c. Which has a larger capacity, a quart or a liter?

9. You've now had some experience working with both the US measurement system and the metric measurement system. Which system do you think is easier to work with? Write an explanation for your choice.

Skill Check

Go to Software Work through Practice in Lesson A.3 of the software before attempting the following exercises.

Convert the units of measurement as indicated. Round to the nearest hundredth

10. 75 °F = ______ °C

11. 5 yd = ______ m

12. 12 cm^2 = ______ $in.^2$

13. 2.5 L = _____ qt

Name: ______________________ Date: ______________________

14. 7 lb = _____ kg

15. 10 m^2 = ______ ft^2

Apply Skills

Work through the problems in this section to apply the skills you have learned related to US to metric conversions.

16. In the United States, cola is sold in 2-liter bottles. How many fluid ounces are in each 2-liter bottle? Round your answers to the nearest hundredth.

a. How many quarts of cola are in each 2-liter bottle?

Lesson Link 

Equivalent measurements in the US Customary System were introduced in Section A.1.

b. Use the answer from part **a.** to determine how many pints of cola are in each 2-liter bottle.

c. Use the answer from part **b.** to determine how many cups of cola are in each 2-liter bottle.

d. Use the answer from part **c.** to determine how many fluid ounces of cola are in each 2-liter bottle.

e. Use the method of dimensional analysis to combine the multiplication or division by equivalent measurements from parts **a.** through **d.** into one step. Perform the conversion. Does dimensional analysis give you the same result as part **d.**?

17. In the United States, sugar is typically sold in 5-pound bags. How many grams are in each 5-pound bag of sugar? Round your answers to the nearest hundredth.

a. How many kilograms are in each 5-pound bag?

b. Use the answer from part **a.** to determine how many grams of sugar are in each 5-pound bag.

18. The 2014 Winter Olympics were held in Sochi, Russia. Due to two consecutive mild winters, Sochi had to store approximately 588,000 cubic yards of snow under insulating blankets on nearby mountains to ensure that they had enough snow for the skiing and snowboarding events.

a. What is the equivalent measurement for yards to meters?

b. Cube each side of the equivalent measurement from part **a.** to find the equivalent measurement for cubic yards to cubic meters. Round your answers to the nearest thousandth.

c. Use the equivalent measurement from part **b.** to determine how many cubic meters of snow were stored. Round your answer to the nearest meter.

Name: Date:

A.4 Absolute Value Equations and Inequalities

Objectives

Solve absolute value equations.

Solve equations with two absolute value expressions.

Solve absolute value inequalities.

Success Strategy

Remember that when solving absolute value equations or inequalities, there will always be two equations or inequalities to solve because the expression inside the absolute value bars can be positive or negative.

Understand Concepts

Go to Software First, read through Learn in Lesson A.4 of the software. Then, work through the problems in this section to expand your understanding of the concepts related to absolute value equations and inequalities.

Absolute value was introduced in Section 7.1. In this section, we will discuss how to solve absolute value equations and inequalities.

Absolute Value

The **absolute value** of a number, written with the notation $|a|$, represents the distance between the number a and 0 on the number line. This value is nonnegative, meaning it is either positive or equal to zero.

$|a| = a$ when $a \geq 0$. For example, $|12| = 12$.

$|a| = -a$ when $a < 0$. For example, $|-7| = 7$.

Quick Tip

The prefix *non-* means *not*, so the word *nonnegative* means *not negative*.

Quick Tip

When working with an equation of the form $n|ax+b|+m=c$, isolate the absolute value expression on one side of the equation before removing the absolute value bars.

Solving Equations with One Absolute Value Expression

For $c > 0$,

1. If $|x| = c$, then $x = c$ or $x = -c$.
2. If $|ax+b| = c$, then $ax+b = c$ or $ax+b = -c$.

For $c < 0$, there are no solutions.

1. Why are there no solutions to the absolute value inequality $|x| = c$ when c is negative (that is, $c < 0$)?

Determine if any mistakes were made while solving the absolute value equations. If any mistakes were made, describe the mistake and then correctly solve the absolute value equation to find the actual result.

2. $|3x-4| = 5$

$3x-4=5$ or $3x-4=-5$

$3x=9$ $\quad$ $3x=-9$

$x=3$ $\quad$ $x=-3$

3. $5|3x+17|-4=51$

$$
\begin{aligned}
5|3x+17| &= 55 \\
|3x+17| &= 11 \\
3x+17 = 11 \quad &\text{or} \quad 3x+17 = -11 \\
3x = -6 \quad &\quad\quad 3x = -28 \\
x = -2 \quad &\quad\quad x = -9\frac{1}{3}
\end{aligned}
$$

Occasionally equations will contain two absolute value expressions. The solving process is similar to equations with only one absolute value expression.

Quick Tip

If two absolute value expressions are equal to each other, then the expressions inside the absolute value bars are either the same or the opposite of one another.

Solving Equations with Two Absolute Value Expressions

If $|a|=|b|$, then either $a=b$ or $a=-b$.

If $|ax+b|=|cx+d|$, then either $ax+b=cx+d$ or $ax+b=-(cx+d)$.

4. Solve the absolute value equation $|x-8|=|x+4|$.

   a. Write an equation of the form $ax+b=cx+d$.

   b. Solve the equation from part **a.** If there is no solution, write "no solution".

   c. Write an equation of the form $ax+b=-(cx+d)$.

   d. Solve the equation from part **c.** If there is no solution, write "no solution".

   e. Check the solutions from parts **b.** and **d.** in the original equation. What are the solution(s) to $|x-8|=|x+4|$?

Name: Date:

Determine if any mistakes were made while solving the absolute value equations. If any mistakes were made, describe the mistake and then correctly solve the absolute value equation to find the actual result.

5. $|3x+1| = |4-x|$

$$\begin{aligned} 3x+1 &= 4-x \\ 3x &= 3-x \\ 4x &= 3 \\ x &= \frac{3}{4} \end{aligned} \quad \text{or} \quad \begin{aligned} 3x+1 &= x-4 \\ 3x &= x-5 \\ 2x &= -5 \\ x &= -\frac{5}{2} \end{aligned}$$

6. $|x+3| = |x-5|$

$$\begin{aligned} x+3 &= x-5 \\ x &= x-8 \\ 2x &= -8 \\ x &= -4 \end{aligned} \quad \text{or} \quad \begin{aligned} x+3 &= -(x-5) \\ x+3 &= -x+5 \\ x &= -x+2 \\ 2x &= 2 \\ x &= 1 \end{aligned}$$

Quick Tip

These pairs of inequalities are often referred to as **compound "and" inequalities** because both inequalities must be true.

Solving Absolute Value Inequalities with < or ≤

For $c > 0$,

1. If $|x| < c$, then $-c < x$ and $x < c$.
2. If $|ax+b| < c$, then $-c < ax+b$ and $ax+b < c$.

7. Solve $|x+3| < 2$.

a. Write an inequality of the form $-c < ax + b$.

b. Solve the inequality from part **a.**

c. Write an inequality of the form $ax + b < c$.

d. Solve the inequality from part **c.**

Quick Tip

When graphing the solution to an inequality on a number line, remember that ( ) are used when the end points of the interval are not solutions and [ ] are used when the end points of the interval are solutions. This notation is the same as that used for interval notation.

e. Graph the combined solutions from parts **b.** and **c.** on the number line.

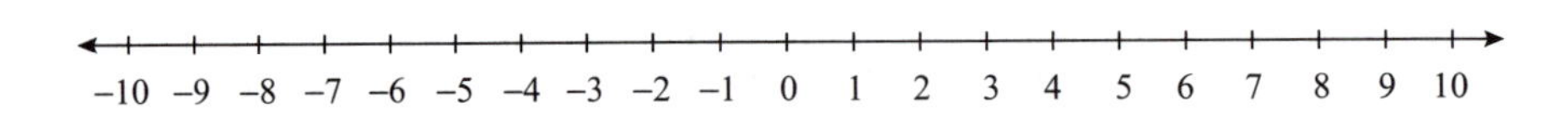

Lesson Link

Interval notation was introduced in Section 8.7.

f. The overlapping part of the solution sets determines the solution of $|x+3|<2$. Write the solution in interval notation.

Quick Tip

These pairs of inequalities are often referred to as **compound "or" inequalities** because only one of the inequalities has to be true.

Solving Absolute Value Inequalities with $>$ or $\geq$

For $c>0$,

1. If $|x|>c$, then $x>c$ or $x<-c$.
2. If $|ax+b|>c$, then $ax+b>c$ or $ax+b<-c$.

8. Solve $|4x-3|>2$.

a. Write an inequality of the form $ax+b>c$.

b. Solve the inequality from part **a.**

c. Write an inequality of the form $ax+b<-c$.

d. Solve the inequality from part **c.**

e. Graph the solution on the number line.

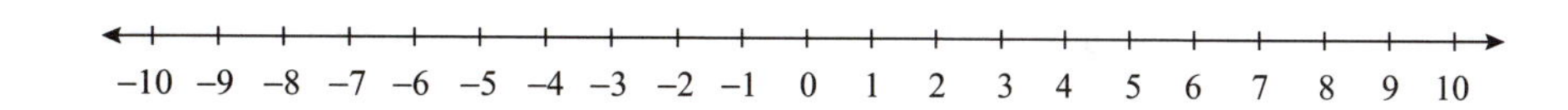

f. The union of the two solution sets determines the solution of $|4x-3|>2$. Write the solution in interval notation.

Skill Check

Go to Software Work through Practice in Lesson A.4 of the software before attempting the following exercises.

Solve for the variable.

9. $|t|=5$

10. $|2x-4|=8$

Solve for the variable. Write the solution sets in interval notation.

11. $|y-4|<5$

12. $|2x-1|\geq 2$

13. $|3x+4|-1<0$

14. $3|4x+5|-5>10$

Apply Skills

Work through the problems in this section to apply the skills you have learned related to absolute value equations and inequalities.

15. In manufacturing, a tolerance is specified to allow the final product to vary slightly without decreasing performance quality. The length x of a screw has a tolerance range defined by $|32 - x| < 2$, where 32 is the target length. What is the tolerance range for the screw length?

16. Gallup performs a daily poll of American adults to determine their mood. On February 26, 2014, 48% of responders reported feeling "a lot of happiness/enjoyment" and 12% of responders reported feeling "a lot of stress/worry." The margin of error for this survey was ±5 percentage points.

a. Determine the range of responders that reported feeling happiness or enjoyment by solving the inequality $|48 - h| \leq 5$.

b. Determine the range of responders that reported feeling stressful or worried by solving the inequality $|12 - w| \leq 5$.

c. Go to http://www.gallup.com/poll/106915/Gallup-Daily-US-Mood.aspx and determine the ranges of happiness and stress for the most recent date reported.

Name: Date:

A.5 Synthetic Division and the Remainder Theorem

Objectives

Divide polynomials using synthetic division.

Use the remainder theorem to find the value of a polynomial at a specific value of x.

Success Strategy

Synthetic division cannot be used for all polynomial division problems. The divisor must have the form $x - c$. Review polynomial long division in Section 11.7 before starting this section.

Understand Concepts

Go to Software First, read through Learn in Lesson A.5 of the software. Then, work through the problems in this section to expand your understanding of the concepts related to synthetic division and the remainder theorem.

Polynomial division was introduced in Section 11.7 in the form of long division. If the divisor in a polynomial division problem has the form $x - c$, then a method called **synthetic division** can be used. Synthetic division is a quick way to divide the polynomials using only the coefficients of the variables.

Synthetic Division

1. Write both the divisor polynomial and dividend polynomial in descending order.
2. Write the **constant divisor**, which is the value of c from the divisor $x - c$, in a partial box. To the right of the constant divisor write the coefficients of the dividend. Use a 0 coefficient to represent any missing terms.
3. Under the coefficients of the dividend, leave space for a row of numbers and draw a horizontal line.
4. Write the leading coefficient of the dividend below the line under the leading coefficient.
5. Multiply the constant divisor by the number that was just written below the line. Write this product above the line and under the next coefficient to the right.
6. Find the sum of the coefficient and the new value from Step 5. Write the sum below the line.
7. Repeat Steps 5 and 6 until all columns are filled in. The final sum is the remainder.
8. Use the bottom row of sums to write a polynomial to represent the quotient. This polynomial will have a degree that is 1 less than the degree of the dividend.

Quick Tip

Writing a polynomial in descending order of degree means that the terms are written with the exponents in decreasing order from left to right.

For example, dividing $x+3\overline{)5x^3+11x^2-3x+1}$ by synthetic division results in the following setup.

$$\begin{array}{r|rrrr} -3 & 5 & 11 & -3 & 1 \\ & & -15 & 12 & -27 \\ \hline & 5 & -4 & 9 & -26 \end{array}$$

Quick Tip

Notice that the constant divisor is negative. This is because the binomial divisor should be written as $x - c$. In this case, we would have $x + 3 = x - (-3)$, so $c = -3$.

Since the dividend is a cubic polynomial and the divisor is a first degree polynomial, the quotient of the division problem will be one degree less. For our example, the quotient is $5x^2 - 4x + 9 - \frac{26}{x+3}$.

1. The polynomial division problem $x+8\overline{)x^3 + 4x^2 + x - 1}$ for synthetic division is set up as follows.

$$\begin{array}{r|rrrr} -8 & 1 & 4 & 1 & -1 \\ & & & & \\ \hline & & & & \end{array}$$

a. Bring the first coefficient 1 down below the line.

b. Multiply the coefficient that was just brought down by the constant divisor –8. Place the result under the second coefficient.

c. Add the second coefficient 4 and the product from part **b.** Write the sum below the line.

d. Multiply the sum from part **c.** by the constant divisor. Place the result under the third coefficient 1.

e. Add the third coefficient and the product from part **d.** Write the sum below the line.

f. Multiply the sum from part **e.** by the constant divisor. Place the result under the fourth coefficient –1.

g. Add the fourth coefficient and the product from part **f.** Place the result below the line. This sum is the remainder.

h. Use the numbers below the line to fill in the boxes in the quotient of the division problem. Each number is the coefficient of a term in a polynomial of degree 2 written in decreasing order. The remainder is written in fraction form with the remainder as the numerator and the divisor $x + 8$ as the denominator.

$$\square x^2 + \square x + \square - \frac{\square}{x+8}$$

2. Redo the division from Problem 1 using long division. You should get the same answer.

3. Which method do you prefer, long division or synthetic division? Write an explanation for your choice.

Name: Date:

Lesson Link

The notation $P(x)$ represents a function. The notation $P(c)$ indicates that the function is evaluated at the value $x = c$. Function notation was introduced in Section 9.5.

The Remainder Theorem

If a polynomial $P(x)$ is divided by $(x - c)$, then the remainder will be $P(c)$.

For example, when $P(x) = 3x^2 - 4x + 7$ is divided by $(x - 2)$, the remainder will be $P(2) = 3(2)^2 - 4(2) + 7 = 11$.

4. The remainder theorem can be used to check if you made any errors while performing synthetic division. Let's check the result from Problem 1.

a. Determine the value of c in the expression $x+8\overline{)x^3 + 4x^2 + x - 1}$.

b. Use the remainder theorem to determine what the remainder will be for $x+8\overline{)x^3 + 4x^2 + x - 1}$ by evaluating the dividend with the value of c found in part **a.**

c. Does this remainder match the remainder you found in part **g.** of Problem 1?

Skill Check

Go to Software Work through Practice in Lesson A.5 of the software before attempting the following exercises.

Divide using synthetic division. Write the quotient polynomial. Check your answer using the remainder theorem.

5. $\dfrac{x^2 - 12x + 27}{x - 3}$

6. $(2x^3 - 4x^2 - 9) \div (x + 3)$

7. $\dfrac{x^4 + 2x^2 - 3x + 5}{x - 2}$

8. $(x^5 - 1) \div (x - 1)$

Apply Skills

Work through the problems in this section to apply the skills you have learned related to synthetic division and the remainder theorem.

9. A moving company uses a box that has a volume of $x^3 + 7x^2 - 6x - 72$ cubic inches.

   a. If the height of the box is $x + 4$, what is the area of the base of the box?

   b. If the height of the box is $x - 3$, what is the area of the base of the box?

Name: Date:

A.6 Graphing Systems of Linear Inequalities

Objectives

Solve systems of linear inequalities graphically.

Use a graphing calculator to graph systems of linear inequalities. (Software only)

Success Strategy

Even if you have a graphing calculator, it is recommended that you solve a few systems of linear inequalities by hand. Using different colored pencils for each region will make it easier to determine the solution, which is where the regions overlap.

Understand Concepts

Go to Software First, read through Learn in Lesson A.6 of the software. Then, work through the problems in this section to expand your understanding of the concepts related to graphing systems of linear inequalities.

Solving a System of Two Linear Inequalities

1. For each inequality, graph the boundary line and shade the appropriate half-plane. Be sure the boundary line is solid for $\leq$ and $\geq$, and dashed for $<$ and $>$.
2. Determine the region of the graph that is common to both half-planes (the region where the shading overlaps).
3. To check the solution set, pick a test-point in the intersection that is not on a boundary line and verify that it satisfies both inequalities.

Lesson Link

The method of solving a system of linear inequalities by graphing was introduced in Section 9.6.

Quick Tip

When colored pencils are unavailable, different shading techniques can be used, such as those shown here.

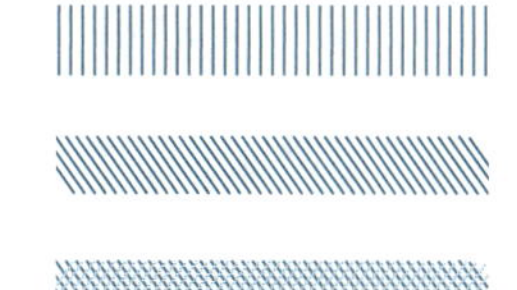

1. Solve the system of linear inequalities $\begin{cases} x \leq 2 \\ y \geq -x+1 \end{cases}$

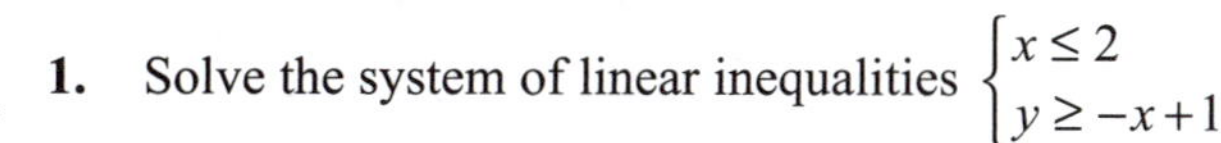

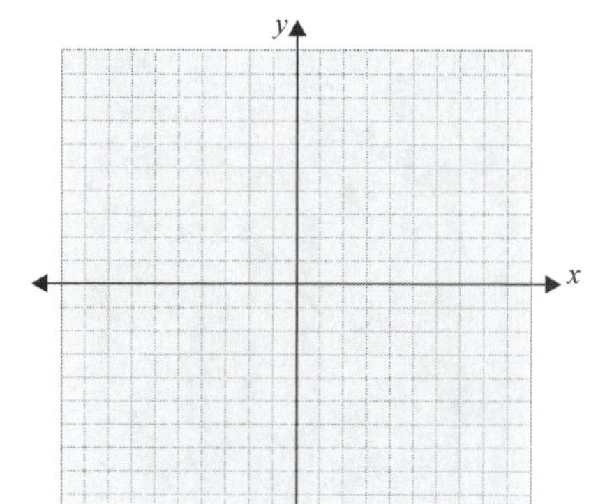

a. Graph $x \leq 2$. Use a colored pencil, or shading technique, to show the solution set.

b. Graph $y \geq -x + 1$. Use a different colored pencil, or different shading technique, to show the solution set.

c. Choose a point from the overlapping solution sets that is *not* on a boundary line. Test this point in both inequalities. Does this point satisfy both inequalities?

2. Match the graph of the solution set to its corresponding system of linear inequalities.

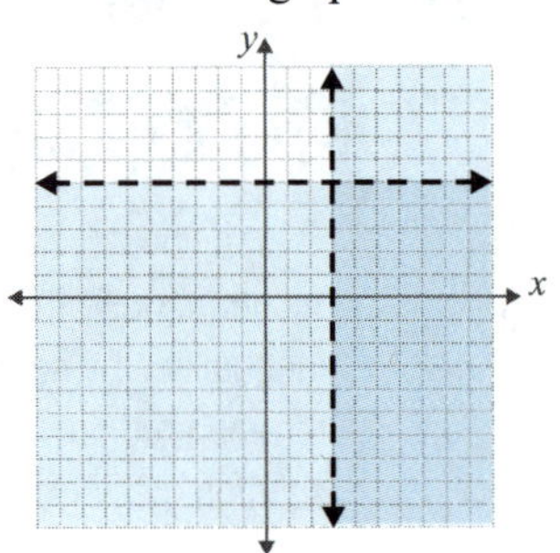

Graph A

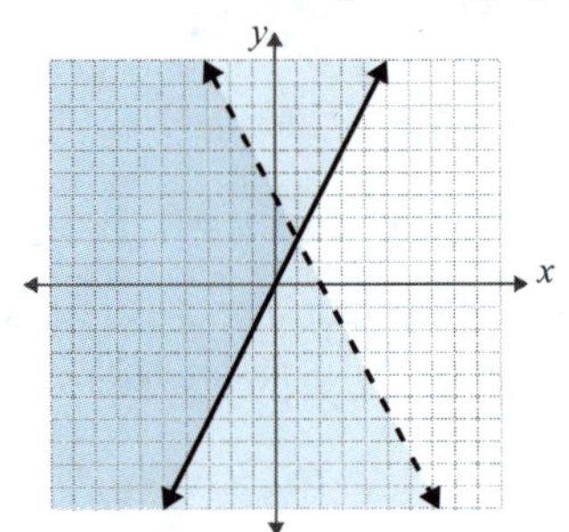

Graph B

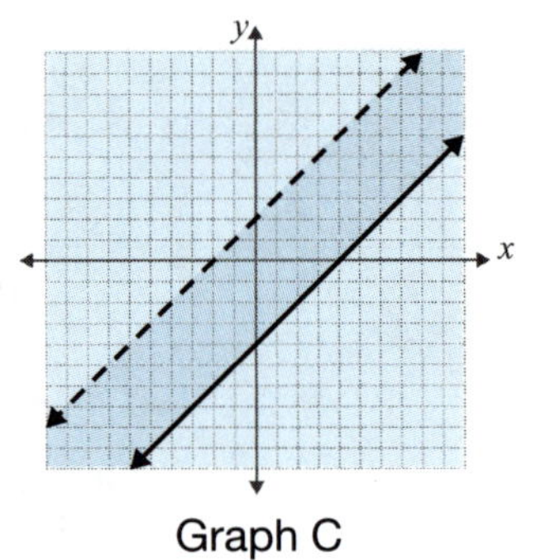

Graph C

a. $\begin{cases} y \geq x - 4 \\ y < x + 2 \end{cases}$

b. $\begin{cases} 2x + y < 4 \\ 2x - y \leq 0 \end{cases}$

c. $\begin{cases} x > 3 \\ y < 5 \end{cases}$

Skill Check

Go to Software Work through Practice in Lesson A.6 of the software before attempting the following exercises.

Solve the systems by graphing. Write the equation for each boundary line and determine if it is a part of the solution set.

3. $\begin{cases} y > 2 \\ x \geq -3 \end{cases}$

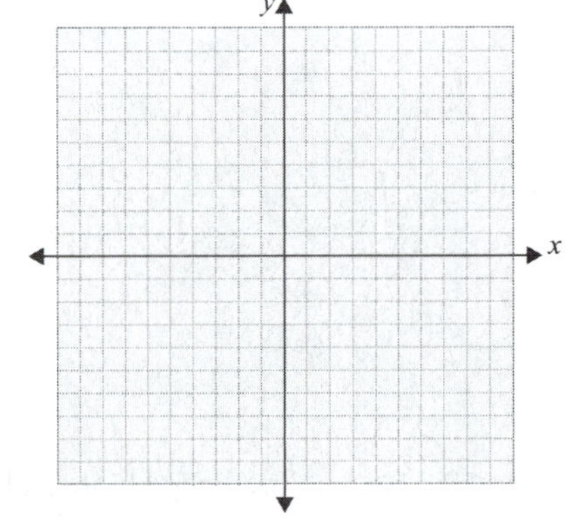

4. $\begin{cases} x + y < 4 \\ 2x - 3y < 3 \end{cases}$

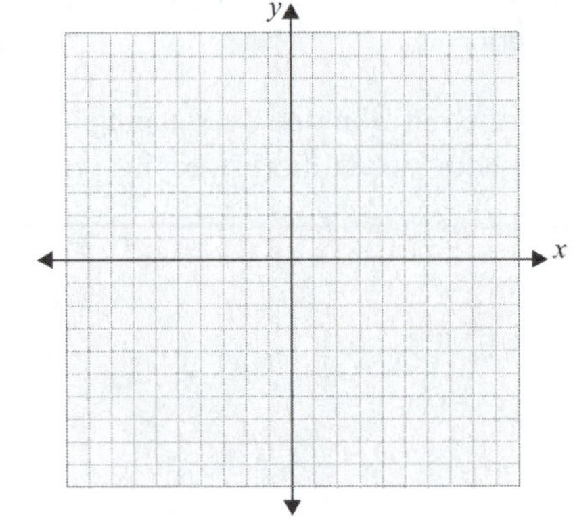

Name: ______________________________ Date: ______________________________

Apply Skills

Work through the problems in this section to apply the skills you have learned related to graphing systems of linear inequalities.

5. Barbara's Bombtastic Bakery sells cookie bouquets where the price depends on the arrangement. Each completed bouquet arrangement needs to weigh less than 5 pounds for shipping purposes. The small cookies weigh 0.1 pounds and the large cookies weigh 0.3 pounds. The flower pot and Styrofoam weigh 1.2 pounds. The cost of each arrangement needs to be less than $30. The small cookies cost $1 each and the large cookies cost $2 each. (The cost of the flower pot and foam are included in the cookie prices.)

a. Write two linear inequalities to describe the situation. Use the variable x to represent the number of small cookies and the variable y to represent the number of large cookies in a bouquet.

b. Graph the two linear inequalities on the same coordinate plane.

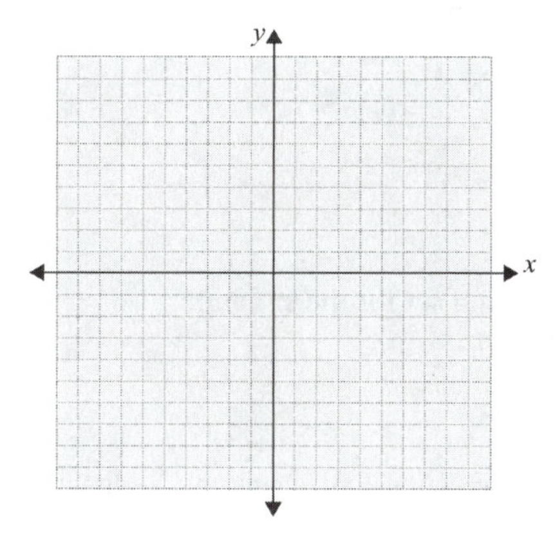

c. Describe the solution set for the situation.

d. Do any of the values in the solution set not make sense in the context of the problem? Explain why or why not.

Quick Tip TECH

For problems that involve large numbers or numbers that are difficult to work with, try using a software program, such as Excel, to graph the equations.

6. Robin is planning a charity ball to raise money for her favorite charity. There are two different ticket options. The VIP option includes dinner, dancing, and cocktails for $150 per ticket. The regular option includes dancing and cocktails for $75 per ticket. Robin wants to make at least $14,000 in ticket sales. The ballroom that is being used for the charity event has a maximum capacity of 150 people.

a. Write two linear inequalities to describe the situation. Let the variable x represent the number of VIP tickets sold and let the variable y represent the number of regular tickets sold.

b. Graph the two linear inequalities on the same coordinate plane.

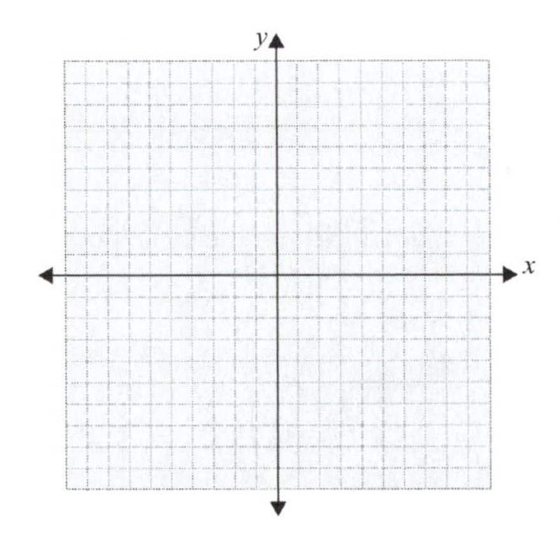

c. Describe the solution set for the situation.

d. Can Robin reach her sales goal if she only sells tickets for the regular option? Explain why or why not.

Name: Date:

A.7 Systems of Linear Equations in Three Variables

Objectives

Solve systems of linear equations in three variables.

Solve applied problems by using systems of linear equations in three variables.

Success Strategy

The addition method and the substitution method for solving linear equations should be used when solving linear equations with three variables. Review these methods in Sections 10.2 and 10.3 for the two-variable case before attempting the three-variable case.

Understand Concepts

Go to Software First, read through Learn in Lesson A.7 of the software. Then, work through the problems in this section to expand your understanding of the concepts related to solving systems of linear equations in three variables.

The general form of an equation in three variables is

$$ax + by + cz = d,$$

where a, b, c, and d are nonzero real numbers and x, y, and z are variables. An equation with three variables represents a plane in 3-dimensional space. A portion of the plane in a 3-dimensional space is shown here. Keep in mind that the plane continues to infinity in the directions of the lines.

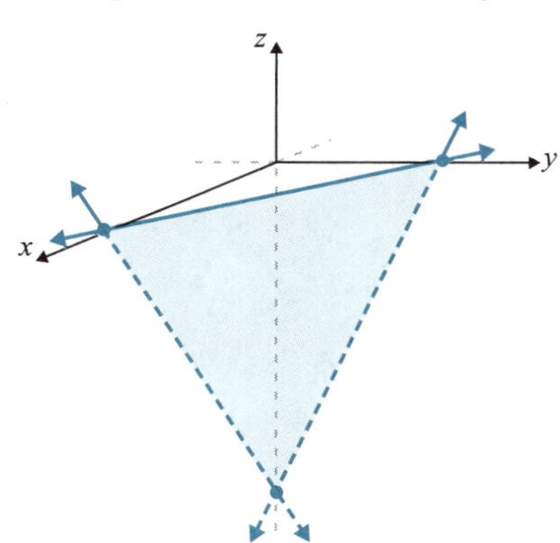

In Chapter 10, you learned that two equations that involve the same two variables are needed to solve a system of linear equations in two variables. In this section, we will discuss how to solve a system of linear equations in three variables. Three equations involving the same set of three variables are needed to solve a system of linear equations in three variables.

The solution to a system of linear equations in two variables is an ordered pair of the form (x, y) that satisfies both equations in the system. A solution for a system of linear equations in three variables is an ordered triple of the form (x, y, z). This ordered triple satisfies all three equations in the system and represents a point on a 3-dimensional plane.

Possible Solutions for a System of Linear Equations with Three Variables

Consistent System: There is at least one solution. The planes intersect in exactly one point or a line.

Inconsistent System: There are no solutions. The three planes do not have a common intersection.

Dependent System: There are an infinite number of solutions. The planes intersect in a line or all three equations represent the same plane.

Independent System: The planes intersect at a single point.

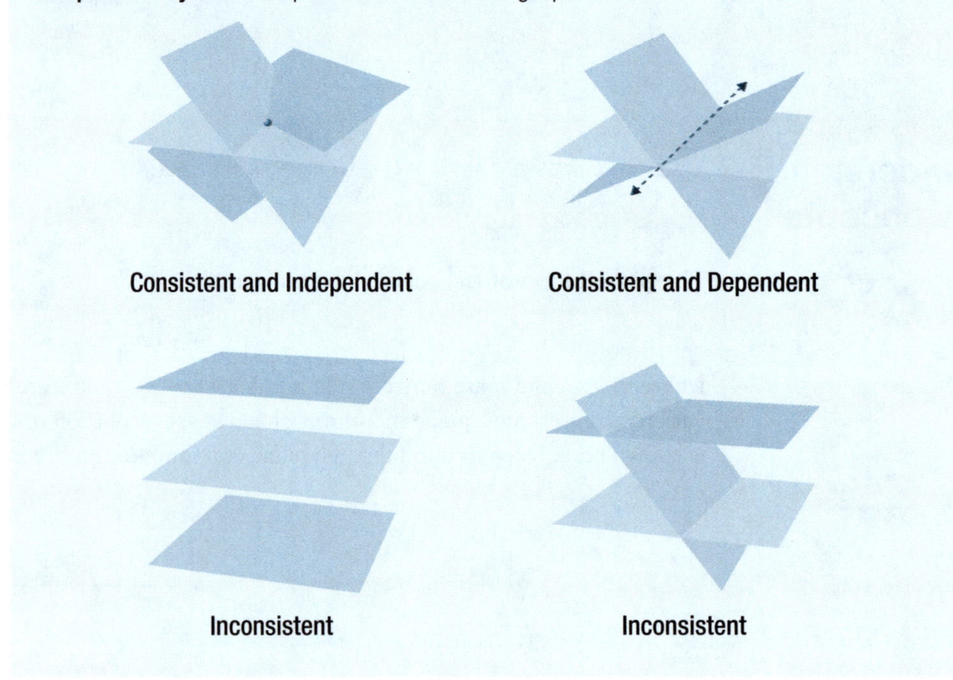

True or False: Determine whether each statement is true or false. Rewrite any false statement so that it is true. (There may be more than one correct new statement.)

1. A linear system in three variables with a solution set of $(1, 2, 3)$ is an inconsistent and independent system.

Lesson Link 

A solution with variables included was introduced for systems of linear equations in two variables in Section 10.2.

2. A linear system in three variables with a solution set of $(x, x+3, 5)$ is a consistent and independent system. (**Hint:** Consider how many possible solutions this system has.)

3. A linear system in three variables with no solutions is an inconsistent system.

Name: Date:

Quick Tip

Back substitution is the method of substituting the values of the solved variables into one of the original equations to find the value of the remaining variable.

Solving Systems of Linear Equations in Three Variables

1. Select two of the three equations and eliminate one of the variables using the addition method.
2. Select a different pair of equations and use the addition method to eliminate the same variable as was eliminated in Step 1.
3. Use the addition or substitution method to solve the system of two linear equations in two variables created by Steps 1 and 2.
4. Back substitute the values found for the two variables in Step 3 into any one of the original equations and solve for the remaining variable.
5. Check the solution (if one exists) in all three of the original equations.

4. Solve the system $\begin{cases} 2x+3y-z=16 \\ x-y+3z=-9 \\ 5x+2y-z=15 \end{cases}$

a. Using the first and second equation, eliminate the variable y with the addition method.

b. Using the second and third equation, eliminate the variable y with the addition method.

c. Solve the system of linear equations in x and z created by parts **a.** and **b.** with either the addition method or the substitution method.

d. Use back substitution in any one of the original equations to solve for y.

Quick Tip

The solution to a system of linear equations in two variables is an ordered pair. The solution to a system of linear equations in three variables is an **ordered triple**, such as (x, y, z).

e. What is the solution to this system of equations?

f. Check that the solution satisfies all three original equations.

5. Solve the system $\begin{cases} y+z=2 \\ x+z=5 \\ x+y=5 \end{cases}$

a. Using the first and second equation, eliminate the variable z with the addition method.

b. Use the equation from part **a.** and the third equation to eliminate the variable x using either the addition method or the substitution method and solve for y.

c. Use back substitution in the third equation to solve for x.

d. Use back substitution in one of the first two original equations to solve for z.

e. What is the solution to this system of equations?

f. Check that the solution satisfies all three original equations.

6. Solve the system $\begin{cases} 3x+y+4z=-6 \\ 2x+3y-z=2 \\ 5x+4y+3z=2 \end{cases}$

a. Using the first and the second equation, eliminate the variable z with the addition method.

b. Using the second and third equation, eliminate the variable z with the addition method.

c. Using the equations from parts **a.** and **b.**, solve for one of the variables with either the addition method or the substitution method.

d. Does the system have a solution? Explain why or why not.

Name: Date:

Skill Check

Go to Software Work through Practice in Lesson A.7 of the software before attempting the following exercises.

Solve. If a system has no solution write "no solution" and if a system has an infinite number of solutions write "infinite number of solutions".

7. $\begin{cases} x+y-z=0 \\ 3x+2y+z=4 \\ x-3y+4z=5 \end{cases}$

8. $\begin{cases} x-y-2z=3 \\ x+2y+z=1 \\ 3y+3z=-2 \end{cases}$

9. $\begin{cases} 2x-2y+3z=4 \\ x-3y+2z=2 \\ x+y+z=1 \end{cases}$

Apply Skills

Work through the problems in this section to apply the skills you have learned related to solving systems of linear equations in three variables.

10. A florist is creating the bridesmaids' bouquets for a wedding. Each bouquet will cost \$92 and have a mixture of 16 flowers consisting of tiger lilies, cream roses, and white daisies. The lilies cost \$10 each, the roses cost \$6 each, and the daisies cost \$4 each. If each bouquet will have as many daisies as it has roses and lilies combined, how many of each type of flower will be in the bouquet?

a. Write a system of linear equations in three variables to describe the situation. Remember that you will have three equations. Use the variable x to represent the number of lilies, the variable y to represent the number of roses, and the variable z to represent the number of daisies.

b. Solve the system of equations from part **a.**

c. Use the solution from part **b.** to answer the question from the problem statement. Write a complete sentence.

11. A theater has seats for a theatrical production located on the main floor, the balcony, and the mezzanine. For an upcoming musical, main floor tickets cost \$60 each, balcony tickets cost \$45 each, and mezzanine tickets cost \$30 each. On opening night, the ticket sales totaled \$27,600. The box office sold 20 more tickets for the main floor than they did for the balcony and mezzanine combined. The number of tickets sold for the mezzanine was 40 more than twice the number of tickets sold for the balcony. How many of each type of ticket did the box office sell?

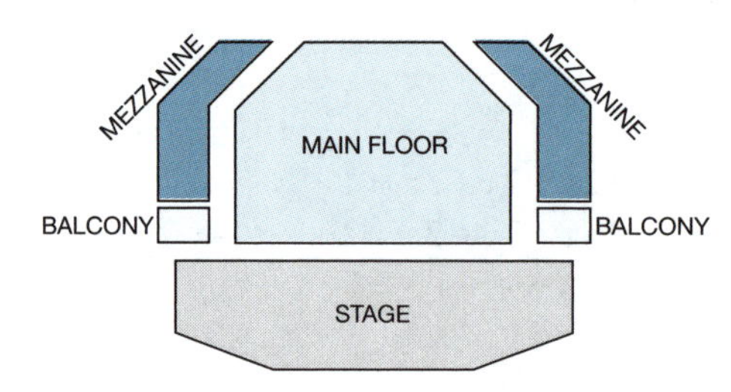

a. Write a system of linear equations in three variables to describe the situation. Remember that you will have three equations. Use the variable x to represent the number of main floor tickets sold, the variable y to represent the number of balcony tickets sold, and the variable z to represent the number of mezzanine tickets sold.

b. Solve the system of equations from part **a.**

c. Use the solution from part **b.** to answer the question from the problem statement. Write a complete sentence.

Name: Date:

A.8 Introduction to Complex Numbers

Objectives

Simplify square roots of negative numbers.

Identify the real parts and the imaginary parts of complex numbers.

Solve equations with complex numbers by setting the real parts and the imaginary parts equal to each other.

Add and subtract with complex numbers.

Success Strategy

There will be a lot of new terminology in this section. Be sure to write down the new terms in your notebook or use the Frayer Model introduced in Chapter 2.

Understand Concepts

Go to Software First, read through Learn in Lesson A.8 of the software. Then, work through the problems in this section to expand your understanding of the concepts related to complex numbers.

Throughout this workbook you have been told that the square root of a negative number is not a real number. This section will introduce you to a new type of number called a **complex number**. The entire set of complex numbers is based on the value of $\sqrt{-1}$.

The Number *i*

$i=\sqrt{-1}$ and $i^2=\left(\sqrt{-1}\right)^2=-1$

Quick Tip

Imaginary numbers were seen as useless and unimportant from their discovery in ancient Greece through the 1700s. René Descartes gave them the name *imaginary* because he thought they had no purpose in mathematics. As new mathematics were discovered, the imaginary numbers became more useful, but the name stuck.

The number i is used to construct the set of imaginary numbers.

Imaginary Numbers

Imaginary numbers are created by taking the square root of negative numbers

$$\sqrt{-a}=\sqrt{a}\cdot\sqrt{-1}=\sqrt{a}\,i=i\sqrt{a}$$

or formed from the product or quotient of a real number and i, such as $4i$ or $\frac{i}{2}$.

Note: When writing radical expressions that contain i, the i can be written before or after the radical sign. However, to avoid confusion, the i is commonly written in front of the radical sign.

True or False: Determine whether each statement is true or false. Rewrite any false statement so that it is true. (There may be more than one correct new statement.)

1. $\frac{4}{3i}$ is an imaginary number.

2. $\sqrt{-25}=5i$

Complex Numbers

Complex numbers are the sum of a real number and an imaginary number. The general form of a complex number is

$$a + bi,$$

where a and b are real numbers.

$$\underset{\text{real part}}{\overset{\uparrow}{a}} + \underset{\text{imaginary part}}{\overset{\uparrow}{bi}}$$

a is considered the **real part** of the complex number and bi is considered the **imaginary part**.

Equality of Complex Numbers

Two complex numbers are equal to each other if the real parts are equal and the imaginary parts are equal. In mathematical notation, for complex numbers $a + bi$ and $c + di$, if $a + bi = c + di$, then $a = c$ and $b = d$.

3. Solve the equation $2x - 8i = -2 + 4yi$ for x and y.

a. Write an equation that sets the real parts of the complex numbers equal to each other.

b. Solve the equation for x.

c. Write an equation that sets the imaginary parts of the complex numbers equal to each other.

Quick Tip

Remember that a number divided by itself is equal to 1. This means that $\frac{i}{i} = 1$.

d. Solve the equation for y.

e. Substitute the values of x and y into the equation $2x - 8i = -2 + 4yi$ and simplify each side. Is the equation true?

Name: Date:

Addition and Subtraction with Complex Numbers

To add (or subtract) two complex numbers, add (or subtract) the real parts and add (or subtract) the imaginary parts. In mathematical notation, for complex numbers $a+bi$ and $c+di$,

$(a+bi)+(c+di)=(a+c)+(b+d)i$ and $(a+bi)-(c+di)=(a-c)+(b-d)i$.

4. Simplify $(6+3i)+(4-5i)$.

a. Find the sum of the real parts.

b. Find the sum of the imaginary parts.

c. Write the sum of the answers from parts **a.** and **b.** as a complex number.

5. Simplify $\left(1\frac{3}{4}+5i\right)-\left(\frac{1}{2}+2i\right)$.

a. Find the difference of the real parts.

b. Find the difference of the imaginary parts.

c. Write the sum of the answers from parts **a.** and **b.** as a complex number.

Skill Check

Go to Software Work through Practice in Lesson A.8 of the software before attempting the following exercises.

Label the real part and the imaginary part of each complex number.

6. $4-3i$

7. $\frac{4-7i}{5}$

Solve for the variables.

8. $x+3i=6-yi$

9. $3x+2-7i=i-2yi+5$

Simplify.

10. $(2+3i)+(4-i)$

11. $(\sqrt{3}+i\sqrt{2})-(5+i\sqrt{2})$

Apply Skills

Applications with complex numbers are beyond the scope of this workbook. If the career pathway you choose requires you to take higher level math courses, chances are that you will eventually use complex numbers. For example, several fields of study based on physics make use of complex numbers. These fields include engineering, optics, and electronics. To learn more about fields of study and career pathways that use complex numbers, try searching on the Internet or asking your instructor for resources.

Name: Date:

A.9 Multiplication and Division with Complex Numbers

Objectives

Multiply with complex numbers.

Divide with complex numbers.

Simplify powers of *i*.

Success Strategy

Remember that a lot of mathematics involves patterns. To simplify higher powers of *i*, you need to understand the established pattern. Do extra practice in the software to help you with this concept.

Understand Concepts

Go to Software First, read through Learn in Lesson A.9 of the software. Then, work through the problems in this section to expand your understanding of the concepts related to multiplication and division with complex numbers.

Quick Tip

When simplifying products of imaginary numbers, the exponent on *i* in the simplified answer should never be larger than 1.

Multiplication of complex numbers follows the same method as multiplying two polynomials. The key thing to keep in mind is that $i^2 = \left(\sqrt{-1}\right)^2 = -1$.

1. Simplify $(3i)(4i)$.

a. Rewrite the product using the commutative and associative properties so the real numbers are next to each other and the *i*'s are next to each other.

b. Simplify the expression from part **a.** Remember, *i* should be treated just like a variable when multiplying and using exponent rules.

2. Simplify $(3i)(2-7i)$.

a. Distribute $3i$.

b. Simplify the expression from part **a.**

3. Simplify $\left(\frac{1}{2}+4i\right)\left(\frac{1}{2}-2i\right)$.

a. Multiply using the FOIL method.

b. Simplify the expression from part **a.**

Determine if any mistakes were made while multiplying complex numbers. If any mistakes were made, describe the mistake and then correctly multiply the complex numbers to find the actual result.

Lesson Link

Always convert square roots of negative numbers to imaginary numbers first and then use the product property for radicals, which was introduced in Section 14.2.

4. $$\begin{aligned}\sqrt{-6}\cdot\sqrt{-2} &= i\sqrt{6}\cdot i\sqrt{2}\\ &= i^2\sqrt{12}\\ &= (-1)2\sqrt{3}\\ &= -2\sqrt{3}\end{aligned}$$

5. $$\begin{aligned}\sqrt{-10}\cdot\sqrt{-8} &= \sqrt{80}\\ &= 4\sqrt{5}\end{aligned}$$

Lesson Link

Rationalizing the denominator of a rational expression was introduced in Section 14.4.

Dividing by a complex number is similar to rationalizing the denominator of a radical expression with a binomial denominator. Both methods use the conjugate of the denominator. After dividing by a complex number, no imaginary numbers should remain in the denominator.

Complex Conjugates

The **conjugate** of the complex number $a + bi$ is $a - bi$. Notice that both the real number and the imaginary number remain the same, only the sign between them change.

For example, the conjugate of $2 + 4i$ is $2 - 4i$.

6. Simplify $(1+i)\div(2-3i)$.

a. Write the division problem in fraction form.

b. What is the conjugate of the denominator from part **a.**?

Name: Date:

c. Multiply the numerator and denominator of the fraction from part **a.** by the complex conjugate from part **b.** Do not simplify.

d. Simplify the result from part **c.**

Skill Check

Go to Software Work through Practice in Lesson A.9 of the software before attempting the following exercises.

Simplify.

7. $8(2+3i)$

8. $i\sqrt{3}\left(2-i\sqrt{3}\right)$

9. $(-2+5i)(i-1)$

10. $\dfrac{2+i}{-4i}$

11. $\dfrac{2i}{5-i}$

12. $\dfrac{6+i}{3-4i}$

Apply Skills

Work through the problems in this section to apply the skills you have learned related to multiplication and division with complex numbers.

13. Work through this problem to discover the pattern for the powers of i.

a. Fill in column two of the table.

i^n	Simplified
i^0	1
i^1	i
i^2	
i^3	
i^4	
i^5	
i^6	
i^7	
i^8	
i^9	
i^{10}	
i^{11}	
i^{12}	
i^{13}	

b. There is a pattern in column two of the table. Describe the pattern.

Name: Date:

A.10 Standard Deviation and z-Scores

Objectives

Calculate the standard deviation of a data set.

Calculate and interpret z-scores.

Success Strategy

The standard deviation can be easily calculated on most scientific calculators or using spreadsheet software, such as Excel, that have common statistical functions. However, it is recommended that you calculate the standard deviation by hand a few times to get a better understanding of the steps involved.

Understand Concepts

Read through the concepts presented in this section, then work through the problems to expand your understanding of standard deviation and z-scores.

In Section 6.1, we discussed the **range**, which is the difference between the largest value and the smallest value of a data set. The range gives you an idea of how spread out the data set is, but it only uses two data items from the data set. A better measure of the spread of a data set is the standard deviation, which measures how much each item in the data set differs from the mean of the data set. That is, the standard deviation tells us how the data items of a data set are spread out around the mean.

Quick Tip

The method presented here is used to calculate the standard deviation of a sample taken from a population.

Finding the Standard Deviation of a Data Set

1. Find the mean of the data set.
2. Subtract the mean from each item in the data set.
3. Square each of the differences from Step 2.
4. Find the sum of the squared differences from Step 3.
5. Divide the sum from Step 4 by $n-1$, where n is the number of items in the data set.
6. Find the square root of the result from Step 5.

While calculating the standard deviation of a set, it's important to keep your work organized because there are several calculations involved. A table can be useful to help organize your work, as shown in Problem 1.

Quick Tip TECH

For large data sets, computer software, such as Excel, and graphing calculators can be used to find the standard deviation.

1. Find the standard deviation of the data set: 15, 16, 19, 22, 28.

a. First, calculate the mean of the data set.

b. Subtract the mean from the value of each data item. Place the differences in column two of the table.

x	$x -$ mean	$(x -$ mean$)^2$
15		
16		
19		
22		
28		

c. Square each difference from part **b.** Place the squared differences in column three of the table.

d. Find the sum of the squared values from column three of the table.

Quick Tip

The mean is considered to be the **balance point** of the data set. As a result, the sum of the differences in column two of the table is equal to zero because the positive and negative differences from the mean "cancel" each other out. This is why the differences need to be squared to find the standard deviation.

e. How many data items are in the set?

f. Divide the sum from part **d.** by one less than the number of data items in the set.

g. Find the square root of the quotient from part **f.** Round your answer to the nearest hundredth.

Quick Tip

The importance of taking the square root is so that the units of the standard deviation are in the same units as the mean. Remember that the differences are squared in Step 3 of calculating the standard deviation, which means that the units are also squared.

In the future, if you take a math course that covers statistics more in depth, you will probably be given the standard deviation formula, which appears in the box below. When using this formula to calculate the standard deviation, you would follow the same steps introduced earlier.

The Standard Deviation Formula

The formula for finding the standard deviation is

$$\sqrt{\frac{\Sigma(x-\text{mean})^2}{n-1}},$$

where x represents each data item in the set, "mean" is the mean of the data set, and n is the number of items in the data set.

Quick Tip

The mathematical formula for finding the standard deviation uses the summation operator Σ, which is used when adding all items in a series or list of numbers.

2. Use the three data sets Q, R, and S to answer the following questions.

Q: 2, 2, 3, 3, 4, 5, 7

R: 2, 2, 2, 2, 2, 2, 7

S: 1, 2, 3, 4, 5, 6, 7

a. Which data set appears to be more spread out than the other data sets?

b. Calculate the standard deviation for data set Q. Round your answer to the nearest hundredth.

c. Calculate the standard deviation for data set R. Round your answer to the nearest hundredth.

d. Calculate the standard deviation for data set S. Round your answer to the nearest hundredth.

e. Do the values of the standard deviations from parts **b.** through **d.** support your answer from part **a.**? Explain why or why not.

The mean and standard deviation are very useful measures of a data set. To compare data items from different data sets, we can use what is called a **standard score**, or ***z*-score**. A z-score tells how many standard deviations a data item is away from the mean.

z-Score

To find the z-score of a data item, subtract the mean from the data item and divide this difference by the standard deviation of the data set. In mathematical notation,

$$z = \frac{\text{data value} - \text{mean}}{\text{standard deviation}}$$

If $z > 0$, the data item lies above the mean.

If $z = 0$, the data item is equal to the mean.

If $z < 0$, the data item lies below the mean.

3. Three friends took the same Introduction to Philosophy class, but they each took a section of the class taught by different instructors. They want to compare how they did on the final exam, but each instructor gave a different exam. Amber scored 75 on the exam where the class mean was 87 and the standard deviation was 4. Bill scored 80 on the exam where the class mean was 70 and the standard deviation was 5. Chris scored 88 on the exam where the class mean was 85 and the standard deviation was 3.

a. Find Amber's z-score.

b. Find Bill's z-score.

c. Find Chris's z-score.

d. Based on the z-scores, who had a better score on the final exam compared to their friends? Explain how you know.

Skill Check

Work through the following exercises to test the skills that you learned in this section.

Find the standard deviation of each data set. Round your answers to the nearest hundredth.

4. Number of text messages per day: 50, 45, 20, 36, 49, 40

Number of Text Messages	x – mean	$(x - \text{mean})^2$

5. ACT scores for a group of students: 28, 20, 30, 31, 24, 34, 15

ACT score	x – mean	$(x - \text{mean})^2$

6. A data set has a mean of 15 and a standard deviation of 1.5. Find the z-score of each data item, to the nearest hundredth.

a. 10

b. 16

c. 18

d. 15

Name: ____________________ Date: ____________________

Apply Skills

Work through the problems in this section to apply the skills you have learned related to standard deviation and *z*-scores.

7. Evan wants to invest a portion of his savings into the stock market. While researching different stock companies, he finds the following standard deviations of the daily stock closing prices for two different companies over the last year.

Company A: Standard deviation of stock prices = $2.57

Company B: Standard deviation of stock prices = $7.48

a. How do the two standard deviations of the stock prices compare?

b. Which is the more stable company to invest in?

8. A food manufacturer creates and packages a variety of individual frozen dinners. The specifications for the production line that produces lasagna frozen dinners are that the dinners produced have an average weight of 10.75 ounces and an allowable standard deviation of 0.15 ounces. The specifications for the production line that produces meatloaf frozen dinners are that the dinners produced have an average weight of 9.85 ounces with an allowable standard deviation of 0.10 ounces. A factory worker randomly selects a completed meal from the lasagna dinner production line and determines that it has a weight of 10.83 ounces. The worker then takes a completed meal from the meatloaf dinner production line and determines that it has a weight of 9.79 ounces.

a. Calculate the z-score of the randomly selected lasagna dinner. Round your answer to the nearest hundredth.

b. Calculate the z-score of the randomly selected meatloaf dinner. Round your answer to the nearest hundredth.

c. Which production line is closer to specifications? Explain your reasoning.

Name: Date:

A.11 Mathematical Modeling

Objectives

Create and interpret scatter plots.

Identify and use linear models.

Calculate and interpret linear correlation.

Identify and use quadratic models.

Identify and use square root models.

Identify and use exponential models.

Use models for prediction.

Success Strategy

Do a brief review of Sections 9.1 through 9.5 on linear equations in two variables and functions in order to be prepared for the material presented in this section.

Understand Concepts

Read through the concepts presented in this section, then work through the problems to expand your understanding of using algebraic functions to model real-world phenomena.

Mathematical modeling uses algebraic functions to represent the relationships that exist between variables in real-world data. These models are not exact and usually involve some error from the actual system being modeled. Mathematical models are used in many different fields, including physics, biology, and economics. These models are used to explain the relationships that exist in a physical system and to predict the behavior of the system as variables change.

Lesson Link

Linear functions were introduced in Section 9.5 and quadratic functions were introduced in Section 15.5.

To this point in the course, you have learned about two important functions, linear functions and quadratic functions. Remember that a relationship between two variables is a function if each element of the domain corresponds to exactly one value in the range. This is an important characteristic needed in mathematical modeling for predicting values of y. Remember that $y = f(x)$ when using function notation. The following table gives a brief review of these two functions.

Quick Tip

Other forms of linear models we have studied include **standard form** $Ax + By = C$ and **point-slope form** $y - y_1 = m(x - x_1)$.

Linear and Quadratic Functions

Model	Graph	Characteristics
Linear: Slope-intercept form: $f(x) = mx + b$, where m and b are real numbers.		m is the slope of the line. $(0, b)$ is the y-intercept of the line. Domain: $(-\infty, \infty)$
Quadratic: $f(x) = ax^2 + bx + c$, where a, b, and c are real numbers with $a \neq 0$.	$a > 0$ $a < 0$	If $a > 0$, the parabola opens up. If $a < 0$, the parabola opens down. Vertex occurs when $x = \frac{-b}{2a}$. Line of symmetry is $x = \frac{-b}{2a}$. $(0, c)$ is the y-intercept. Domain: $(-\infty, \infty)$

Lesson Link

Square root functions were introduced in Section 14.7.

Any function whose graph is not a line is called a **nonlinear function**. A quadratic function is an example of a nonlinear function since its graph is a parabola, which is a curve. Two other nonlinear functions that are useful in modeling are the **square root function** and the **exponential function**.

Lesson Link

Remember from Section 9.5 that the **domain of a function** is the set of all real x-values for which the function is defined.

Keep in mind that the domain of the standard square root function $f(x) = \sqrt{x}$ is $[0, \infty)$ since you can't take the square root of a negative number in the real number system. As a result, the value of the expression under the square root sign must always be greater than or equal to zero.

1. Let's look at a table of values for the function $f(x) = \sqrt{x}$. Calculate the decimal approximations for the square roots in column two using a calculator. Round your answer to the nearest thousandth.

x	$f(x) = \sqrt{x}$
1	1
2	$\sqrt{2} \approx$ ___
3	$\sqrt{3} \approx$ ___
4	2
5	$\sqrt{5} \approx$ ___

Increasing and Decreasing Functions

An **increasing function** is a function where the values of y get *larger* as the corresponding values of x get larger.

A **decreasing function** is a function where the values of y get *smaller* as the corresponding values of x get larger.

2. Looking at the table of values from Problem 1, is the function $f(x) = \sqrt{x}$ an increasing or decreasing function?

Read the following information about exponential functions and work through the problems.

In Chapter 11, you worked with exponents on constants and variables. Another type of function that is used in mathematical modeling is the **exponential function**. In an exponential function, the exponent is a variable and has the general form $f(x) = Cb^x$, where C is a real number such that $C \neq 0$ and b is a positive real number such that $b \neq 1$. Two examples of exponential functions are graphed on the next page. Notice how the value of b affects the shape of the graph.

Exponential Models

In the exponential function $f(x) = Cb^x$,

1. If $0 < b < 1$ then the values of y are decreasing as x increases. (See Graph A.)
2. If $b > 1$ then the values of y are increasing as x increases. (See Graph B.)

Name: Date:

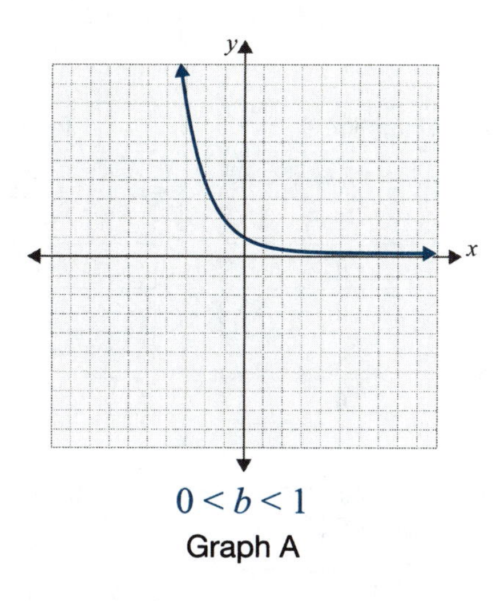

Graph A

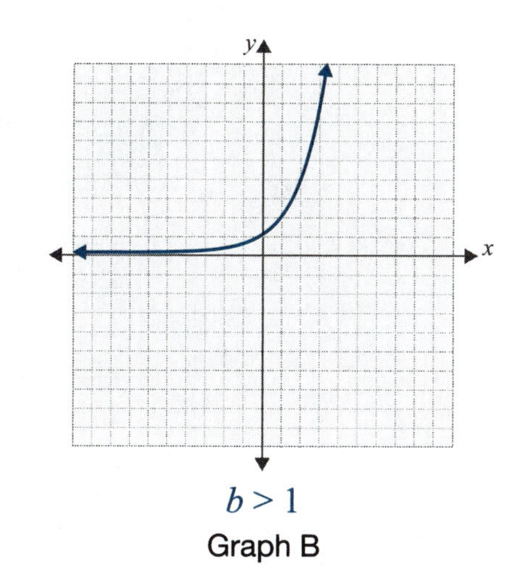

Graph B

3. Let's explore why b shouldn't be equal to 1 in an exponential model.

a. Fill in the y-values in the table for the exponential model $y = 1^x$.

x	$y = 1^x$
–1	
0	
1	
2	
3	

b. Is this a very interesting model? Explain your reasoning.

c. Can you think of any real-world situation that can be modeled by this function? If yes, provide an example. If no, explain why.

d. The data in the table can be modeled using a simple linear equation. Write a linear equation to represent this data.

Four tables of values are provided below, where each one represents a different type of function. Use the tables to work through Problems 4 and 5.

Table A: Linear Function	
x	$y = 2x$
–1	–2
0	0
1	2
2	4
3	6

Table B: Quadratic Function	
x	$y = x^2$
–1	1
0	0
1	1
2	4
3	9

Table C: Square Root Function	
x	$y = \sqrt{x}$
–1	Undefined
0	0
1	1
2	1.414
3	1.732

Table D: Exponential Function	
x	$y = 2^x$
–1	0.5
0	1
1	2
2	4
3	8

4. Use the data in Tables A, B, C, and D to determine if there is a pattern in the rates of change for any of the functions.

a. Calculate the rate of change, or slope, between each pair of consecutive points in Tables A through D. Write the rates of change in the table below.

x-values	A: Linear	B: Quadratic	C: Square Root	D: Exponential
–1 to 0				
0 to 1				
1 to 2				
2 to 3				

b. Is there a pattern in the rates of change for any of the functions? If yes, describe it.

5. Use the data in Tables A, B, C, and D to determine if there is a pattern in the consecutive ratios for any of the functions.

a. Calculate the ratio of each pair of consecutive y-values for each pair of points in Tables A through D. Write the ratios in the table.

x-values	A: Linear	B: Quadratic	C: Square Root	D: Exponential
–1 to 0				
0 to 1				
1 to 2				
2 to 3				

b. Is there a pattern in the ratios for any of the functions? If yes, describe it.

Linear and Exponential Growth and Decay

If the relationship between two variables is **linear**, the average rate of change, or slope, is a constant value. If the linear function is increasing we can say there is **linear growth**. If the linear function is decreasing then we can say there is **linear decay**.

If the relationship between two variables is **exponential**, the ratio of consecutive *y*-values is a constant value. This constant ratio is equal to the base *b*.

If the exponential function is increasing then we can say there is **exponential growth** and *b* is called the **growth factor**. If the exponential function is decreasing then we can say there is **exponential decay** and *b* is called the **decay factor**.

For actual data that exhibit a linear pattern, the slope between each pair of points will not be exactly the same value. However, the slopes should be approximately the same. Likewise, for actual data that exhibit an exponential pattern, the ratio of consecutive *y*-values will not be exactly the same value. However, the ratios should be approximately the same.

6. The following table of values has an exponential pattern.

x	y
–1	$\frac{1}{8}$
0	$\frac{1}{2}$
1	2
2	8
3	32

a. In the exponential function $f(x) = Cb^x$, the value of C is the y-value when $x = 0$. Looking at the table, what is the value of C?

b Calculate the ratio of consecutive y-values for each pair of points.

x-values	Ratio
–1 to 0	
0 to 1	
1 to 2	
2 to 3	

c. The value of b in the exponential function is the ratio of consecutive y-values. What is this value?

d. Write the function that represents the table of values in the form $f(x) = Cb^x$.

e. Verify that your function $f(x)$ from part **d.** is correct by substituting each value of x from the table into your equation to verify that it gives the same y-value as the table.

Read the following information about mathematical models and work through the problems.

When determining a mathematical model to describe the relationship between two variables for actual data, you should begin by graphing a **scatter plot** of the data. A scatter plot is a graph of ordered pairs of data on a coordinate plane, where the points are not connected. One of the variables will correspond to the x-axis and is called the **independent variable**. The other variable will correspond to the y-axis and is called the **dependent variable**. The dependent variable is the variable whose value depends on the other variable, when this makes sense. For example, if our two variables were the temperature outside and the amount of snowfall, the dependent variable would be the amount of snowfall since it depends on the temperature. In some cases, it is not clear which variable is the dependent variable and which is the independent variable. For modeling purposes, a mathematical model can be determined regardless of which variable is chosen to be the dependent variable and which is chosen to be the independent variable.

Creating a Scatter Plot

On a coordinate plane,

1. Determine which variable is the independent variable and label the x-axis to reflect this.
2. Determine which variable is the dependent variable and label the y-axis to reflect this.
3. Determine the scales of both axes by looking at the range of data values for each variable.
4. Plot the data as ordered pairs (x, y), where x is the value for the independent variable and y is the corresponding value for the dependent variable.

7. Draw a scatter plot of the three sets of data below and decide which of the four functions fits the data best: linear, quadratic, square root, or exponential. If none of them fit the data, then write "other".

a.

Data A	
x	y
0.5	0.7
2	1.5
3	1.8
4	2.0
5	2.3

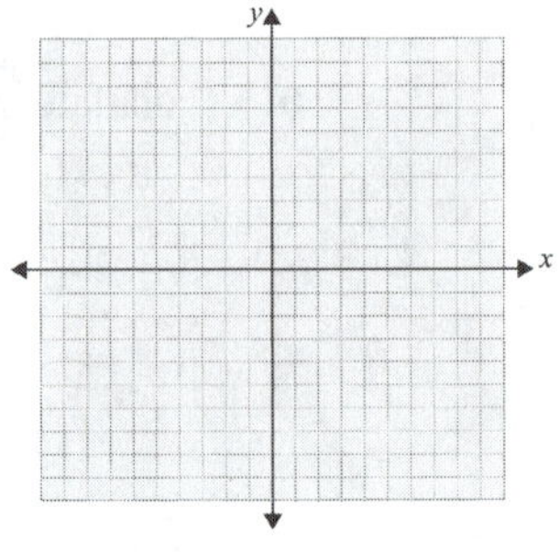

Name: Date:

b.

Data B	
x	**y**
–1	0.9
0	3.1
1	4.9
2	7.2
3	8.8

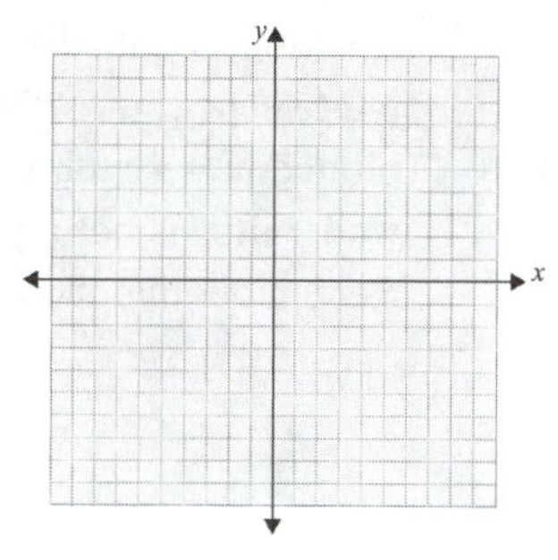

c.

Data C	
x	**y**
–1	1.3
0	1.1
1	0.8
2	0.5
3	0.4

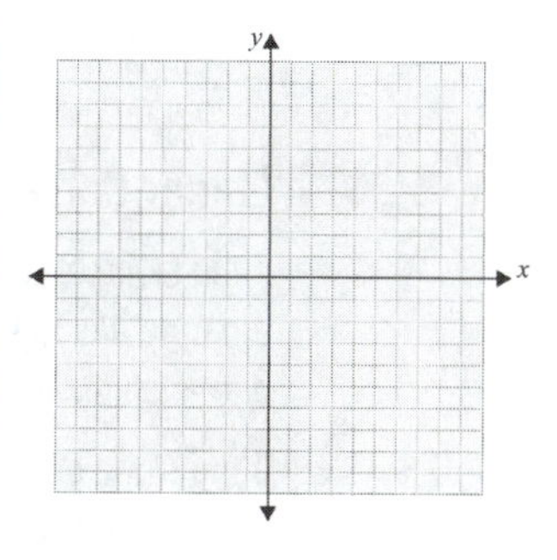

Read the following information about regression analysis and work through the problems.

Quick Tip

Regression analysis is a statistical method for estimating the relationships among variables.

From the scatter plot of the data, you will determine which of the four functions fits the data best based on the shape of their graphs. In other words, which function has a shape similar to the scatter plot of the data you just created? Once you have decided which function to use for your model, you can then determine the actual model in two ways: using the mathematics learned in this course or using **regression analysis**. Excel and most graphing calculators have the capability to perform a regression analysis. If these are unavailable to you, the website Wolfram|Alpha can be used to perform a regression analysis.

Follow along with the steps to use Wolfram|Alpha to fit a line to the given data.

x	y
–2	–3.7
–1	–2.8
0	–2.0
1	–1.2
2	0
3	1.1

Quick Tip TECH

Wolfram|Alpha can also fit quadratic, cubic, exponential, and logarithmic regression models as well. For exponential models, the base used is the irrational number $e \approx 2.71828$.

To perform a linear regression analysis using Wolfram|Alpha,

1. Go to the website http://www.wolframalpha.com.
2. In the input line of Wolfram|Alpha, type "linear fit $\{-2, -3.7\}$, $\{-1, -2.8\}$, $\{0, -2\}$, $\{1, -1.2\}$, $\{2, 0\}$, $\{3, 1.1\}$". (Notice that braces $\{\ \}$ are used instead of parentheses ( ) for the ordered pairs.)
3. Click the "=" button to the right.

WolframAlpha computational... knowledge engine

linear fit {-2, -3.7},{-1, -2.8}, {0, -2}, {1, -1.2}, {2, 0}, {3, 1.1}

Examples Random

Quick Tip

The regression line is often referred to as the **least-squares regression line** due to the statistical method used to obtain it. This line is also called the line of best fit.

Wolfram|Alpha gives the following linear regression equation: $y = 0.948571x - 1.90762$. This equation is called the **regression line**. It has an estimated slope of 0.948571 and an estimated y-intercept of -1.90762. The output from Wolfram|Alpha also provides a graph with the data points plotted along with the regression line. You can see from the graph that the line fits the data very well, but it is not a perfect model since the actual data values do not all lie on the regression line. Remember that there will always be some error involved when modeling actual data.

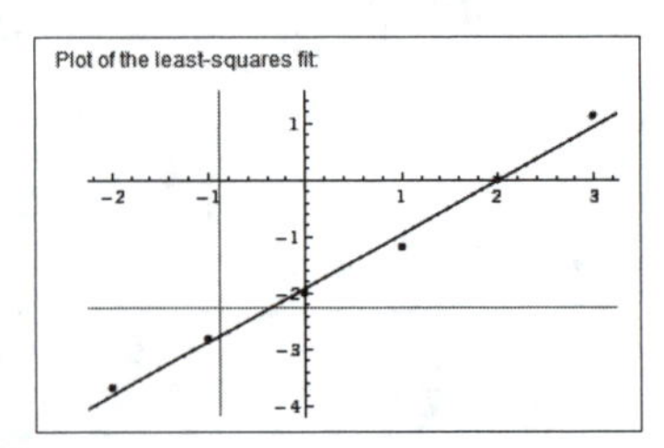

Linear Correlation

If the slope of a regression line is positive (that is, the regression line rises from left to right), the data has a **positive correlation** and the variables x and y are **positively correlated** with one another.

If the slope of a regression line is negative (that is, the regression line falls from left to right), the data has a **negative correlation** and the variables x and y are **negatively correlated** with one another.

The results from Wolfram|Alpha also provide a value denoted by R^2 that is called the **coefficient of determination**. This value describes how well the linear regression fits the data. The coefficient of determination will take on values between 0 and 1, inclusive. The closer the R^2 value is to 1, the better the regression line fits the data. An R^2 value of 1 indicates that the regression line is a perfect fit to the data. An R^2 value of 0 indicates that a linear relationship does not exist between the two variables.

8. The regression analysis gives $R^2 = 0.993248$. How well does the regression line fit the data?

Name: Date:

9. Another way to create a linear model for this data is to "eyeball" a line that you think fits the data and use the point-slope form of a linear equation to write an equation for the line.

a. Plot the points from the table that we used for the regression analysis on the coordinate plane provided. Use the data points $(0,-2)$ and $(2,0)$ and draw a line of "best" fit through these two points. (**Note:** These points were chosen to make the calculations easier in part **b.**)

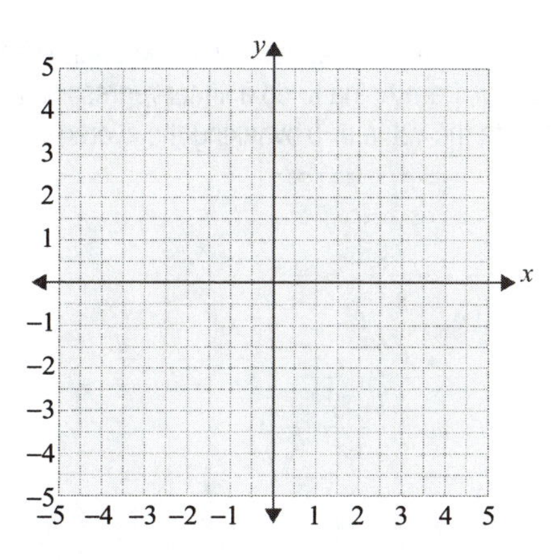

Lesson Link

The point-slope form of an equation was introduced in Section 9.4.

b. Use the point-slope form of an equation to construct an equation for this line using the two specified points in part **a.**

c. Solve the equation in part **b.** for y so that it is in slope-intercept form.

d. Compare the slopes and y-intercepts from the regression line obtained from Wolfram|Alpha and the line from part **c.** Are they similar?

e. The importance of having a mathematical expression that models the data well is that this model can then be used for prediction. Predict the y-value for this data when $x = 3$ using the regression line obtained from Wolfram|Alpha and the line from part **c.**

Quick Tip

A regression analysis minimizes the sum of the squares of the residuals, or errors, to determine a line of best fit for the data. This line of best fit is the **least squares regression line**.

f. Calculate the difference between the predicted values from part **e.** and the actual value of y from the table of values. (The difference between the actual observed value from the data and the predicted value from the model is called a **residual**.) Which predicted value is closest to the actual value?

Skill Check

Work through the following exercises to test the skills that you learned in this section.

For each of the table of values below,

a. Draw a scatter plot of the data.

b. Decide which model would fit the data best: linear or exponential.

c. If the data are linear, determine if the correlation between the variables is positive or negative, then "eyeball" a line of best fit or use WolframlAplha to do a linear regression analysis and calculate the slope. If the data are exponential, estimate the value of b by taking the average of the ratios of consecutive y-values.

10.

x	y
–3	9.8
–2	8.1
–1	6.2
0	3.7
1	2.3
2	0.1
3	–2.2

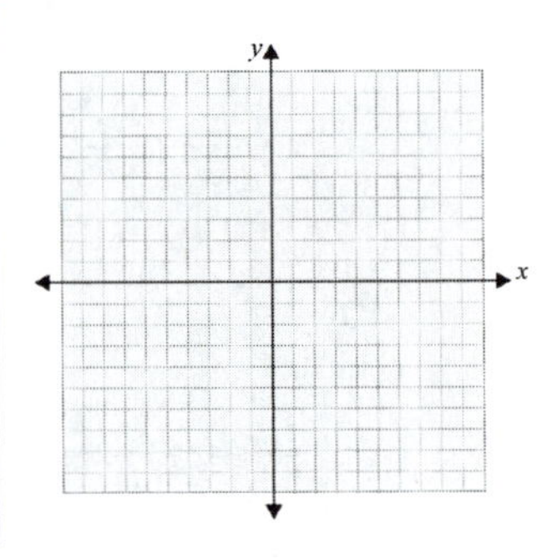

11.

x	y
–3	0.3
–2	0.5
–1	0.7
0	1.1
1	1.6
2	2.4
3	3.5

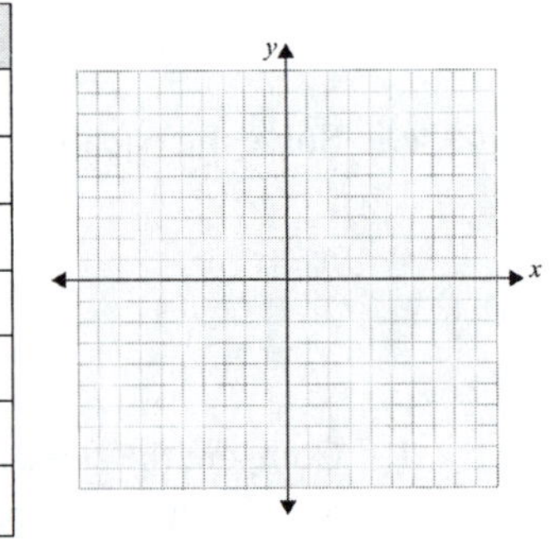

12.

x	y
–3	–7.9
–2	–6.1
–1	–4.2
0	–2.0
1	–0.1
2	2.1
3	3.8

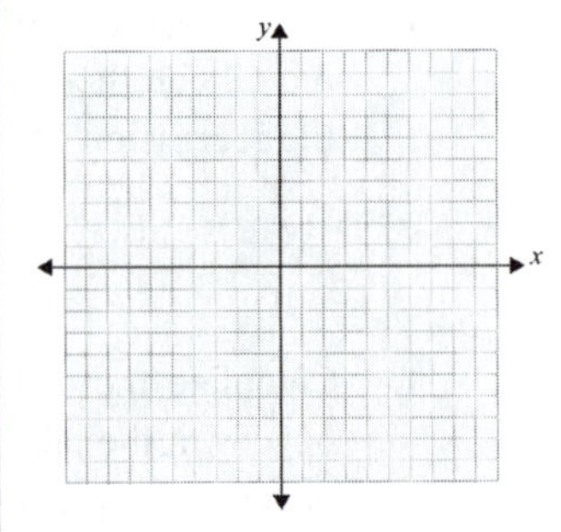

Name: Date:

Apply Skills

Work through the problems in this section to apply the skills you have learned related to using algebraic functions to model real-world phenomena.

13. The following data represent the estimated value of a certain Chevrolet model SUV from the date of purchase (age = 0) to 5 years after purchase. The y-values decrease due to the depreciation of the value of the car over time. Notice the large decrease in value from the time of purchase to 1 year of age.

Age of Car (in years)	Car Value ($)
0	25,000
1	16,915
2	14,713
3	12,555
4	10,440
5	8368

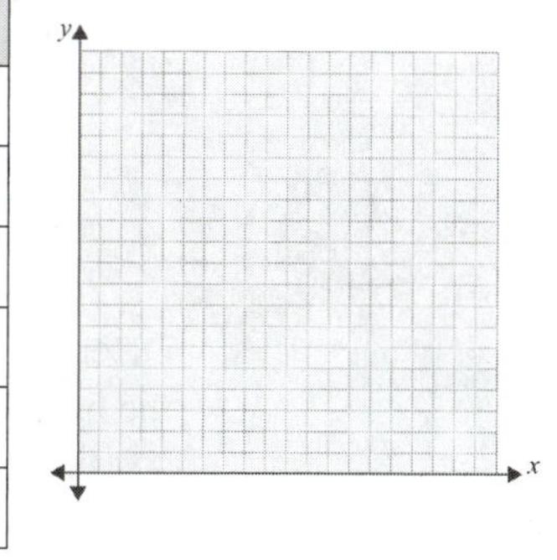

a. Determine which variable should be the dependent variable and which variable should be the independent variable for this problem and label the axes of the graph accordingly.

b. Draw a scatter plot of the data on the coordinate plane.

Quick Tip

Mathematical models don't always include all of the data values. This problem excludes the data point for age 0 due to the large decrease in value from age 0 to age 1.

c. If you only look at the data values from age 1 to 5 years (excluding the data point for age 0), what type of model would fit this data?

d. Fit the model type specified in part **c.** to the data from age 1 to 5 years.

e. Determine the rate of decrease in value of the SUV per year (age 1 to 5 years only). (**Hint:** This will be a value taken from the model in part **d.**)

Quick Tip

In general, if the mathematical model created was not a function, then a specific y-value could not be predicted for a given x-value.

f. Use the model from part **d.** to predict the value of the car at age 6 years. Round your answer to the nearest dollar.

14. A professor at a local university created a free phone app to help his students learn Spanish. The following number of downloads of the app took place over the first 6 weeks after release.

Weeks after Release	Number of Downloads
0	4
1	5
2	6
3	8
4	10
5	12
6	15

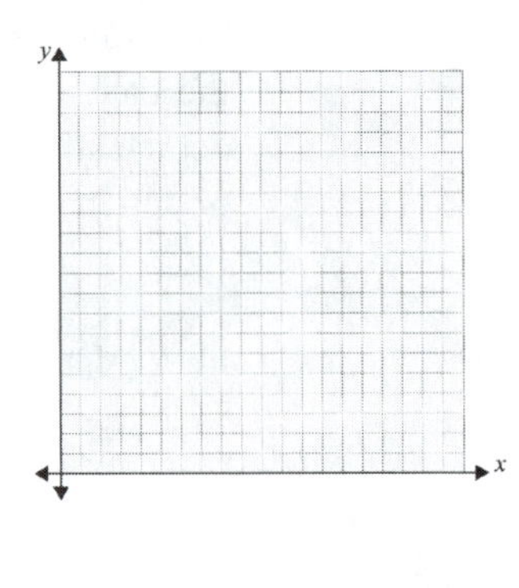

a. Determine which variable should be the dependent variable and which variable should be the independent variable for this problem and label the axes of the graph accordingly.

b. Draw a scatter plot of the data on the coordinate plane.

c. Fit an exponential model $f(x) = Cb^x$ to this data.

Quick Tip

Remember that C is the y-value when $x = 0$ and b is the ratio of consecutive y-values.

d. Since this data is increasing (that is, $b \geq 1$), b is referred to as the growth factor. This growth factor is often interpreted as a percent, which allows us to calculate the percent increase from week to week. Calculate the percent increase in the data from week to week by converting the base b to a percent and then subtracting 100% from this value.

Lesson Link

Percent increase was introduced in Section 4.7.

e. Predict the number of downloads of the app 7 weeks after release using the exponential model from part **c.** Round to the nearest whole number since you can't have a partial download.

f. Predict the number of downloads of the app 7 weeks after release by multiplying the number of downloads from Week 6 by the value of b. Round to the nearest whole number.

g. How do the results from parts **e.** and **f.** compare?

h. Assuming the rate of growth in the number of downloads continues at the current rate, predict the number of downloads of the app 6 months after release. Assume 1 month = 4 weeks. Round to the nearest whole number.

i. If the number of downloads doubled each week, what would the value of the growth factor be?

j. What percent increase would the value from part **i.** correspond to?

Name: Date:

Appendix Projects

Project A: Smartphones and Survey Errors

An activity to demonstrate the use of solving absolute value inequalities in real life

In a nationwide survey conducted in 2013 by Grunwald Associates, in conjunction with the Learning First Alliance, it is clear that mobile devices have become an essential part of our lives. At the high school level, 51% of all students carry a smartphone to school with them. Almost one-fourth of all US students in kindergarten through grade 12 carry a smartphone, and 8% of students in grades 3 through 5 bring a smartphone to school. Source: http://grunwald.com/pdfs/Grunwald%20Mobile%20Study%20infographic.pdf

1. In this study, Grunwald Associates reported that the margin of error for this study was about 3%. What this means is that there may be some error in the percentages reported due to the fact that only a representative sample of parents from across the United States were surveyed and not the entire population of parents in the United States. For example, for the 51% of high school students who carry a smartphone to school, the "true" population percentage of high school students who carry a smartphone to school is within a 3% margin of error of 51%. This can be determined by solving the absolute value inequality $|x-51|\leq 3$.

a. When the absolute value bars are removed from this absolute value inequality, does it result in a compound "and" or "or" inequality?

b. Solve the absolute value inequality for x.

c. Write the solution from part **b.** in interval notation.

d. Graph the solution to the absolute value inequality on the real number line.

e. Does the solution represent the intersection or the union of two intervals?

f. Interpret the meaning of this solution set in the context of the problem.

2. The margin of error in a survey applies to all the percentages reported in the survey. Determine the range of values for the "true" population percentage of students in kindergarten through grade 12 who carry a smartphone.

   a. Write $\frac{1}{4}$ as a percent.

   b. Write an inequality similar to the one described in Problem 1 to represent the margin of error around the percent in part **a.** Let the variable x represent the percent of students who carry a smartphone in kindergarten through grade 12.

   c. Solve the inequality from part **b.** for x.

   d. Write the solution from part **c.** in interval notation.

   e. Graph the solution to the absolute value inequality on the real number line.

   f. Interpret the meaning of this solution set in the context of the problem.

Name: Date:

Project B: A Diversified Portfolio

An activity to demonstrate the use of linear systems of three equations in real life

Diversifying your investments by splitting up your money into two or more funds may keep you from suffering significant losses if one of the funds performs poorly.

For this activity, if you need help understanding some of the investment terms, use the following link as a resource. http://www.investopedia.com/

Let's suppose that you have saved $7500 and want to invest this money in three funds paying 5%, 6%, and 8% simple interest annually. Remember that interest rates fluctuate as the economy changes and there are no guarantees on the amount of interest you will actually earn on your investment. Also, notice that higher rates of interest typically indicate a higher risk on your investment.

1. If you want to earn $460 total in interest on your investments this year and you want the amount of principal invested in Fund 2 to be equal to half the amount invested in the other two funds combined, how much money would you need to invest in each fund? Let the variable x represent the amount invested in Fund 1, the variable y represent the amount invested in Fund 2, and the variable z represent the amount invested in Fund 3. Recall that to calculate simple interest on an investment, you use the formula $I = Prt$, where P is the principal or amount invested, r is the annual interest rate, and t is the amount of time the money is invested, which for our problem will be 1 year $(t = 1)$. Use the table below to help you organize this information. Notice that interest rates must be converted to decimals before using them in an equation.

	Principal	Interest Rate	Interest
Fund 1	x	5%	$0.05x$
Fund 2	y	6%	$0.06y$
Fund 3	z	8%	$0.08z$
Total	**a.**		**b.**

a. Fill in the total amount available for investment in the bottom row of the table.

b. Fill in the total amount of interest desired in the bottom row of the table.

c. What does $0.05x$ represent in the context of this problem?

d. What does $0.08z$ represent in the context of this problem?

e. Using the principal column of the table, write an equation in standard form involving the variables x, y, and z for the total amount available for investment.

f. Using the interest column of the table, write an equation in standard form involving the variables x, y, and z for the total amount of interest desired.

g. Using the information that the amount of principal invested in Fund 2 should be equal to half the amount invested in the other two funds combined, write an equation in standard form involving the variables x, y, and z.

h. Solve the linear system of three equations created in parts **e.**, **f.**, and **g.** You can use either the substitution method or the addition method to solve the system of equations.

i. Check to make sure that your solution to the system is correct by substituting the values for x, y, and z into all three equations and verify that the equations are true statements.

j. How much money should be invested into each fund to earn $460 in interest? Write a complete sentence.

Name: Date:

Final Projects

Project A: To Buy or Not to Buy?

According to the 2010 census figures, the American dream of homeownership experienced its biggest drop since the Great Depression of the 1930s. The homeownership rate fell to 65.1 percent in 2010 from 66.2 percent in 2000. Some of the reasons for this include tighter credit by lending institutions, prolonged job losses and reduced government involvement. Patrick Newport, an economist with IHS Global Insight stated that "While 10 years ago owning a home was the American Dream, I'm not sure a lot of people still think that way." With Congress considering the idea of eliminating the federal tax deduction for home mortgage interest and given depressed housing values in many areas of the country, it may make more sense to rent than own.

1. To purchase a home, you must have a minimum down payment of 5% to 10% of the purchase price of the home, depending on the lending institution and your credit rating. For a $312,000 home what is the minimum down payment you should have?

2. If you buy a home with less than a 20% down payment, your lender typically will require you to purchase private mortgage insurance, or PMI, which protects the lender in case you default on the loan.

a. To avoid paying PMI on a home purchase of $312,000, how much of a down payment will you have to have?

b. The annual premium (yearly cost of the insurance) that you pay for PMI will vary but typically it is around 0.5% of the total amount you borrow (the loan amount). If you have a 10% down payment on a $312,000 home, how much will your annual PMI premium be?

c. If you pay the annual premium from part **b.** on a monthly basis, how much will the monthly PMI expense add to your monthly mortgage payment?

3. You can usually cancel your PMI when the balance of the mortgage is paid down to 80% of either your home's original purchase price or its appraised value at the time you took out the loan. If you forget to cancel your PMI, your lender is required by federal law to end the insurance once your outstanding balance reaches 78% of the original purchase price or appraised value at the time of purchase. For a home originally priced at $312,000, calculate the loan balance at which the lender will be required by law to stop charging you for PMI.

4. Annual property taxes for a $200,000 house in Charleston County, South Carolina, (after tax credits) amount to $1094.60.

   a. Assuming the same property tax rate applies to all homes in the county, how much property tax would you pay on a $312,000 home? (**Hint:** Set up a proportion since the same tax rate applies to both houses.) Round to the nearest cent if necessary.

   b. If you pay the annual taxes from part **a.** on a monthly basis, how much will the monthly tax expense add to your monthly mortgage payment? Round to the nearest cent if necessary.

5. If you purchase a 2400-square-foot home for $295,000, what is the price per square foot of the home? Round to the nearest cent if necessary.

6. If you purchase an 1850-square-foot home for $235,000, what is the price per square foot of the home? Round to the nearest cent if necessary.

7. Which of the two houses in Problems 5 and 6 is the best buy based on price per square foot?

8. What other considerations would be important to take into account when buying a home besides the cost per square foot?

9. Suppose that you plan to live in your current house for 5 years and then sell it. You expect the home to appreciate and hope that the profit made on the house will be enough to use as a down payment for a larger home.

   a. To list your house with a real estate agency, you will typically pay a 6% commission on the sales price to the realty company for marketing and advertising your home. How much commission would be charged if your home was priced to sell for $250,000?

   b. Suppose that you still have an outstanding loan balance of $168,000 on your $250,000 home. What would your net profit be on the sale of the house after paying the commission to the real estate company and paying the bank the outstanding loan balance?

   c. Will your net profit from part **b.** be enough to make a down payment on a home priced at $312,000 and allow you to avoid paying private mortgage insurance or PMI? (See Problem 2.)

Name: Date:

Final Project B: Taking the Bus

Justin recently retired from his job and decided to purchase a charter bus and offer bus excursions between his home in Savannah, Georgia, and Disney World in Orlando, Florida. The bus will seat 56 people. He has been trying to determine how much to charge for the round trip and has decided to charge \$425 per person, which includes the price of a three-day pass to Disney World. As an incentive, if a person brings a friend, he will give them both a \$5 discount. If they bring a third person, they will each get a \$10 discount, and so on. So Justin's revenue function for a group of friends would look like this:

1 person	$1\cdot(425-0)=1\cdot(425-0\cdot5)$
1 person + 1 friend	$2\cdot(425-5)=2\cdot(425-1\cdot5)$
1 person + 2 friends	$3\cdot(425-10)=3\cdot(425-2\cdot5)$
1 person + 3 friends	$4\cdot(425-15)=4\cdot(425-3\cdot5)$

In general, the revenue function for x number of people riding the bus who are all friends is given by

$$R(x)=x(425-(x-1)\cdot5).$$

Before advertising his incentive plan, Justin wants to make sure he will make a profit, so he is trying to figure out how many people with friends he can carry on the bus before he starts to lose money.

1. Write the revenue function $R(x)$ in standard form.

2. Does this quadratic function have a minimum or maximum? Explain your reasoning.

3. Find the zeros of the revenue function by setting $R(x)$ equal to zero and solving by factoring.

4. How much revenue will there be at the zeros of the function?

5. Find $R(32)$ and explain what it represents.

6. Determine the number of people that will maximize Justin's revenue function by finding the x-coordinate of the vertex.

7. Another way to find the x-coordinate of the vertex of a quadratic function is to find the **midpoint** of the two zeros. **The midpoint is the number that is exactly halfway between two numbers.** You can find the midpoint of two numbers by calculating their mean. Find the midpoint of the zeros of the quadratic function (found in Problem 3) and compare it to the result in Problem 6. Are the results the same? Why do you think this happens?

8. Will Justin be able to fill the bus with people before he starts losing money? Why or why not?

9. What is the maximum revenue he could earn using this incentive plan? (**Hint:** It is the y-coordinate of the vertex.)

10. At the maximum revenue point, how much will each person pay for their ticket, assuming they are all friends?

Name: Date:

11. What will Justin's revenue be if the bus is **full** of people who are friends with one another?

12. How much money will Justin lose by filling up the bus with people who are all friends as compared to the maximum revenue?

13. If the bus is full of people who are all friends, how much will each one pay for their ticket?

14. Suppose that the bus is filled with people who are all friends.

a. How much will each person **save** off the regular price of $425 for the trip to Orlando?

b. What is the percent decrease in price (round to the nearest tenth of a percent)?

To this point we have only talked about the revenue that Justin will bring in. His ultimate concern is how much profit he will make. Profit is what's left over from revenue after you pay all your expenses. In other words, *Profit = Revenue – Expense*.

15. Justin will have to pay for gas and for the 3-day Disney park passes out of his revenue. From Savannah to Orlando, the mileage is approximately 280 miles and the bus only gets about 9 miles to a gallon of gas. Using the price of gas in your local area, calculate the cost of gas for the round-trip bus ride from Savannah to Orlando. (**Hint:** Use unit fractions from Appendix A.1 to help you calculate the cost.)

16. Disney World has offered Justin a discount on the price of the 3-day park passes as long as he brings a certain minimum number of visitors to the park over the course of the next year. The price of the passes, assuming he meets this quota, will be $120. Calculate Justin's profit, assuming he makes the maximum revenue (see Problems 6 and 9) by deducting the cost of gas (see Problem 15) and his cost for the 3-day park passes.

17. Now calculate Justin's profit based on the bus being full of people who are friends. (See Problem 11 for his revenue in this situation.)

18. Based on your profit calculations in Problems 16 and 17

a. What do you think of Justin's business plan for chartering trips to Disney World? Is it a good plan? Explain your answer.

b. Is there anything you would do differently if you were Justin to ensure a good profit margin? Explain your answer.

c. Can you think of a better incentive plan? Explain your plan in depth.

Index

A

Absolute value(s) 251, 621
  solving equations with 621–622
  solving inequalities with 623–624

ac-method for factoring a trinomial 470, 575

Addition
  addend 7
  carrying 8
  sum 7
  with complex numbers 643
  with fractions
    with different denominators 75
    with the same denominator 73
  with mixed numbers 79
  with polynomials 433
  with rational expressions
    with different denominators 505
    with the same denominator 503
  with real numbers 255
  with whole numbers 7

Addition principle of equality 295

Additive inverse 259, 273

Adjacent angles 176

Algebraic expression(s) 277
  evaluating 279

And 4, 97

Angle(s)
  acute 176
  adjacent 176
  alternate interior 178
  bisect 180
  complementary 177
  congruent 176
  corresponding 178
  created by transversals 178
  definition 175
  naming 177
  obtuse 176
  right 176
  straight 176
  supplementary 177
  symbol for
    angle 176
    degree(s) 176
    measure of 176
  vertical 177

Appreciation. *See* Percent(s): percent increase

Area 12, 194
  enclosed by a circle 190
  formulas 12, 185
  square unit 14
  with cut out sections 186

Associative property of addition 11, 273

Associative property of multiplication 11, 273

Average. *See also* Statistics: mean
  definition 33
  of whole numbers 33
  problems
    solving with linear equations 319

B

Back substitution 389

Balance point of a data set 650

Bar graph. *See* Graph(s): bar graph

Base of an exponent 27

Base ten system 4

Binomial 429

Boundary line 368

C

Capacity 607

Cartesian coordinate system 337
  horizontal axis 337
  ordered pairs 338
  origin 338
  plotting points 338
  quadrants 337–338
  vertical axis 337
  x-axis 337
  y-axis 337

Cartesian plane. *See* Cartesian coordinate system

Change in value 260

Circle graphs. *See* Graph(s): circle graph

Circles. *See* Geometry: circles

Circumference 189

Closed plane 368

Coefficient 277, 429

Combined variation 522

Commission 159

Commutative property of addition 11, 273

Commutative property of multiplication 11, 273

Complementary angles 177

Completing the square 579

Complex fractions 84, 509
  simplifying 84

Complex number(s)
  addition with 643
  definition 642
  division with 646
  equality of 642
  imaginary part 642
  multiplication with 645
  real part 642
  subtraction with 643

Complex rational expressions 509. *See also* Rational expression(s): complex

Composite number 43

Compound linear inequalities 325

Conditional equation 306

Congruent angles 176

Congruent triangles 201

Conjugate(s)
  of complex numbers 646
  of radical expressions 550

Consecutive integers 309

Consistent system of equations 383

Constant monomial 497

Constant of variation 521

Contradiction equation 306

Coordinate plane. *See* Cartesian coordinate system

Cost problems 400

Counterexample 40

Counting numbers. *See* Natural numbers

Cross multiplying. *See* Cross products

Cross products 130

Cube root(s) 533

D

Decimal number(s)
  addition with 103
  comparing 99
    with fractions 117
  decimal fraction 99
  decimal point 97
  definition 99
  division with 111
    by powers of 10 112
  estimating 104, 108, 112
  fraction part of 97
  history of 99
  multiplication with 107
    by powers of 10 107
  nonterminating 116
    nonrepeating 116
    repeating 116
  plot on a number line 99
  reading 97
  rounding 100, 104
  subtraction with 103
  terminating 116
  writing 97
  writing a fraction as a 115
  writing a percent as a 139
  writing as a percent 139
  writing fractions as 115

Decimal system 4. *See also* Base ten system

Decreasing function 656

Degree of a polynomial 429

Dependent system of equations 383

Dependent variable 660

Depreciation. *See* Percent(s): percent decrease

Diameter 189

Difference of two cubes 475

Difference of two squares 442, 473

Digit(s) 3

Dimensions of space 194

Direct variation 521

Discount 155

Distance-rate-time problems
- solving with linear equations 317
- solving with systems of linear equations 399
- with quadratic equations 589
- with rational expressions 518

Distributive property 11, 273

Divisibility
- divides 39
- divisible by 39
- tests for 39

Division
- by powers of 10 112
- by zero 17
- dividend 17, 111
- divisor 17, 111
- quotient 17, 111
- synthetic 627
- with complex numbers 646
- with decimal numbers 111
- with fractions 62
- with monomials 447
- with polynomials 447
- with rational expressions 501
- with real numbers 263
- with scientific notation 426
- with whole numbers 17

Division algorithm 447

Division principle of equality 295

Domain 362, 563

E

Element of a set 323

Ellipsis 249

Empty set 324

Equation(s)
- $ax + b = c$ 301
  - solving 302
- $ax + b = cx + d$ 305
  - solving 306
- $ax = c$ 295
  - solving 297
- conditional 306
- contradiction 306
- definition 255, 295
- identity 306
- linear 295
  - general form 305
  - solving 306
- one-step 295
- point-slope form 356
- roots of 482
  - extraneous 513
- slope-intercept form 352
- solution 255, 295
  - extraneous 513
- two-step 301
- variable 12, 255
- verifying 256
- $x + b = c$ 295
  - solving 297
- zeros of 482

Estimating
- methods 21
- overestimating 22
- underestimating 22
- with decimal numbers 108, 112
- with whole numbers 21

Evaluating expressions 269

Event 237

Excluded value 497

Expanded form 98

Experiment 237

Exponential decay 659

Exponential expression(s) 27

Exponential growth 659

Exponent(s)
- base 27
- cubed number 533
- definition 27
- exponential expressions 27
- exponent of 0 27, 419
- exponent of 1 27
- negative 419
- power 27
- rational 559
  - n^{th} roots 559
- rules for
  - power of a product 424
  - power of a quotient 424
  - power rule 423
  - product rule 417–418
  - quotient rule 418
  - summary of 560
- squared number 533

Expressions
- algebraic 277
  - coefficient 277
  - complex 510
  - degree 277
  - evaluating 279
  - like terms 277
  - term 277
  - variable 277
- definition 30
- exponential expressions 27
- simplifying 116

Extraneous information 65

Extraneous solutions 513

F

Factoring by grouping 462

Factoring polynomials
- *ac*-method 470, 575
- by grouping 462
- difference of two cubes 475
- difference of two squares 473
- factoring out the GCF 461
- guidelines for 477–479
- not factorable 462
- perfect square trinomials 474
  - square of a binomial difference 474
  - square of a binomial sum 474
- solving quadratic equations by 481
- special factoring techniques 473
- sum of two cubes 475
- sum of two squares 474
- trial-and-error method 465, 469
  - patterns 466, 470

Factors 11

Factor theorem 482

Finite 249, 324

FOIL method 441

Formulas
- area 185
  - enclosed by a circle 190
- circumference of a circle 190
- compound interest 162
- distance traveled 313
- force 313
- lateral surface area of cylinder 313
- perimeter 181
- simple interest 162
- slope 350
- solving for a variable 314
- surface area 193
- temperature 615
- volume 193
- working with 313

Fraction(s) 55
- addition with
  - with different denominators 75
  - with the same denominator 73
- borrowing a 1 79
- comparing 83
- complex 84, 509
  - simplifying 84
- denominator 55, 75
- division with 62
- equivalent 57
- history of 70
- improper 56
- invisible denominator 64
- least common denominator 74
- linear 655
- mixed number 56
- multiplication with 55
- numerator 55, 75
- plot on a number line 58
- proper 56
- quadratic 655
- reciprocal of a 62
- reducing 59, 144
- subtraction with
  - with different denominators 75
  - with the same denominator 73
- unit fractions 608
- writing a decimal number as a 115
- writing a percent as a 143
- writing as a decimal number 115
- writing as a percent 143

Frayer Model 54

Function(s) 362, 563. *See also* Relation
- decreasing 656
- exponential 655
  - decay 659
  - growth 659
- increasing 656
- linear 363
  - decay 659
  - growth 659
- nonlinear 655
- notation 364
- review of 563
- square root 655
- vertical line test 362

G

GCF. *See* Greatest common factor

Geometric shapes 181

Geometric solids 193

Geometry
- circles
  - area 190
  - center 189
  - circumference 189–190
  - definition 189
  - diameter 189
  - radius 189
  - semicircle 190
- cube 197
- height 186
- line 175
- line segment 175
- plane 175
  - boundary line 368
  - closed half-plane 368
  - half-plane 368
  - open half-plane 368
- point 175
- polygons 181
  - parallelogram 181
  - rectangle 181
  - regular 184
  - square 181
  - trapezoid 181
  - triangle 181
  - with more than 4 sides 181
- ray 175
- solids
  - rectangular pyramid 193
  - rectangular solid 193
  - right circular cone 193
  - right circular cylinder 193
  - sphere 193

Grade 349

Graph(s)
- bar graph 37, 223
  - break symbol 158
  - creating 231
  - interpreting 216
- circle graph 223
  - creating 232
- Excel directions for creating 233
- histogram 223
  - class 225
  - class boundary 225
  - class width 225
  - frequency 225
- line graph 223
- Pareto chart 244
- pictograph 226
- properties of 223
- with negative values 265

Greatest common factor
- factoring out of a polynomial 461
- of a set of terms 460
- of integers 459

Grid reference 340

H

Half-plane 368

Histogram 223. *See also* Geometry

Horizontal lines 351

Hyphens 5

Hypotenuse 206, 577

I

i 641

Identity equation 306

Identity property of addition 11, 273

Identity property of multiplication 11, 273

Imaginary numbers 641
- *i* 641

Improper fraction 56

Income tax 156

Inconsistent system of equations 383

Increasing function 656

Independent system of equations 383

Independent variable 660

Index of a radical expression 533

Inequalities 251
- compound "and" 623
- compound linear 325
- compound "or" 624
- graph 252
- linear 323, 325
  - solving 326
- symbols 251
- verifying 251

Infinite 249, 324

Integers 249
- consecutive 309
- greatest common factor 459

Interest
- compound 162
- effective interest rate 165
- simple 162

Interest problems
- solving with linear equations 318
- with systems of equations 403

Interval(s) 325
- closed 325
- half-open 325
- open 325

Inverse variation 522

Invisible one 278

Irrational numbers 205, 249, 549

J

Joint variation 522

K

Key words 34
- for addition 9
- for division 19
- for multiplication 15
- for subtraction 9
- for translating English words into algebraic expressions 283

L

LCD. *See* Least common denominator

LCM. *See* Least common multiple

Leading coefficient 429

Least common denominator 74

Least common multiple 67
- of a set of polynomials 504

Like radicals 543
- combining 543

Like terms 277
- combining 278

Line graph 223

Line of symmetry 591

Linear decay 659

Linear equation(s) 295
- general form 305
- in one variable. *See* Equation(s)
- in three variables
  - solutions 635
  - standard form 635
- in two variables 338
  - dependent variable 339
  - graphing 344–345, 355
  - graph of 339, 343
  - independent variable 339
  - slope 349
  - solutions 338–339
  - standard form 343
  - summary of 358
  - *x*-intercept 345
  - *y*-intercept 345
- solving 297, 302, 306

Linear function(s) 363

Linear growth 659

Linear inequalities 325, 367
- graphing 369
- standard form 367

Line(s)
- horizontal 351
- intersection 178
- parallel 178, 357
- perpendicular 178, 357
- segment 175
- slope of a 349
- transversal 178
- vertical 351

M

Mathematical model(s)
- definition 655
- exponential 656
- regression analysis 661
  - coefficient of determination 662
  - negative correlation 662
  - positive correlation 662
  - regression line 662

Maximum value 594

Mean 218

Measures of center 218

Median 218

Method of addition for solving systems 395

Method of elimination for solving systems 395

Method of substitution for solving systems 389

Metric system
- capacity 611
- converting measurements within 612–613
- converting to US system 615
- equivalent measurements 611
- history of 611
- length 611
- liquid volume 611
- mass 611
- temperature 615

Minimum value 594

Mixed number(s) 56
- addition with 79
- changing from an improper fraction 56
- changing to an improper fraction 57
- multiplication with 61
- subtraction with 79

Mixture problems
- with systems of equations 404

Mode 219

Model(s) 343. *See also* Linear equation(s)

Monomials 429

Multiples 67

Multiplication
- by a binomial 438
- by a monomial 437
- by a polynomial 437
- by powers of 10 107
- factor 11
- product 11
- with complex numbers 645
- with decimal numbers 107
- with fractions 55
- with mixed numbers 61
- with polynomials 437
- with radical terms 545
- with rational expressions 500
- with real numbers 263
- with scientific notation 426
- with whole numbers 11

Multiplication principle of equality 295

Multiplicative inverse 273

N

Natural numbers 3

Negative correlation 662

Negative exponents 419

n^{th} roots 559

Null set 324

Number problems
- with equations 309
- with rational expressions 517
- with systems of equations 399
- with whole numbers 35

O

Open plane 368

Opposite 259

Ordered pair(s) 338

Ordered triple 635

Order of operations
- with decimal numbers 113
- with fractions 84
- with polynomial expressions 434
- with real numbers 269
- with whole numbers 29

Origin 338

Outcome 237

P

Parabola 591

Parallel lines 178, 357

Pareto chart 244

Percent(s) 139
- decrease 157
- equation 151
- equivalent decimal forms 140, 141
- increase 157
- of profit 161
  - based on cost 160, 161
  - based on selling price 160, 161
- proportion 147
- symbol 139
- writing a decimal number as a 139
- writing a fraction as a 143
- writing as a decimal number 139
- writing as a fraction 143

Perfect cube 549

Perfect square 205, 549

Perimeter 194
- definition 8
- formulas 181

Perpendicular lines 178, 357

Pi 189
- history of 189
- Pi Day 189

Pictograph 226

Pitch 349

Place value system
- for decimal numbers 97
- for whole numbers 3

Plane 175

Point 175

Point-slope form 356, 655
- graphing with 355

Polygons 181

Polynomial(s)
- addition with 433
- binomial 429
- degree of a 429
- division with 447
  - by a monomial 447
  - division algorithm 447
- factoring. *See* Factoring polynomials
- leading coefficient 429
- monomial 429
  - constant 497
- multiplication with 437
  - binomial by a binomial 438
  - difference of two squares 442
  - FOIL method 441
  - monomial by a polynomial 437
  - perfect square trinomials 444
  - polynomial by a polynomial 438
- opposite 499
- prime 462
- quadratic 481
- special products of binomials 441–444
- subtraction with 434
- trinomial 429

Population density 169

Positive correlation 662

Power of a product rule 424

Power of a quotient rule 424

Power rule for exponents 423

Prime factorization
- division by primes 44
- factor tree 44

Prime numbers 43

Prime polynomials 462

Principles of Equality
- addition principle 295
- division principle 295
- multiplication principle 295
- subtraction principle 295

Probability 237
- chance 145
- event 237
- experiment 237
- experimental 237
- of an event 238
- outcome 237
- sample space 237
  - tree diagram 237
- theoretical 237

Problem solving
- basic strategy 34
- Pólya's approach xviii
- RIDGES strategy 485
- SOLVE strategy 485
- S.T.A.R. strategy 486

Product property of radicals 537

Product rule for exponents 417–418

Proper fraction 56

Properties
- additive inverse 259, 273
- distributive property 11, 273
- factor theorem 482
- multiplicative inverse 13, 273
- of addition
  - associative property 7, 273
  - commutative property 7, 273
  - identity property 7, 12, 273
- of multiplication
  - associative property 11, 273
  - commutative property 11, 273
  - identity property 11, 273
  - zero-factor law 11, 273
  - zero-factor property 481

of radicals
product property 537
quotient property 537
of real numbers 273
of triangles 200

Proportions 129
nursing notation 135
extremes 135
means 135
solving with 135
setting up 133–134
solving 133
verifying 129
with cross products 130

Pythagorean Theorem 206, 577

Pythagorean triple 206–207

Q

Quadratic equation(s)
definition 481
solving
by completing the square 580
with the quadratic formula 583
with the square root method 575–576
solving by factoring 481

Quadratic Formula 583

Quadratic function(s)
description of 591
discriminant 583, 593
graphing 593
line of symmetry 591
maximum value 594
minimum value 594
vertex 591
x-intercept 591
y-intercept 591

Quadratic polynomials 481

Quotient property of radicals 537

Quotient rule for exponents 418

R

Radical expression(s)
rationalizing denominators
with a binomial 551
with a single term 549
solving equations with 555

Radical function(s)
definition 564
domain of 564
graphing 564

Radical(s)
expression 533
index 533
like radicals 543
combining 543
multiplication with 545
properties of 537
radicand 533
simplest form 537
simplifying 538–539
symbol 533

Radicand 533

Radius 189

Range 217, 362, 563, 649

Rate of change 350

Rates 129
unit rate 132

Rational exponents 559

Rational expression(s)
addition with
with different denominators 505
with the same denominator 503
complex 509
simplifying 509–510
definition 447, 497
division with 501
fundamental principle of 499
multiplication with 500
opposites 499
solving equations with 513
subtraction with
with different denominators 505
with the same denominator 503

Rational function(s) 505
restriction on the domain 497
finding 498

Rationalizing a denominator 549, 551

Rational number(s) 249
arithmetic rules for 498
definition 498
finite terminating 249
infinite, repeating 249

Ratios 129

Ray 175

Real number(s) 249
addition with 255
division with 263
line 249
multiplication with 263
nonnegative 251
subtraction with 259

Reciprocal 62

Reference value(s) 160

Regression analysis 661

Relation 361, 563
domain 362
mapping diagram 361
range 362

Remainder theorem 629

René Descartes 337

Restriction 497–498

Roster form 323

Rounding
decimal numbers 100
rules for 21
whole numbers 21

S

Sales tax 156

Sample space 237

Scatter Plot(s) 660

Scientific notation 425–426
dividing with 426
multiplying with 426
writing a number in 426

Semicircle 190

Set-builder notation 324

Set(s) 249, 323, 361
element 323, 361
empty 324
finite 249, 324
infinite 249, 324
intervals
closed 325
half-open 325
open 325
null. *See* Set(s): empty
roster form 323
set-builder notation 324
subset 250

Similar triangles 201

Slope 349
calculating 350
finding from a graph 350
formulas 350
rise 349
run 349

Slope-intercept form 352

Solution of an equation 255, 295, 338

Square of a binomial difference 444, 474

Square of a binomial sum 444, 474

Square root method 575

Square root(s) 205, 533
negative square root 534
principal square root 534
simplifying 537

Standard deviation 649

Statistics
definition 217
mean 218
finding. *See* Average
with negative values 264
measures of center 218
median 218
finding 218
mode 218
finding 219
range 217
regression analysis 661
coefficient of determination 662
negative correlation 662
positive correlation 662
regression line 662
standard deviation 649
formula 650
weighted mean 220
z-score 651

Substitute 181

Subtraction
borrowing 7, 8
with mixed numbers 79
change in value 260
difference 7
minuend 7
subtrahend 7
with complex numbers 643
with decimal numbers 103
with fractions
with different denominators 75

with the same denominator 73
with mixed numbers 79
with polynomials 434
with rational expressions
with different denominators 505
with the same denominator 503
with real numbers 259
with whole numbers 7

Subtraction principle of equality 295

Sum of two cubes 475

Sum of two squares 474

Supplementary angles 177

Surface area
formulas 193

Synthetic division 627

Systems of linear equations
consistent system 383, 636
definition 383
dependent system 383, 636
inconsistent system 383, 636
independent system 383, 636
solution of 383
solving
by graphing 384
in three variables 637
method of addition 395
method of elimination 395
method of substitution 389

Systems of linear inequalities
solving 631

T

Temperature 615

Term 277

Tests for divisibility 39

Three-dimensional space 635
ordered triple 635, 637

Transversal lines 178

Trial-and-error method of a trinomial 465, 469

Triangle(s)
acute 199
congruent 201
angle-side-angle (ASA) 201
side-angle-side (SAS) 201
side-side-side (SSS) 201
equilateral 199
hypotenuse 206
isosceles 199
obtuse 199
properties of 200
right 199
legs of 206
scalene 199
similar 201
symbol for 200

Trinomial 429

U

Undefined
definition 17
division by zero 17

Uniqueness
definition 13
of the additive identity 13
of the multiplicative identity 13

Unit fractions 608
converting measurements with 609

Unit rate 132

Units 13

US measurement system
capacity 607
common abbreviations 607
converting measurements within
using division 608
using multiplication 608
using unit fractions 609
converting to metric 615–617
equivalent measurements 607
history of 607
length 607
temperature 615
time 607
weight 607

V

Variable 255

Variation
combined 522
constant of 521
direct 521
inverse 522
joint 522
solving 521

Venn diagram 68
to find GCF 460
to find LCM 68

Vertex 591

Vertical angles 177

Vertical lines 351

Vertical line test 362, 563

Volume 194
formulas 193

W

Weighted mean 220

Whole numbers
addition with 7
definition 3
division with 17
estimating 21
multiplication with 11
rounding 21
subtraction with 7

Work problems
with quadratic equations 588
with rational expressions 518

X

x-axis 337

x-intercept 345

Y

y-axis 337

y-intercept 345

Z

Zero
division by 17
history of 18

Zero-factor law 11, 273

Zero-factor property 481

z-score 651